Contemporary Ergonomics 2001

Edited by

Margaret A. Hanson

Institute of Occupational Medicine, Edinburgh, UK

THE **Ergonomics**
society

First published 2001
by Taylor & Francis
11 New Fetter Lane, London EC4P 4EE

Simultaneously published in the USA and Canada
by Taylor & Francis Inc,
29 West 35th Street, New York, NY 10001

Taylor & Francis is an imprint of the Taylor & Francis Group

The design of alerts for future platform management systems
Michael A. Tainsh
© DERA 2001

Printed and bound in Great Britain by TJ International Ltd, Padstow, Cornwall

British Library Cataloguing in Publication Data
A catalogue record for this book is available from the British Library

Library of Congress Cataloging in Publication Data
A catalogue record for this book has been requested

ISBN 0-415-25073-0

CONTENTS

UPPER LIMBS

DISPLAY SCREEN EQUIPMENT

MANUAL HANDLING

MANAGING BACK PAIN

HEALTH AND SAFETY

OCCUPATIONAL HAZARDS

ALARMS AND WARNINGS

HCI

COGNITIVE ERGONOMICS

SYSTEM DESIGN

METHODOLOGY

EDUCATION
(Also see pages 549-560)

SELLING AND COMMUNICATING ERGONOMICS
(Also see pages 561-566)

USABILITY

DESIGN

THE OFFSHORE INDUSTRY

GENERAL ERGONOMICS

ADDITIONAL PAPERS

PREFACE

Contemporary Ergonomics 2001 is the proceedings of the Annual Conference of the Ergonomics Society, held in April 2001 at the Royal Agricultural College, Cirencester, UK. The conference is a major international event for Ergonomists and Human Factors Specialists, and attracts contributions from around the world.

Papers are chosen by a selection panel from abstracts submitted in the autumn of the previous year and the selected papers are published in *Contemporary Ergonomics*. Papers are submitted as camera ready copy prior to the conference. Each author is responsible for the presentation of their paper. Details of the submission procedure may be obtained from the Ergonomics Society.

The Ergonomics Society is the professional body for ergonomists and human factors specialists based in the United Kingdom. It also attracts members throughout the world and is affiliated to the International Ergonomics Association. It provides recognition of competence of its members through its Professional Register. For further details contact:

The Ergonomics Society
Devonshire House
Devonshire Square
Loughborough
Leicestershire
LE11 3DW
UK

Tel: (+44) 1509 234 904
Fax: (+44) 1509 235 666

Email: ergsoc@ergonomics.org.uk
Web page: http://www.ergonomics.org.uk

APPLIED PHYSIOLOGY

THE EXTENT AND EFFECTS OF DEHYDRATION AMONG LOGGERS

Liz Ashby, Richard Parker, Graham Bates[1]

Centre for Human Factors and Ergonomics (COHFE)
Private Bag 3020, Rotorua
New Zealand
liz.ashby@forestresearch.co.nz

[1]*School of Public Health*
Curtin University, Western Australia

Forest harvesting (logging) is well recognised as being a hazardous occupation. There have been many efforts to try and reduce fatalities and injuries, but in New Zealand much of the work is still physically demanding, in harsh environmental conditions. One area which has produced much interest is that of dehydration and fluid intake of workers. Dehydration is associated with fatigue, poor concentration, unsafe work practice and longterm health problems. Although there have been efforts to encourage sufficient fluid intake among workers, there has been little research to determine the extent of the problem. This pilot study aimed to investigate the level of hydration among a sample of loggers and establish whether further work was needed to clarify the status of workers and to implement and measure solutions.

Introduction

The forest industry and in particular harvesting (logging) is associated with high rates of injury and fatalities. There have been many efforts to address the risks associated with logging, for example through increasing mechanisation, design and use of personal protective equipment and improved hazard identification. However, in New Zealand the difficult terrain precludes mechanised harvesting in many areas, leaving motor-manual techniques (a worker with a chainsaw) as the only means of harvesting plantation trees in these areas. The difficult terrain and extreme weather conditions place heavy physical and physiological demands on workers.

Alongside addressing the hazards and risks associated with logging, attempts have been made to educate workers to ensure they are equipped to deal with these physical demands, for example in terms of their fitness, level of hydration and nutrition, and fatigue (Paterson and Kirk, 1997; Kirk *et al*, 1996; Cummins and Kirk, 1999).

However, there has been little work to establish the actual level of hydration of workers to establish the extent of the problem, and whether means of addressing it, such as educational programmes, are being effective.

The human body maintains a constant core temperature via complex mechanisms, with sweating being the most effective means of cooling. Heat is produced within the body through metabolism and muscle activity, and so will increase with escalating physical activity, and sweating allows dispersion of this heat. Many factors influence the effectiveness of heat balance within the body and effectiveness of sweating, including clothing, air temperature, humidity, air movement and radiant heat.

Whilst sweating is effective in helping to maintain a constant core temperature, it does mean that fluid is constantly being lost. The response of the body to dehydration is thirst, but this mechanism is not experienced until the body is already somewhat dehydrated. The tasks associated with logging can result in large volumes of sweat loss because of high physical demands, high humidity and temperatures, carriage of equipment and clothing (such as chainsaw, and personal protective equipment) and uneven and steep terrain. This can mean a loss of around 3 litres a day, with up to 6.2 litres being recorded during work in high temperatures (Paterson, 1997).

If the fluid loss is not replaced, the body will become increasingly under (hypo) hydrated. The effects of dehydration are well documented, and include physical changes, mental deterioration and long-term health effects.

Physical work capacity reduces with a fluid loss of 1% to 2% body weight and a loss of 4% in a hot environment can cause approximately 50% reduction in physical work capacity (Bates and Matthew, 1996). This has safety implications where physiological demands on the body are high. Mental deterioration can be expected with hypohydration, which can have serious consequences where the safety of an operation may be associated with complex tasks and decision making. Long-term, there may be health consequences of inadequate fluid intake, for example with the development of kidney stones.

Dehydration was observed in a previous study involving New Zealand loggers, indicating they had inadequate fluid intake to replace their sweat loss (Paterson, 1997). The study also found that loggers who worked under a regular fluid regime (i.e. had sufficient fluid intake) worked with lower heart rate and lower relative workload. These subjects reported feeling fresher, stronger and more awake when fluid was taken regularly.

This current work was a pilot study, to obtain a 'snapshot' of hydration status of a sample of logging workers. It is envisaged that this pilot work will be part of a larger more detailed study to investigate a wider sample of workers in variable conditions such as seasonal variations, and to assess the effectiveness of interventions such as educational programmes.

Method

Thirty-one loggers were included in the sample and were randomly selected from available logging crews. All gave consent to the study and were able to withdraw at any time. Loggers were given a specimen jar and asked to provide a small urine sample. One drop of the urine sample was placed onto a refractometer viewing platform (Atago Instruments) using a pipette, and the specific gravity of the sample was recorded. At the same time, the loggers answered questions about the quantity of fluid consumed so far that day, and what sort of drinks they had consumed. Researchers also noted age, height and weight of the worker along with a description of their clothing. Environmental conditions were recorded at the time of sampling. Informal interviews with loggers supplemented data through obtaining subjective opinions on fluid intake problems and issues.

Results

Urine samples were provided by 31 loggers, representative of New Zealand workforce in terms of age, physical characteristics (see Table 1) and clothing (which included high visibility and protective clothing, hard hat and leather or rubber chainsaw cut resistant boots). Environmental conditions were variable, reflecting the New Zealand climate.

Table 1. Loggers' physical characteristics

	Mean	SD	Min	Max
Age (years)	34	9	17	53
Height (cm)	178	10	153	191
Weight (kg)	89	12	70	113

The specific gravity of the samples ranged from 1.012 to 1.031 g/ml with an average of 1.022 g/ml (see Table 2). This indicated that most of the loggers were under, or hypohydrated (1.015 to 1.030 g/ml). Those with a value over 1.030 could be described as dehydrated at the time of the sample. Figure 1 shows the distribution of specific gravity among the loggers sampled.

Table 2. Fluid intake and specific gravity

	Mean	SD	Min	Max
Fluid intake (ml)	1124	822	0	3000
Specific gravity (g/ml)	1.022	0.005	1.012	1.031

The type and amount of fluid consumed also varied. One logger had consumed no fluid at all and 23% had only had caffeinated drinks. A similar proportion had drunk only water. Informal interviews with the loggers resulted in opinions on factors that influence their fluid status. These included:

- Physical difficulties associated with fluid consumption such as equipment in which to carry it. Loggers carry equipment with them such as chainsaws, wedges, petrol cans, mallet etc, and fluid bottles add to the bulk and weight carried;

- Loggers had difficulty understanding what is best to drink, having had conflicting information from different sources. For example pros and cons of soft drinks, expensive sports drinks, water, and tea and coffee;
- Keeping drinks cold whilst working in the bush all day;
- Water containers can get contaminated with petrol and oil;
- Social influences which may limit fluid consumption, such as reluctance to drink so as not to have to stop a machine to pass urine; lack of conveniences for example for female workers.

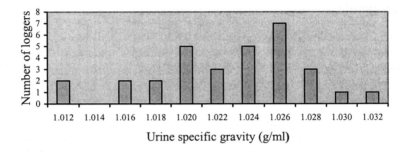

Figure 1. Distribution of specific gravity results

Conclusions

This pilot study used measurement of specific gravity as an effective tool to estimate fluid status of workers in the field. The test does have limitations, the primary one being the influence of caffeine and other substances over specific gravity. A high caffeine intake will produce a lower specific gravity result than that which might represent the true fluid status of the worker. This is because the diuretic effect of caffeine causes excretion of fluid, diluting the urine sample.

However, as a field-test, and if influential factors are considered alongside results, it provides a useful indication of fluid status, and also provides potential for training in the form of biofeedback to loggers. Loggers are able to see their specific gravity, and colour of urine associated with that value, and so regular testing could help them determine the effects of drinking increased quantities of fluid. This could be helpful where wide individual and task variations makes it difficult to advise on specific volumes to consume. The method of field testing specific gravity in this way has been used successfully in the Australian mining industry (Bates, personal communication).

The study determined that there are complex issues associated with drinking sufficiently to maintain fluid balance. An overall ergonomics approach is needed to ensure contributory factors are accounted for when devising means of implementing sufficient fluid intake. Such factors include methods of carrying fluid, how to keep it cool and types of convenience facilities, for example to cater for female workers.

A positive culture needs to be in place so that workers do not limit their fluid intake to avoid the need to pass urine – this was demonstrated by machine operators who did not want to be noticed by surrounding workers if they stopped their machines.

The type of fluid consumed is also an issue. If large quantities are required during physical work, to maintain a fluid balance, a variety may be indicated so those workers do not get bored of drinking water. They should be encouraged not to consume large quantities of bottled drinks such as cola drinks or lemonade – these can contain up to 10% sugar resulting in high sugar consumption per day. High caffeine drinks will increase fluid loss rather than aid replacement. There is anecdotal evidence of contractors supplying cold cordial to their crews, which is a positive way of encouraging sufficient and appropriate fluid intake.

Further progress in this work is taking the form of more detailed screening of a larger sample of workers. Specific gravity, alongside the other measures described, is being measured in winter and summer conditions, to establish seasonal variations in drinking patterns as well as climatic influences. Workers are providing samples over a four day period, three times a day.

Different equipment designs are also being explored, such as the use of fluid backpacks. Further consideration will be given to what additional educational and training information should be used and what the effects of any of these implementations are.

References

Bates, G. and Matthew, B. 1996. *A new approach to measuring heat stress in the workplace*. Occupational Hygiene Solutions. Proc 15th Annual Conference of the Australian Institute of Occupational Hygiene: 265-267. Perth, 30th November to 4th December.

Cummins, T. and Kirk, P. 1999. *Sleep loss: a deadly issue!!* Liro Report volume 24, 3. Liro Forestry Solutions, Forest Research, Rotorua, New Zealand.

Kirk, P., Gilbert, T., Darry, K. 1996. *Increased safety and performance through "smart food"*. Liro Report volume 21, 26. Liro Forestry Solutions, Forest Research, Rotorua, New Zealand.

Paterson, T. 1997. *Effect of fluid intake on the physical and mental performance of forest workers*. Project Report 66, Liro Forestry Solutions, Forest Research, Rotorua, New Zealand.

Paterson, T. and Kirk, P. 1997. *Fluid and Energy for Forest Workers*. Liro Report, volume 22, 8. Forest Research, Rotorua.

Physiological aspects of hill-walking

P.N. Ainslie[1], I.T. Campbell[2], K.N. Frayn[3], S.M. Humphreys[3], D.P.M. MacLaren[1], and T. Reilly[1]

[1]*Research Institute for Sport and Exercise Sciences, Liverpool John Moores University, Liverpool, L3 2ET;* [2]*University Department of Anaesthesia, University Hospitals of South Manchester, Withington Hospital, Manchester M20 2LR; and* [3]*Oxford Lipid Metabolism Group, Radcliffe Infirmary, Oxford OX2 6HE*

The aim of the current study was to investigate the physiological responses to a typical hill-walking event. Thirteen subjects (11 male and 2 female) participated in the study during the months February to March. On separate occasions subjects completed a self-paced 12 km hill-walk, varying in elevation from 100 m to 902 m above sea level consisting of a range of gradients and terrain typical of a mountainous hill-walk. During the hill-walk, continuous measurements of oxygen consumption and rectal temperature were made. During the first 5 km of the walk (100 - 902 m) rectal temperature increased (36.9 ± 0.2 to 38.5 ± 0.4°C). Subjects operated at approximately 50% of $VO_{2\,peak}$, with a average heart rate (HR) of 148 ± 8 (b.min^{-1}) during this first part of the walk. Rectal temperature decreased by approximately 1.5°C during a 30-min stop for lunch, and continued to decrease a further 0.5°C after walking recommenced. During the final 6 km, subjects descended from 902 - 100 m; over this period, VO_2 was maintained at approximately 30% peak, and HR averaged 126 ± 5 (b.min^{-1}). Practical recommendations to enable both hill-walkers and leaders to be better equipped for the mountainous environment are considered.

Introduction

The varied terrain, typical of a hill-walk, may place a high physiological and psychological strain on the enthusiast. The 'stresses' of hill-walkers have not been studied systematically despite the potentially deleterious physiological and psychological consequences of activity sustained over a whole day, sometimes in adverse climatic conditions.

Hill-walkers can be caught unexpectedly and unprepared when rain and wind accompany outdoor activities in cool weather. Decreased thermal insulation of wet clothing presents a serious challenge to body temperature regulation, which can be compounded by fatigue associated with prolonged exercise such as hill walking (Pugh, 1966a, 1967; Noakes, 2000).

Pugh (1966b) proposed that maintaining a VO_2 of 2-2.5 l·min^{-1} or 50-60% maximum oxygen uptake ($VO_{2\,max}$) would offset heat loss and combat the debilitating effects of the cold, wet and windy environment. Pugh (1966b) and Weller *et al*, (1997) showed that when exercise metabolism is reduced, the increase in shivering may be insufficient to prevent a decrease in deep body temperature. Weller *et al*, (1997) reported that rectal temperature (Trec) responses to an initial 120-min phase of exercise at approximately

60% of peak VO_2 (VO_{2peak}) were not influenced by the cold stress of a wet and windy environment. During a subsequent 240-min phase of exercise at 30% VO_2 peak, Trec was lowered by 0.6°C.

The studies reviewed have been limited to simulated conditions. Consequently the influence of a 'typical' day's hill-walk on a range of physiological and subjective variables, in the field, has not been established. Given such limitations, the aim of the current study was to investigate selected physiological and subjective responses to a typical hill-walking event.

Methods

Subjects: Thirteen subjects, 11 male and 2 female, participated in this study which was reviewed and approved by the Human Ethics Committee of Liverpool John Moores University. The physical characteristics, mean and range, were 26 (18-32) years of age, body mass 73 (55-82) kg, body fat 17 (10-30) %, and maximal oxygen consumption 59 (48-67) $ml\,kg^{-1}min^{-1}$. The majority of the subjects were active and experienced hill-walkers.

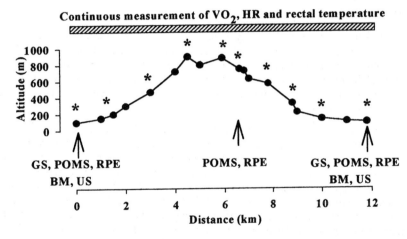

Figure 1. Illustrated profile of hill walk and associated measurements. *, measurements of wind chill index, GS, grip strength, POMS, Profile of Mood State, BM, body mass, US, urine osmolality .

Protocol and procedures: On separate occasions subjects completed a 12-km (8 mile) hill-walk. The course varied in elevation from 100 m to 902 m above sea level and consisted of a range gradient and terrain typical of a mountainous hill-walk. Subjects woke each morning between 05:00 and 05:30 hours and completed the preliminary experiments prior to the hill walk, shown in Figure 1. Self-paced walking began each day between 07:00–08:00 hours. Experiments were conducted from January through March. Prior to the walk and upon its completion, the subjects weighed themselves nude. Following the initial weighing the participants inserted a rectal temperature probe to a depth of 10 cm beyond the anal sphincter. The rectal probe was connected to a data logger (Squirrel meter 1000, Grant Instruments, Cambridge, UK) which recorded data every 6 min. On the walk a rest period of approximately 1 – 3 min was allowed every time thermal measurements were made, and 30 min for lunch (Figure 1.). Subjects were permitted fluid and food *ad libitum* which were pre-weighed allowing the total energy intake to be subsequently determined.

During the hill-walk, heart rate and VO_2 were measured continuously using a short-range radio telemetry for heart rate (PE3000 Sports Tester, Kempele, Finland) and with a portable respiratory gas analysis system (Metamax, Cortex Biophsik GmbH, Borsdorf) respectively.

Subjective and motor measurements: The Profile of Mood State (POMS) was used to measure fluctuating affective states of depression, tension, anger, confusion, fatigue and vigor (McNair et al, 1992). In addition, the subjects were asked to rate their overall perceived exertion (RPE) on a 6 - 20 scale (Borg, 1970) from the start of the walk until the lunch stop, and from the lunch stop until the completion of the walk.

Statistical Analysis: Data were analysed by repeated-measures analysis of variance (ANOVA). Post hoc tests (Tukey) were performed to isolate any significant differences. Student's paired t-tests ascertained between-condition differences when a variable was measured once. Statistical significance was set at $P \leq 0.05$ for all statistical tests.

Results and Discussion

Energy balance: The relative energy intake, during the hill-walk, was 5.6 ± 0.7 MJ relative to energy expenditure 12.5 ± 0.5 MJ (P<0.001), in addition to a decrease of ~2 kg in nude body mass was observed. This relatively high-energy expenditure serves to reflect the high energetic cost of hill walking, even when pursued over a relatively short duration, as well as the need for the hill walkers to increase their energy intake to help to maintain energy balance.

Metabolic, heart rate and thermoregulation: Pugh (1966b) described a VO_2 'cut-off' point, above which individuals exercising in a cold, wet and windy environment would not experience any influence on the physiological responses to exercise i.e., drop in core temperature, mental impairment, extreme fatigue and exhaustion. Below this point there would be an obligatory increase in energy expenditure and subnormal Trec and muscle temperatures. The present observations highlight the variability in VO_2. It was only during the high intensity part of the walk that subjects reached this cut-off point where they were operating at ~2.0 l·min^{-1} (Figure 2). Since the hill-walkers in this study walked at their own pace, it could be cautiously concluded that hill-walkers do not consistently operate at, or above this proposed 'cut-off' level.

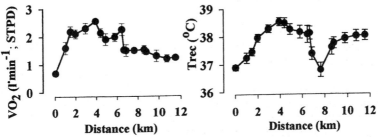

Figure 2. Oxygen consumption and rectal temperature results from the hill walk. Values are mean (SE).

The average HR during the first part if the walk until lunch was 148 ± 8 (b.min⁻¹). From the lunch stop until the completion of the walk, a average HR of 126 ± 5 (b.min⁻¹) was shown. The analysis of heart rate responses during the hill walk revealed that hill-walking in the present study requires moderate to high levels of exercise intensity. According to American College of Sports Medicine (1986), moderate intensity exercise and an appropriate training zone for cardiovascular fitness consists of 60-90% of maximal heart rate. The average heart rate during the hill-walk fell within this target zone, suggesting the possible advantageous health aspects of hill-walking.

Five of the walks monitored were conducted in cold, wet and windy weather. Snow and ice were regularly encountered, along with high winds, as reflected by a high wind chill index; these factors represent walking in very demanding climatic conditions. When the subjects stopped for lunch and measurements mid-walk for approximately 30 min, the exercise hyperthermia was cancelled out by the decreased heat production and increased heat loss through conduction and radiation, as highlighted in the decrease in Trec at this time point (Figure 2). The initial physiological responses to cold exposure to maintain core temperature in the cold are peripheral vasoconstriction to reduce heat loss and shivering to generate heat. Although shivering was not formally quantified directly, pronounced shivering was noted in four of the subjects. Once peripheral vasoconstriction is maximised, core temperature can only be maintained by an increased heat production, i.e., shivering which is thought to be the major contributor to the cold induced increase in heat production in the cold (Doubt, 1991). The core temperature continued to fall in a number of subjects when they began walking following lunch (Figure 3). This temperature 'after-drop' has been reported in a number of cold water immersion studies (Golden and Hervey, 1977), but to our knowledge, has not been reported in situations such as the present. Even though Trec did not drop below 35.9°C, this after-drop may describe the reason for hill-walkers slipping into the early stages of hypothermia after stopping for a rest. This continued decrease in Trec may underpin some of the changes in behavior observed when walking commenced after lunch. A number of subjects tended to withdraw, walking with their heads lowered, eyes ground-ward, apparently inattentive to their surroundings.

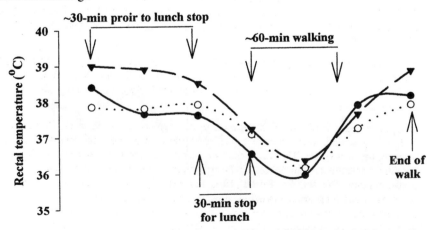

Figure 3. The apparent temperature 'afterdrop' which was evident in 3 subjects after the 30-min stop for lunch.

Subjective and motor measurements: An increased in tension and confusion was observed prior to the walk ($P<0.05$) in the POMS, additionally there was an increase in fatigue ($P<0.05$) post-walk relative to both before and at mid-walk. Overall RPE from the start of the walk to the lunch stop, and the final part of the walk was 15±2 and 13±3, respectively. Urine osmolality increased pre-walk to post-walk (603±86 to 744±71 mosmol; $P<0.05$), indicative of dehydration. A decrease in grip strength (45.4±2.7 to 43.5±2.8 kg·m²; $P<0.05$) was recorded from pre-walk to post-walk. Both the dehydration and decrease in motor performance may impair the hill-walker and leader in the event of an accident and/or emergency when either the ability to climb or use rope-rescue techniques may be required.

In summary, the relatively high work-rates during the hill-walk induced a heat production sufficient to offset heat loss, apart from during the lunch break and the subsequent ~30 min walking after this point. In addition, hill-walkers operating in adverse weather conditions may be affected by high subjective ratings of fatigue and decreases in motor performance.

Practical Implications: The reported findings have some practical applications, some of which are summarised here:

- Fluid intake: In order to avoid dehydration, as much fluid should be consumed as possible. Walkers should make a conscious effort to be fully hydrated prior to the hill walk. Monitoring of 'urine' colour should be made as an index of hydration (Shirreffs, 2000) – darker the colour, the more fluid required. Problems of stopping in serve weather, and subsequent cooling, may be avoiding by the use of an 'camel pack' or other 'bladder' systems designed to give easy access to fluid.

- Food intake: Regular snacks are ideal. Snacks that can be taken on the move are clearly beneficial in order to increase energy intake and avoid the subsequent cooling of a lunch stop if operating in adverse weather conditions.

- Physique and fitness: For the reasons described, leaders should be particular aware of both the very 'lean' and seemingly 'unfit' members of their party (Freeman and Pugh, 1968). These individuals should be observed with caution. If possible, the intensity of the walk should be related to the fitness level of the party. Grouping by exercise history is perhaps the simplest safeguard.

- Subjective and motor performance: Marked feelings of fatigue, along with subsequent decreases in motor performance may affect the ability to climb and or initiate emergency rescue techniques, even over relatively short walking distances in adverse weather.

- Stopping times: Clearly, lunch stops are part of a typical day's hill-walk. However, in bad weather conditions, even if lunch is in a survival shelter, decreases in core temperature will occur. This decrease in core temperature may continue to fall after walking recommences. In the hill-walking scenario, this means that although walkers may complain of feeling 'very cold', tired and possibly start to withdraw, they should on most occasions be counselled not to panic because they think they have hypothermia. They should, at this point, be encouraged to keep moving to enable an increase in heat generation. Caution is needed when subjects have stopped for prolonged periods of time in the stressful environment (> 30 min), these people could indeed be slipping into hypothermia. Therefore, lunch stops should be kept to a minimum in adverse weather conditions.

- Weather conditions and walking intensity: Walkers and leaders should be especially aware of cold, wet and windy conditions, if they plan to spend prolonged periods of time in the hills. In addition, care should be taken during the 'low intensity' part of the walk. At this point heat loss may be greater than heat production. This imbalance may lead to compromises in physiological and psychological functioning.

In conclusion, it is hoped that the current observations will extend the knowledge of both hill-walkers and leaders, enabling them to be better equipped for the mountainous environment. Information regarding the physiological responses to a 'stressful' hill-walking event may be used to facilitate decision making for enhancing survival and the outcome of some of the potential risks of activity sustained over a whole day, often in adverse conditions.

Acknowledgements:

The work was supported by Mars Incorporated.

References

American College of Sports Medicine. 1986, *Graded Exercise Testing and Exercise Prescription* (Philadelphia, PA: Lea and Febiger

Borg, G.A.V. 1970, Perceived exertion as an indicator of somatic stress, *Scandinavian Journal of Rehabilitation Medicine*, **2**, 92-98

Doubt, T.J. 1991, Physiology of exercise in the cold, *Sports Medicine*, **11**, 367-381

Freeman, J., and Pugh, L.G.C.E. 1968, Hypothermia in mountain accidents, *International Anaesthesiology Clinical*, **7**, 997-1007

Golden, F,StC, and Hervey, G.R. 1977, The mechanism of the 'after-drop' following immersion hypothermia in pigs. *Journal of Applied Physiology*, **227**, 35-36

McNair, D.M., Lorr, M., and Droppleman, L.F. 1992, *Edits Manual For The Profile Of Mood States*, (Educational and Industrial Testing Service, San Diego)

Noakes, T.D. 2000, Exercise and the cold, *Ergonomics*, **43**, 1461-1479

Pugh, L.G.C.E. 1964, Deaths from exposure on Four Inns Walking Competition, March 14-15, *Lancet*, **1**, 1281 - 1286

Pugh, L.G.C.E. 1966a, Clothing insulation and accidental hypothermia in youth, *Nature (London)*, **209**, 1281-1286

Pugh, L.G.C.E. 1966b, Accidental hypothermia in walkers, climbers, and campers: report to the Medical Commission on Accident Prevention, *British Medical Journal*, **1**, 123-129

Pugh, L.G.C.E. 1967, Cold stress and muscular exercise, with special reference to accidental hypothermia, *British Medical Journal*, **2**, 333-337

Shirreffs, S.M. 2000, Markers of hydration status, *Journal of Sports Medicine and Physical Fitness*, **40**, 80-4

Weller, A.S., Millard, C.E., Stroud, M.A., Greenhaff, P.L., and MacDonald, I.A. 1997, Physiological responses to a cold, wet, and windy environment during prolonged intermittent walking, *American Journal of Physiology*, **272**, R226-R233

PSYCHOSOCIAL FACTORS AND MUSCULOSKELETAL DISORDERS

PSYCHO-SOCIAL INFLUENCES ON REPORTING OF WORK RELATED MUSCULOSKELETAL DISORDERS - THE NEED FOR A GROUNDED THEORY APPROACH

Andrew Weyman & Mark Boocock

Health & Safety Laboratory
Sheffield S3 7HQ, UK

Despite widespread calls for multi-disciplinary approaches to the study of psycho-social influences on WRMSDs the majority of studies remain entrenched within a single perspective and reflect extensions of the traditional individual centred, bio-medical model. While having recognised the importance of social phenomena on reporting of WRMSDs, many studies remain fundamentally asocial in their method of enquiry. Reported here are insights from a pilot study of sewing machinists at four clothing manufacturing plants. Although well matched in terms of basic ergonomics variables, with respect to levels of technology and methods of production, there was a wide disparity in the incidence of reporting of WRMSDs. Fundamentally qualitative insights highlight the role of cultural influences associated with social and institutional support / legitimisation of WRMSD related ill-health reporting and management commitment.

Introduction.

Despite widespread calls for the application of multi-disciplinary, multi-method approaches to the study of work related musculoskeletal disorders (WRMSDs) the majority of studies to date remain entrenched within a single perspective (see, Norman, 1994; Sauter & Wanson, 1996 and Fordyce, 1996). A further criticism which might be levelled at research in this area relates to an apparent over emphasis on reductionist approaches, based upon extensions of the traditional biomedical model (Hockling, 1996). It is perhaps surprising that, while there is increasing recognition of the importance of social phenomena within this context, so many studies remain fundamentally asocial in their method, failing to take the unit of study beyond the individual.

Evidence of the presence of apparent epidemics in the reporting of WRMSD symptoms and their relative absence in ostensibly equivalent workplace environments has highlighted the potential importance of social context and raised questions regarding the utility of approaches which have sought to typify the personality profile of WRMSD *sufferers*, to the extent that such approaches have largely been abandoned (see Fordyce, 1996). Potentially relevant influences identified in the literature include: low levels of job-satisfaction; high job demands; low social support; and limited control and autonomy over physical, social and environmental conditions. The literature even contains reports of associations between specific musculoskeletal symptoms and questionnaire based measures of the above variables. Ahlberj-Hulter et al (1995), for example, report associations between 'social support at work' and shoulder / neck strain, similarly back pain with job strain. It is difficult to conceive that such findings reflect preordained experimental hypotheses, leaving the impression that they are at risk of identifying statistical artefacts arising from assumptions of causality on the basis of correlation.

A further influence that seems apparent as the research base in this area expands is that the range of potentially relevant variables also expands, despite limited commonality on findings. Rather than taking this as indicative of the importance of context and devising methods of enquiry well suited to the investigation of contextual influences, the majority of researchers have reacted to this by adding to the number of measures they include in their studies. This trend raises a number of issues: firstly, as the number of variables increases the potential for spurious relationships between them increases; secondly, there is a potential contradiction in using measures at the level of individual attitude to measure social phenomena, particularly if used in isolation; thirdly, hypotheses tend to remain at best vague and very loosely defined; and fourthly there is a tendency, frequently exhibited, to select generic measures of the variables of interest. It is perhaps this last issue which is at greatest contradiction with the object of social enquiry, there being a significant risk that generic measures merely measure generic things, where what is needed are measures which provide us with insight into the social environment which drives behaviour in a specific context. It is further the case that, while quantifiable measures, at the level of attitude, have potential to provide useful insight regarding relevant factors, they tell us rather less about why and in what way they are important.

What is grounded theory?

Grounded theory has its roots in sociology and social psychology, and is an approach which typifies much of the practice of contemporary qualitative researchers. At its essence is the concept of understanding, rather than attempting to quantify per se. Its central premise is that there has been an overemphasis in social science method on the verification of theory, rather than the logically prior process of discovering what concepts and hypotheses are relevant in a given context, by directly interacting with the subject matter. This interaction typically takes the form of broad based freely associative elicitation techniques with members of the population(s) of interest. Building upon these insights the objective is to identify hypotheses for future testing and verification. In essence the approach seems eminently sensible (for a detailed discussion of the approach see Glaser & Strauss, 1967).

Why is it then that so few studies seem to take the time to gain insight at this level, before plunging headlong into the quantification phase based upon non-contextualised *understandings*? In short, the psychology in the majority of academic psycho-social investigations into WRMSDs is open to criticism on the grounds that it has not been particularly social in terms of its method and, as a consequence, is at risk of underplaying the role of context.

Study design

The study reported upon here represents a pilot investigation into the basis for wide disparities in WRMSD reporting rates amongst sewing machinists at four garment manufacturing plants belonging to a common parent company. The investigation, conducted by a multi-disciplinary research team, comprising an ergonomist, an occupational psychologist and a biomechanist, sought to gain an insight in to the relative importance and interplay of psychological, physiological and environmental influences which might account for the observable differences in reporting rates.

From the outset it was known that one of the four sites had a considerably higher reporting rate than the other three, although the incidence of chronic disability apparent was limited to two or three cases annually at each. A principal advantage of the investigation was considered to be the presence of four well matched organisations, in terms of the tasks performed, the design and layout of workstations and the organisational structures present. In essence this study population had the appeal of offering an element of experimental control frequently absent in such studies, i.e.: if it could be established that differences in terms of the above criteria were minor or effectively absent then these variables might reasonably be excluded from the equation as determinants of the apparent disparities in reporting rates between sites. Two of the sites were in the North East, and the remaining two in the East Midlands.

A common approach was adopted at each of the sites, where attempts were made to:
• establish the extent and nature of musculoskeletal symptoms recorded in the established company reporting system;
• conduct an appraisal of the design and variability present in workstations;
• gain an appreciation of the physical nature of the work processes and tasks performed and the degree of variability present;
• address cultural influences by conducting individual and small group discussions with a range of personnel by taking a cross sectional slice through the organisation
• other variables appraised included hours of work, sickness absence and staff; turnover rates, methods of payment, work rate targets; levels of unionisation; and opportunities for alternative employment in each locality.

Discussion of findings.

Reference to table 1 reveals, perhaps foreseeably given the nature of the machinists' task, that reports of upper limb symptoms predominate. The relative epidemic is apparent in the wide disparity in reporting rates between Factory A and the other three. Significantly, a inverse association seems apparent between reporting rate and staff turnover, at face value this being suggestive of an association between symptom reporting and exposure.

Table 1. Nature and frequency of musculoskeletal complaints recorded 1990-7.

Recording period 1990 - 1997

Factory	Mean annual reporting rate	Most frequently reported complaints related to	Monthly mean absence rate all causes	Mean annual staff turnover
A	14%	hand, wrist; forearm & upper arm; upper back	11%	2%
B	1%	hand & wrist, upper back	10%	9%
C	<1%	*insufficient data*	4%	10%
D	3%	right hand, wrist & forearm	7%	6%

An initial appraisal of workstation designs revealed that there existed a high level of commonality at each site in terms of their principal dimensions. Variations were apparent, however, regarding the operating features, some being of two and others of three pedal operation. As far as could be ascertained the mix and distribution of make and model of machine were comparable at each site. The scope for adjustment to workstations was limited, principally relating to the position of foot controls, regarding which operatives expressed a range of preferences. At the beginning of the study all workstations were designed for seated operation, however, during the course of the study each site underwent a (parent company led) transition to standing workstations.

At the beginning of the study productive activity was organised in the manner of traditional line working, with each individual receiving payment in direct proportion to their output. The division of labour was high, with each machinist completing a brief discrete task. During the course of the investigation the company changed the mode of production to autonomous work groups, of between four and six members, under the title of 'teamworking'. In this orientation, machinists were required to be multi-skilled, and interdependent, the output based remuneration they received being dependent upon the productivity of the team. The transition to team working was further associated with the introduction of standing work, although some teams remained seated.

Individual and small group discussions with senior managers, supervisors and shop floor staff were conducted at each site, but were of greater frequency at sites A and C, these two sites constituting the extremes in terms of reporting rates. Discussions with supervisory and shop floor staff at Factory A revealed that a high priority seemed to be placed upon the reporting of musculoskeletal symptom. This level of priority appeared to be reflected in the presence of a site safety officer with ergonomics training. Supervisory staff also appeared proactive in monitoring staff symptoms, which, when reported, resulted in prompt attention from the site safety officer. Appraisal of staff musculoskeletal symptom reports at Factory A revealed that most were discrete, and nonrecurring, suggesting that the 'ergonomics' intervention had been successful. However, it was apparent that the scope for ergonomics interventions to 'modify' workstations was rather limited, amounting to little more than minor repositioning of foot pedals, or the substitution of a chair for one of the range of types available at the site, none of which appeared to possess any outstanding features, good or bad.

By contrast interviews at Factory C revealed a much lower emphasis being placed up -on the reporting of symptoms of discomfort. Supervisors in this context appeared to be rather dismissive of machinists complaints, to the extent that they seemed to constitute a significant barrier to reporting. It was not possible to establish the basis for this apparent impasse, although, speculatively, it might be viewed as consistent with the cultural legitimacy of reporting and level of priority ascribed to musculoskeletal stress within the management chain. Of note appeared to be the presence of a coping strategy amongst machinists, where reports indicated that modifications to their workstations, of an ostensibly equivalent nature to those at Factory A, were performed as a result of direct inter- action between machinists and site maintenance staff, i.e.: effectively circumventing the official reporting system. Given the limited scope for ergonomics intervention, the nature and quality of modifications made appeared indistinguishable from those at Factory A.

With regard to hours of work, each site operated a 39 hour basic working week, spread over 4.5 days in the case of Factory's A, C and D, and four days, in the case of Factory B. As noted above methods of payment during the early phase of the study were ascribed on an individual basis, but following the move to team working were made on a group basis, bonuses being achieved on the realisation of prescribed targets. Some variability in basic rates of pay and bonus payments were apparent between sites by geographical region. Exploration of issues surrounding alternative employment opportunities revealed that these were greater at sites B, C, and D, which it seems may, in part, account for the greater labour turnover at these sites. Reports of alternative employment in the North East (Factory C) were not expected in the economic climate of the mid 1990's. However, it was apparent that a significant proportion of machinists were motivated to migrate between a number of local sites as labour demand and rates of pay fluctuated with each sites acquisition of orders for garments. A differential in basic pay of around 50 pence per hour appeared sufficient to motivate a change of employer for some. Of further potential salience, within this context, was the finding that leavers frequently left to take up employment as machinists elsewhere rather than leaving the industry, and often appeared equally likely to return at some later date. This finding raises questions over the level of confidence we might have in labour turnover accounting for the disparities between sites in terms of variable symptom reporting rates

Conclusions.

The fact that during the course of the study each site went through a transition resulting in significant changes to the organisation of the productive process might be viewed as unwelcome from a traditional experimental perspective However, mapping this change arguably provided further insight. Central to many study findings in this area are issues surrounding the influence of job satisfaction, where the merits of task rotation and job enlargement tend to be stressed. Similarly, benefits are frequently considered likely to accrue from increased levels of autonomy and control over pace of work, all of which might reasonably be considered likely to be enhanced by the transition from traditional line working to autonomous teamworking in the current context. There was, however, no discernible impact on the relative magnitudes or differentials in reporting of symptoms between the four sites, and no significant reduction at Factory A. It is perhaps of note that, although limited data was available, the move to standing work at Factory A hinted at the presence of a shift in reported symptoms from upper to lower limbs. However, it would be unwise to draw any conclusions on the basis of the available evidence.

From the insights gained from this pilot study, it would seem not unreasonable to hypothesise that the observable disparities in reporting rates between the sites represented differences in the manner in which the system of reporting and recording of musculoskeletal symptoms was applied. Specifically, the available evidence seems suggestive of both attenuative and amplification effects. With regard to the basis for these effects, it seems apparent that the level of 'social' and 'institutional legitimacy' ascribed to reporting was variable between the sites. At Factory A it seemed apparent that reporting of symptoms had entered the social / cultural discourse to the extent that this was seen as a legitimate source of concern, which required managerial intervention. The cultural profile of the reporting of symptoms at the remaining sites appeared appreciably lower.

Of further note was the presence of a number of disbenefits associated with the move to teamworking. Reports from supervisory staff highlighted significant increases in levels of psychological stress, attributable to the need for greater involvement in human relations issues resulting from the significantly greater co-operation necessary for team working. The change to team working reflected a major cultural change for machinists, with the potential for antipathy between team members being enhanced by the need for greater co-operation in order to achieve performance targets and bonus payments. An inevitable outcome of the move to team working was that a performance league table resulted, with *effective* high performing teams typically exhibiting high levels of co-operation and multi-skilling. At the other end of the spectrum were less productive teams, comprised of individuals who felt less able to co-operate with each other. Here multi-skilling was effectively absent, team members appearing to revert to their previous orientation to the extent that their activity reflected line based production in microcosm. Other negative effects included the need for multi-skilling of machinists, and the financial disbenefits associated with greater time necessary for staff training, a particularly important issue in an industry with a relatively high staff turnover.

Based upon the insights gained from the fundamentally qualitative approach adopted, it would seem not unreasonable to hypothesise that within the current context primary variables worthy of further investigation relate to 'management & supervisory commitment' and 'social and institutional legitimisation' of musculoskeletal symptoms. With regard to the method of investigating these variables the authors would advocate methodological triangulation involving a contextualised psychometric approach.

The principal benefit of adopting a grounded approach for the investigation was considered to be that it provided a degree of insight that would likely have been absent from a less contextualised approach, particularly one restricted to the use of traditional survey tools alone. Taking the time to gain insight into the nature and salience of variables with potential to impact upon behaviour within a given context permits a more hypothesis driven, and scientific, approach to our investigations. Indeed, the monstrous and unwieldy survey tools which frequently result, in the absence of a hypothesis led approach, not only risk further muddying the waters, but arguably raise important ethical questions relating to the redundancy of unnecessarily collected data.

The issues discussed in this paper should not be interpreted as seeking to explain away reports of WRMSDs, but as highlighting the need to take due cognisance of the role of context in studies in this area.

References

Ahlberj-Hulter, G; Thorell, T & Sigala, F (1995) Social support, Job strain and Musculoskeletal Strain Among Female Health Care Personnel. *Scand. Journal of Env. Health;* 21 435-439

Fordyce, WE (1996) A Psychosocial Analysis of Cumulative Trauma Disorders; In Moon, D & Sauter, SL *Beyond Biomechanics; Psychosocial aspects of musculoskeletal disorders in office work.* Taylor & Francis.

Gazer B. and Strauss A. 1967 *The discovery of grounded theory: strategies for qualitative research*, Aldine de Gruyter.

Norman, RW (1994) Occupational Injury: Is it a psychosocial or a biomedical issue? *IEA, International Perspectives on Ergonomics*, Vol1

Sauter, SL Swanson, NG (1996) Work organisation, stress and cumulative trauma disorder, In Moon, D & Sauter, SL *Beyond Biomechanics - Psychosocial aspects of musculoskeletal disorders in office work.* Taylor & Francis.

ARE OCCUPATIONAL PSYCHOSOCIAL FACTORS RELATED TO BACK PAIN AND SICKNESS ABSENCE?

Serena Bartys[1], Malcolm Tillotson[1], Kim Burton[1], Chris Main[2], Paul Watson[2], Ian Wright[3], Colin MacKay[4]

[1]*Spinal Research Unit, University of Huddersfield, 30 Queen Street, Huddersfield, HD1 2SP*
[2]*Dept of Behavioural Medicine, Hope Hospital, Salford*
[3]*GlaxoSmithKline, UK*
[4]*Health and Safety Executive, Bootle*

There is increasing interest in occupational psychosocial factors and their association with back pain and disability. The workforce of a large multi-site company was invited to complete a booklet of psychosocial questionnaires, along with self-reports of back pain. Absence data were extracted from company records. Cut-off points for the psychosocial variables of distress, job satisfaction, social support, attribution (of back pain to work) and perceptions of organisational climate were established. Odds-ratios were then calculated for the back pain variables and certificated absence. The psychosocial factors (with the exception of attribution) were statistically significantly related both to reported back pain and absence in the previous 12 months. These results offer further support for the suggestion that psychosocial aspects of work should be addressed in clinical intervention for back pain.

Introduction

Musculoskeletal complaints are the most frequently reported work-related illness in the UK, are often attributed to work and are regularly associated with work absence (Hodgson *et al*, 1993). Of these complaints, that of low back pain (LBP) is the most common, and it is now established that disability due to LBP (reflected in the high level of social security benefits and lost production) has been increasing exponentially over the past few decades (Waddell, 1998). This is in spite of purported improvements in work environments and medical management.

The social and individual cost of musculoskeletal disorders (MSDs) is not a result of increased incidence (or prevalence); these rates have remained unchanged. A recent survey under the Occupational Physicians' Reporting Activity surveillance scheme reported that over a 4 year period 43,764 new cases of work-related disease have been reported, of which MSDs make up nearly half of all these cases (Cherry *et al*, 2000). However, only a small number of individuals contribute to the disability statistics and it is, therefore, a small number of individuals that account for the substantial costs to industry and the state. Additionally, though, there is also a group of individuals who take short, recurrent absences and also incur substantial costs to industry in terms of lost production. There is a clear need to investigate the factors which may predict, (and/or contribute to), a progression to chronicity or recurrent absence.

It is now established that psychosocial factors are major risk factors for chronicity in LBP, and the effects of certain clinical psychosocial factors, such as distress and somatisation, are well documented (Pincus *et al*, 2001). In addition to the established clinical psychosocial factors, the increasing amount of literature which documents the influence of work argues for a greater understanding of occupational psychosocial factors. The existing organisational literature emphasises a number of ways in which the individual may respond to their environment, but the mechanisms through which such psychosocial interactions might constitute obstacles to recovery from MSDs are not yet clearly understood. A literature review by Davis and Heaney (2000) summarises the hypotheses to date about how these mechanisms may operate: Firstly, it has been hypothesised that psychosocial factors are directly related to MSDs by influencing the perceptions of loading on the body or muscles. Another possible mechanism is based on psychosocial factors influencing various chemical reactions in the body that take place during the performance of job tasks, i.e. increased tension caused by fear of over-exertion or injury. This tension in turn restricts the blood flow and causes discomfort - the resulting pain then seems to justify the fear. Lastly, a mechanism that has an effect on the reporting of an injury is hypothesised to be a psychosocial influence that alters tolerance to pain. It is this final hypothesis that is of most interest here, as it suggests that differing responses to (perceptions of) environmental factors influence how the individual may accept and cope with pain or injury.

A broader understanding of how these factors operate, and how the individual perceives and interacts with the work environment would provide a potentially important expansion to further research investigating obstacles to recovery. There is still much that is unknown about the causes and the course of MSDs, and the role of psychosocial factors as obstacles to recovery in occupational back pain. Clinical guidance focuses on so-called 'yellow flags' (Kendall *et al*, 1997), but the clinical and occupational factors have not been disentangled. It has recently been suggested that psychosocial factors related to work may have a profound influence on occupational LBP (Burton and Main, 1999).

Traditionally, occupational health literature tends to focus on the effects that workplace factors have on biological processes and the immune system (i.e. stress). Acknowledging occupational psychosocial factors introduces the notion that work has social consequences and can place certain constraints on the individual. The concept also proposes that it is important, in terms of health outcomes, to investigate how the worker interacts with and perceives these factors. Thus, identifying obstacles to recovery from work-related MSDs is likely to require recognition of psychosocial influences that arise as a consequence of being a worker, and those which comprise individual experiences and beliefs.

To further investigate this dimension a workforce survey has been undertaken specifically to explore the relationships between musculoskeletal symptoms, related sick leave and psychosocial factors (both occupational and individual).

Methods

The workforce of a large multi-site pharmaceutical company (n=7,500) was invited to complete a booklet of questionnaires. Self-report data on musculoskeletal disorders (using an abbreviated version of the Nordic questionnaire), and data on clinical and occupational psychosocial factors were collected. The occupational psychosocial factors included job satisfaction, social support, attributing back pain to work, sources of pressure at work, and elements from the demand/control model. Clinical psychosocial factors included distress (measured by the General Health Questionnaire (GHQ)) and beliefs about the nature and progression of musculoskeletal disorders. Company sickness absence data were also collected for a 12-month period. The response rate for this survey was 62% (n=4,637).

Results

In order to look at the distribution of occupational psychosocial factors in this cohort, five occupational psychosocial factors were chosen for this analysis: job satisfaction, perceptions of social support at work, organisational climate and of control over the work situation, and also attribution (of back pain to work).

Statistical procedures were used to establish cut-off points (for each questionnaire) that indicated a statistically significant relationship with LBP. Table 1 illustrates the frequencies with which these occupational factors are displayed in the study population. Sixty-five percent did not score above the cut-off points for any of the occupational psychosocial factors, and the results shown in Table 1 relate to the 35% of the population that did display these scores. From these figures, the factor that is most prominent in the population is attribution of back pain to work factors (e.g. heavy lifting).

Table 1. Proportion of survey population with scores above the cut-off point on each occupational psychosocial factor (n=1,506)

Psychosocial variable	Proportion of population
Job satisfaction	27.1%
Social support	30.5%
Attribution of low back pain to work	50.0%
Control over the work situation	27.9%
Organisational climate	12.9%

Further exploration of the data involved determining statistically significant relationships between psychosocial scores and both back pain history and absence. Using the established cut-off points, odds-ratios were calculated for the outcome variables of self-reported back pain in the previous 12 months and previous 7 days, self-reported disability in the previous 12 months, and certificated absence in the previous 12 months. For each variable, the cut-off point is given in Table 2 along with an indication that this score (or a score above or below, depending on the scale direction indicated by the arrow) is associated with reports of back pain and absence.

Psychological distress is a well-established clinical psychosocial factor when looking at low back pain, but interestingly most of the occupational psychosocial factors here are also statistically significantly related both to reported back pain and absence. The effect size is less than that for distress in respect of self-reported back pain, but variously higher and lower for certificated absence. The variables that show the weakest association are attribution of low back pain to work, and organisational climate. Whilst the attribution variable showed to was the most prominent in the survey population, the results in Table 2 indicate that this belief does not necessarily have a strong association with the reporting of back pain or taking of absence.

Table 2. The association between cut-off points and self-reported low back pain (LBP) and certificated absence, expressed as odds-ratios(OR)

Psychosocial variable	Cut-off point	LBP 12m (OR)	LBP 7d (OR)	LBP disability 12m (OR)	Absence 12m (OR)
Distress (GHQ)	14↑	1.9	1.9	1.9	2.4
Job satisfaction	16↓	1.4	1.6	1.6	3.2
Social support	11↓	1.5	1.7	1.5	2.9
Attribution of LBP to work	41↑	ns	1.3	1.2	1.8
Control over work	11↓	1.5	1.7	1.4	2.1
Organisational climate	20↑	1.5	1.5	1.3	2.0

[Nordic questionnaire responses: LBP 12m = low back pain in the previous 12 months; LBP 7d = low back pain in the previous 7 days; LBP disability 12m = self-reported disability due to LBP in previous 12 months. Absence 12m = recorded absence for LBP in previous 12 months]

Although Table 2 indicates that certain occupational psychosocial factors do have associations with low back pain and absence, those results are for single psychosocial factors only. What is perhaps more interesting is to look at the effect of multiple psychosocial factors. Therefore, considering only those workers who have experienced low back pain in the past 12 months, Table 3 shows the percentages that took absence over the same period. Using the same cut-off points, odds ratios for 0, 1 or 2 or more occupational psychosocial factors, along with the GHQ factor were calculated.

Table 3. Percentage of workers reporting low back pain in the past 12 months who took absence (n=134)

	0 occupational factors	1 occupational factor	2 or more occupational factors
GHQ factor present	2.5%	3.5%	7.8%
GHQ factor not present	4.0%	5.8%	9.8%

There are two issues worthy of special note: firstly, the psychosocial factor effect seems to be cumulative, and secondly, the effect of any one occupational psychosocial factor alone is similar to one clinical psychosocial factor alone (in this case, distress) measured by the GHQ.

Matching the psychosocial data with the company absence data, it was found that 58% of absentees had scores on the occupational psychosocial variables that exceeded the cut-off points. Of those, 71% had scores that exceeded the cut-off point on more than one psychosocial variable.

Discussion

A recent review of biopsychosocial determinants of non-return to work following LBP (Truchon and Fillion, 2000) concluded that the role of psychological variables is emerging, but further investigation is required to specify the nature of the inter-relationships among them. The present study aims to address this issue, where the analysis presented here is a preliminary explanation of those relationships. It is recognised that the data are cross-sectional and it is not possible from these to determine the direction of influence. Nevertheless, statistically significant association between clinical and occupational psychosocial factors and both back pain and related absence were observed.

Numerous prevention and rehabilitation programs have targeted those factors assumed to be obstacles either to recovery or work retention, but few are successful (Waddell, 1998). This limited success could be due to the fact that such interventions have addressed factors with only a minor influence (such as education about the mechanics of the back), and have not addressed important psychosocial occupational factors, and therefore the effectiveness of treatment is considerably reduced.

There is some (albeit limited) evidence to suggest that a workplace educational program targeting beliefs about low back pain can be effective for reducing sickness absence (Symonds *et al*, 1995), and the results from the present survey offer support for the suggestion that perceptions about work and the causes of low back pain usefully could be addressed in a clinical intervention.

The recent Occupational Health Guidelines (Carter and Birrell, 2000) recognise both individual and work-related psychosocial factors as obstacles to recovery from low back pain but, apart from some general management principles, they found no scientific evidence on specific interventions. The guidelines call for further studies of innovative approaches to management specifically designed to overcome psychosocial obstacles to recovery, ideally, perhaps, within the occupational health environment. Following on from the workforce survey reported here, we are currently conducting a trial of an occupational health psychosocial intervention designed to target the factors found to be potentially important as obstacles to recovery; and hope to be able to report the findings in due course.

References

Burton, A. K. and Main, C. J. 1999, Obstacles to recovery from work-related musculoskeletal disorders, in W. Karwowski (eds), *International Encyclopedia of Ergonomics and Human Factors,* (Taylor & Francis, Inc)

Carter, J. T. and Birrell, L. N. 2000, *Occupational health guidelines for the management of low back pain at work - principal recommendations*, (Faculty of Occupational Medicine, London)

Cherry, N. M., Meyer, J. D., Holt, D. L., Chen, Y. and McDonald, J. C. 2000, Surveillance of work-related diseases by occupational physicians in the UK: OPRA 1996-1999, *Occupational Medicine*, **50**, 496-503

Davis, K. G. and Heaney, C. A. 2000, The relationship between psychosocial work characteristics and low back pain: underlying methodological issues, *Clinical Biomechanics*, **15**, 389-406

Hodgson, J. T., Jones, J. R., Elliott, R. C. and Osman, J. 1993, *Self-reported work-related illness. Results from a trailer questionnaire on the 1990 Labour Force Survey in England and Wales*, (HMSO, Norwich)

Kendall, N. A. S., Linton, S. J. and Main, C. J. 1997, *Guide to assessing psychological yellow flags in acute low back pain: Risk factors for long-term disability and work loss* (Accident Rehabilitation & Compensation Insurance Corporation of New Zealand and the National Health Committee, Wellington, NZ)

Pincus, T., Burton, A. K., Vogel, S. and Field, A. P. 2001, A systematic review of psychosocial factors as predictors of chronicity/disability in prospective cohorts of low back pain, *Spine*, (in press)

Symonds, T. L., Burton, A. K., Tillotson, K. M. and Main, C. J. 1995, Absence resulting from low back trouble can be reduced by psychosocial intervention at the work place, *Spine*, **20**, 2738-2745

Truchon, M. and Fillion, L. 2000, Biopsychosocial determinants of chronic disability and low-back pain: a review, *Journal of Occupational Rehabilitation*, **10**, 117-142

Waddell, G. 1998, *The Back Pain Revolution,* (Churchill Livingstone, Edinburgh)

MUSCULOSKELETAL DISORDERS

MUSCULOSKELETAL PROBLEMS IN THE NEW ZEALAND FOREST INDUSTRY

Liz Ashby, Richard Parker, Tim Bentley

Centre for Human Factors and Ergonomics (COHFE)
Private Bag 3020, Rotorua
New Zealand
Liz.ashby@forestresearch.co.nz

The forest industry is recognised as being hazardous, with high rates of injury. The gradual increase in logging mechanisation (which removes the need for manual felling) has helped reduce the risk of some of the serious injuries, but may be introducing alternative risks, for example due to long periods of time seated in machines. In New Zealand, musculoskeletal disorders have had little attention, as injuries of a more serious nature have been the focus of ergonomics safety and health research and intervention. There are many tasks in logging (harvesting) and silviculture (planting and tending trees) which contain factors known to be associated with musculoskeletal disorders. This study is the first part of work which aims to evaluate the extent and nature of musculoskeletal disorders among forest workers and implement effective injury prevention measures.

Introduction

Work in the forestry industry involves many tasks that are associated with risk of injury and is considered as a hazardous occupation. Logging in particular is associated with serious injuries and fatalities (Carruthers et al, 2000), and much work in the area of ergonomics and health and safety has necessarily focused on reducing these risks. In silviculture, where injuries are generally less severe, there has been little research in injury prevention strategies.

All major forest companies in New Zealand contribute to an industry Accident Reporting Scheme (ARS) run by the Centre for Human Factors and Ergonomics (COHFE) who use the reported injury information to determine trends, guide research efforts and measure effects of implementations. However, although the ARS provides extremely useful information, it is limited by the constraints common to reporting schemes, for example a reliance on willingness and ability to report and unknown accuracy of information.

It is particularly difficult to establish the extent of musculoskeletal disorders (MSD) among the workforce due to the cumulative nature of many of these disorders, as workers tend to report acute events more readily than gradual onset of discomfort or pain. This study aimed to establish an estimate of the extent and nature of MSD among silvicultural and logging workers in New Zealand using direct interviewing, alongside other supplementary sources of information, including the ARS.

Method

The project stretches out initially over a two-year period and will look at specific tasks and interventions to address MSD. Figure 1 illustrates the project plan.

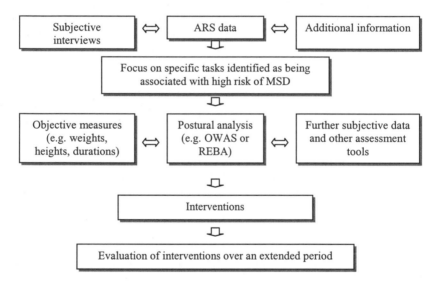

Figure 1. Plan for addressing MSD in the forest industry

The first part of the study involved gathering subjective information from a sample of workers and reviewing this alongside other data, including information held by government organisations. The subjective information was gathered through semi-structured interviews with silvicultural and logging workers. There is difficulty associated with gathering such information in both groups. Firstly, the literacy of forest workers is generally poor. A study carried out in 1999 found that 58% of workers tested with reading comprehension test had a reading ability below the adult level, with an average reading age of 11.5 years among this group (Cummins and Sullman, 1999). Therefore, it would be difficult to collect rich data using a written questionnaire. An alternative method of gathering information was used, which involved interviewing workers using a semi-structured questionnaire. This method still had associated difficulties. New Zealand has a diverse geography, with a widely distributed workforce and carrying out interviews face-to-face limited the locations that would be covered by the survey.

The lack of a workforce database means subjects tend to be selected from geographically accessible regions. Interviews were therefore carried out on a convenience basis, using company contacts to access crews from a variety of locations. The crews included silvicultural and logging workers, and within each group, subgroups were identified to include samples of main crew functions. For example, silvicultural crews are generally either thin-to-waste (cutting down small surplus trees) or pruning and planting, each having different risk factors associated with them. Logging crews include motor-manual crews (using a chainsaw to fell trees) or mechanised (where workers operate harvesting machines to fell the trees).

The interviews were optional, and the purpose of the study explained. Names were not recorded, and information was kept confidential. The questionnaire was comprised of three sections. A general information section recorded the age, height and weight of the worker. A measuring pole and scales were used and where measures were not practical, estimates were recorded. Details were taken on the length of employment in their current job, in the industry as a whole, and their previous job. The hours worked each day and number and length of breaks was recorded and workers gave an indication of their level of job satisfaction. A Nordic questionnaire-based section was completed to establish subjective reports of MSD. For each body part where the subject described a problem, further questions were asked. If the worker perceived the problem as being work related, details on the condition and associated tasks were asked. This included establishing from them which elements of the task they associated with either causing or aggravating their problem.

To supplement the gathered information, ARS data from a five-year period (1995-1999) was interrogated. One of the ARS fields, 'strains and sprains' captures reported musculoskeletal injuries, which are either acute injuries (such as a fall from a height or a slip) or gradual process or cumulative problems. The narrative description provided within each report enabled a judgement to be made on the type of injury and cause. Data were analysed to establish trends and common associated tasks.

Additional information was supplied by an Accident Compensation and Rehabilitation Corporation (ACC) industry report (ACC Injury Prevention Unit, 2000). In New Zealand any injury (whether work related or not) that requires treatment is covered by ACC, removing the litigation potential. ACC collects baseline information from any claimant, and a summary of this data relating to the forest industry was considered alongside the ARS and subjective findings.

Results

Subjective interviews
At the time of writing, 119 silvicultural and logging workers had been interviewed over a four-month period. Table 1 summarises the job category and characteristics of the workers interviewed. Of the interviewed workers, 77% reported pain or discomfort, during the 12-month period up to the interview, in one or more parts of their body, with a total of 246 reports from 92 workers. Of these, 60% of the reports were perceived as having a work-related cause. Of the 113 days reported lost over the previous twelve months, 87 were described as having a work-related cause.

Table 1. Worker characteristics

	Logging	Silviculture	Trainee (silviculture)	Trainer (silviculture)
Number of workers (n=119)	52	58	8	1
Mean height (cm)	180.1	174.1	177.3	178.0
Mean weight (kg)	89.7	80.7	80.8	95
Mean age (years)	33.8	28.7	24.5	33
Mean job experience (yrs)	4.83	4.19	0.02	0.17
Mean industry experience (yrs)	9.52	5.68	0.02	13

Figure 2 shows the reports described as caused by work, according to part of body. Low back problems accounted for the highest proportion of reported MSD, with 52% of the workers reporting problems. Of these, two-thirds were perceived as being work-related by the worker. Pruning in silviculture, and machine operating in logging were the two main tasks highlighted. The wrist and hands were the next largest group, the majority of these work-related reports being associated with use of the chainsaw, with the weight, vibration and wrist position being cited as contributory to pain and discomfort.

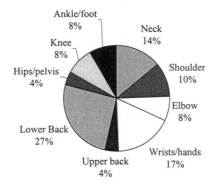

Figure 2 Work-related reports according to body part

The tasks associated with the work-related reports tended to relate to the current jobs (e.g. silvicultural interviews were with either pruning crews, who plant outside the pruning season, or thin-to-waste crews and this was reflected in their reported problems). Pruning workers mostly reported pruning actions and falls (either on rough terrain or from pruning ladders) as contributing to their reported MSD. They specifically identified high forces (e.g. large branches), overexertion (such as returning after a holiday), and overreaching (having to prune with outstretched arms). Thin to waste workers, who fell small trees using chainsaws, tended to describe similar causative factors to logging fallers such as slips, trips and falls, alongside chainsaw related factors like vibration, postural extremes and weight of the saw to carry around.

In logging, manual felling was associated with neck problems, perceived as relating to end of range movements and awkward postures when looking up at the top of the tree. Work involving chainsaw use, including skid site work and felling, produced accounts of upper limb problems.

Other tasks on the skid-site, like log making and trimming, were associated with discomfort from bending for long periods, as the logs are placed at ground level for processing.

ARS data

The ARS data indicated that both silviculture and logging were associated with reports of MSD. Many of the 'strains and sprains' reported to the logging and silviculture ARS were associated with slips, trips and falls rather than overuse or overexertion, and this is another area of for concern. However it is unclear whether the predominance of slips, trips and falls is due to the frequency of occurrence, or ease of reporting a specific event as opposed to a developing discomfort.

In silviculture, ARS reports were predominantly associated with pruning activities, whereby the worker uses either a chainsaw or pruning loppers to remove branches on young trees. The three main causes were identified as falling off ladders, injuries during pruning activities and slipping while walking between trees. In logging, tasks associated with skidwork (where initial processing of trees into logs takes place) are where most MSD are reported to the logging ARS. Recent analysis of skidwork showed that 'struck by/against' injuries were most predominant (72%), but were followed by slips, trips and falls (on the level or from a height, 20%) and manual handling (2%) (Parker and Bentley, 2000). Data from the ARS indicate that MSD are significant and increasing in proportion (Leov, 2000). Aside from slips, trips and falls, pruning activities and skidwork are the primary areas of concern.

ACC Information

ACC compiled data summaries on injury claims related to the forest industry. Job definitions in the ACC data do not allow comparison with ARS information. ACC data (1994/5 to 1998/9) indicated that the back/spine accounted for at least 17% of claims in the industry as a whole. Gradual process injuries accounted for 6% but with other soft tissue injuries (which will include a mixture of acute and ongoing injuries) accounting for 50%. 'Loss of balance' and 'lifting/carrying/strain' were among the main injury causes but specific tasks were not identified within the data. Although the data in the summary report was limited, it still indicated a need to further investigate MSD in the industry, to establish more detail on exactly which tasks contribute and why.

Conclusion

The study highlighted the need to use a variety of sources to establish an idea of injury extent and nature, particularly in an industry where gathering information is constrained. This first stage of the research (whilst not complete), provides information rich in detail. The high percentage of reported injuries indicates the need to address the physical demands placed on forestry workers, particularly as many reports were of an 'unknown' cause and could also be work-related in origin. Much of the work carried out by cross industry organisations is general and relies on raising awareness and educating in prevention. Within the industry there is scope to address specific tasks to reduce exposure of the workers to factors associated with MSD.

Priority tasks to consider are pruning, felling trees and other activities associated with use of the chainsaw, particularly where adverse postures are encouraged by task design. Workers provide valuable clues as to the harmful elements of the tasks that they perform.

Interventions need to focus on the primary factors associated with a given task, and establish ways to remove or avoid the risk. In many cases this will be using or redesigning equipment, for example improved pruning tools to reduce joint range of movement and force required, or better machine suspension and design. Where the risk cannot be further reduced, then organisation of the work, such as task rotation, planning to ensure adequate rest breaks are accounted for and encouraged, and training and education to assist workers to reduce the risks where it is feasible for them to do so.

At the time of writing the paper, crew visits were being planned to obtain further information about the highlighted tasks. This will involve obtaining baseline objective information (such as recording weights, forces, distances etc) and evaluating the tasks with a view to intervening. Other objective information will include postural analysis of primary tasks, using methods such as OWAS and REBA. Identifying specific problem areas will help focus interventions and design of new tasks, equipment or training.

Some intervention measures have already been investigated alongside the described study. For example Electromyography biofeedback (EMG biofeedback) has been trialed by a group of machine operators, as a possible training tool to reduce static work during normal machine operating activities (Parker and Wright, 1999). This trial has had positive results and will be investigated further.

References

ACC Injury Prevention Unit, 2000. *Analysis of claims data for the period 1994/95 to 1998/99 for (i) forestry and logging & (ii) log sawmilling and wood product manufacturing*. Unpublished Report.

Carruthers, P., Parker, R. and Ashby, L. 2000. *Analysis of lost time injuries – 1999*. COHFE Report, volume 1, number 2.

Cummins, T. and Sullman, M. 1999. *Reading comprehension levels in the New Zealand Forest Industry*. Liro Report Volume 24, number 12.

Leov, M. 2000. *A study investigating the factors contributing to musculoskeletal injuries in harvesting and silvicultural operations*. A dissertation submitted in partial fulfilment of the requirements for the Degree of Bachelor of Forestry Science at the University of Canterbury. University of Canterbury, 2000.

Parker, R. and Wright, L. 1999, EMG-biofeedback for forest harvesting machine operators. In *New Zealand Ergonomics Society Conference Proceedings*, Christchurch, NZ, September 1999.

Parker, R. and Bentley, T. 2000. *Skid Work Injuries 1995 – 1999*. COHFE Report Volume 1, number 4. COHFE, Rotorua, New Zealand.

ADDRESSING MUSCULOSKELETAL RISK FACTORS IN THE COMPLEX WORKING ENVIRONMENT OF TREE HARVESTING; INITIAL FINDINGS FROM MSD SURVEY AND FIELD STUDIES.

Clare Lawton, Ed Milnes, David Riley

Health and Safety Laboratory, Broad Lane, Sheffield, S3 7HQ, UK

In drafting practical guidance on ergonomics issues it is important to consider all aspects of the work task. However, in doing so there is a risk of being overwhelmed by the number of variables and volume of data that such an approach can generate. This paper discusses the initial stages of a research study investigating musculoskeletal disorders among tree harvester operators in which numerous potential risk factors are present. A broad approach is initially taken to formulate an appropriate research strategy. Consideration of the wider work context aims to assist in the identification of adverse interactions between organisational and technological design factors, intervention strategies can then be tailored accordingly.

Introduction

This paper reports the initial results from a major research project involving the UK forestry industry. This has been funded by the Health and Saftey Executive to investigate the prevalence of musculoskeletal disorders (MSDs) among tree harvester operators and the risk factors involved. Evidence of problems within the UK forestry industry has come to the fore over the last few years, with concerns being raised by the industry and unions. Technological advances have transformed tree harvesting from a heavy manual task into a relatively sedentary occupation seated at the controls of a mechanical harvester. The development of multiprocessing harvester heads has created a new generation of forestry vehicles which can fell, delimb, cut and stack a tree in a continuous sequence of operations. A positive effect of the change has been the decline of major injuries such as broken legs, amputations and lacerations. Unfortunately, such incidences have been replaced by reports of more subtle musculoskeletal injuries, the causes of which are harder to identify.

An initial industry survey suggested that over one third of the harvester operator work force had complaints regarding the upper limbs. MSDs within the forestry industry have received particular attention abroad, primarily in Scandinavia and Canada. In a study conducted by Axlesson and Ponten (1990), health investigations of 1,174 machine operators indicated musculoskeletal problems in 50% of the operators, mainly characterised by neck and shoulder complaints.

Research into tree harvesting has typically taken a microergonomics approach in tackling a specific and isolated area of interest: for example by specifically focusing on hand control design, cab design or vibration attenuation. However, despite significant developments in machine design and the incorporation of ergonomics design principles, there appears to have been little reduction in MSD prevalence (Axlesson and Ponten, 1990). This suggests that ergonomics studies/interventions in isolated and specific areas of machine design have not improved the situation as might have been expected.

Study context

Two types of machine exist: Excavator Based machines (Figure 1) which are tracked excavators commonly used for digging which have been adapted for tree harvesting purposes and, Purpose Built machines (Figure 2) which are wheeled machines designed specifically for the tree harvesting task.

Figure 1. Excavator Based Harvester   **Figure 2. Purpose Built Harvester**

The introduction of technological advances has led to a number of changes in the harvesting task. The main effects of these technological advances are as follows:

- Machines have become more expensive to buy which has resulted in operators working longer hours to make their machines profitable;
- Machines are more reliable reducing the time spent conducting maintenance tasks, however this is at the expense of increasing task invariability and isolated working;
- Internal computer systems provide continual feedback about performance and provide information which increases the cognitive demands on the operator;
- Machine functioning is faster creating highly repetitive and monotonous work tasks;
- Work pressures to match the high output capabilities of the modern machines are also significantly increased.

Tree harvesting occurs in a complex working environment with many interacting factors. These include:

Interacting Risk Factors

Psychosocial Issues	Organisational issues:-	Physical design factors:-	Environmental factors
Work pressures Isolation Unsociable hours Job security.	Payment systems Shift lengths Task variability Team work.	Hand controls Seat design Vibration, shocks, jolts Speed and precision of function	Terrain Temperature, Isolation, Rest facilities.

The approach

Initially a broad approach was taken to formulate an appropriate research strategy that would: 1. Enable the identification of adverse interactions between risk factors in the working context, and; 2. Identify areas of particular concern that require further investigation while taking account of the wider context. This paper reports on stages one and two; operator surveys and field studies respectively.

Stage 1 - Operator surveys
To gain further information on the prevalence of MSDs among tree harvester operators and possible causes, two operator surveys were designed and administered. These surveys incorporated the HSE version of the Nordic musculoskeletal questionnaire and questions covering psychosocial and physical design issues. Out of a national tree harvester operator population of approximately 350, a total of 129 completed surveys .were received. Findings indicated that the hands and wrists develop significantly higher levels of musculoskeletal problems in harvester operators than in a comparable working population ($p<0.01$). Compared to the general population harvester operators show significantly higher prevalence of trouble in all upper body areas (Figure 3).

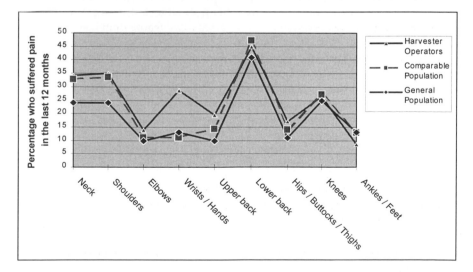

Figure 3.   12 month prevalence of MSDs by body area

Findings also indicated that Purpose Built machine operators showed higher pain prevalence than Excavator Based machine operators. This result was surprising as recent developments in the design of Purpose Built machines place greater emphasis on the importance of operator comfort. Results from other questions within the survey provided some insights into why differences between the two operator populations may exist. These include past health, psychosocial issues, organisational issues and physical differences in machine design.

Past health
It is possible that operators experiencing muscular pains from previous work such as chain saw use or operating Excavator Based machines are transferred to operate Purpose Built machines on the premise that these machines offer a less demanding method of working. It is likely that if the problem was serious enough it would continue to be painful whilst operating Purpose Built machines.

Psychosocial issues
Subjective psychosocial data indicate that Purpose Built machine operators feel more busy, more pressurised and find the work more mentally and physically demanding than Excavator Based machine operators. Even though Purpose Built operators do not differ significantly in the number of breaks taken during the day or hours worked during an average week. Purpose Built machines offer a more productive way of working. This may have a knock on effect on work pressures and working pace. Expectations for increased output from these machines are placed on the operators to meet new and higher targets. In addition, the machines are more expensive to buy, to justify the costs machine owners come under greater work pressures to achieve the extra output.

The perception of how busy you are, work pressures, and the effects of mental and physical fatigue can affect stress, muscle tension and fatigue. These psychosocial factors can increase the risk of MSDs and may provide an insight into why a slightly higher percentage of Purpose Built operators suffer from musculoskeletal complaints.

Organisational issues
The underlying cause of several of the psychosocial issues may be associated with organisational issues. The introduction of different work regimes have been developed within different regions / companies. Two person operating teams have been established in which tasks outside the operation of a harvesting machine are evenly distributed between each operator. Shifts of continuous operation of harvesting machines are shortened and are conducted alternately between other tasks such as site planning, product selection, maintenance, etc. Other methods of team working include operators working alternately between tree harvesting machines and forwarders (log loaders). Changes in work patterns and team functioning aim to reduce the stresses and work pressures placed on the individual. However, work patterns within such an environment have in some places proved difficult to sustain, either through reducing profits, or timing and location of machine usage. Again these are areas that require further investigation.

Physical design
Body regions affected by pain were also found to differ between the two operator populations. A higher percentage of Excavator Based machine operators complained of

shoulder problems than Purpose Built machine operators. Whereas a higher percentage of Purpose Built machine operators complained of hand and wrist problems.

Hands and wrists

Survey findings indicate that these differences may be related to hand control design. The majority of purpose built operators with hand and wrist complaints operated machines fitted with bear grip controls, whereas the majority of Excavator Based machine operators used hand levers. These findings although not conclusive, as they do not take into account previous operator occupations or previous hand control usage, do support earlier findings by Grevsten and Sjogren (1996). Survey results also indicate that Purpose Built machines are perceived by the operators as requiring greater precision and control. If machines require greater precision this may affect the extent of muscle tension to control fine hand movements within an environment subject to vibration and jolting movement. The survey highlights the need for further investigation in this area.

Neck and shoulders

Neck and shoulders problems were greater in the excavator based machine population, a higher percentage of these operators stated that their pains got worse due to awkward body postures. Survey questions on visibility, back and arm rest usage did not provide conclusive results regarding a relationship between these factors and neck and shoulder problems, further investigation into this area is also required.

Stage 2- Field studies.

Findings from the surveys provided the foundation for a series of field studies. Results indicated that perceived exertion and the physical design factors accounting for terrain, visibility, hand, wrist, neck and shoulder complaints required further investigation within the real working context. The field studies were conducted in the summer of 2000. Data was collected throughout an operator's normal working day. Six different machines were studied, two excavator based and four purpose built machines. The studies aimed to provide data on the interaction between machine design characteristics, the physical environment and the effects on working posture.

Perceived exertion of the operators was recorded using the Borg RPE Scale. Pressure pads were placed on the seat pan, arm rests and back rests to gather data on their usage. Inclinometers were fitted to the cab floor to gain data on cab inclination with respect to terrain and boom shock. Two digital video cameras were mounted inside the cab, one focused on the operator to gain posture information and the second was mounted to record the view of the operator. In addition, the actions of the machine were recorded from a safe distance. Hand posture data were recorded in two planes; flexion/extension and ulnar/radial deviation using goniometers. Production data were also gathered, on type of trees, the terrain, and the time spent on each task element. All data capturing devices were synchronised to enable each set of data to be analysed in relation to one another and interactions to be analysed. Analysis of the data is still on going, however initial findings regarding goniometer data indicate that different hand control designs do vary significantly in the postures that they require the user to adopt.

Bear grip controls (Figure 4) seem to require fewer movements at the extreme ranges of motion however, it should be noted that bear grips require pronated hand postures at all times. Previous studies indicate that pronated wrist postures are more likely to result in muscular complaints (Grevsten and Sjogren (1996), Putz-Anderson (1988)). Hand levers require semi-pronated postures requiring more gross movements in flexion and extension but require little movement in the ulnar/radial plane. Mini- levers require smaller movements of the wrist, with fine finger movements manipulating the control. Future analysis will be done investigating, percentage of time spent in various wrist angles ranges for each control and which task elements / design aspects of the controls create poor working postures.

Bear grip   Hand lever   Mini joystick

Figure 4. Three different types of tree harvester hand controls.

Conclusion

Our findings initially indicate that machines can still benefit from further ergonomic input in the physical design of machines, e.g. hand controls, visibility, etc. However, by considering the broader issues it can be recognised that the production of more ergonomically designed controls/cabs may not only increase operator comfort but may also further increase work efficiency and pace. This in itself may therefore counteract the potential improvement of such interventions. Taking account of the wider context directs the study towards aspects that require more detailed research and yet also emphasises the role of and the interaction of psychosocial and organisational issues. A broad approach enables intervention strategies to be tailored to meet the specific needs and constraints of the industry. Providing empirical evidence supporting a set of recommendations will encourage the industry in the adoption of any new guidance / recommendations that arise from the results of this study, particularly when changes in work patterns and organisation maybe required.

References:

Axlesson S.A. and Ponten B. 1990, New Ergonomic Problems in Mechanized Logging Operations, *International Journal of Industrial Ergonomics* 5(3), 267-273

Grevsten S. and Sjogren B. 1996, Symptoms and Sickleave among Forestry Machine Operators Working with Pronated Hands, *Applied Ergonomics*, **27**(4), 277-280

Putz-Anderson V. 1988. *Cumulative Trauma Disorders: A Manual for Musculoskeletal Diseases of the Upper Limbs.* (Taylor & Francis, London)

POSTURE AND DISCOMFORT

The use of the Portable Ergonomic Observation Method (PEO) to monitor the sitting posture of schoolchildren in the classroom

Sam Murphy and Peter Buckle

Robens Centre for Health Ergonomics
University of Surrey
Guildford GU2 7TE
s.murphy@surrey.ac.uk

Contrary to common belief, back pain amongst young people is a frequent phenomenon. Epidemiological studies have found high prevalence rates of back pain. The study reported here aims to identify the extent of back pain experienced by 11 to 14 year old schoolchildren, and establish the intensity, duration and frequency of exposure to physical risk factors present in schools. This paper considers the sitting postures of schoolchildren in the classroom. The sitting postures of eighteen children were recorded using three methods, the Portable Ergonomic Observation Method (PEO), video analysis and self-report. The three methods were compared using SPSS. PEO was significantly correlated with video analysis of the sitting postures after development of the method. Self-report was not significantly correlated with video analysis of the sitting postures. Therefore PEO was selected as the main observation tool in further analysis of children's sitting posture in this large UK based survey.

Introduction

Back pain is a significant burden on industrialised countries. If the symptoms and causes of back pain could be identified at an early stage the opportunity for remedial action would be improved. It has been shown that a strong predictor of having future back pain is a previous history of such symptoms, (Troup et al. 1987). A large portion of adult sufferers report a first onset of back pain in their early teenage years or in their twenties, (Papageorgiou, et al. 1996). It is commonly perceived that back pain amongst young people is uncommon. However epidemiological studies have found high prevalence rates of back pain, (Brattberg and Wickman, 1992, Troussier et al. 1994). The Robens Institute of Health Ergonomics is conducting a three-year study at the University of Surrey in conjunction with the Arthritis and Rheumatism Councils' Epidemiology Research Unit at the University of Manchester.

Objectives of the study
This study aims to:
- Identify the extent of back pain experienced by schoolchildren, aged 11 to 14.
- Establish any physical risk factors, which may be present in schools.

- Provide advice to prevent problems arising in the future.

Direct observation of children in the classroom was considered the most suitable method to use in schools to record posture, (Murphy and Buckle 2000). This paper describes the calibration of the observer and the development of the Portable Ergonomic Observation method PEO (Fransson-Hall et al. 1992), for use in the classroom to observe the sitting posture of school children.

Ethics
Permission was granted from the director of Education for Surrey, the Ethics committee of the University of Surrey and the Head Teacher of the school involved. The parents and children were each sent a consent letter informing them of the study with the option to withdraw at any stage.

Calibration of the observer

This calibration work involved direct measures in the laboratory. Measurements were taken of a group of subjects in the laboratory using the Lumbar Motion Monitor, (Marras et al. 1992). This equipment was used to set reference angles. The angles were those used in the PEO system. The observer spent 10 hours in the laboratory observing a subject bending while wearing the LMM to trunk flexion of 20° and 45°. Six female and five male students at the University of Surrey were then observed standing while wearing the LMM. The students were asked to bend forward by the observer to 20° and 45° flexion of the trunk, a total of 110 observations 60 at 20° trunk flexion and 50 at 45° trunk flexion were observed. Another observer recorded the readings from the LMM.

Calibration Results
The observations showed very good agreement with the reference measures. (Pearson's r = .817, p< .01). This was considered to be sufficiently good to enable the researcher to use the PEO based direct observations in the classroom environment.

Methods

18 schoolchildren aged 11 to 14 at a large secondary school in Surrey were studied, posture was recorded using both PEO and Video. At the end of each class the children were given a short questionnaire relating to sitting posture during the lesson. The lessons recorded included: Textiles, English, History, Geography, Science (computer based), Biology, Religious Education and Art. Recording took place in both morning and afternoon lessons. The video and computer equipment were set up at the back of the classroom and focused on one pupil per 30-minute session. The video camera was set between 1.5 and 2.5 meters from the subject in the sagittal view.

PEO
Observations of body postures were made in real time directly in the classroom using a Viglen Dossier 486 laptop computer. 18 children were recorded for 30 minutes each. The PEO screen set-up and the postures recorded were as follows:

Portable Ergonomic Observationmethod	Posture <w> Backseat <q> Frontseat	Activities
Post&Act	Trunk <e> Truflex > 20° <r> Truflex > 45° <t> Truflex > 60° <y> Trurot > 45°	<k> Workdesk <l> Look <m> Read <,> Talking
File 20 TIME		
CLOCK 20:28:14 (hh:mm:ss) Obs-time: 00:02:16 Obs-time: 154	Neck <d> Neckflex > 20° <f> Neckflex > 45° <g> Neckrot > 45° <s> Standing <a> Supported <z> Unsupported	<;> Listen
<Ctrl-Q> - Quit		

The categories included in the PEO system are selected according to risk factors in the literature, (Fransson-Hall et al. 1994, pp 97). Trunk flexion > 45° was added as a medium range measure for sitting as during pilot studies children had been observed sitting at this angle of flexion. Neck flexion > 30° was changed to neck flexion > 20° as this was the angle used during the calibration for trunk posture. All postures were recorded in relation to an upright sitting posture, i.e. trunk flexion > 20° was activated when the subject's torso was at an angle of 20° or more from vertical. Recording started around ten minutes after the lesson began to allow the children to settle down and become accustomed to the presence of the researcher. When the observer presses the pre-defined keys, continuous visual feedback is provided on screen by the posture being highlighted, the start of the event is recorded and when the same key or a mutually exclusive key is pressed the end of the event is recorded. At the end of the observation the PEO software gives a read out of the percentage of time spent in the recorded postures, the number of registrations of the postures, and the start and finish of each event.

Video analysis
The video was stopped at 15-second intervals and measurements taken directly from the screen. This was judged to be sufficient for postures that were held for longer periods and registrations that did not involve the estimation of angles. For postures that were held for shorter periods the time spent in the posture was measured from the time the subject moved into the posture until they moved out of the posture. This was calculated for one body area at a time (in seconds). During the observation neck flexion was recorded as one category regardless of the angle of flexion. This was due to the difficulty of judging this angle in real time recording.

Questionnaire
The questionnaire was distributed during the last ten minutes of the lesson. The questionnaire was concerned with the duration of sitting postures. The children were asked to look at drawings of people sitting at desks in three positions: Upright, 20° trunk flexion and 45° trunk flexion and asked whether they sat on the front or back of their seat. They were then asked which of the pictures were similar to the way they had been

sitting while working at the desk during the lesson and for how long in minutes they had been sitting that way, (a). They were also asked if they had to twist their back or neck, or bend their neck forward, to see the blackboard, teacher or to see their work during the lesson and to estimate the length of time they did each for that lesson, (b). The total usable subjects recorded by PEO, video and completing the questionnaire was 14.

            (a)   (b)

Results stage 1.

Percentage of time spent in postures
PEO showed very good agreement with the video analysis for trunk flexion > 20° and for sitting at the front of the seat, back of seat, arms supported and unsupported. Trunk flexion > 45°, trunk rotation > 45°, neck flexion > 20° and neck rotation > 45° showed poor agreement with the video analysis, (table 1.).

PEO and video analysis (table 1.)

Observation	Pearson's correlation
Trunk flexion > 20° (time)	r = .926**
Trunk flexion > 45° (time)	r = .450
Trunk rotation > 45° (time)	r = .241
Neck flexion > 20° (time)	r = .469
Neck rotation > 45° (time)	r = .402
Sitting at the front of seat (time sample)	r = .768**
Sitting at the back of seat (time sample)	r = .834**
Arms supported on desk (time sample)	r = .808**
Arms unsupported on desk (time sample)	r = .735**

**Correlation is significant at the 0.01 level

Number of posture registrations
PEO showed very good agreement with video analysis for trunk flexion > 20°, trunk rotation > 45° and neck rotation > 45° and good agreement with neck flexion > 20°. Trunk flexion > 45° showed poor agreement with the video analysis, (table 2.)

PEO and video analysis (table 2.)

Observation	Pearson's correlation
Trunk flexion > 20°	r = .831**
Trunk flexion > 45°	r = .445
Trunk rotation > 45°	r = .803**
Neck flexion > 20°	r = .644**
Neck rotation > 45°	r = .816**

**Correlation is significant at the 0.01 level

Time spent in postures
None of the questionnaire data showed agreement with the video analysis, (table 3.)

Questionnaire and video analysis (table 3.)

Observation	Pearson's correlation
Trunk flexion > 20°	r = -.327
Trunk flexion > 45°	r = .009
Trunk rotation > 45°	r = -.006
Neck flexion > 20°	r = .061
Neck rotation > 45°	r = -.145

**Correlation is significant at the 0.01 level

Development of the method

The recording of trunk flexion > 20° is sufficiently accurate to use in this study but the other flexion and rotation categories are not. This could be due to the overload of the observer (i.e. trying to record too many events at the same time). It was decided to reduce the number of categories. Sitting at the back of the seat, arms supported on the desk, trunk flexion > 60° and neck flexion > 45° were omitted. Trunk flexion > 60° was not used during the observation and the differences in neck posture were too difficult to judge in a real time continuous recording. The videotapes were then re-coded using PEO.

Results 2nd stage

Time spent in postures
PEO showed excellent agreement with video analysis for trunk flexion > 20° and trunk flexion > 45° and very good agreement with trunk rotation > 45° and neck flexion > 20°. Neck rotation > 45° showed poor agreement with video analysis, (table 4.).

PEO and video analysis (table 4.)

Observation	Pearson's correlation
Trunk flexion > 20°	r = .934**
Trunk flexion > 45°	r = 1.000**
Trunk rotation > 45°	r = .742**
Neck flexion > 20°	r = .780**
Neck rotation > 45°	r = .590

**Correlation is significant at the 0.01 level

Number of posture registrations
PEO showed excellent to very good agreement with video analysis for trunk flexion > 20°, trunk flexion > 45°, trunk rotation > 45°, neck flexion > 20° and Neck rotation > 45°, (table 5.)

PEO and video analysis (table 5.)

Observation	Pearson's correlation
Trunk flexion > 20°	r = .955**
Trunk flexion > 45°	r = .985**
Trunk rotation > 45°	r = .901**
Neck flexion > 20°	r = .858**
Neck rotation > 45°	r = .894**

**Correlation is significant at the 0.01 level

Discussion

All of the categories showed an improvement at the second stage. This could be due to the reduced number of categories, continued learning and the improved angle from recording from the videotapes. The observer and the video camera were observing the same events from the same angle. The closer agreement for the rotation registrations and video analysis than the time spent in the rotated postures may be due to the difficulty judging the position of the trunk, neck and head when the subject is turning. The registration key may be hit either slightly before or after the head has rotated back to a neutral position. The observer knew that the rotation had occurred but had difficulty judging exactly when the rotation began and ended. The calibration of the flexion categories improved the accuracy of the results for time spent in the posture and the number of registrations. As no calibration was used for the rotation categories the exact cut off points for registrations were not as clear and were left to the observer's judgment. Observation of females with long hair also made it difficult to see the neck. PEO used in the classroom does give an acceptable measure of the intensity, duration and frequency of sitting posture. The method does not measure the shape of the spine but gives an accurate measure of gross trunk posture.

Conclusions

PEO meets the requirements set out at the start of this paper. The method can be used to record the intensity, duration and frequency of sitting posture in the classroom and will be used in a further large-scale study to identify different sitting behaviours.

References

Brattberg, G. and Wickman, V. 1992, Prevalence of back pain and headache in Swedish school children: A questionnaire survey. *The Pain Clinic*, **5**, 211-220.

Fransson-Hall, C., Gloria, R., Kilbom, Å., Winkel, J., Karlqvist, L., Wiktorin, C. 1995, A portable ergonomic observation method (PEO) for computerized on-line recording of postures and manual handling. *Applied Ergonomics*, **26**, 93-100.

Marras, W.S., Fathallah, F.A., Miller, R.J., Davis, S.W. and Mirka, G.A. 1992. Accuracy of a three-dimensional lumbar motion monitor for recording dynamic trunk motion characteristics. *International Journal of Industrial Ergonomics*, **9**, 75-87.

Murphy, S.D. and Buckle, P. 2000. The occurrence of back pain in school children and triennial Congress of the International Ergonomics Association and 44th meeting of the Human Factors and Ergonomic Society, July 29-August 4 San Diego, California, 5-549-552.

Papageorgiou, A.C., Croft, P.R., Thomas, E., Ferry, S., Jayson, M.I.V., and Silman, A.J. 1996. Influence of previous pain experience on the episode incidence of low back pain: results from the South Manchester Back Pain Study. *Pain*, **66**, 181-185.

Troup,J.D.G., Foreman, T.K., Baxter, C.E., and Brown, D. 1987. The perception of back pain and the role of psychophysical tests of lifting capacity. *Spine*, **12**, 645-57.

Troussier, B., Davione, P., deGaudemaris, R., Fauconnier, J., Phelip, X. 1994. Back pain in school children - A study among 1178 pupils. *Scandinavian Journal of Rehabilitation Medicine*, **26**, 143-146.

ANTHROPOMETRY

THE USE OF 3D COMPUTER-GENERATED MANNEQUINS IN DESIGN INDUSTRY AND DESIGN EDUCATION.

Steve Rutherford[1] and Ken Newton[2]

[1]Senior Lecturer, Design Department,
Nottingham Trent University, Burton Street, Nottingham, UK.
[2]Senior Lecturer, School of Law, Arts and Humanities,
University of Teesside, Middlesbrough, Cleveland, UK.

In the field of computer-aided design (CAD) the use of 3D versions of the human form to assist in the design process has had a slow gestation. Ten years ago there were several systems available of varying cost. However, compared to other aspects of information technology, there has been little change in what is on offer over the past decade.

This work looks at the issues surrounding the use of 3D CAD human models, the poor uptake of them in industry and the teaching of these methods to designers. The paper examines a successful approach to the use of 3D computer-generated mannequins within a degree level industrial design course and, in an experimental situation with design students, compares the use of 3D CAD models with more traditional methods. The implications for the teaching of ergonomics to design students, specifically the use of 3D computer models, are discussed.

Introduction

The concept of accurate representations of the human form in CAD models of products and environments has been with us for several decades. In the life or death situations of military or space use the technology has enabled complex interaction to be tested more fully than ever before. Systems in use in these areas were not commercial, i.e. they were not widely available in the marketplace, but were normally the product of in-house research and development programmes. The use of 3D computer mannequins then spread. Systems became more commercial and available, principally with SAMMIE and Mannequin. However, although used in many high-profile projects the software has never in any major sense permeated down to the street level of smaller design companies. The lack of availability is one factor, but the lack of understanding of ergonomics and the irregularity of the methods used by designers is another (see Erbug, 1999 and Ruiter, 1999).

The concept and use of CAD mannequins will not be explained here, as it has been explained very well previously by Porter *et al* (1990). Despite this excellent advert for the use of systems such as these, the ergonomics methods used by designers, even today, are often woeful. Some projects not only require good ergonomics input but also lend themselves to the use of 3D CAD mannequins, but the uptake of professional ergonomics input is still very low. A recent survey has put the number of design consultancies and manufacturing companies using ergonomists at less than 30% (Rutherford, 1999).

With this level of mis-use or lack of use of ergonomics we should look to the education system to examine the methods employed by design educators to right this situation. There are examples of good practice in education at the present time. However, for the subject to be treated in so derisory a fashion in industry today there must be a history of a lack of tuition and expertise in ergonomics within the design education system.

The use of 3D CAD human models in design education

More and more design courses in areas such as industrial, product, transport and furniture design are including ergonomists as part of the course teams. This is improving the appreciation of ergonomics by designers. An institution benefiting from this is the University of Teesside, where Mannequin ergonomics software has been in use on their Industrial Design courses for over 10 years. It is successful in that many students use it at various stages of projects. The programme is largely project based, with modules tailored to the typical design challenges encountered in the real world. The first stage of the programme lays the foundations of industrial design, introducing students to the philosophy of design and exposing them to generic design problems at a variety of physical scales. The first year includes a specific module, Human Factors, which introduces students to the theoretical underpinning to related issues. Using a project – these have included VR simulators and more recently a single seat helicopter cockpit – students investigate issues relating to body size, joint movement, fields of vision and how best to communicate them.

Anthropometric data is acquired through published data (Pheasant, 1986) and digital databases such as People Size (Friendly Systems, 1994). Students use this information to establish worst case scenarios for clearance and reach, which are factored into a theoretical solution that represents a design envelope. Conflicts are then resolved which may include the need for adjustability. This theoretical arrangement is then tested through the mocking up at full size and running fitting trials with appropriate user groups. In tandem with this approach, students are introduced to the use of 3D CAD packages as an alternative or complimentary approach. Working in small groups, students are formally taught using exercises, which expose them to the functionality available. As part of the module submission students must show evidence of their use of all approaches and explain in a report how and why they differ. Initially, students find it difficult to grasp the constraints the human body actually has and find it frustrating that the computer model appears to be at variance with their understanding. As an example, when wanting to rotate the palm of the hand it is the forearm that must be manipulated rather than the hand itself, which is constrained by the wrist joint.

In the third year most students would use Mannequin in their final projects. Interestingly, the projects in which it has been most successfully used are where it provides story board characters, demonstrating how to perform certain tasks when it would be difficult to photograph someone performing them, underwater for instance - swimming aid, scuba equipment, etc. Mannequin also has a sighting capability which has been used in a number of projects to demonstrate how seating / screen arrangements have improved visibility.

Strengths and weaknesses of this approach

The integrated use of Mannequin as described above is encouraging. There would seem to be many instances of the use of the system highlighting problem areas or at least making designers more aware of the different factors involved. In contrast there seems to be quite a degree of freedom available to the students regarding the amount of use they make of the system or whether they use the system at all. In defence of this it should be added that during much of the time this system has been in use at the University of Teesside its use has been overseen by an ergonomist - one of the full-time course staff. What it has done very successfully is increase awareness, involving designers in a more intuitive way with the issues and aiding communication of the issues.

From the designers' point of view, they feel that there is no reflection of the true variability of the human frame and the difficulties of manipulation of the mannequin make it often difficult to mimic real situations. They seem to be aware of the limitations of the Mannequin system and are encouraged to follow this work up with full-size mock-ups. However, we know that the use of these systems is still not widespread. Are more traditional methods a substitute for the use of software such as Mannequin?

Comparison of 3D CAD mannequin use with more conventional methods of anthropometric analysis

If the use of 3D CAD mannequins is not widespread can we rely on traditional methods to offer similar benefits? To explore the use of basic ergonomics methods by designers an experiment was set up to compare 3 methods of anthropometric analysis. First year undergraduate product design students were introduced to the ergonomics methods and asked to complete a simple exercise -

"You are required to produce, to one-tenth scale, sketch, plan and elevation views for people working at a desk. You should communicate the 'zone of convenient reach' dimensions sideways, upwards and away from the body of someone sitting at an ordinary desk. Your aim is to define a zone that the majority of people can reach."

The following methods were used:
1. *Mannequin ergonomics software.*
2. *Bodyspace anthropometric tables.*
3. *Real people.*

All students were given some background to the proper use of body data for ergonomics design – awareness of percentile variation across the population and within individual's bodies; the use of worst case scenarios to predict problems. Three groups of 3 to 4 students used each of the three methods described. This enabled some comparisons to be made and added some validity to the results.

The exercises provided very graphic, immediate evidence of the way designers work. The groups using Mannequin were engrossed by the technology. The groups using the anthropometric data tables were not amused at having to deal with numbers. The groups using real people were enthusiastic but not prepared to search the building at length for more subjects for their exercise. This highlighted several important factors:

- The use of 3D CAD mannequins is a visual way of working which accords well with designers' other working methods.
- The difficulty of manipulating the mannequins, i.e. the unfriendliness of the software, limits its usefulness.
- The use of data tables is very unfamiliar to many designers, leading them to look for shortcuts or other means to justify their proposals.
- Getting real people for fitting trials is difficult and the variability of the measuring process added to the 'fuzziness' of the results.
- The lack of real people at the lower end of the percentile scale led to an underestimation of the minimum size of the zone of convenient reach in comparison to the other methods, although the resultant horizontal span of 1697mm still accords well with an average recommendation of 1650mm taken from 3 key text books (Sanders & McCormack, 1992; Kroemer & Grandjean, 1997; Pheasant, 1986).

In conclusion, the zones of convenient reach determined by the groups using anthropometric data and the groups using CAD mannequins accorded well. Both could be used as first attempt estimations for the building of test rigs.

The advantages and disadvantages of 3D CAD mannequins during the design process

There is a strong case for the use of 3D CAD mannequins in many areas of industrial design. Although they are flawed in well documented ways (Ruiter, 1999), attempts are being made to standardise their use and explain to users the limitations of future systems (Launis *et al*, 1999). In the future their use should improve, but it would still seem to be a long process. The disadvantages appear to be quite serious – lack of intricate manipulation of the manikins; lack of variability of percentiles within standard body frames; rigidity of the postures compared to real peoples' adjustments and body movement.

Ergonomists appreciate the complexity of human-machine interaction but 3D CAD mannequins may never mimic this complexity over the full range of uses people put products through. There is also a danger with computer systems which aid ergonomic design, in that they provide precise answers, to the nearest millimetre, which designers may be tempted to believe.

The advantages however are great, given the lack of appreciation of ergonomics within the design industry. As seen in their use at Teesside and other enlightened institutions, the visual communication aspects tie in with designers' working methods. This will help the designer working in a less serious area of design than aerospace or transport, such as furniture, to be more in control of the situation. It could also be a focal point for communication with ergonomists in more serious design projects.

The comparative exercise with the students obviously and immediately heightened their awareness of the issues. It would seem to be easy to communicate to design students the importance of getting ergonomics issues right in a design project. This is not simply stating an obvious fact though. Designers, and indeed the public at large, have a simplistic idea of ergonomics as a subject. The contrast that is important here is with these students' awareness and use of the term 'ergonomics' *before* the exercise took place. Before the exercise these students would speak with authority on a subject, ergonomics, that they plainly thought was theirs. This is an attitude very prevalent in industry today.

In the author's opinion, an important part of the communication to designers should be an awareness of the importance of product and professional liability in law, and the need for important decisions to be backed up by professional consultation. Designers must be made aware of the need for professional ergonomics input. Possibly, they should be dissuaded from all but the most elementary ergonomics methods in non-serious areas of design.

References

Erbug, C, 1999, Use of computers to teach ergonomics to designers, in: Mondelo, P., Mattila, M. and Karwowski, W. (eds.), *Proceedings of the International Conference on Computer-Aided Ergonomics and Safety CAES'99.* CD-ROM. Universitat Politechnica de Catalunya, Barcelona

Friendly Systems Limited, 1994, PeoplcSize, Anthropometric database software. Loughborough

Kroemer, K. H. E. & Grandjean, E., 1997, *Fitting the Task to the Human.* London; Taylor & Francis

Launis, Jones & Örtengren, 1999, A European and International Standard on the Anthropometric Characteristics of Computer Manikins and Body Templates, in Mondelo, P., Mattila, M. and Karwowski, W. (eds.), *Proceedings of the International Conference on Computer-Aided Ergonomics and Safety CAES'99.* CD-ROM. Universitat Politechnica de Catalunya, Barcelona

Pheasant, S., 1986. *Bodyspace.* London: Taylor & Francis

Porter, J. M., Case, K. & Bonney, M., 1990, Computer workspace modelling. *In:* Wilson, J. R. & Corlett, N. E., (eds): *Evaluation of Human Work.* London: Taylor & Francis, 472-499

Ruiter, I. A., 1999, *Anthropometric man-models, handle with care*, in: Mondelo, P., Mattila, M. and Karwowski, W. (eds.), *Proceedings of the International Conference on Computer-Aided Ergonomics and Safety CAES'99.* CD-ROM. Universitat Politechnica de Catalunya, Barcelona

Rutherford, S., 1999, The User Culture - User as Designer? The User-Centred Design Process – a Survey of its use and a Case Study, in *Design Cultures, Proceedings of the Third European Academy of Design Conference.* European Academy of Design, Sheffield.

Sanders, M. S. & McCormick, E.J., 1992, *Human Factors in Engineering Design.* New York: McGraw - Hill.

GETTING STUCK – THE ANTHROPOMETRY OF YOUNG CHILDREN'S FINGERS

Mic L. Porter

School of Design
University of Northumbria at Newcastle
Newcastle-upon-Tyne
NE1 8ST
mic.porter@unn.ac.uk

An incident occurred in a "Community Infant School" when a child, aged 3+, trapped a finger while playing with a water-bath. The bath was found to conform to the relevant safety standards and this apparent incompatibility acted as the *trigger* for this study. The study sought to obtain anthropometric data, graded by age, of the finger sizes to be found in children living in the North East and attending similar schools to the one which had experienced the incident. This data is outlined and compared with various published sources. Guidance is also given for Designers of consumer and other products with which young children might come into contact.

Introduction and background

A child attending a "Community Infant School" was able to trap a finger in the drain hole of the water-bath with which they had been playing. This was, obviously distressing and painful for the child and of concern to the School. The School referred the matter to the local Trading Standards Department which reported that the bath complied with the relevant safety standards. As this was clearly an undesirable situation, an anthropometric data collection project was started.

The finger length was measured, palm upward, with a metal ruler marked in half millimetres. The larger finger joint diameters were measured with standard UK Jewellers ring "sizer" gauges and a special set made to extended the range downwards and thus accommodate the smallest fingers. The traditional "sizers" are of known diameter and produced in steps of, approximately, 0.2 mm. These will yield more precise data than would be collectable with the commonly used drilled (in 1 mm steps) plastic sheet. Examples of the infant "sizers" are shown (overleaf) in Figure 1. Further details of the measurements techniques adopted can be found in Porter (2000).

The data reported in this paper has been revised from that reported previously (Porter 2000) to include subjects from another geographical area and especially to enlarge the sample taken from the cohort of pupils who would, shortly, transfer, at 7+, to a "Community Junior School".

The "sizer" ring diameter recorded was the smallest through which the finger would, without the application of force, freely pass. During the trial the child was instructed not to push or twist their fingers in an attempt to "make it fit". The experimenter was also vigilant to ensure an unforced measurement and thus avoid the possibility of a trapped finger. The diameter through which the finger would not pass ("just binds") was taken to be one size smaller in diameter.

Figure 1. Examples of the "sizers" produced in the School of Design

Three caveats to the study should be noted. The sample, like the population in the North East was largely *white* and included only about 3% of *visibly non European* children. Ethnic diversity issues have not, therefore, been investigated in this study.

It should also be noted that fingers are ellipsoidal (not circular) in cross section and deformable soft tissue overlies the structure of the finger. Thus care must be taken by a Designer considering non circular holes.

The soft tissue of the finger does deform when forced through a tight but thin hole and, for reasons not yet understood by the author, would appear to "flow" more easily when the ring "sizer" is inserted than when it is removed. No fingers were trapped during the data collection but the possibility was considered and removal strategies planned. It is expected that data collected with a drilled thin plastic sheet would have similar risks.

Results

The anthropometric data collected is available from the author. This database consists of finger lengths and joint diameters for children aged 6 months to 7½ years in six month sets. The data referred to is presented, in table 1.

Table 1. Distal joint diameter and (aged 7 only) finger length; both sexes in mm

Mean(SD)	Thumb	Index Finger	Middle Finger	Ring Finger	Little Finger	Criteria
6 - <8 mth. n = 20	10.1 (1.0)	7.7 (0.4)	8.1 (0.4)	7.8 (0.4)	7.1 (0.3)	"just binds"
30 - <36 mth. n = 20	12.6 (0.4)	10.0 (0.6)	10.3 (0.7)	10.0 (0.5)	8.8 (0.3)	"just passes"
3½ - <4 yrs. N = 51	13.3 (0.8)	10.5 (1.1)	10.7 (0.8)	10.3 (0.8)	9.1 (0.6)	"just binds"
7 - <7½ yrs. n = 31	15.6 (1.0)	12.2 (0.9)	12.4 (0.8)	11.7 (0.7)	10.7 (0.7)	"just passes"
	43.6 (4.2)	51.7 (3.2)	56.7 (3.9)	51.1 (4.6)	43.0 (3.9)	

Discussion and guidance for designers

Nearly twenty five years ago in the USA Snyder *et al* collected and published two sets (1975 and 1977) of anthropometric data for infants and children. Comparison of the data for the middle finger shows that the latter sample generally found slightly smaller values. The data published by the German Standards Institute (DIN 1981) is, where comparisons are possible, slightly larger than Snyder *et al* samples. However, direct comparison is impossible because of variations in the age bands reported. Norris and Wilson (1995) also quote limited details from Steenbekkers (1993) who reports longer, but much thinner fingers than the other samples.

In comparison to these data (table 2 overleaf) this survey is also in general agreement albeit with finger lengths generally larger and diameters generally smaller. Again, however, precise comparisons are often impossible due to the differences in age categories. This sample, however, does contain a greater quantity of hand anthropometric data collected from a sample of children under 7½ years old.

The "European Normalised" British Standard relevant to the safety of toys (BS 1998) has developed from BS 5665 (BS 1988) and the sections referring to holes are, essentially, unchanged. The test appears to be mainly concerned with preventing the child pushing their finger through the hole and coming into contact with something undesirable rather than simply becoming stuck. The Standards specify an "articulated accessibility probe" which comes in two sizes for products to be used by children under 36 months and another for between 36 months and 14 years. The Standard requires that both probes should be applied if the product will be used by all children under 14.

Table 2. Comparative data (both sexes) from published sources (mm)

Dimension and source	Thumb	Index finger	Middle finger	Ring finger	Little finger
Snyder *et al* (1975)					
7 - 9 mth. Length			36 (4)		30 (5)
distal joint diameter			8.4 (0.5)		7.4 (0.5)
31 - 36 mth. Length			45 (4)		33 (3)
distal joint diameter			10 (0.7)		8.6 (0.5)
79 - 84 mth. Length			56 (4)		41 (04)
distal joint diameter			11.9 (0.8)		10.1(0.7)
8 years Length			58 (4)		43 (4)
distal joint diameter			12.2 (0.8)		10.3 (0.7)
Snyder *et al* (1977)					
6 - 8 mth. Length	9.9 (0.7)		8.3 (0.7)		
2 - 3½ yr. Length	36 (4)	41 (4)	44 (4)		
distal joint diameter	12.4 (0.9)	9.6 (0.6)	9.9 (0.6)		
6½ - 7½ yr. Length	47 (4)	52 (3)	58 (4)		
distal joint diameter	15.0 (1.0)	11.7 (0.8)	12.0 (0.9)		
DIN (1981) (All standard deviations estimated from range data)					
3 years Length	37 (2.1)	42 (2.4)		45 (2.7)	36 (2.7)
7 years Length	46 (3.3)	54 (3.3)		55 (3.3)	44 (3.6)
8 years Length	47 (3.6)	55 (3.9)		57 (3.9)	45 (3.9)
Steenbekkeers (1993) (Standard deviations for diameter estimated from range data)					
3 years Length			49 (4)		
7 years Length			59 (4)		
0 - 2 mth. distal joint					6.5 (1.0)
3 - 5 yrs. distal joint					6.6 (0.5)
6 - 8 yrs. distal joint					7 (0.3)

The end of the smaller probe is 5.6 mm in diameter and 44 mm long. This would safely satisfy the data collected in this study. However, it is clear that the larger probe does expose children to the risk of trapped fingers if they are only a little older than 3 years but intent on "exploring". The child whose incident triggered this study fell into this category. For the older age group the data is 8.6 mm diameter and 57.9 mm long. The length standard is acceptable given that children whose fingers are this long will be prevented from insertion due to the hole diameter.

However, the Standard will expose some children to the risk of trapping fingers. For example, the data given in table 1 for mean and standard deviations of the 3½-4 year old child would imply approximately 4% of children risk jamming their Index finger and about 1.75% their ring finger into a hole that would satisfy the Standard. Of course, the risk of being able to jam the little finger is even greater; approximately 20 of that population.

There are two approaches to ensuring that fingers cannot get stuck in orifices or holes. The Index is to ensure that the hole is too small for the finger to penetrate and thus the critical dimension is for the youngest child's smallest finger. The data collected suggests that the critical hole diameter should be less than 5.9 mm (mean minus four standard deviations). Alternatively a less secure/satisfactory standard would be 6.2 mm diameter (mean minus three standard deviations). (Table 3, below)

Table 3. Data calculated from the means and standard deviations (SD)

	Thumb	Index Finger	Middle Finger	Ring Finger	Little Finger
6 - <8 months					
Mean - 3SD	7.1 mm	6.5 mm	6.9 mm	6.6 mm	6.2mm
Mean - 4SD	6.1 mm	6.1 mm	6.5 mm	6.2 mm	5.9mm
30 - <36 months					
Mean + 3SD	13.8 mm	11.8 mm	12.4 mm	11.5 mm	9.7 mm
Mean + 4SD	14.2 mm	12.4 mm	13.1 mm	12.0 mm	10.0 mm
7 - 7½ years					
Mean + 3SD	18.6 mm	14.9 mm	14.8 mm	13.8 mm	12.8 mm
Mean + 4SD	19.6 mm	15.8 mm	15.6 mm	14.5 mm	13.5 mm

The second, alternative, approach is to ensure that any finger (or the thumb) can pass freely through the hole without the risk of becoming trapped or of scratching against the edge. This general standard would need to be based upon the largest diameter finger on the largest child to be considered. Thus the critical dimension would be the mean plus three standard deviations or the much safer plus four standard deviations. For small items the general guidance is for the under 36 months. For this group a hole diameter of 14.2 mm or 13.8 mm respectively is implied. However, a hole of this diameter will have the potential to trap the fingers and thumbs of some of the older children. In the case of the 7 -7½ age group a 14.2 mm diameter hole can be calculated to be, approximately, equivalent to a 8th percentile thumb joint and thus a potential trap. The 13.8 mm hole could also trap fingers. The diameter of the desirable "clearance" hole for all children (7 -7½) may be calculated to be 18.6 mm (mean plus 3SD) or 19.6 mm (mean plus 4SD).

In the case of this second criterion, a further factor may need consideration from the point of view of safety. Is it possible for a finger inserted in the hole to come to harm from what is behind; either by physical contact or by sufficient proximity to an electrical source that a spark may jump the remaining gap. The length of fingers for the 7 -7½ age group is given in table 1. The longest finger is the middle and consideration of which suggests a clearance between the base of the finger and the risk should be 68.4 mm (mean plus 3SD) or the safer 72.3 mm (mean plus 4SD).

Conclusions

Children whose age is near to the bottom of the 36 month – 14 year range could, potentially, get their fingers caught in holes that met the Standards for which revision would, thus, be desirable. A trapped finger is unlikely to be "life threatening" but will be distressing and, potentially painful.

The mechanism by which soft tissue can roll and "squirm" through very thin objects, such as used in this and other studies, increases this risk and should be investigated further. It would appear that the flexibility of the soft tissue is different in each direction, generally it is easier to force the finger through the hole than it is to extract it!

The data collected in this study is more complete that that published in other sources and thus should aid the Designer, especially those working on consumer products.

Acknowledgements

Acknowledgement must be made to all the children who agreed to be measured, to their parents/guardians who gave the required permission and to the staff of the schools and nurseries involved in this study. To Derek Anderson, Jewellery Technician, within the School of Design who expertly produced the infant ring "sizers" essential to this data collection exercise and to Katherine Johnston who under took the pilot project.

References

BS, 1989, *BS 5665: Safety of Toys Part 1,Specification for Mechanical and Physical Properties*, (British Standards, London)

BS, 1998, *EN BS 71-1: Safety of Toys Part 1, Mechanical and Physical Properties*, (British Standards, London)

DIN, 1981, *DIN 33402:Body dimensions of people, June 1981*, (Deutsches Institut fur Normung (DIN) e V, Berlin) (Quoted in Norris and Wilson 1995)

Norris B. and Wilson J.R., 1995, *Childata*, URN 95/681, (Consumer Safety Unit, Department of Trade, London)

Snyder, R.G. Spencer M.L. Owings C.L. and Schneider L.W. 1975, *Anthropometry of US Infants and Children (SP394)*, (Society of Automotive Engineers(SAE), Washington, DC)

Synder, R.G., Schneider L.W., Owings C.L., Golomb D.H. and Schork M.A. 1977, *Anthropometry of infants, children and youths to age 18 for product safety design*, Report# UM-HSRI-77, (Consumer Product Safety Commission, Washington, DC) (Quoted in Norris and Wilson 1995)

Porter, M.L. 2000, *The Anthropometry of the fingers of Children*, In the proceedings of the IEA2000/HFES Congress, San Diego, (Human Factors and Ergonomics Society, Santa Monica)

Steenbekkers, L.P.A., 1993, *Child development design implication & accident prevention*, No 1 in Physical Ergonomics Series, (Delft University of Technology (TU Delft), The Netherlands)

BODY ASYMMETRY - AN OBSERVATIONAL STUDY

Bernard Masters[1]

Centre for Complementary Health Studies[2]
University of Exeter
BERNIE_MASTERS@HOTMAIL.COM

Anthropometric measurements were taken of healthy, pain-free sports female athletes, (n=71). Body symmetry is used to assess the horse, but the same criterion is not applied to the human. Neuro-physiology suggests that one side of the body can have a more efficient nervous, muscular or cardiac system, (a reflection of asymmetrical neural control?). Anthropometric tables do not detail the differences of human symmetry. Using simple measuring devices, the two sides of the body were compared, weight & shape, standing & sitting. Descriptive statistics were employed ($\alpha=0.01$). Unexpected results suggest that there are complex patterns of body shape/weight distribution with individuals having an unique *weight signature*. Further research into performance of the human vehicle and the implications of functional anthropometrics should be undertaken.

Introduction

As part of a much larger ergonomic trial focussing on equestrian saddle design, the work presented here relates to some of the preliminary anthropometric data. Testing seven different saddle shapes, and having assessed 71 athletes, (mean age 21) the study was halted for ethical reasons, the non-riders suffering pudendal and obturator nerve compression with numbness and muscle cramps in the legs (Masters 1999).

It is presumed that any human vehicle that is eccentrically laden will be subjected to accumulative stress producing uneven structural strain, leading to premature system failure. Logically, the greater the asymmetry, the faster the fatigue. Ergonomic tables rarely describe the differences between the two sides of the body, (Pheasant 1988, Bullock 1990, Dul & Weerdmeester 1993), yet there has been a limited interest in comparing performance between the two sides (Stefanyshyn & Engsberg 1993, Steele & Mays, 1995). There has been little research into the comparative function of the two sides of the body.

[1]*In private practice, specialising in chiropractic neurology.* [2]*This PhD research degree was supported by the Anning-Morgan Trust and the European Chiropractic Union.*

Literature review

Eccentric body function suggests that some individuals may have a more efficient nervous, muscular or circulatory system on one side or the other, (Annett 1973). This may be exacerbated by eccentric mechanical loading from handedness (Steele and Mays 1995), for 'overuse' of one limb could be an indicator of cerebral dominance (Balogun & Onigbinde, 1992, Bishop 1990a).   Breedlove (1997) contentiously proposed that differences in activity cause, rather than are caused by, differences in brain structure. One sided body usage may compound any adaptive changes in peripheral and central neural tissue, reflected in cerebral asymmetry or cerebral dominance, (Milner *et al.* 1966, Bishop 1990b).

Following injury, cerebral asymmetry - a product of brain plasticity - can produce subtle responses in the entire neurological system through gene expression (Lenn 1992). The resultant neuroplasticity allows cells to survive, sometimes with deleterious changes [de-afferentation or abnormal firing pattern].   In contrast, Morgan & Fields (1993) suggested that simple environmental stimuli can reduce pain, ('rub it better').

Le May (1992) related handedness to cerebral asymmetry, influenced by gender, race, position and other uterine factors (Galaburda *et al.* 1987), with certain diseases preferring the left side of the body (Bradshaw & Nettleton, 1983).   The growth patterns of embryological and foetal development (Bareggi *et al.,* 1994) showed differences in right and left long bone length.   Jee and Frost (1992) found many defects were due to skeletal tissue adaptation to mechanical usage, whereas, Puustjärvi *et al.* (1994) induced fibre type muscle changes by increasing activity.

Newman & Ainscough-Potts (1997) considered limb usage effect on spinal muscles, implying that over-use of one side affected the axial spine, (Bishop 1990c).   As the majority of people use the right side of the body for executing tasks, over-development of one side of the body could increase spinal muscle development and uneven body shape. Wall *et al.'s*, (1993) research on neural damage found sensory changes to be reflexogenic both centrally and in peripheral nerve activity, producing asymmetry in the motor cortex, (Acquadro & Montgomery 1996), thus altering the nervous system's ability to handle information, so injury to a single nerve is reflected in the non-injured peripheral nerve tissue.

Mankind's upright flexible structure would suggest a whole body interaction, controlled from the eyes downward, (Berthoz & Pozzo, 1988, Roberts 1995).   Butler & Major (1992) saw segmental control learned from repeated action whereas, Leonard (1998) viewed postural equilibrium to be maintained in part by the vestibulo-occular reflexes, with cerebellar feed-forward, feed-back, efference copy and multiple loops (Guyton 1992).   The loss of symmetry in the thoracic system may effect the efficiency of the cardio-respiratory pump (Mitchell, 1984, Lamb 1984, Schafer 1987), producing long term effects on respiration and cardiac output, particularly as these mechanisms are impeded by gravity (D'Arcy Thompson, 1992).   Siclare (1993) postulated Repetitive Postural Stress Patterns in the human frame and suggested that habitual muscle contractions may create asymmetrical pattern generators in the spinal cord.

The human body with an upright posture has preferred limb usage and this may affect the central nervous system, leading to greater efficiency on one side.   Task adaptation and pain avoidance may cause postural compensations simultaneously resisting gravity.   In order to establish the normal variations of shape and weight between right and left side of the body, an observational study was undertaken; 125 pain-free, healthy female athletes were invited to participate (α=.01), 75 non-riding athletes assigned to the control group and 50 professional horsewomen to the tested group.

Method

The athletes were measured and weighed in their 'natural' sitting and standing stance, the right and left sides were recorded. These naïve volunteers were measured only once, (assuming that learning of their postural differences could cause conscious re-adjust of learned patterns of movement, Guyton 1992b). Measurement using callipers (tested for intra-examiner reliability recorded the volunteers' shoulders and hips, both standing and sitting. Weight distributions were compared using two calibrated weight scales. Descriptive statistics were employed ($\alpha = 0.01$), with Spearman's correlation, SPSS computer programme was used to process the data.

Results

Table 1. Anthropometric measurement of standing & seated heights (mm)

	mean	median	mode	std dev	vari	range	mini	max	$25^{th}p$	$50^{th}p$	$75^{th}p$
body height	1643	1640	1650	49.0	2404	225	1525	1750	1615	1640	1670
left shoulder	1349	1350	1320	50.0	2506	220	1230	1450	1320	1350	1380
right shoulder	1350	1350	1380	49.5	2452	235	1220	1455	1320	1350	1380
shoulder difference	**1**	**0**	**60**	**0.5**	**54**	**15**	**10**	**5**	**0**	**0**	**0**
left hip height	1000	1000	970	39.6	1570	160	910	1070	970	1000	1030
right hip	1000	1000	970a	40.5	1640	180	920	1100	970	1000	1030
hip height difference	**0**	**0** *(a = multiple modes)*	**0**	**0.9**	**70**	**20**	**10**	**30**	**0**	**0**	**0**

	mean	median	mode	st dev	vari	range	mini	maxi	$25^{th}p$	$50^{th}p$	$75^{th}p$
seated height	871	870	880	27.8	775.0	155	805	960	855	870	887
left shoulder	597	600	600	28.9	837.5	120	540	660	570	600	620
right shoulder	593	590	570	27.9	780.5	120	540	660	570	590	610
shoulder ht differences	**4**	**10**	**30**	**1**	**57**	**0**	**0**	**0**	**0**	**10**	**10**
left hip height	235	240	230a	17.0	291	70	200	270	220	240	250
right hip height	235	240	220	16.0	258	70	200	270	220	240	250
seated hip height differences	**0**	**0** (a = multiple modes)	**10**	**1**	**33**	**0**	**0**	**0**	**0**	**0**	**0**

N.B. Of 71 subjects, 65 were right handed, and 6 left handed, 5 were left legged and 65 right legged, with one claiming *cross dominance,* left handed but right legged.

Table 2. Standing & Seated Body Weight on two scales [kilos]

	mean	median	mode	st dev	vari	range	mini	max	25thp	50thp	75thp
total weight	61.3	60	65	7.8	61.4	42	48	90	55	60	67
left leg wt	30.9	30	30	4.8	23.0	24	20	44	28	30	34
right leg wt	30.2	30	30	4.3	18.9	24	22	46	28	30	34
standing wt difference	0.7	0	0	0. 5	4.1	0	2	2	0	0	0
left seated wt	28.6	28	25	4.5	20.3	25	20	45	25	28	32
right seated wt	28.6	28	25	4.99	24.9	25	20	45	25	28	32
seated wt differences	0	0	0	0.49	4.6	0	0	0	0	0	0

Table 3. Correlation of weight & shape[rho =0 .01] (ρ^2=coefficient of determination)

	R seated wt	L seated wt	R pelvic height	L pelvic height
R sat wt	1.000	.691* (48%)	.296* (09%)	.359* (13%)
L sat wt	.691* (48%)	1.000	.204 (04%)	.249 (06%)
R pelvic ht	.296* (09%)	.204 (04%)	1.000	.926* (86%)
L pelvic ht	.359* (13%)	.249 (06%)	.926* (86%)	1.000

	Handedness	R shoulder stood	L shldr stood	L shdr sat	R shldr sat
Handedness	1.000	.018 (00%)	.057 (00%)	.118 (01%)	.041 (00%)
R shdr stood	.018 (00%)	1.000	.952*(91%)	.590*(35%)	.662*(44%)
L shdr stood	.057 (00%)	.952*(91%)	1.000	.636*(40%)	.649*(42%)
Lt shdr sat	.118 (01%)	.590*(35%)	.636*(40%)	1.000	.899*(81%)
R shldr sat	.041 (00%)	.662*(44%)	.649*(42%)	.899*(81%)	1.000

(*= statistically significant at 0.01)

Discussion

In spite of the fact that all the volunteers were young, fit, healthy, pain-free athletes, the results were unexpected, (see tables 1-3). Postural strategies and weight distribution may be a reflection of asymmetrical neurological function, and it was concluded that the individual's weight and shape strategies may be a unique neural record of their past activities and/or damage.

Each person appeared to have an individual weight distribution. Control subjects' *weight signature* appeared to be constant, (sometimes becoming more symmetrical when in pain). The weight signature was not always found to be synonymous with the individual's symmetry of body shape. Lopsided posture and eccentric weight distribution of the individual may contribute to an eventual physical breakdown. Body shape/weight indices could be regarded as an indicator of neurological integrity.

The relationship between the two sides of the body may be a reflection of the complicated mechanisms of standing and sitting posture. Upright posture may put high demands on blood supply for oxygen delivery to tissues. Some postural habits could reduce the efficiency of the musculo-skeletal system, impairing respiration and thus putting further demands on the arterial system. This may produce plastic changes in the nervous system accommodating to the learned posture.

There was poor correlation between reported handedness/leggedness, (see table 3) and between weight and shape. The relative shoulder heights and weights were not constant as the individual changed from the sitting to standing posture. Postural mechanisms may have been affected by previous body usage, possibly contributing to the individual's *weight signature*.

Postural mechanisms allow the human frame to compromise between flexibility and rigidity in order to keep upright and yet still perform tasks. Eccentric usage may allow one half of the nervous system to become more efficient. Implicitly, the less used body half may be compromised, and postural strategies may have subtle long term health consequences. Eccentricity of body shape could be used as an index to predict future physical health. With so few a number of subjects, this study has only just begun to describe the dynamic differences of shape and weight loading between the two sides of the human frame. Further work is needed to explore postural mechanisms as a branch of functional anthropometrics of the human vehicle.

Conclusions

These 'dynamic' records compared the right and left sides of the body with respect to weight and shape. The majority (*90%*) were right-handed implying that humans arc naturally *anti-symmetrical,* and that *dys-symmetry* is the norm. The data suggests that:-
(a) the human frame is flexible with complex patterns of body shape/weight distribution:
(b) The changes in relative shoulder and hip heights, from the seated to the standing position, were unexpected, but did not apparently correlate to handedness.
(c) The symmetry of shape was poorly related to symmetry of weight, those who stood evenly were not necessarily the same people who sat evenly.
(d) The subjects appeared to have an individual *weight signature.*
(e) The constancy of the weight signature is yet unknown.
(f) The long term consequence of asymmetrical loading needs investigation.

References

D'Arcy Thompson, W., (1917) & (1992) *On Growth and Form,* this edition 1992, Canto, Cambridge University Press, ISBN 0 521 43776 8, p. 288.

Acquadro M.A., Montgomery W.W., (1996) Treatment of chronic paranasal sinus pain with minimal sinus disease, *Annals of Otology Rhinolology Laryngelology,* **105** pp. 607-614.

Annett M., (1972) The distribution of manual asymmetry, *British Journal of Psychology,* **63**, 3, pp. 343-358.

Balogun J.A., Onigbinde A.T., (1992) Hand and leg dominance: do they really affect limb muscle strength? *Physiotherapy Theory and Practice,* **8,** 89-96.

Bareggi R., Grill V., Zweyer M., Sandrucci M.A., Narducci P. & Forabosco A., (1994) The growth of long bones in human embryological and fetal upper limbs and its relationship to other developmental patterns, *Anatomy and Embryology,* **189:** 19-24.

Berthoz A., Pozzo T., (1988) Intermittent head stabilisation during postural and locomotion tasks in humans. In: Amblard B., Berthoz A., Clarac F., (eds) *Posture and gait: Development, adaptation and modulation,* Elsevier, Amsterdam, pp. 198-198.

Bishop D.V.M., (1990) *Handedness and Developmental Disorder,* Lawrence Erlbaum Associates, Hove, ISBN 0-86377-288-9 (PBK).(a, p.70) (b, p. 18) (c, p.79).

Bradshaw J.L. & Nettleton N.C. (1983) *Human Cerebral Asymmetry,* Prentice-Hall Inc., New Jersey, ISBN 0-12-444646-1.

Breedlove S.M. (1997) Sex on the Brain, *Nature*, **Vol. 389**, 23 October, p. 801.

Bullock M. I., (1992) *ERGONOMICS,* The Physiotherapist in the Workplace, (Editor) Churchill Livingston, Edinburgh ISBN 0 443 03612 8.

Butler P.B. & Major R.E., (1992) Biomechanics of postural control and derived management principles, *Physiotherapy Theory and Practice*, **8**, 183-184.

Dul J., & Weerdmeester B., (1993), *Ergonomics for Beginners, A Quick Reference Guide,* Taylor & Francis London, ISBN 0 7484 0076 6

Galaburda A.M., Corsiglia J., Rosen G.D. & Sherman G.F., (1987) Planum temporale asymmetry, reappraisal since Geschwind and Levistky, *Neuropsychologia* **vol. 25**, no. 6, 853-868.

Guyton A. C. (1992) *Basic Neuroscience Anatomy & Physiology*, 2nd Ed., W.B. Saunders Company, Philadelphia, ISBN 0 7216 3993 3 (a p.224-239) (b, p.235-237).

Jee W.S. & Frost H.M., (1992) Skeletal adaptations during growth, *Triangle* **31**, 2/3 pp. 77 - 87.

Lamb D.R., (1984), *Physiology of Exercise, Responses and Adaptations*, 2nd edition, Macmillan, New York, ISBN 0-02-367210-2, p. 149, p.285.

Leonard C.T., (1998) *The Neuroscience of Human Movement*, Mosby, St Louis, ISBN 0 8151 5371 6, pp. 34-43 & 53.

Le May M., (1992) Left-right Dissymmetry, Handedness, *American Journal of Neuroradiology,* **vol. 13,** 493-504.

Lenn N.J., (1992) Brain Plasticity and regeneration, *American Journal of Neuroradiology,* **13:** 505-515.

Masters B., (1999) *An examination of the neuro-musculo-skeletal health of the horsewoman in relationship to saddle design*, vols. I & II, University of Exeter, PhD Thesis.

Milner B., Branch C., Rassmussen C., (1966) Evidence for bilateral speech representation in some non right-handers, *Transactions of the American Neurological Association,* **91**, 306-308.

Mitchell F.J. Jr., (1984) Respiratory-Circulatory Model: *Concepts and Applications of Neuromuscular Functions,* An international Conference on Concepts and Mechanisms of Neuromuscular Functions, Ph. E. Greenman Editor, Springer Verlag, Berlin, ISBN 3-540-13470-0.

Morgan M.M., & Fields H.L., (1993) Activity of nociceptive modulatory neurons in the rostral ventromedial medulla associated with volume expansion-induced antinociception, *Pain*, **vol. 52,** 1-9.

Newman D.J. & Ainscough-Potts A.M., (1997) Musculoskeletal basis for movement, in Trew M, Everett T., eds *Human Movement, An Introductory Text,* Churchill Livingstone, New York ISBN 0 443 04441 4, p. 55.

Pheasant S., (1988) *Bodyspace, Anthropometry, Ergonomics and Design,* Taylor & Francis London, pp 91-109, ISBN 0 85066 440 7

Puustjärvi K., Tammi M., Reinikainen M., Helminen H.J., Paljärvi L., (1994) Running training alters fibre type composition in spinal muscles, *European Spine* **3:** 17-21.

Roberts T.D.M., (1978) *Neurophysiology of Postural Mechanisms,* 2nd edition Butterworths, London, p. 251

Schafer R.C., (1980) *Chiropractic Physical and Spinal Diagnosis* Associated Chiropractic Academic Press, Oklahoma, ISBN 0-936948-00-0, p. 438.

Siclare R., (1993) Repetitive Posture Stress Patterns (RPSP), *Motion Palpation Institute,* Dec 17th.

Steele C., (1992). The assessment of sensory interaction in single leg stance: A pilot study, *Physiotherapy, Theory and Practice*, **8.** 175-177.

Steele J. & Mays S., (1995) Handedeness and Directional Asymmetry in the Long Bones of the Human Upper Limb, *International Journal of Osteoarchaeology* **vol. 5:** 39-49.

Stefanyshyn D. & Engsberg J.R., (1994), Right and left differences in the ankle joint complex range of motion, *Medicine in Science and Sport and Exercise*, **vol. 26,** No. 5, 551-555.

Wall J.T., Nepomuceno V. & Rasey, S.K, (1993) Nerve Innervation of the Hand and Associated Nerve Dominance Aggregates in the Somatosenory Cortex of a Primate (Squirrel Monkey). *Journal of Comparative Neurology, 337;* 191-207.

UPPER LIMBS

"COULD YOU OPEN THIS JAR FOR ME PLEASF STUDY OF THE PHYSICAL NATURE OF JAR Of ⌐

GE Torrens[1], G Williams[2] and R Huxley[3]

[1]Hand Performance Research Group, [2]Dexterity Research Limited,
Department of Design and Technology, Loughborough University, Loughborough
Leicestershire, LE11 3TU
[3]Department of Art & Design, Staffordshire University, Stoke on Trent,
Staffordshire
g.e.torrens@lboro.ac.uk

The Department of Trade and Industry have recently published a report highlighting the problems of opening foodstuff packaging. The aim of this pilot study was to observe and document the different techniques used by a range of male and female subjects as they tried to open one of the packaging types highlighted in the report as difficult to open, a vacuum-sealed fruit jam jar. Three types of fruit jam jars were assessed with four subjects, two males and two females. Each subject was physically characterised. During the task performance upper limb posture and the forces applied through the jam jar were recorded using a video camera, a CODA motion capture system and a universal grip dynamometer (UGD). The relationship between the physical characteristics and task performance will be discussed in relation to vacuum jar packaging design, and the appropriateness of the assessment methods used.

Introduction

The Department of Trade and Industry (DTI) indicates that shape, size, weight, surface finish, visuals and opening devices all play a part in the ease of package opening for all age groups. (DTI, 2000) The three physiological functions used in opening packs are visual, cognitive and manual (muscle). Visual function is employed in reading and in inspecting and identifying the mode of opening, cognitive function relates to understanding and adopting to perhaps unfamiliar mechanisms whilst the manual and muscular function relates to the force required to open the pack.

The DTI assessment of broad age-related issues for packaging states that after the age of 25-30 years there is a gradual reduction in the power and speed of muscular contraction, together with decreased capacity for sustained muscular effort. The report also states that if muscle strength deteriorates then there may also be a reduction in the relative accuracy of the movements. A 70 year is as weak as a 10 year old and only 65% as strong as a 20 year old. If the person also has a condition in the wrist and fingers such as arthritis then they will have further difficulties with packaging. The difficulties may be due to a lack of overall strength, weak grip, inability to squeeze, inability to press or lack of manual dexterity. (DTI, 1999)

There are no formal guidelines that lay down opening torque requirements for the full range of jar/bottle diameters; industry is generally governed by the loose 'rule of thumb'. It is suggested that the vacuum accounts for 80% of the torque required to open this pack type. (DTI, 2000)   Recent tests by the DTI have indicated that the 'rule of thumb' significantly overestimated the torque capabilities of consumers, especially the elderly and disabled.

This pilot study is contributing to a European Commission sponsored project named PACKAGE. The aims of the project are to improve upon the range of opening devices already available on the market and to provide suggestions minor modifications to packaging to assist in the ease of opening. However, if the redesign of packaging is to be successful it is essential to observe the ways in which people currently open jars and to examine the various physical techniques applied.

As the title of the paper suggests, many people have to ask other people to assist with opening and in the case of disabled and elderly people living alone, they quite often just have to wait.

Method

The assessments were carried out rooms with ambient temperature and limited direct sunlight. Four subjects, two female and two male, were chosen to provide a wide range of physical differences from a small sample population.   The test procedure was discussed with each subject.   Subjects were asked to confirm that they had no history of neuromuscular or musculoskeletal injuries, or diseases.   They were also asked if they had a skin condition that might be affected by cleansing of the finger prior to the assessments. The assessment of each subject was divided into two phases; (i) physical characterisation, and (ii) task performance.

(i) Physical characterisation

The physical characterisation of each subject was assessed using a previously documented series of methods (Torrens and Gyi, 2000).   The measurements taken included:

- Stature
- Weight
- Fingertip arm length
- Fingertip to first wrist crease
- Hand width (across MCP joints of digits 2-5)
- Hand depth
- Fingertip to first crease (DIP, digit 2)

- Fingertip depth
- Fingertip width
- Finger vertical compliance (vertical displacement of ungula proximal pulp between two parallel platens under a 10N force) using a prototype device developed in the Department.
- Grip strength
- Pinch strength

The results from the measurement of each subject's anthropometrics were related to a United Kingdom population from the software package PEOPLESIZE (Open Ergonomics, 1999). Grip and pinch strength, finger compliance and weight were related to unpublished data collected from students within the department of Design and Technology, Loughborough University, over the last four years.   The comparison provided some context for the assessments of the sample group within a larger U.K. population, to enable a more meaningful discussion of the outcomes.

(ii) Task performance

The methods used in the recording of task performance were based on those documented in previous studies (Torrens and Gyi, 2000, Torrens and Newman, 2000). The methods included:

- Finger friction (coefficient) when using three different jar top samples.
- Force and torque data capture using a Universal Grip Dynamometer (a prototype unit developed within Loughborough University).

- Motion capture using a CODA mpx30 system supplied by Charnwood Dynamics Limited, Leicester, UK. (http://www.charndyn.com)
- Grip pattern observation through video recording.

Finger friction

The coefficient of friction was taken from each subject's finger using a finger friction meter (a prototype device developed in the Department). The three sample tops came from proprietary brands of: (A) Apple, 66mm dia. Smooth edged top, 364gm; (B) Sandwich spread, 60.2mm dia. Smooth edged top; and, (C) Jam, 66.2mm dia. Knurled edged top, 650gm.. All jar tops were of a metal lug type, with a polymer-coated surface. A section of each of the jar tops had been mounted in turn, curved edge upwards, upon a fibreboard base using double-sided tape. The tape enabled a quick changeover of top sections between each sample testing.

Force and torque data capture

The subject was asked to open three types of unopened jar (A, B and C described above), repeating this task with the universal grip dynamometer (UGD) attached to the glass base of the jar. The dynamometer, with bracket attachment, weighed 4.5 Kg. Whilst the extra weight would affect how the base of the jar was held during opening, it was envisaged that it would not significantly affect the nature of the forces used to open the jar top.

Motion capture

The motion-capture system employed used infrared emitting markers that were placed over anatomical reference points on the upper limbs and head, including:

- Supraorbital foramen (right and left)
- Mandible (at the midline)
- Acromium point (right and left)
- Humerus (at the lateral epicondyle, right and left)
- Radius (at the styloid process, posterior, right and left)

- Ulna (at the styloid process, posterior, right and left)
- Digit 2 (at the Metacarpophalangeal joint, right and left)
- Digit 5 (at the Metacarpophalangeal joint, right and left)

The motion capture system was used primarily to record elbow flexion/extension and forearm/hand position through the wrist. There are many studies that indicate wrist deviation from a neutral position affects grip performance, notably Pryce (1980).

The distal phalanges were not marked for motion capture, due to the marker size in relation to the smallest fingers of the sample subjects and that the markers would be out of view of the single CODA system when the fingers curled around an object during a task. A three-system CODA set-up would be required (with CODA monitoring units set at right angles to each other in three axes) to ensure the majority of the movement of the phalanges was recorded. Pre-pilot assessments showed a jar opening task to task lest than two seconds if the subject could open it.

Grip pattern observation

   A video recording ensured that the phalange positions within grip patterns used by each subject would still be documented. The changeover between each jar sample enabled each subject to rest for a period of five minutes, to recover from muscle fatigue and enable the re-inflation of soft tissues in the hand through blood pressure.

Results and discussion

   The results of each subject's physical characterisation are shown in Tables 1 and 2. The results of the task performance are shown in Tables 3 and 4. A summary description of the processed results from the motion capture recordings and grip pattern observation follows Table 2. The total time taken to process sections (i) and (ii) of the trials was calculated to be two hours per subject, involving three operators. The time taken to process the physical characterisation and task performance results from the four subjects was approximately 8 hours.

   The comparison of stature of the subjects and U.K. data through PEOPLESIZE (Open Ergonomics, 1999) indicates that the males in the sample group are at the extremes of stature scale and the female at the mid to high percentiles of the same dimension. The grip strength of the larger male related to the higher values recorded from Design and Technology students at Loughborough University (20Kg-40+Kg). The grip strength of the smaller male relates to a lower range value within the student sample population. The vertical finger compliance from both females and males fell within the boundaries of expected values ranging between 1mm-4mm vertical displacement for females and 2mm-5mm for males. The results from the finger friction comparison of three jar tops are shown in Table 3. The performance results from the jar-opening task are shown in Table 4. The friction assessments were repeated due to a loss of data during processing, with the three subjects who were still available. These results are shown in Table 3.

Table 1. Subject physical characteristics

Subject	Gender	Stature mm	Percentile equivalent PEOPLESIZE	Weight Kg	Fingertip to elbow mm	Hand width MCP joints mm	Hand Length mm
1	M	1950	>99	96	525	93	210
2	F	1664	80	52	415	70	161
3	F	1775	>99	101	452	85	185
4	M	1601	1	79.9	428	85	172

Table 2. Subject physical characteristics

Subject	Grip strength Kg	Pinch strength Kg	Dominant hand	Fingertip Length mm	Fingertip width mm	Fingertip depth mm	Finger temperature C°	Finger vertical compliance mm
1	50	7.5	Right	30.34	18.3	16.19	30	2.84
2	26	2.65	Right	25.09	14.19	11.51	26	2.95
3	20	5.5	Right	26.38	16.03	13.65	34	2.88
4	33	4.25	Right	27.18	16.28	13.81	27	2.96

Table 3. Coefficient of friction values of right-hand digit 2 using three different jar tops

Subject	Finger temperature C°	Jar A	Jar B	Jar C
1	23	0.54	0.50	0.50
2	27	0.69	0.47	0.57
4	35	0.60	0.65	0.50

Table 4. Peak force (Kg) value from X, Y or Z axis for each of 4 subjects when opening three types of jar top with other axes values and x, y, z axes torques (Kg m) from same point in time. Positive force values are upward, positive torque values are clockwise.

Sample A	Force (x)	Force (y)	Force (z)	Torque (x)	Torque (y)	Torque (z)
1	-1.95	1.70	1.70	0.02	-0.10	0.40
2	-0.50	0.43	-0.10	0.00	0.06	0.03
3	-0.77	1.31	1.03	0.02	-0.07	0.07
4	2.78	-0.57	-12.6	0.03	0.26	0.90

Sample B	Force (x)	Force (y)	Force (z)	Torque (x)	Torque (y)	Torque (z)
1	0.11	0.48	0.64	0.01	0.04	0.05
2	-0.11	0.66	1.30	0.01	-0.08	0.10
3	0.08	-1.80	-0.94	-0.04	0.07	-0.44
4	0.50	0.90	-4.1	0.03	-0.15	0.43

Sample C	Force (x)	Force (y)	Force (z)	Torque (x)	Torque (y)	Torque (z)
1	-0.32	2.62	-0.80	0.08	0.00	0.97
2	0.21	0.48	1.4	0.00	-0.15	0.08
3	0.57	1.08	-3.78	0.02	0.29	0.20
4	0.10	-0.90	-11.43	0.07	-0.29	1.72

The finger friction results did indicate that more friction for grip on to the jar top was available using sample A. However, further investigation is required to provide more robust evidence. The coating on the surface of the jar top could change the friction qualities. It was not obvious if there was any difference in surface coating.

The grip patterns used in the task by all four subjects involved a clamping power grip in a mid-supinated position and a manipulating grip with the other. The non-dominant (left) hand in three of the four subjects was used as the clamp hand. All subjects used a composite grip of a power and lateral pinch grip with their manipulating hand. The extent to which the grip pattern could be identified as a lateral pinch came with increased difficulty of the female subjects in opening the jar or gripping the jar top size. The motion capture data indicated an increase in flexion in both hands when the subject found it difficult to open the jar.

The force data indicates that most of the forces and torques were low, under +/-2 Kg and +/-2Kg m respectively. The force values correspond to static torque dynamometer measurements taken by one of the authors. The forces and torques that exceeded these values were in the z (vertical) axis of the jar body and top. The largest force was −12.6 Kg exerted by subject 4 (small male), indicating some downward force. However, subject one (large male) produced the highest y-axis forces 1.70 Kg, indicating some side force application. The apparent difficulty of opening a jar compared to the forces measured seems incongruous. However, this phenomenon has been previously identified when assessing the difficulty of cutting meat with subject who had limited grip strength. This pilot study has raised issue that require further investigation.

References

Department of Trade and Industry, 1999, Assessment of Broad Age-Related Issues for Package Opening. Government Consumer Safety Research – Department of Trade and Industry Research commissioned by Consumer Affairs Directorate, (HMSO, London)

Department of Trade and Industry, 2000, A Study of the Difficulties that Disabled People have when Using Everyday Consumer Products' Government Consumer Safety Research – Department of Trade and Industry Research commissioned by Consumer Affairs Directorate, (HMSO, London)

Open Ergonomics Limited, 1999 PEOPLESIZE software, Loughborough

Pryce J.C., 1980, The wrist position between neutral and ulnar deviation that facilitates the maximum power grip strength. *Journal of Biomechanics*, **13**, 505-511

Torrens G.E. and Gyi D., 1999, Towards the integrated measurement of hand and object interaction. *7th International Conference on Product Safety Research*, European Consumer Safety Association, U.S. (Consumer Product Safety Commission, Washington D.C.), 217-226

Torrens G.E. and Newman A., 2000, The measurement of range of movement of the wrist: man or machine? In P.T. McCabe, M.A. Hanson, S.A. Robertson (ed.), *Contemporary Ergonomics*, (Taylor & Francis, London)

Forearm Discomfort for Repeated Isometric Torque Exertions in Pronation and Supination

L.W. O'Sullivan and T.J. Gallwey

Ergonomics Research Center, University of Limerick,
Plassy Technological Park, Limerick, Ireland

This study investigated forearm torque strength and forearm discomfort for right arm torque exertions. The first part of the study examined maximum forearm torque in both the clockwise and anti-clockwise direction at different forearm joint rotation angles. The second part of the study involved subjects exerting intermittent isometric torques at 20% of MVC in both directions at eleven pronation and supination joint angles. The results show that clockwise torque was strongest and that maximum torque in both directions were significantly affected by forearm joint angle. Discomfort scores from the intermittent exercises revealed that anti-clockwise exertions had considerably higher discomfort scores overall, and also, that exertions in both directions were significantly affected by forearm joint angle. Regression equations were developed that predict these relationships.

Introduction

There is growing concern about the lack of epidemiological studies investigating common forearm and elbow injuries such as pronator teres syndrome, lateral epicondylitis (tennis elbow), medial epicondylitis, and supinator (radial tunnel) syndrome amongst others. Based on a review of over 600 Work-related Musculo Skeletal Disorder (WMSD) studies of the neck, upper limb and back, Bernard (1997) indicated that there are fewer epidemiological studies addressing workplace risk factors of the elbow than for other WMSDs. One of the first references to WMSDs of the elbow is that of Cyriax (1936) via Bernard (1997) who defined a "chronic occupational" variety of lateral epicondylitis that appeared in patients who completed work involving repeated pronation and supination forearm movements with the elbow almost fully extended. Similarly Hughes et al. (1997) found a very strong relationship between elbow forearm disorders and the number of years of forearm twisting. In addition, Sinclair (1964) found a strong relationship for occupations that included the gripping of tools with repeated forearm supination and pronation.

A number of studies have documented tasks that result in cases of forearm/elbow injuries, the majority of which include forms of epicondylitis. For example, in the study of an engineering plant Dimberg (1987) diagnosed lateral epicondylitis in 7.4% of 540 workers surveyed. Silverstein (1998) associated epicondylitis with tasks requiring forceful laborious work, e.g. wallboard installation, roofing, masonry, foundries, building construction, furniture making, wood frame construction, paper products manufacturing, meat dealers, and concrete construction. All occupations involved repetitive, forceful work that required pronation and supination.

Bernard (1997) notes that overall, the majority of related epidemiological studies are supportive of the hypothesis of an increased risk of epicondylitis for occupations that involve forceful and repetitive work, frequent extension, flexion, supination and pronation of the hand and forearm. However, little is known of the relationships of forearm rotation and torque exertion direction on the propagation of injuries. The lack of epidemiological studies defining dose-response relationships between the various disorders and the levels and combinations of factors causing them, is a major impediment to their prevention.

In this study, forearm torque strength in both the clockwise and anti-clockwise directions and for various forearm rotations were evaluated. Equations were also developed that predict forearm discomfort in both directions in relation to percentage of forearm Range Of Motion (ROM).

Method

Experimental design

Twenty-two right-handed University students participated in a study that lasted 4 hours. Maximum torque strength in both the clockwise and anti-clockwise directions was measured at five forearm rotation joint angles, i.e. 75% and 30% of pronation ROM, neutral, 30% and 75% of supination ROM. Subjects also completed a total of 22 intermittent isometric trials of duration five minutes each, i.e. clockwise and anti-clockwise directions in 11 forearm rotation joint angles from 75% pronation ROM to 75% supination ROM. In each trial, subjects exerted torque equivalent to 20% of the maximum torque in the specified direction for one-second duration every five seconds i.e. 10 repetitions per minute. Forearm discomfort was recorded on a 100 mm visual analog scale (0 –10).

Apparatus

A Penny and Giles Biometrics electo-goniometer model Z180 was used to measure forearm ROM. Signals from the goniometers were amplified and interfaced with the PC (333 MHz) using a National Instruments data acquisition and A/D converter board (model PCI-MIO-16XE-50) with a BNC adapter board (model BNC2090). The forearm torque meter consisted a purpose built T-bar mounted with strain gauges. Voltage signals from the strain gauges were also interfaced with the PC using the BNC adapter board. Virtual Instruments (VIs) were written using G code in LabVIEW V5.0 to control the experiment. A series of separate VIs were coded for each part of the experiment and loaded dynamically into memory. The electro-goniometer and torque signals were configured within LabVIEW and displayed the readings in real time on the VDU for the VIs.

Procedure

For the duration of the experiment, subjects were positioned in front of the VDU at an adjustable height table set such that the included elbow flexion angle was 90^0. The experiment commenced with the measurement of maximum pronation and supination ROM followed by the measurement of maximum forearm torque in the specified postures. For the main part of the experiment, subjects were presented with a VI containing a six second cycle analog clock and a torque force meter that measured the subjects exertion as a percentage of their maximum values in that direction. Subjects were requested to maintain an exertion of 20% in the direction specified by the software for one second every six-second cycle. Each exercise was five minutes in duration followed by a one-minute rest before testing in the next posture. The discomfort scores were standardised for each subjects using the following procedure

$$\text{standardised value} = \frac{(\text{raw data} - \text{minimum data})}{(\text{maximum data} - \text{minimum data})} \times 10$$

Results

Repeated measures ANOVA was used to test if forearm joint angle (%ROM) and torque direction (clockwise/anti-clockwise) affected maximum torque exertion. The ANOVA results (Table 1) indicate that forearm joint angle had a highly significant effect on maximum torque (p<0.001). Both the direction main effect and the direction*ROM 2-way interaction also had a significant effect on max torque (p<0.05). The plot of the mean maximum torque strength values (Figure 1, including regression equations) indicates that the clockwise direction had the higher torque strength overall (15.4 Nm clockwise versus 12.8 anti-clockwise). The significant main effect for joint angle identified in the ANOVA is clearly evident in the plot of the mean torque values, as are the difference in mean strength values for both directions. The significant 2-way interaction is also supported by the plot, as clockwise strength diminished considerably for postures between neutral and 75% maximum supination ROM while the anti-clockwise values were affected only slightly.

Table 1 ANOVA results for effects of ROM and direction on maximum torque values

	SS	df	MS	F	Sig.
ROM	284.06	4	71.01	8.1	p<0.001
Direction	144.66	1	144.66	4.8	p<0.05
Direction*ROM	127.65	4	31.91	2.8	p<0.05
error	2393	210	11.39		

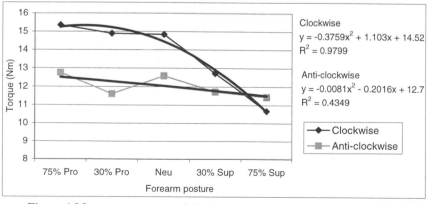

Clockwise
$y = -0.3759x^2 + 1.103x + 14.52$
$R^2 = 0.9799$

Anti-clockwise
$y = -0.0081x^2 - 0.2016x + 12.7$
$R^2 = 0.4349$

Figure 1 Mean torque strength for both directions and forearm joint angles

Repeated measures ANOVA was used to test if forearm joint angle (ROM, 11 levels) and direction significantly affected the standardised discomfort scores. The results (Table 2) indicate that both main effects had a highly significant effect on the discomfort scores (p<0.001) while the 2-way interaction was not significant (p>0.05). Figure 2 contains a profile plot of the mean standardised discomfort scores across all subjects including regression lines predicting discomfort for both torque directions. Figure 2 indicates that overall, anti-clockwise torque exertion had discomfort scores a lot higher than clockwise, with anti-clockwise discomfort scores for the neutral forearm joint angle similar to the clockwise values at 75% of pronation ROM.

Table 2 ANOVA results for effects of ROM and direction on discomfort

	SS	df	MS	F	Sig.
ROM	251.74	10	25.17	4.24	p<0.001
Direction	247.96	1	247.96	18.99	p<0.001
Direction*ROM	71.27	10	7.13	1.22	p=0.27
error	2725	462	5.9		

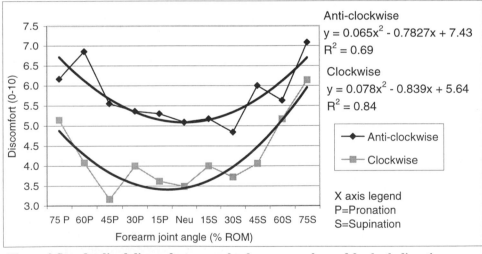

Figure 2 Standardised discomfort scores for forearm angles and for both directions

Both discomfort regression equations had reasonably good R^2 values (Figure 2, anti-clockwise 0.69, clockwise 0.84). While the discomfort lines were parallel for pronation postures, the discomfort lines for supination postures were not, in spite of a non-significant 2-way interaction in the ANOVA (Table 2). The non-significant 2-way interaction was unexpected and therefore examined in further depth. For each subject, a natural log transformation was applied to the original discomfort scores. The slope of the line through the transformed discomfort scores for both directions was obtained. Paired sample t-test was used to test if the slopes of the log discomfort lines were parallel for pronation and supination postures separately. The results indicated that for pronation postures, the lines were parallel but for the supination postures the lines were not (Table 3 t= 3.07, p<0.01).

Table 3 Paired sample t-test for 2-way interaction of direction and ROM

	t-stat	df	Sig.
Pronation	1.57	21	p=0.12
Supination	3.07	21	p<0.01

Discussion

The results show that forearm torque was strongest in the clockwise direction, but as the forearm joint angle tends towards the extreme supination ROM, the strength in both directions decreased with a considerable reduction for clockwise torque exertions. This suggests that at minimum, pronation and neutral forearm joint angles are preferable for tasks, especially those requiring forceful exertions such as screwdriver work and valve twisting operations. Data from the maximum torque strength exertions and the discomfort profiles both indicate that, irrespective of forearm joint angle, anti-clockwise exertions for the right arm were weaker than clockwise by approximately 20% in a neutral forearm posture. Also, anti-clockwise exertions resulted in discomfort scores approximately 40% greater than clockwise when tested at 50% MVC in their respective directions. Therefore it is fortunate that many mechanical controls and screws require right-handed threads i.e. clockwise twisting. But if the same operations are reversed, this may result in a considerably higher susceptibility to injury due to the weaker anti-clockwise musculature. It should also be recalled that clockwise exertions when completed by left-handed individuals involve the opposite weaker anti-clockwise musculature.

Although the ANOVA (Table 2) did not indicate a significant 2-way interaction for ROM*direction on discomfort, the 2-way interaction for the strength values in Table 1, and analysis of the discomfort scores separately for pronation and supination ROMs in Table 3, indicate that there was a combined effect for ROM*direction for supination postures only with greater increasing discomfort for clockwise exertions as the forearm joint angle tends towards the extreme of supination ROM. This may be due to the majority of the muscles controlling clockwise exertions crossing the elbow joint, and due to their lengths being greater affected by forearm rotation joint angle than for anti-clockwise exerting muscles. Even though the discomfort values for the clockwise exertions were lower than anti-clockwise, the 2-way interaction may be indicative of a greater susceptibility to injuries when in extreme supination postures. This is supported by the considerable reduction in maximum torque in the clockwise direction for supination ROM, and the significant two-way interaction identified in the t-tests.

Conclusions

1. Both forearm joint angle and direction significantly affected maximum forearm torque.
2. Maximum clockwise torque was affected more by joint angle than was anti-clockwise.
3. Intermittent torque exertions at 50% of MVC in the anti-clockwise direction resulted in considerably higher forearm discomfort (40% greater at neutral forearm angle).
4. The significant effect of joint angle on discomfort was modelled with regression equations.
5. As supination joint angle increased, discomfort increased more for clockwise exertions than anti-clockwise.

References

Bernard, B.P., 1997, Musculoskeletal disorders and workplace factors, NIOSH, Cincinnati, Ohio.

Dimberg, L., 1987, The prevalence and causation of tennis elbow (lateral epicondylitis) in a population of workers in an engineering industry, Ergonomics, 30, 573-580.

Cyriax, J.H., 1936, The pathology and treatment of tennis elbow, Journal of Bone and Joint Surgery, 18, 921-940.

Hughes, R.E. Silverstein, B.A. and Evanoff, B.A., 1997, Risk factors for work-related musculoskeletal disorders in an aluminium smelter, American Journal of Industrial Medicine, 32, 66-75.

Sinclair, A., 1964, Tennis elbow in industry, British Journal of Industrial Medicine, 22,144-148.

Silverstein, B.A., 1998, Claims incidence of work related disorders of the upper extremities: Washington State 1987-1995, Public Health, 88, 1827-1833.

PILOT STRENGTH AND SIDESTICK CONTROL FORCES IN MILITARY COCKPITS

Sue Hodgson

Human Factors Engineer, BAE SYSTEMS
Warton Aerodrome (W427E), Warton, Preston,
Lancashire, PR4 1AX, UK.
sue.hodgson@baesystems.com

A critical driver in the design of modern combat aircraft cockpits is to ensure that the target population is catered for. In particular, the introduction of women into the military cockpit has given rise to smaller pilots with reduced strength capability. The aim of this study was to investigate whether one sidestick force gradient would accommodate an extended population, or whether multiple models are needed. A flight simulation study was conducted in the BAE SYSTEMS flight simulation facility at Warton using subjects fulfilling the extended anthropometric requirements and ranging in strength from weak to strong. The results obtained show that one single sidestick force configuration could accommodate the extended population without compromising pilot performance.

Introduction

A key factor in the design of modern combat aircraft cockpits is ensuring that the target population can be fully accommodated. The cost of training, the scarcity of potential pilots, and the drive for maximising the potential for export have led to the expansion of the aircrew anthropometric population at both ends of the distribution. In particular, the introduction of women into the military cockpit has given rise to smaller pilots with reduced strength capability.

It is of prime importance that all pilots are able to attain displacement of the stick throughout its full range of movement whilst maintaining necessary precision. Failure to do so could compromise the aircraft mission and endanger the pilot. However, while stick forces must be low enough for the smaller and weaker pilots to handle, they must be high enough for the stronger aircrew to maintain a good feel of the aircraft, without the risk of over-controlling. The issue of strength in cockpit design is an important one, and one which has been largely overlooked. However the expansion of the aircrew population has given rise to the need for extensive research into this area.

The aim of the study was to evaluate whether one sidestick force model is able to accommodate an extended anthropometric range or whether multiple models, such as can be provided by active stick technology, are required. The intention was not to develop or define the actual sidestick force-displacement models, but to examine how performance with different sidestick force configurations is affected by a pilot's strength.

Method

Subjects
The experimental design required a combination of fast-jet test pilots, female commercial pilots and non-pilots. The non-pilot subjects were used to ensure that the required population strength/anthropometry range was represented. A total of thirty-one subjects (20 male and 11 female) participated in the simulation trials. These comprised six male BAE SYSTEMS pilots, four male subjects who were simulator proficient, nine British Airways pilots (2 male and 7 female), and twelve non-pilots (8 male and 4 female).

Strength measurements
It was important to the study that an extended population was represented in terms of both strength and anthropometry. There is no direct relationship between strength and anthropometry, which meant that the required subject strength range could not be determined from anthropometric databases, and so the population strength data used had to be specific to stick-based strength. Strength data measured in a study by McDaniel (1995) was provided to BAE SYSTEMS. The stick-related strength measurements done by McDaniel were in a centre stick position, whereas this study was related to sidestick forces. However, this was the most comprehensive stick based data available at the time, and is considered to be suitable for use in this study. Strength measurements were taken in both a centre stick position (in the same geometry as that of the McDaniel study), and in a sidestick position.

Isometric strength measurements were taken on the lateral and longitudinal axes for both the centre and sidestick measurements. The measurements were taken in the BAE SYSTEMS Multifunction Anthropometry Cockpit using a dynamometer with a stick handle attached. Analysis showed a highly significant correlation between the centre and sidestick measurements. This allowed the required strength population and percentile ranges for this study to be predicted and defined using the McDaniel population data. The anthropometric measurements taken were weight, stature, sitting height, shoulder-elbow length, elbow-fingertip length, hand length, bideltoid breadth, biceps circumference, and forearm circumference.

Simulation Tasks
The simulation tasks comprised fine and gross manoeuvring tasks, and included simple aircraft tracking tasks to be done by all subjects, and more complex tasks, including an air combat task, to be done by the pilots and simulator proficient subjects only. The simple flying tasks were designed for the non-pilots, to minimise training time (and the associated fatigue effects), and performance inconsistencies. The simulation tasks took place in the Flight Simulation department at BAE SYSTEMS Warton, and all tasks used the same Head Up Display (HUD) symbology, as shown in Figure 1. Nine different sidestick force-displacement gradients were used, and are detailed in Table 1.

Control of the aircraft in all tasks, other than the air combat, was by stick only, and no use of any other cockpit controls was required. The relative distance of the target aircraft in the tracking tasks was automatically controlled, to simplify and standardise the tasks.

The throttle was necessary in the combat task to allow pilots to allow better control in the engagement. Each task was repeated for all of the stick gradients.

The aim of both the fine and gross manoeuvring tasks was to keep the target aircraft image inside of the target circle. If possible the aiming cross at the centre of the circle was to be kept over the dot in the centre of the aircraft target symbol, which followed and overlaid the target aircraft image.

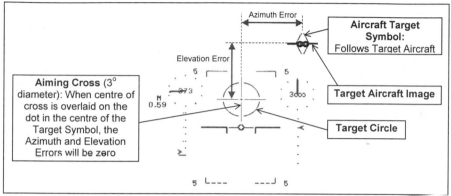

Figure 1: The HUD symbology used in the simulation tasks

The simple fine manoeuvring task involved precision movements of the stick to assess the feel qualities of each gradient. This task involved the target aircraft doing a series of gentle (1.5 g) turns every 5, 10 or 15 seconds. Each turn was a smooth transition in which the aircraft remained at the same height. The task lasted for 90 seconds. The more complex fine manoeuvring task for pilots was exactly the same, with turbulence added to make the task more difficult and increase the need for fine stick adjustments.

The simple gross manoeuvring task was designed to require large displacement of the stick for prolonged periods with shifts between axes, to address any strength and fatigue issues associated with the stick. The task started with the aircraft flying straight and level, followed by a series of 5.5g turns, and lasted for 90 seconds.

The pilot only gross manoeuvring task was a typical air-to-air combat task. The aim of the task was to obtain a lock on the target aircraft. This was achieved by keeping the centre of the target aircraft within the aiming circle for two seconds. The task ended when a lock was achieved, or at 120 seconds, depending on which occurred first.

Stick force gradients
Nine sidestick force-displacement gradients were used, comprising combinations of low, medium and high pitch and roll forces, the maximum values of which (at full displacement) are shown in Table 1. The forces used were based on flight control system experience the strength measurements taken, strength theory (e.g. Kroemer, 1970), and a simulation shakedown.

Design
The trials were of a mixed experimental design. The within subjects factors were the sidestick gradients, and pitch and roll levels. The between-subjects factor was subject strength. The independent variable in the trial was the sidestick force gradient for each

task. The dependent variables (metrics) were as follows:
- RMS (root mean square) error from the centre of the aiming circle. The RMS measurements comprise azimuth, elevation and radial error.
- The percentage of time spent with the target aircraft outside of the aiming circle.
- Time to achieve a target lock in the combat task
- Borg's Rating of Perceived Exertion (RPE) scale (Borg, 1982), used to assess perceived effort for each stick force gradient.

Counterbalancing was used in the ordering of the sidestick force gradients for each task to reduce fatigue and learning effects.

Table 1: Maximum longitudinal and lateral sidestick forces

Gradient	Longitudinal force (lbf)			Lateral force (lbf)		
	Level	Push	Pull	Level	Left	Right
1	Low	-10.2	18	Low	-9	7.41
2	Low	-10.2	18	Medium	-12	9.88
3	Low	-10.2	18	High	-14	11.53
4	Medium	-13.6	24	Low	-9	7.41
5	Medium	-13.6	24	Medium	-12	9.88
6	Medium	-13.6	24	High	-14	11.53
7	High	-17	30	Low	-9	7.41
8	High	-17	30	Medium	-12	9.88
9	High	-17	30	High	-14	11.53

Results

Repeated measures ANOVAs were used to test the effects of gradient on radial error and percent time out for all simulation tasks. These ANOVAs were of a one-factor design with nine levels (one for each gradient). Repeated measures ANOVAs were also employed to test the effects of pitch and roll levels on radial error and percent time out for all simulation tasks. The ANOVAs were of a two-factor design, the two factors being pitch and roll. Each factor had three levels – low, medium and high. Tukey's Honest Significant Difference (HSD) was used in both cases for post-hoc comparisons of any significant effects.

A Friedman ANOVA was used for the RPE ratings. This non-parametric test is only available in a one-factor design, so only the main effect of gradient could be examined. There are also no post-hoc tests associated with the Friedman ANOVA.

An analysis was done with regard to subject strength, to examine how performance with different stick gradients was affected by strength. The population was divided into weak, medium and strong, and grouped according to the average strength percentile for pulling back and lateral forces exerted on the stick. The weak and strong subjects were considered to be those at the extremes of the population, as shown below:
- Weak subjects: Subjects of ≤10th percentile strength
- Medium subjects: Subjects of 11th to 79th percentile strength
- Strong subjects: Subjects of ≥80th percentile strength

A repeated measures ANOVA was done for the radial error on the tracking tasks, with the addition of a between subjects factor for strength.

Discussion of Results

The analysis of radial error for the simple fine manoeuvring task suggested that it was not the actual stick forces that were an issue, but the way in which the pitch and roll forces interacted. A low roll force appeared to make the stick too sensitive for the fine movements required of this task, unless the pitch forces were matched and were also light. In terms of the simple fine manoeuvring task, the radial error and % time out statistics indicated that gradients 4 and 7 were undesirable, with gradients 5, 6 and 9 the most desirable in terms of performance.

The analysis of data for performance on the turbulence fine manoeuvring task did not yield significant results (p>.05), but appeared to concur with the results of the simple fine tracking task, in that low roll gradients resulted in higher errors. In addition, high pitch forces also seemed to result in increased errors, but these are just observations as the differences were not statistically significant. Again, this suggests that gradient 7 is undesirable. The RPE ratings for both the simple fine tracking and the turbulence tasks suggest that fatigue and physical effort are not an issue for these task types.

The analysis of radial error for the simple gross manoeuvring task showed that, although not statistically significant, gradients 4, 7, 8 and 9 had the highest error and gradients 2, 3 and 5 the lowest, suggesting that high pitch forces were an issue in addition to low roll forces. An analysis of pitch forces showed a significant main effect of pitch (p<.05) with low pitch forces resulting in significantly lower radial error. The analysis of % time out showed a significant main effect of gradient (p<.005), and that gradient 7 was significantly worse that most other gradients. It was also found that low and medium pitch forces yield significantly better performance than high pitch forces (p<.0005), and that low roll results in significantly worse performance than both medium and high roll (p<.005). The RPE ratings for this task suggest that exertion was an issue for gradients 7, 8 and 9 (those with high pitch forces).

The analysis for the combat task, although not statistically significant, was in general agreement with the gross manoeuvring task, in that low roll forces and high pitch forces yielded the highest error. No significant effects were found for % time out, since this was high for all gradients due to the nature of the task. The time to achieve a lock was not significantly affected by gradient. The RPE ratings showed significant effects, suggesting that exertion was an issue for gradient 9, which had high pitch and high roll forces.

The analysis of strength for the fine manoeuvring task showed no significant differences between the three strength groups, suggesting that strength is not an issue for this task, and that overcontrolling due to the low forces did not appear to be an issue for the 'strong' pilots (other than with low roll forces, which have the same effect for all three groups). It was found that for the gross manoeuvring task, weak subjects performed best with gradients 1, 2 and 3, and worst with gradients 8 and 9. Performance was much more consistent over the nine gradients with the medium and strong groups, suggesting that high pitch forces are an issue for the weaker subjects.

Observations and recommendations
Generally speaking, the results suggest that high pitch forces are an issue, especially for the weaker pilots, who are more likely to become fatigued or are unable to exert the required forces. Low roll forces are also undesirable, due to oversensitivity of the stick with low roll forces, and this appears to be the case regardless of strength. Although there was a concern with regard to the possibility of stronger pilots overcontrolling, low pitch forces do not appear to be an issue with regard to stick sensitivity.

For the fine manoeuvring tasks (both simple fine tracking and turbulence), it appears that stick forces are not an issue because of the relatively small stick movements required. This means that the subject is rarely, if ever, operating the stick at maximum force levels or having to make large or frequent shifts between axes. The only gradient that was actually undesirable for these tasks was gradient 7, with the high pitch and low roll force.

For the exertion tasks (both gross tracking and combat) there was a more pronounced distinction between the different stick force gradients. Pitch force was the pertinent factor in these tasks, especially for the weaker pilots. Low pitch forces appeared to be desirable, but not with low roll forces.

From the results obtained, it appears that an extended population can be accommodated by one set of stick force gradients. However, although gradients 1, 2, 3 and 5 resulted in the best performance overall, it does not mean that these configurations are optimal for all aircraft. The reason is that the stick characteristics and feel are not just a function of stick force and displacement, but also of factors like the flight control system and the response of the aircraft to stick movements. From the results obtained, it can be concluded that the preferred gradient is in the order of 10-13 lbf push/18-24 lbf pull for the longitudinal axis, and 12-14 lbf left/10-11.5 lbf right for the lateral axis.

Conclusion

The aim of this study was to evaluate whether one stick force model is able to accommodate an extended anthropometric range or whether multiple models, such as that provided by active stick technology, are required. The results obtained show that one stick force configuration will accommodate the entire proposed population without compromising pilot performance. The flight simulation trials revealed that the main issue is that of the weaker pilots becoming quickly fatigued, or even not being able to exert the required forces. The concern regarding the possibility of the stronger pilots overcontrolling due to low stick forces was unsubstantiated by the results.

Although absolute stick forces cannot be defined as a result of the study, due to the complex relationship between the stick and flight control system, it is recommended that pitch forces should be low, in the order of 10-13 lbf push and 18-24 lbf pull, while the roll forces should be more substantial, in the order of 12-14 lbf left and 10-11.5 lbf right.

Acknowledgements
Thanks go to Dr. Mike Llewellyn from DERA for the loan of the strength measuring equipment and advice on the study; Dr. Joe McDaniel for the strength/anthropometry data and advice on strength research; and British Airways pilots for their time and enthusiasm.

References

Borg, G.A.V. 1982, Psychophysical bases of perceived exertion, *Medicine and Science in Sports and Exercise*, **14**, 377-381

Kroemer, K.H.E. 1970, Human strength: terminology, measurement, and interpretation of data, *Human Factors*, **12**, 297-313

McDaniel, J. 1995, Strength capability for operating aircraft controls, *Safe Journal*, **25**, 28-34

DISPLAY SCREEN EQUIPMENT

THE ERGONOMIC RAMIFICATIONS OF LAPTOP COMPUTER USE

Steven Wilson

Ergonomic Work Systems Ltd
1 Hop Garden Way, Canterbury, Kent, CT1 3SH

This study examined the working situations of 48 laptop computer (LC) users. The most notable disadvantage reported was that of musculoskeletal discomfort, mainly in the neck. Ten subjects who used their LC both on a desk and on their laps were observed, photographed, and their postural angles measured. Sign tests demonstrated significantly greater neck, shoulder, and viewing angles when work was performed with the LC set up on the lap. Viewing distances were also significantly shorter when LCs were used on laps. Discomfort-time plots were compared for 6 subjects who completed two ninety-minute work periods, with the LC used on a desk and with the LC used on the lap. The most frequently reported areas of discomfort were the neck and thoracic spine. Regression analysis indicated that LC work on the lap is indeed less comfortable than LC work on a desk.

Introduction

Portable computers have been commercially available since the early 1980s. Their emergence has been the product of both advancements in computer technology and a drive toward devices that are smaller and more easily transportable. Apart from their characteristic portability, LCs have added advantage of being both space saving and energy saving [Saito et al 1997]. These characteristics made them perfectly suited for the small 'onsite' tasks for which they were initially intended [Straker et al 1997].

However the nature of many work environments is changing and many companies are seeing the advantages of a more mobile workforce. Indeed it is forecast that companies will increase investment to support these dispersed workplaces from less than 5% of their budgets to greater than 30% [Gartner Group, cited by Nunn, 1999]. The result is an expected boom in the portable computer market over the next five years [Nunn, 1999].

Laptop computers as a workplace hazard
However LCs might present hazards for users. The keys, keyboard, and screen are all smaller than a desktop computer (DC) and the screen is also hinged to the keyboard. Therefore there is little scope for adjustment, particularly in terms of screen height and distance. The height and distance of the screen impacts on the user's head and neck posture and the height and distance of the keyboard affects neck, shoulder, arm, and trunk posture. The result is that LC users might need to assume awkward postures. With advertising showing a trend to replace DCs with LCs [Saito et al 1997], it is possible that the growth in the LC market might be accompanied by an increase in LC related musculoskeletal complaints.

Laptop computer - background
The literature has hitherto focused on the ubiquitous desktop computer and the physical implications of their use are well documented. However there is a dearth of comparable research for laptop computers. Harbison & Forrester [1995] investigated the working environment of five accounting staff who used LCs 'in the field' for an average of 22 hours/week. The results showed considerable forward head inclination (average 45 degrees), and the highest ratings of discomfort for the neck and upper thoracic regions.

Straker et al [1997] found a 6.6-degree increase in mean neck flexion angle and an 11.5-degree difference in head tilt angle when using a LC compared to a DC. 75% of subjects also reported visual tiredness after using the LC for only 20 minutes. There was however no significant difference in discomfort ratings though it is possible that if the work task had lasted longer subjects might have developed more appreciable differences in reported discomfort levels. Saito et al, [1997] observed significant differences between laptop computers and desktop computers in terms of viewing distance, viewing angle and head angle.

Aims and Objectives of Present Study
Part 1 of this study examined the *working conditions* of LC users. Part 2 explored the postural ramifications of LC use through comparison of working postures of ten subjects who frequently used LCs, both on a desk and on their laps. The third part of the study explored the effects on perceived discomfort through time for two laptop set-ups, LC-desk and LC-lap. Previous studies have looked only at LCs set up on a desk and not for longer than 20 minutes.

Part 1: Interviews and focus groups

Method
Semi-structured interviews were conducted with a convenience sample of 15 laptop users. Focus groups were subsequently organised for three user groups, university students, National Health Service (NHS) employees, and managers from an insurance company. Respondents were asked about their perceptions of the advantages/disadvantages of LCs. Other information gathered included the type of work undertaken, the length of time LCs were used, and any physical consequences of their use. Those interviewees who reported physical symptoms of discomfort associated with LC use were asked to define the area of their symptoms on a body map. All interviews were recorded. Transcripts were analysed for common themes (advantage and disadvantage) that were, when identified, codified, and in the case of physical symptoms, counted.

Results

Twenty-eight subjects (13 males and 15 females) took part in five NHS focus group discussions, and two subjects (2 females) participated in a University student 'focus group'. The insurance company focus group was abandoned and individual interviews conducted instead. The mean age of subjects was 35.1 years, (19-55 years). The mean minimum time that subjects used LCs was 17 minutes (5 minutes – 1 hour). The mean maximum time per uninterrupted session was 2 hours, 41 minutes (20 minutes – 8 hours). The mean estimates for time used per week was 19 hours, 48 minutes (5 hours - 40 hours). The *advantage themes* that emerged were *portability* and *space saving, effective use of time*, and *family/social interaction*. However respondents also generally felt that the presence of a LC meant that there was often *no respite from work*. Users also complained of *hardware/expense* problems and the *postural and physical consequences* of LC use.

Part 2 - Postural Measurements

Methods
Ten subjects who regularly performed LC work on a desk and on their laps agreed to be observed while working in their normal workplace. Bony landmarks (C7 spinous process, acromion process, mid iliac crest, lateral humeral epicondyle, ulnar styloid process) were highlighted with adhesive markers and a spirit level was used to define a vertical reference point at each workplace. Subjects were given a settling in time of 5-10 minutes and photographs were taken every five minutes for 20 minutes with a camera positioned to the side of the subject. Postural angles were measured and averaged from the five photographs per subject. The data were analysed using SPSS software.

Results – Part 2
Mean age was 31.9 years. Sign tests showed that four of the six pairs (neck angle, shoulder angle, viewing angle, and viewing distance) varied significantly from the desk set-up to the lap set-up (at $p = .05$). Trunk and elbow angles were not significant.

Table 1. Results of Sign Test for Postural Angles

	Neck Angle (lap) – Neck angle (desk)	Trunk Angle (lap) – Trunk angle (desk)	Elbow angle (lap) – Elbow angle (desk)	Shoulder Angle (lap) – Shoulder angle (desk)	Viewing angle (lap) – Viewing angle (desk)	Viewing distance (lap) – Viewing distance (desk)
Exact sig. (2-tailed)	.004	1.000	.344	.021	.021	.021

Part 3: Discomfort-time

Method
Six subjects were recruited through an e-mail message to NHS employees. Candidates were excluded if they had experienced neck or upper limb discomfort in the previous three months that had resulted in treatment or time off work. Each subject was required to complete two 90-minute sessions at a LC, the second a day after the first. The order of

use was balanced across the group. The desk height was 720mm and the subjects were permitted to adjust the chair to their preferred settings. Efforts were made to ensure that all aspects of the environment were the same for both sessions. Subjects were asked to perform a series of tasks of the type described in the focus groups. At ten-minute intervals the subjects were presented with a body map and asked to define the whereabouts of any areas of current discomfort. They were also asked to rate the levels of that discomfort on a visual analogue scale. Prior to this subjects were asked for an assessment of overall discomfort using the same scale.

Results - Part 3.

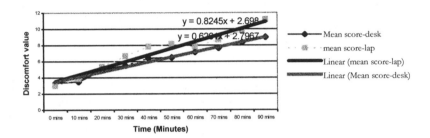

Figure 1: Discomfort-time Comparison

The mean age of subjects was 34.17 years. The figure above shows that the levels of discomfort increased with time and that the rate of increase in discomfort was greater for the LC-lap set up. A regression model was also explored using a logarithmic transformation of the discomfort data.

Table 2: Regression analysis of logarithmic transformation of discomfort data

Regression Statistics	
Multiple R	0.7688
R Square	0.5910
Adjusted R Square	0.5840
Standard Error	0.2938
Observations	120

ANOVA	Df	SS	MS	F	Significance F				
Regression	2	14.5975	7.2987	84.5326	1.9305E-23				
Residual	117	10.1020	0.0863						
Total	119	24.6995							
	Coefficients	Standard Error	t Stat	P-value	Lower 95%	Upper 95%	Lower 95.0%	Upper 95.0%
Intercept	1.3310	0.0566	23.5093	3.8305E-46	1.2188	1.4431	1.2188	1.4431
TIME	0.0120	9.3389E-04	12.8163	4.9821E-24	0.0101	0.0138	0.0101	0.0138
DESK	-0.1176	0.0536	-2.1926	0.0303	-0.2239	-0.0114	-0.2239	-0.0114

Discussion

Part 1: Interviews/Focus Groups
The most notable theme was the *postural/physical consequences*. The most common position for the LC to be used was on a desk though *90%* of users reported using their LC at some time on their lap. 45.8% of subjects reported musculoskeletal symptoms. This figure seems high although it is lower than the 60% reported in a group of Australian school children [Harris & Straker 1999]. The most common site of symptoms was the neck (35.4%). This is similar to the >38% figure reported by Harris & Straker [1999]. Disturbingly, many subjects stated that they ignored discomfort and continued work. An explanation for this behaviour might be found in the attitudes toward health and safety literature that accompanies most laptop computers. Only 7 of 18 interview subjects could say for certain that their computer came with safety and comfort guidelines and only four had read them. Despite all the problems mentioned 86.7% of subjects believed that the advantages of LCs outweighed the disadvantages.

Part 2: Postural angles
Mean *neck angles* for both set ups were well outside the limit of 15 degrees suggested by previous guidelines [Harbison & Forrester, 1995], LC-lap - 58.5°, LC-desk – 51.7°.

Mean *viewing angle* (LC-desk) was 35.8 degrees against 47.3 degrees (LC-lap). It has been suggested that monitors should be positioned at eye level to minimise the forces that need to be generated by the neck muscles [De Wall, 1992]. However Pheasant [1996] stated that the preferred zone of sight extends from the horizontal to 30 degrees below the horizontal and that a further 15 degrees is perhaps permissible. The mean viewing angle for the LC-desk falls inside this range while the mean viewing angle for the LC-lap set up remains outside.

The differences in viewing distance data were also significant illustrating a tendency towards shorter viewing distances with the LC-lap set-up (LC-desk-52.7cm, LC-lap-48.6cm). Pheasant [1996] regarded 50 cm as the absolute minimum viewing distance. Therefore it seems that in terms of viewing distance also both LC set-ups could be considered less than salubrious. Furthermore the evidence of this study suggests that the LC-lap set-up is perhaps the more hazardous.

There were also significantly greater shoulder angles for the LC-lap set-up (mean values 73.6°-LC-desk and 84.1°-LC-lap). This result could probably have been anticipated because the LC-lap position brings the device closer to the user's body, with the shoulders compensating by moving into extension. Nevertheless, it is not anticipated that this is especially pertinent. An angle closer to 90 degrees means that the arm is held closer to the vertical. This should be less hazardous because less muscular effort is required to support the arm in a vertical position.

Part 3. Discomfort-time plots
Both regression models took the form $y = a \times time + b \times desk + c + error$, where y was discomfort rating and desk was a 'dummy' variable that took the value 1 for LC-desk and 0 for LC-lap. It was expected that 'a' would be positive and 'b' would be negative. The models give 'a' (coefficient of time) positive and 'b' (coefficient of desk) negative as required. The p-values are all less than 0.05 so the values a and b are unlikely to have the signs they have by chance and therefore it can be reasonably claimed that it is more comfortable to use a laptop on a desk.

This much was perhaps evident from inspection of the 'trendlines' of the mean ratings of discomfort through time (figure 1). However the trend line might be misleading. A logarithmic transformation of discomfort data was required to eliminate *heteroskedasticity* when regression residuals were plotted. Therefore the relationship between discomfort and time might not necessarily be linear.

No statistical analyses were conducted on individual areas of discomfort. However it was noted that subjects tended to experience discomfort in the neck and thoracic spine, the same areas reported in part 1. Inspection of the raw data also revealed that subjects tended to experience discomfort in more areas throughout the course of the sessions.

Conclusion

Most users like LCs. They provide workplace flexibility and versatility. However it appears they can also be a workplace hazard. The remedial measures suggested by Harbison & Forrester [1995] seem appropriate; i.e. the provision of detachable screens that can be positioned independent of the keyboard. The subjects in part 1 of this study concurred with this suggestion with one caveat: the additions should not reduce portability.

DCs should be favoured, where possible. However, if portability is essential, then judicious use of laptops might be indicated, but only for incidental use and certainly not for large scale word processing tasks. If possible the LC should be docked with monitors and keyboards. When docking is not possible laptop devices should only be used for short periods of time, probably for no more than 15-30 minutes per session.

References

DeWall, M., Van Riel, M., and Snuders, C. 1992. Improving the sitting posture of CAD/CAM workers by increasing VDU monitor working height. Ergonomics, 35, (4), 427-436.

Harbison, S. and Forrester, C. 1995. The ergonomics of notebook computers: problems or just progress? Journal of Occupational Health and Safety-Australia/New Zealand, 11, (5), 481-487.

Harris, C. and Straker, L. 1999. Survey of the physical ergonomic issues associated with school children's use of laptop computers. Correspondence with the author.

Nunn, J. 1999. "Upwardly Mobile" –CBI News (The magazine of the Confederation of British Industry). January edition. London, 40-42.

Pheasant, S. 1999. Bodyspace. Anthropometry, Ergonomics and the Design of Work. Taylor & Francis, London, 2nd Edition, 93-104.

Saito, S., Miyao, M., Kondi, T., Sakakibara, H., and Toyoshima, H. 1997. Ergonomic Evaluation of Working Posture of VDT Operation Using Personal Computer with flat panel display. Industrial Health, 35, 264-270.

Straker, L., Jones, K.J. and Miller, J. 1997. A Comparison of the postures assumed when using laptop computers and desktop computers. Applied Ergonomics, 28, (4), 263-268.

COMPARING THE USE OF DIFFERENT DESIGNS OF COMPUTER INPUT DEVICES IN A REAL LIFE WORKING ENVIRONMENT

Luke Wardell[1] and Paul Mrozowski[2]

*[1]Department of Human Sciences, Loughborough University,
Loughborough, Leicestershire, LE11 3TU, UK
L.E.Wardell-97@student.lboro.ac.uk*

[2]Nortel Networks, Maidenhead, Berkshire, SL6 3QH, UK

As the popularity of personal computers continues to increase, we can also see the substantial widening of supporting products and accessories. This study takes just one element in this rapidly developing technology, the computer mouse and compares the use of a typically standard style of device with four other mouse styles. Measurements are taken of experienced users at their own workstations and findings indicate that though the different designs and operational requirements do contribute towards postural differences in particular for wrists and fingers, it is the physical position on the desk that can have the most varied impact.

Introduction

Since its invention in the mid 1960's the computer mouse has become well renown as a satisfactory and effective tool that enables users to interact with a computer. It is by far the most common form of pointing device on the market at present (Greenstein, 1997). It is defined as *a hand held device consisting of a small plastic box that can fit under the palm of finger tips attached to the computer by a wire,* and its functions as, *movement on a flat surface is used to generate cursor movement. Mice have one to three buttons that may be pressed to perform such functions as changing menus drawing lines or confirming inputs. Movement of the mouse is detected mechanically or optically* (Arnaut, 1998). Despite these relatively simple explanations, rapid technological advancements in the computer industry have already enabled wireless solutions to be developed and a wide variety of different styles created in an attempt to improve performance.

In recent years there have been a small number of studies that have investigated some of these changes. Variation in upper limb posture and movement during word processing with and without mouse use (Karlquvist et al, 1994), Computer mouse and trackball operations: Similarities and differences in posture, muscular load and perceived exertion (Karlquvist et al, 1999) and a Comparative study of two computer mouse designs (Hedge et al, 1999), to name but a few. All of these use similar methods to evaluate the interaction between mouse and user. Their results go some way towards generating awareness of the impacts of the use of this device both in the scientific community and in the public domain. A simple search on the Internet can reveal the extent of this awareness. Two examples, the first, an article found in Science Daily Magazine, claiming that a Larger Adjustable Computer mouse could risk of Wrist Injury, Cornell Study Finds and the second titled Mouse Menace – What you don't know about point and click can hurt you (Cohen, 1999)

It is well known that there is an increasing number of computer users as well as a wider spread through the population. Children sending homework into their schools via e-mail or the elderly generation searching for travel information, the diversity of today's user knows no limits. Consequently, as the number of users increases, so does the concern for the potential health risks associated with computer use, in particular for this small plastic hand held device. Comparison of basic anthropometric data measured from the above mentioned groups can reveal significant physical differences, yet often, they could well be operating the same style of device. Not surprisingly then, there is growing concern that the design and operational requirements of such devices are significantly contributing to the increasing numbers of compensation claims from users (Burgess-Limerick et al, 1999) and work into researching the extent of use and problems associated with non-keyboard input devices (Hastings et al, 2000).

In this field based study the company involved provides a Microsoft mouse as a standard piece of equipment with the computer that an employee will operate. In a few cases where it is thought that this mouse is not suitable, usually through personal preference, any alternative style of mouse can be purchased. This study attempts to evaluate the differences in the use of these individual devices comparing them to the 'standard'.

Materials and Methods

Subjects

12 subjects were recruited, 4 males and 8 females. The participants were all employees of a technologically based company and were full time computer users and had been chosen according to the type of device they used. 8 used the Standard Microsoft mouse, one used a Logitech Finger Tracker ball, another used an Ergo-View Whale (Flat) mouse, another a standard Compaq mouse and finally one used an Anir Joystick mouse. All participants were advised of the methods that were going to be used. The participants were all deemed to be experienced users of computer mice, the least being 6 months experience, they also reported that they usually operated the mouse with their right hand. Each subject reported that they had set up their workstation to suit their own individual needs.

Apparatus

Measurement of wrist flexion/extension and radial/ulnar deviation was taken via a Biometrics Goniometer (Type number XM65) and converting box units which were connected to a PC compatible Laptop computer. A Panasonic standard VHS camcorder, (model number NV-M50B), was set up on a tripod to the rear of the participants. A tape measure was used for all workstation measurements.

A set task was devised based on observation of a number of other users prior to the assessment and the experimenter's 5 year experience of mouse usage. The majority of typical mouse movements and actions were incorporated using these two knowledge bases, (clicking, scrolling dragging, copying and pasting for example) and the task involved navigating a submitable form on the company's intranet.

Procedure

The goniometer was placed on the mouse hand in accordance with the guidelines set out in the biometrics operating manual. It was attached with double sided medical tape in line with the proximal middle knuckle to the lateral epicondyle, while in neutral position and with 90° arm abduction. The goniometer was calibrated as per the Biometrics Laboratory Data Acquisition System Operating Manual and was set to record at 1200 Hz. Additional tape was applied as necessary to the connection wires to hold them together at the wrist end of the goniometer. The participants were then asked to perform their maximum range of movement in the two planes, to allow for percentage comparisons of the data. The video was set to record for the duration of the assessments. Participants were allowed a short period of approximately 5 minutes to demonstrate the typical task that they would perform with the mouse in a normal working day and to become comfortable with the operation of their device with the goniometer attached. Having demonstrated their typical tasks the participants were left to continue with their daily duties whilst the experimenter took anthropometric measurements of the users and physical measurements of their workstations.

Having completed these measurements a note was made of the time reading of the goniometer data and the participants were guided through them to perform the set task. They were advised that neither speed of performance nor accuracy were being measured in this task. On completion of the set task the goniometer data recording was stopped and the video switched off.

The measurements were imported in to an Excel Spreadsheet to convert the raw data in to degrees and then to be graphically represented as well as statistically analysed.

Results

The mean age of the users was 22.3 ± 0.8 years and the range was from 18 to 46 years.

Data from one participant (a standard Microsoft mouse user) was omitted from the study due to complications of the goniometer readings when the set task was interrupted for a lengthy period by their normal course of work, resulting in only half of the video data being gathered. The data for the remaining seven standard Microsoft mouse users were grouped to represent the readings for one type of mouse. This was compared to the readings for the 4 other styles of device. The tables below summaries the findings from the goniometer data:

Table 1. Maximum readings in degrees for wrist movements in the two planes

Types of mouse	Standard	Finger track ball	Compaq	Joystick	Flat mouse
Flexion Max	65.10	63.05	89.10	54.18	74.25
Extension Max	-69.15	-62.19	-69.71	-82.31	-54.41
Radial Max	20.58	21.87	24.75	19.13	29.75
Ulnar Max	-41.22	-27.36	-27.68	-52.20	-31.28

Table 2. Maximum readings in degrees during the set task

Types of mouse	Standard	Finger track ball	Compaq	Joystick	Flat mouse
Flexion Max	50.89	35.78	52.83	50.09	47.84
Extension Max	-5.16	16.61	22.05	-6.26	0.36
Radial Max	10.59	2.03	11.16	16.43	11.16
Ulnar Max	-22.29	-10.80	-11.48	-12.51	-17.24

Table 3. Average readings in degrees during the set task

Types of mouse	Standard	Finger track ball	Compaq	Joystick	Flat mouse
Flex/Ext	27.18	28.48	36.63	5.32	15.90
Standard deviation	13.56	2.24	5.64	3.57	4.67
Radial/Ulnar	-5.68	-5.08	-0.01	6.18	-5.77
Standard deviation	11.24	1.98	5.01	2.64	4.23

Table 4. Representation of the average readings in relation to maximum movements

Standard	Finger track ball	Compaq	Joystick	Flat mouse
Flexed at 41.75%	Flexed at 45.17%	Flexed at 41.11%	Flexed at 9.82%	Flexed at 21.41%
Ulnar Deviated at 13.78 %	Ulnar Deviated at 18.57 %	Ulnar Deviated at 0.0004%	Radially Deviated at 32.31%	Ulnar Deviated at 18.45%

The video data was analysed to support and explain the readings of the goniometer and in establishing the time during the readings that the set task was initiated. No other specific measurements were taken from it.

Discussion and Conclusion

The aim of this study was to compare the use of different styles of input device in a real life working environment. Full time computer users were assessed at their own workstations whilst using their own preferred style of input device, each were measured for their specific range of wrist movement (Table 1). The number of participants was therefore limited by simply the nature of this study. Simple measurements were taken and simple analytical techniques were used. As can be seen from above, descriptive statistics form the basis of this study, no specific significance test were carried out on the data due to the limited amount of participants. These factors should be considered throughout this section and especially when applying the findings in this report.

Overall, the mouse that required the least amount of wrist flexion and deviation to operate was the finger track ball. This had the lowest reading in term of maximum readings during the set task (Table 2). Compared to the Standard Mouse, for flexion this was 15.11 degrees less, extension 21.77 degrees less, for radial deviation this was 8.56 degrees less and for ulnar deviation this was 11.49 degrees less. The maximum readings for the remaining 4 types of mouse were comparable. The video data supported this result when analysed as very little upper limb movement was required to perform the set task. This is understandable when considering the design of the device as it the fingers that perform all the necessary movements for operation. Though this is desirable for maintaining a neutral wrist position, this is an obvious flaw for other carpal problems, like those thought to be associated with repeated depressing of mouse buttons, such as arthritis. For reducing this type of discomfort it is felt that this device would be in effective.

The average readings, in terms of flexion and extension when compared to the use of the standard mouse (Table 3) showed that both the Joystick mouse and the Whale (flat) mouse were operated in a more neutral position during the set task, approximately 22 degrees nearer the horizontal and 11 degrees respectively. The findings for the flat mouse support the findings from the Cornell Study in that a more neutral position of the wrist is adopted compared to a standard Microsoft mouse, however this was only noted in the flexion/extension plane. Considering the design of this device would suggest that this would be the case.

Comparing the percentages of maximum movements the average readings in terms of radial/ulnar deviation for the set task (Table 4), the Compaq mouse faired best at 0.0003% away from neutral whilst the Joystick mouse was recorded as the greatest angle being 32% radially from neutral. There are a number of factors external to the design of the mouse that could account for these. In particular is the mouse software setting i.e. the movement sensitivity, this would determine how far each device would have to be moved to move the cursor. All the users had reported that they were comfortable with the operation of these controls and had adjusted them to met their own need as they deemed necessary. However, they also reported that no guidance had been given as to what setting is preferable.

The video analysis revealed that the design of device has a small effect on movements in the upper limbs, very little movement was needed for finer tracker ball as the device sits stationary on the desk, where as for the others the arm must move to move the cursor. What was more apparent was that the position of the shoulder and upper arm can vary considerably depending on the relative position of the device on the desk. This element is consistent with other studies, in particular those that focus on workstation design, one excerpt being that workstation ergonomics contribute to chronic musculoskeletal symptoms. (Karlqvist, 1999). Further analysis of the video data and a comparison of movements in the shoulders in relation to the device's position on the desk would provide useful supporting data in future work.

The findings of this study go a short way to understating how a different type of device is likely to be used. Understandably no one particular device stood out as being the best type mouse to use, as all require some element of repeated movement by the arm and fingers. As there are many elements to the problems and conditions that may arise from using a computer mouse, no one particular device would greatly reduce the risk of them all.

In the case of limiting problems associated with wrist discomfort it would appear that the finger track ball does effectively reduce the need for bending at the wrist. Similarly either the Whale (flat) mouse, or the Joystick mouse, appear to be good at reducing the need for flexing movements during operation. Therefore, in these cases it could be suggested that wrist problems could be alleviate using any of these three devices.

In conclusion, this study provides an insight in to a method of comparing the use of different styles of input device. Though much of the data can be shown to support findings from other studies, the recordings must be taken in context as they were based on a very small number of participants and it was limited by the popularity of the style of device. To development the study, work performed could include taking physical measurements from the video data to analyse upper limb posture in more detail in relation to device position, taking accuracy or speed measurements for the set task and involving greater number of participants to make more use of statistical significance techniques. To further understand the impacts of different styles of device and the way that they are used in real life environments, it is suggested that this data could be a useful start in gathering a database of information to compare styles on a wider scale. A similar approach could be taken based on the Cornell study, though it should be noted that recording were made in a laboratory. This progression would be advisable as even using just the 24 participants of students and staff went as far as to produce a magazine headline on the Internet. As the popularity of these devices increase it may be necessary to track their impacts and the potential risk associated with them using some of the methods that have been outlined. Awareness of these issues, the physical and operational requirements as well as the potential risks as part of user training or general understanding is essential in reducing the likelihood of health problems that may be associated with computer use.

References

Arnaut L. Y., Joel S. Greenstein, J.S., 1998, Human factors considerations in the design and selection of computer input devices. In M. Helander (ed.) *Handbook of Human-Computer Interaction*, (North-Holland, Amsterdam)

Burgess-Limerick, R., Shemmell J., Scadden,R., Plooy, A.,. 1999, *Wrist Posture during pointing device use,* (Elsvier Science Ltd)

Cohen, S., 1999 *Mouse menace – what you don't know about point and click can hurt you,* http://www.womenconnect.com/LinkTo/aug0599a_biz.htm

Greenstein, J.S., 1997, *Pointing device - beyond the desk top,* Handbook of Human-Computer Interaction, 2nd Edition, (Elsvier Science Ltd)

Hastings, S., Woods, V., Haslam, R. A., and Buckle, P., 1999, Health Risks from Mice and Other Non-Keyboard Input Devices, *Contemporary Ergonomics 2000,* (Taylor and Francis, London)

Hedge, A., Muss, T., Barrero, M., 1999, *Comparative Study of Two Computer Mouse Designs,* (Cornell University, NY)

Karlqvist, L., Bernmark, E., Ekenvall, L., Hagberg, M., Isaksson, A., Rostö, T., 1999, *Computer mouse and track-ball operation: similarities and differences in posture, muscular load and perceived exertion.* (Elsvier Science)

Karlqvist, L., Hagberg, M., and Selin, K., 1994, Variation in upper limb posture and movement during word processing with and without mouse use, *Ergonomics 1994,* vol. 37 no 7 1261-1267

Science Daily Magazine, 1999, *Larger adjustable computer mouse could reduce risk of Wrist injury, Cornell Study finds.*
http://www.sciencedaily.com/releases/1999/12/991223010717.htm

USE OF SUPPORT APPLIANCES WITH COMPUTER INPUT DEVICES

C L Brace, S Hastings, R A Haslam

Health and Safety Ergonomics Unit, Department of Human Sciences, Loughborough University, Loughborough, Leicestershire, LE11 3TU

Long periods of VDU work using the keyboard and other input devices can lead to fatigue and muscle tension. A range of support appliances are available for use with computer input devices, which manufacturers claim are 'ergonomic', 'supportive' and 'comfortable'. Support appliances used within a number of organisations were identified through a questionnaire survey, followed by a more detailed survey of users to identify perceived problems and benefits. The most commonly used equipment was reported to be foam and gel keyboard rests and foam and gel mouse support appliances. Three aids were subsequently evaluated in a laboratory trial using electrogoniometry and subjective comfort ratings. No significant differences were found for wrist angles whilst using the three products, although subjective measures indicated a preference for a gel appliance.

Introduction

There are numerous types of equipment that can be classified as support appliances, available to assist with computer input work. Such devices are primarily designed for use with keyboards and mice. Although the mouse is used extensively, and there have been suggestions of problems resulting from its use, little research has been carried out to investigate support appliances specifically designed for mouse users. From those studies that have been reported there is the suggestion that use of a mouse support appliance can result in a reduction in muscle strain and exertion in the muscles of the hand-arm system (Paul & Nair, 1996; Strasser & Keller, 1997) and shoulder region (Feng *et al.*, 1997). Several studies also corroborate these results with supporting subjective responses (Paul & Nair, 1996; Strasser & Keller, 1997; Cook & Kothiyal, 1998). However, other studies have shown mouse support appliances to have no positive effects on wrist posture (Damann & Kroemer, 1995; Hedge, 1995).

The aims of the study reported here were to identify commonly used support appliances, problems and benefits associated with them, and to assess the physical effects of such products using objective and subjective measures.

Method

Questionnaire Surveys

In order to gather information about current use of support appliances with non-keyboard input devices, ten organisations were approached, and their Health and Safety/IT managers asked for information on usage via a questionnaire survey. A second questionnaire was then used to gather information from specific users of support appliances within these organisations, to identify any perceived problems or benefits. The results of the questionnaires were used to inform the design of a laboratory trial.

Laboratory Study

Ten participants, five male and five female, inexperienced in the use of support appliances, were recruited to take part in a laboratory trial. Ages ranged from 18 to 26 (mean 21.7 years). All subjects were right-handed and experienced in keyboard and mouse use (range 5 to 11 years experience, mean 7.6 years). Participants were screened to assure no history of diagnosed work-related upper limb disorders (WRULDs), no wrist or hand pains, and no previous wrist or hand injuries.

Foam and gel mouse support appliances, those most commonly found in use in the organisations surveyed, were evaluated against a control condition (a rectangular mouse mat). The dimensions of the appliances were similar. The mouse used in this study was a standard Apple Desktop Bus Mouse (Family Number M2706). Subjects performed an editing task specifically designed for this experiment using Word 6.0 for Macintosh on a Power Mac 7200/90. A Pentium PC recorded wrist measurements using the Biometrics™ software package and equipment. A biaxial electrogoniometer of a compact and non-restrictive design (Biometrics Ltd., Gwent UK) accurate within ±5%, recorded wrist flexion/extension (F/E) and wrist ulnar/radial deviation (U/R) on the right side. These were measured with sensor model XM65.

The keyboard and mouse mat/support appliances were placed on a non-adjustable surface. They were arranged in a 'typical' workstation set-up with the keyboard placed directly in front of the monitor, and the mouse and mouse mat/support appliance located to the right of the keyboard. A document holder was positioned directly behind the mouse mat/support appliance, adjacent to the monitor. Participants were seated directly in front of the keyboard in an adjustable chair so that the participant's forearm was parallel with the worksurface and thighs parallel with the floor. A footrest was provided for use if required. Desk surface markers were used to denote keyboard and mouse mat/support appliance placement. This ensured consistency of placement between the participants and throughout the assessment period for each participant. Although the appliances were positioned consistently, the participants chose how to use each product, i.e. which part of the upper extremity rested on the appliance, and whether the appliance was used continuously as a support or only during breaks from the task. This decision was left to the participants' discretion and was dependent on their feelings of personal comfort, as no documentation was available with the products regarding their usage. Each subject performed text editing requiring use of the mouse, and occasionally the keyboard, for a duration of thirty minutes, during which measurements of flexion, extension, ulnar deviation and radial deviation were collected, using the electrogoniometer. The Body Part Discomfort Scale (Corlett & Bishop, 1976) was used to assess participants' body part discomfort, immediately before and after each product evaluation period, to evaluate changes in comfort levels. A Product Feature Analysis Questionnaire was administered directly after the use of each product and was later analysed for subjective preferences.

The procedure was repeated for each of the three products. The order of conditions was randomised across participants.

Using previously collected calibration data, minimum, maximum and median (neutral) flexion/extension (F/E) and ulnar/radial (U/R) deviation points were identified for each participant. These data points were then applied to the trial data, so that F/E and U/R deviation could be interpreted. Joint angle data from each channel were averaged so that the mean F/E and U/R deviation per condition and per participant could be summarised and compared to the neutral position. This enabled the products to be compared according to the extent by which wrist posture deviated from neutral. Statistical summary values for joint positions (°) were calculated for each participant over each full data acquisition period. Mean wrist joint postures and within- and between-participant standard deviations were calculated. To examine the differences between products, the results were analysed using two-way analysis of variance (ANOVA). Differences between the conditions were analysed using paired sample t-tests, applying the Bonferroni technique (Norusis, 1997). Relationships for mean wrist joint postures with selected anthropometric measures were evaluated. Statistical significance was defined as $p<0.05$. All statistical analyses were conducted using SPSS™ (Version 9 for Windows™) or Excel 97 for Windows™.

Results

The questionnaire surveys indicated that support appliances are in use in many different environments and for various different tasks. In addition to being found in the standard office environment, they are used by rescue services and other operatives performing a range of work. It was found that the most commonly used support appliances were those for the keyboard and/or mouse. Various problems were identified with supports, including increased discomfort and lack of working space. Benefits were also identified, including improved comfort, posture and reduced pain.

**Table 1. Postural measurements for each product
during laboratory trial found using electrogoniometry**

Product	Postural Measurement	Mean Angle (degrees)	Std. Deviation (degrees)
Foam	Flexion	9.6	6.9
	Extension	32.3	4.9
	Ulnar deviation	12.5	5.0
	Radial deviation	24.1	6.7
Gel	Flexion	6.3	10.4
	Extension	30.1	5.8
	Ulnar deviation	13.2	4.9
	Radial deviation	24.0	6.5
Control	Flexion	9.1	7.0
	Extension	41.2	3.6
	Ulnar deviation	11.6	5.6
	Radial deviation	22.9	6.4

Statistical analysis found no significant difference in the mean wrist postures measured between using a gel and a foam support appliance (see Table 1) at $p< 0.0042$, which was the observed significance level after the application of the Bonferroni Technique. Nor was there a significant difference between using a support appliance and the control condition for the wrist postures of flexion, ulnar deviation and radial deviation. However, there was a significant difference between the joint posture of extension for the foam support appliance and the gel support appliance versus the control condition. (The control was significantly higher than the gel or foam, $p< 0.001$ in both cases.)

Table 2. Mean scores of the products resulting from the subjective assessment during the laboratory trial (1=lowest score, 5=highest score)

FEATURE → / PRODUCT ↓	Length of mouse mat	Length of pillow	Width of mouse mat	Width of pillow	Shape of mouse mat	Shape of pillow	Stability of product	Aesthetic appeal	Functional rating	General comfort	Overall rating
Foam	3.25	3.70	4.10	4.00	3.65	3.80	4.70	3.20	3.60	3.30	**3.30**
Gel	3.10	4.00	3.80	4.30	3.80	4.35	4.60	4.00	4.20	4.20	**3.95**
Control	3.50	-	4.35	-	4.00	-	4.50	2.35	4.10	3.25	**3.45**

Analysis of subjective data indicated that the gel appliance was strongly preferred over the foam appliance and the control (see Table 2). Nine out of ten participants found it the most comfortable appliance to use.

In general, wrist postures and motions were not predictable based on body size, elbow height, hand length, shoulder width, or wrist dimension (wrist width and thickness), as determined by simple statistical correlations.

Discussion

The study demonstrated that the posture adopted whilst manipulating a mouse is characterised by sustained wrist extension and flexion, and wrist ulnar and radial deviation throughout the operating period.

Mean joint postures recorded for each product were between 6 and 9 degrees for flexion, 30 and 41 degrees for extension, 11 and 12 degrees for ulnar deviation, and 22 and 24 degrees for radial deviation. The presence of a gel support appliance decreased flexion and extension, and increased ulnar and radial deviation, compared to the control condition. The presence of a foam support appliance increased flexion, decreased extension, and increased ulnar and radial deviation, compared to the control condition.

When compared with the foam device, the gel support appliance resulted in reduced flexion, extension and radial deviation, but increased ulnar deviation. The increased ulnar deviation during use of the gel appliance was not associated with increased musculoskeletal pain or discomfort in the upper extremity whilst using the keyboard or mouse, as has been found in previous studies (Duncan & Ferguson, 1974; Hunting *et al.*, 1981), but this was a short trial.

One possible source of influence on the study data arises from the positioning of the hand goniometer sensors. Although this may have varied slightly between subjects, it remained constant throughout the different conditions for each participant. Depending on the length of the mousing area, and the anthropometry of the individual, different parts of the wrist and/or arm were supported by the pillow. Wrist and finger posture, as well as the force exerted, affect the carpal canal pressure, which if sufficiently increased can lead to WRULDs (Rempel, 1995). Using the fingers to grip and manoeuvre the mouse as well as using them for mouse clicking, creates a greater risk to the user of the development of WRULDs. It appears that a longer length of mousing area may have a more beneficial effect on the long-term health of the user and perhaps should be recommended.

Another limitation of the study was that during the trial task, periods spent typing became shorter as participants became familiar with the task. Although the experimenter attempted to keep the same speed of instruction throughout, participants often anticipated what was coming next and jumped slightly ahead.

Variation may also have been introduced into the task due to participant error. This probably occurred for most participants at some time or other during each evaluation period. However, counter-balancing the order of the evaluated products should have reduced any bias in results from this source.

The participants in this trial were all university students, which could have affected the results. As individuals differ in their susceptibility to the incidence, severity and aetiology of musculoskeletal disorders with age, a study with older persons, engaged in various occupations could lead to different findings. It is possible to speculate that discomfort levels may increase with age and inexperience of using computer input devices.

Conclusions

The aims of the study have been addressed, with the most commonly used support appliances and problems and benefits associated with them, identified. Physical effects of selected products were then assessed using objective and subjective measures.

The objective and subjective data collected by this study, suggest that users' requirements of support appliances are not currently met with respect to safety, comfort and performance. It appears that more consideration should be given to the long-term health of the user with regard to size of mousing area and documentation for the product (instructions as to how and when they should be used).

Further research is recommended to examine the comfort of a larger sample of participants, more representative of the general population, whilst using the devices. Support appliances for use with other input devices could also be investigated, using a range of methodologies e.g. electromyography, to assess muscle activity/fatigue resulting from their use.

References

Cook, C.J., Kothiyal, K., (1998). Influence of mouse position on muscular activity in the neck, shoulder and arm in computer users. *Applied Ergonomics*, **29** (6), 439– 443.

Corlett, E.N., Bishop, R.P., (1976). A technique for assessing postural discomfort, *Ergonomics*, 19, 175–182.

Damann, E.A., Kroemer, K.H.E., (1995). Wrist posture during computer mouse usage. In: *Proceedings of the Human Factors and Ergonomics Society 39th Annual Meeting – 1995*. Vol.1. 625– 629.

Duncan, J., Ferguson, D., (1974). Keyboard operating posture and symptoms in operating. *Ergonomics*, **17** (5), 651– 662.

Feng, Y., Grooten, W., Wretenberg, P., Arborelius, U.P., (1997). Effects of arm support on shoulder and arm muscle activity during sedentary work. *Ergonomics*, **40** (8), 834– 848.

Hedge, A., Powers, J. (1995). Wrist postures while keyboarding: effects of a negative slope keyboard system and full motion forearm supports. *Ergonomics*, **38** (3), 508– 517.

Hunting, W., Laubli, T.H., Grandjean, E., (1981). Postural and visual loads at VDT workplaces. I. Constrained postures. *Ergonomics*, **24**, 917– 931.

Norusis, M.J., (1997). SPSS® 7.5. Guide to Data Analysis. Prentice-Hall, New Jersey.

Paul, R., Nair, C., (1996). Ergonomic evaluation of keyboard and mouse tray designs. *Proceedings of the Human Factors & Ergonomics Society 40th Annual Meeting – 1996*. Philadelpia, 632– 636.

Rempel, D., (1995). Musculoskeletal loading and carpal tunnel pressure, in Gordon, S., Blair, S., and Fine, L. (eds.). Repetitive Motion Disorders of the Upper Extremity, Rosemont, IL:American Academy of Orthopeadic Surgeons, 123-133.

Strasser, H., Keller, E. (1997). Ergonomic Evaluation of a wrist rest for VDU work via electromyographic methods. In: *Design of Computing Systems: Cognitive Considerations*. Advances in Human Factors/Ergonomics, 21A, Proceedings of the Seventh International Conference on Human-Computer Interaction (HCI International '97), San Fransisco, California, USA, August 24-29. Salvendy, G., Smith, M.J. & Koubek, R.J. (ed.s).

MANUAL HANDLING

An Ergonomics Evaluation of the Spare Wheel Changing Process

Glyn Lawson[1], Paul Herriotts[2], Joanne Crawford[3]

[1] *formerly Industrial Ergonomics Group, School of Manufacturing and Mechanical Engineering, The University of Birmingham*
[2] *Principal Ergonomist, Land Rover Group (formerly BMW Group UK Design)*
[3] *Industrial Ergonomics Group, School of Manufacturing and Mechanical Engineering, The University of Birmingham*

This paper details an ergonomics evaluation of the spare wheel changing process. The study was conducted to provide the designers at Rover Group Ltd. with recommendations for car boot dimensions and tyre weights.

A test rig was set up to represent a car boot. Thirty-two participants adjusted the wheel weight, wheel bay height and boot lip height, according to their capabilities. The results were calculated as percentiles. To be acceptable to at least 95% of men and women, the wheel should weigh less than 6.32kg. If positioned horizontally in the boot, the top surface of the wheel should lie 904-935mm from the ground and there should be no boot lip. In a typical production car, the wheel weight and surface height are 13.20kg and 440mm. Alternative placements for the spare wheel are suggested and a device to help users lift it from the boot is described.

Introduction

During the spare wheel changing process, the user must lift a wheel of considerable weight (9 to 28kg) in a stooped posture, both factors which may contribute to back injury (Health and Safety Executive, 1992; Kroemer and Grandjean, 1997).

Although manual handling has typically received a great deal of attention in ergonomics, few studies have looked specifically at lifting items out of the boot of a car. Karwowski *et al* (1993) attempted to provide ergonomics guidelines for the design of a car boot, based on the results of a psychophysical experiment. The participants lifted six objects of various weights into a test rig to simulate loading a car following a shopping trip. They could change the appropriate dimensions of the test rig until the preferred distances were obtained.

The results were plotted as percentiles, which were used to determine recommendations for the design of a car boot. The authors recommend that the boot lip height should be 600-840mm from the ground, and the boot floor height 510-600mm, if the design is to be acceptable to 90% of the population. However, the participants were studied lifting shopping into the car boot, and therefore did not have to bend as far into the car boot as they would when lifting a wheel from the wheel bay.

Thus, in light of the lack of manual handling studies that have used the specific constraints of a car boot, it was concluded that the ergonomics of lifting the spare wheel out of the boot would be investigated. Data from the study would be used as the basis for recommendations for car boot dimensions and wheel weights.

Method

In order to explore the ergonomics problems in more depth, a car mechanic was observed changing the wheel of a Rover 200. The task was analysed using the Ovako Working Posture Analysing System or OWAS (Long, 1993), the NIOSH equation (Waters *et al*, 1993) and a manual handling risk assessment (Health and Safety Executive, 1992). The results of this pilot study confirmed that lifting the spare wheel out of the boot should be investigated, as this was the task most likely to cause back injury. The risk of injury was caused by the stooped posture that the participant was forced into by the location of the spare wheel. The weight of the wheel (13.20kg for the Rover 200) increased the risk.

Before the main experiment was conducted, the variables required for the development of the design recommendations were identified. These were:

- Wheel weight
- Highest and lowest acceptable spare wheel height from the ground
- Highest acceptable boot lip

Participants
The participants were randomly sampled from staff and students of the University of Birmingham. Thirty-two participants were used in the experiment, seventeen male (mean age=38.06, SD=16.6, range=19-69) and fifteen female (mean age=35.2, SD=14.03, range=22-60).

Apparatus
A 175x65x14 tyre was obtained, which is typical for a Rover 200. Several long bags were filled with sand to weigh 0.5, 1.0 and 2.0kg. These could be placed inside the wheel to provide evenly distributed weight. The wheel was laid on an electric table with a height adjustment range of 205mm to 1400mm. A board, the same width as the tyre, was secured in front of the wheel to represent the top of the wheel bay. This was not removed for any of the trials, as the spare wheel is usually recessed into the boot floor. The board prevented the wheel moving any closer than 155mm towards the edge of the desk (taken from the horizontal distance between the bumper and the edge of the spare wheel in the Rover 200). A pair of stands supported a height adjustable crossbeam, which represented the boot lip.

Procedure
A psychophysical approach was thought to be the best method to use for the experiment. This allowed participants to adjust the variable being measured until they reached the highest, lowest or optimal value. The main advantage of the psychophysical approach is that it takes into account biomechanical and physiological factors, according to the participant's subjective experience of a task (Stevens, 1960; Haslegrave and Corlett,

1995). This approach was particularly beneficial when participants were selecting the heaviest acceptable wheel weight, as they could gradually increase the weight until it was the heaviest they could lift without straining. Thus, they were never faced with lifting a wheel that was too heavy.

Before taking part in the experiment, participants were asked to sign a consent form stating that they were in good health and that they would stop if they felt any pain or discomfort. Each participant was asked to feel the weight of the empty tyre, with the experimenter supporting it, to familiarise them with the weight with which they would start the experiment. They were then asked to raise the height of the table to what they thought would be the optimum height for a heavy lift. This was done with no boot lip. The participant was asked to lift the tyre off the table. They were told to ask for more weights until the tyre was as heavy as they would be able to lift. After each lift, the participant was encouraged to relax for a few minutes.

Once the participant had reached the maximum weight they could manage, the weights were secured in the tyre. They were then asked to lift the wheel off the rig, onto the floor. The participant was asked if their choices were correct. If so, the height of the table and the weight of the wheel were recorded. If not, the process of selecting the wheel weight and table height were repeated.

The participant was then asked to raise the lip of the boot to the maximum acceptable height over which they thought they could lift the wheel. The lip height was recorded at the optimum wheel bay height.

The lip was then removed, and the participant adjusted the table until it reached the lowest acceptable height from which they could comfortably lift the wheel. They were asked to select the maximum lip height, this time at the lowest acceptable table height.

Finally, the participant raised the table to the highest from which they could lift the wheel.

For each variable, the participant was asked to lift the wheel off the rig to check their choice before the value was recorded.

Once measurements had been taken for each of these variables, the whole experiment was repeated to check for consistency. If the values were within 15%, the mean was taken. If this was not achieved, the data was disregarded.

Results

The results of the experiment were analysed using the descriptive statistics function on Minitab 11. As the means and medians were very similar for each variable, it was assumed that the data was normally distributed. For this reason, it was feasible to use the means and standard deviations to calculate the percentile values for each. The 10th, 50th, and 95th percentile values of wheel weight, lowest acceptable wheel bay height, lip height (wheel bay to lip) and lip height (lip to ground) are shown in Table 1. These were used in the development of the design recommendations. The highest acceptable wheel bay height was also used in the design recommendations, but only the 95th percentile female value (935mm). Wheel bay height refers to the top surface of the spare wheel.

The lip heights (lip to ground) taken at the lowest acceptable wheel bay height are shown in Table 1. These are lower than those taken at the optimal wheel bay height, and are therefore safer to use in the design guidelines. However, the lip heights (wheel bay to lip) at the optimum wheel bay height are shown, as these are lower than the lip heights (wheel bay to lip) at the lowest acceptable wheel bay height.

Table 1. 10th, 50th and 95th percentile values for the variables evaluated in the main experiment

	Maximum wheel weight (kg)		Lowest wheel bay height (mm)		Highest lip height (wheel bay-lip) (mm)		Highest lip height (lip-ground) (mm)	
Percentile	Men	Women	Men	Women	Men	Women	Men	Women
95	11.60	6.32	903	857	52	27	629	831
50	15.97	9.08	680	739	173	181	914	983
10	19.38	11.24	505	648	268	301	1136	1102

Discussion

It should be noted that for any particular percentile, the addition of the wheel bay height and lip height (wheel bay to lip) does not equate to the lip height (lip to ground). This is partly due to the low correlation between the variables, and the variation in the results. Additionally, the 95th percentile values for highest lip height (lip to ground) are less than the 95th percentile values for lowest acceptable wheel bay height. This is because the values for lowest wheel bay height are constrained by people bending down, whereas highest lip height (lip to ground) is constrained by the height people can reach. Thus, there is an overlap in the results for these variables. Because the 95th percentile value (female) for highest acceptable wheel bay height is 935mm, it is safe to state that providing there is no lip, a wheel bay height of 904-935mm will be acceptable to at least 95% of the population.

The discrepancies in the results highlight the importance of running user trials using a full-scale mock-up when designing a car boot.

Recommendations

The recommendations for boot dimensions and wheel weights are presented in Table 2, based on the results of the main experiment. They have been categorised into optimal, acceptable, tolerable and unacceptable ranges. This was partly to make them more usable for the designers at Rover Group Ltd., but also because it was recognised that the optimal values would be difficult to achieve. Presenting the different ranges will allow the designers to judge the acceptability of any given design. If a variable is within the optimum range, it will be satisfactory to at least 95% of men and women. Acceptable will be satisfactory for 50-95%, tolerable 10-49% and unacceptable less than 10%.

Wherever any of the discrepancies in the results mentioned in the discussion were encountered, a value that would be satisfactory for the largest number of people was chosen. However, the recommendations must still be used with caution. When designing a car boot, the lip height must be compared to the recommendations for both lip height (wheel bay to lip) and lip height (lip to ground). For example, a lip height (wheel bay to lip) of 173mm would not be classified as "acceptable" if used with a wheel bay height of 903mm. This is because the addition of these values exceeds the upper limit for the lip height (lip to ground).

Table 2. Design recommendations

	Wheel weight (kg)	Wheel bay height (mm)	Lip height (wheel bay-lip) (mm)	Lip height (lip-ground) (mm)
Optimum >95%	**<6.32**	**904 - 935**	**0**	**N/A**
Acceptable 50-95%	6.32 - 9.08	739 - 903	27 - 173	766 - 914
Tolerable 10-49%	9.09 - 11.24	648 - 738	174 - 268	915 - 1102
Unacceptable <10%	>11.24	<648	>268	>1102

Wheel weight is obviously a problem as 95% of the female population tested can only lift 6.32kg, whereas the Rover 200 wheel weighs 13.20kg. It would be difficult to reduce the weight of the wheel from 13.20kg to 6.32kg, although some weight could be reduced by using a space saver wheel. It is unlikely that the weight reduction would be sufficient, however.

Lowest acceptable wheel bay height could also pose a problem to the car designers. To be acceptable to 95% or greater of men and women, the top of the spare wheel must lie between 904-935mm above the ground. This is only feasible for an off-road vehicle.

To be acceptable to 95% or more of the population, the boot should have no lip. This would reduce postural stress as objects could be rolled or slid out of the boot.

If the designers are to provide users with a boot that is acceptable to 95% of the population, it will require considerable re-work. This may not be practical due to other constraints, e.g. aerodynamics, structural engineering or styling. For this reason, it was necessary to consider other methods of improving the ergonomics of the task.

A possible solution is the alternative placement of the spare wheel. In particular, if the wheel were positioned outside the car on the back door, the user would be able to keep their body close to it, thus reducing the strain on their back. This is already used for 4x4 vehicles, such as the Land Rover Freelander and Discovery. However, because of the weight of the wheel this would be impractical for hatchback or saloon cars.

Another alternative wheel placement would be inside the boot, positioned vertically against the side-wall. This would allow users to roll, not lift, the wheel out of the boot. However, locating the wheel vertically in the boot would considerably reduce the available boot space, and is therefore impractical for small to medium sized cars.

These alternative wheel placements do not provide a solution for Rover's current car range. Figure 1 shows a device that could be used in almost any car that has the spare wheel located in a wheel bay in the boot.

Two short straps fit through the gaps in the wheel and are fastened with quick release mechanisms. A third, longer strap joins them. The user would lift the wheel to a vertical position within the boot using the longer strap. This would allow them to maintain a good posture in what would otherwise be the most dangerous operation in the wheel-changing task. Once the wheel was vertical, the user would lift it out of the boot using the two short straps. They would use the short straps to lower the wheel to the floor, and would maintain a good posture throughout the whole process.

The idea was evaluated using the NIOSH equation (Waters *et al*, 1993), and the resulting recommended weight limits did not exceed that of the Rover wheel.

One of the main benefits of this idea is that it could be added to the Rover 200 with very little modification. It might add a little to production costs, but for a simple nylon strap, the ergonomics benefits are huge.

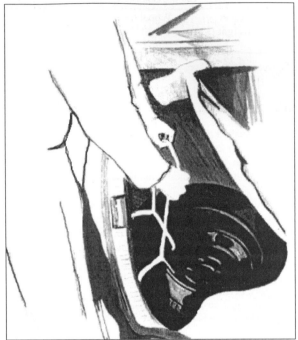

Figure 1. Device to assist the user when lifting the wheel from the wheel bay

References

Haselgrave, C.M. and Corlett, E.N. 1995, Evaluating work conditions and risk of injury - techniques for field surveys. In J.R. Wilson and E.N. Corlett (eds) *Evaluation of Human Work,* Second Edition, (Taylor and Francis, London), 892-920

Health and Safety Executive, 1992, *Manual Handling: Manual handling operations regulations, 1992. Guidance on Regulations,* (HMSO, London)

Karwowski, W., Yates, J.W. and Pongpatana, N. 1993, Ergonomic guidelines for design of a passenger car trunk. In B. Peacock and W. Karwowski (eds) *Automotive ergonomics,* (Taylor and Francis, London), 117-139

Kroemer, K.H.E. and Grandjean, E. 1997, *Fitting the task to the human,* (Taylor and Francis: London)

Long, A. 1993, Overview of the Ovako Working Posture Analysing System (OWAS), *Ergonomics in a changing world, Proceedings of the 29th annual conference of the ergonomics society of Australia, Perth, 1-3 December 1993,* (The Society, Downer), 3-10

Pheasant, S. 1996, *Bodyspace. Anthropometry, Ergonomics and Design,* Second Edition, (Taylor and Francis, London)

Stevens, S.S. 1960, *The psychophysics of sensory function,* American Scientist, **48**, 226-253

Waters, T.R., Putz-Anderson, V., Garg, A. and Fine, L.J. 1993, Revised NIOSH equation for the design and evaluation of manual lifting tasks, *Ergonomics,* **36**, 749-776

ON SITE ANALYSIS OF THE MUSCULAR LOAD WHILE WORKING WITH HOISTS

Jan Seghers, Hans Ponnet , Arthur Spaepen, Hans Lefever

Laboratory of Ergonomics and Occupational Biomechanics, KULeuven
Tervuursevest 101
3001 Heverlee, Belgium

This study describes the on site analysis of the muscular load of workers while they were manipulating three types of hoists during manufacturing DURACELL® batteries. The muscular activity of six persons was analysed while they were performing their normal job. For data-analysis, two subjective screening methods and two objective screening methodes were used. From the results with OWAS and RULA it was concluded that several postures are inappropriate. The electromyographic registrations showed peak efforts between 80 and 90% MVC in the shoulder muscles of the workers while they were operating the hoist. In general, the muscular activity of the Trapezius muscle was higher than the activity of the Deltoideus muscle. Moreover, periods of static effort are much more extended for the Trapezius muscles. The muscular load during other subtasks also consisted of dynamic peak values and periods with static effort. In conclusion, it was clear from the EMG analysis that using material handling devices may even cause a shift in muscular load to other body regions, resulting in a reduction for the back muscles, but not in the shoulder-neck region.

Introduction

Heavy physical demands and biomechanical stresses, imposed by the manual handling and lifting of loads, may cause musculoskeletal disorders of the upper limbs and lower back. One method to reduce the physical load is to use material handling devices. However, Hermans *et al.* (1999) concluded that using such tools did not always decrease muscular activity significantly during the end-assembly of cars. In some situations they found a shift in muscular load to other body regions, resulting in a reduction for the back muscles, but not in the upper limbs. The purpose of this study was to analyse the muscular load in the neck-shoulder region of workers at the Duracell plant while working with three types of hoists during the manufacturing of batteries.

Methods

Working tasks

During the manufacturing of batteries, the batteries themselves are produced fully automatically. To store the batteries, empty boxes are placed in a filler machine. Batteries are automatically organised in this box. The batteries are visually inspected and cleaned if necessary. Finally, the boxes are then transported with a hoist to a pallet. After stocktaking the different types of hoists, three types (CCA 1500 Trio 1, CCA 1500 Trio 4 and DEPO 1300) were selected for further research mainly based on the force production needed to operate the hoist. It was seen that with some types of hoists, the effort needed to operate the hoist was twice greater in the beginning than in the end. Five boxes of 16kg are transported at once. The user interface of the hoist was manipulated above shoulder level, with a relatively high amount of force, and periods with sustained static muscular effort. The hoist was manually pushed to rotate the hoist beam, from the starting position towards the end position.

Materials and methods

For data-analysis, two subjective screening methods and two objective screening methodes were used. These methods are summarised briefly below. The subjective screening methods, *Ovako Working-posture Analysing System (OWAS)* (Karhu *et al.*, 1981) and the *Rapid Upper Limb Assessment (RULA)* (Mc Atamney and Corlett, 1993), were used for task analysis in order to split up the tasks in isolated parts, and quantify the subjective strain for these parts. This study made use of surface-electromyography and video digitalisation for objectively screening the working tasks. From the task analysis it was decided that it was best to monitor the electromyographic activity of M.Trapezius pars descendens and M.Deltoideus pars anterior bilaterally. To obtain a large number of work cycles, continuous EMG registrations lasted approximately 30 minutes per person. The bipolar surface electrodes (type Nikomed, silver-silver chloride, Denmark) were placed according to standard procedures (Zipp, 1982). The surface electrodes were attached to a portable EMG device (ME-3000 Professional, MEGA Electronics Ltd., Kuopio, Finland) which was placed in a small hip bag and worn around the waist by the subject. The sample frequency was 1000 Hz. From the raw EMG-data, a parameter called muscular activity (ACT) was measured. The rationale of the ACT parameter is described elsewhere (Spaepen *et al.*, 1987; Hermans *et al.*, 1999). Before starting each task, an isometric Maximal Voluntary Contraction (MVC) of the shoulder muscles was performed. The recorded EMG signals during the working task were normalized to the MVC value, to analyse the relative muscular effort for each work task and to compare the results. Finally the EMG signals were synchronised with a SONY HI8 video camera. Six persons (two for each type of hoist) were observed while performing their normal job.

Results and discussion

OWAS and RULA

From the results with OWAS and RULA it was concluded that several postures are inappropriate and fell in action categories three and four, indicating that work postures have a harmful effect on the musculoskeletal system, and that the working methods should be changed as soon as possible.

Electromyographic registrations

From the electromyographic registrations it was concluded that not only the operation of the three types of hoists was physically demanding. Figure 1 shows the normalised muscular activity of the registered neck-schoulder muscles while the workers were operating the hoist type CCA 1500 Trio 1. During this working task peaks in shoulder muscular effort between 80 and 90% MVC were observed, and occurred at a rate of 10 to 12 times per hour for the three types of hoists. We can observe that the muscular activity of the right and left Deltoideus was lower than the activity of left and right Trapezius. Also periods of static effort are much more extended for the Trapezius muscles.

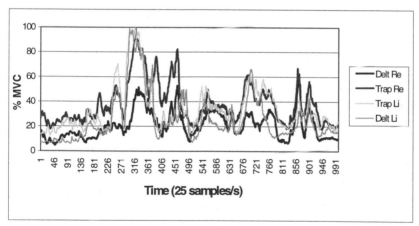

Figure 1. Normalised muscular activity of the left (Trap Li) and right (Trap Re) trapezius and deltoid (Delt Li and Delt Re) muscle while a worker was operating the hoist type CCA 1500 - Trio 1

Several periods with a static effort of more than 20% MVC occur for the right and left Trapezius. The intramuscular pressure during the static contraction prevents the blood flow, and the metabolic byproducts, such as lactic acid, accumulate in the muscle. Under ischemia, oxygen supply to an active muscle is prevented (Masuda *et al.*, 1999). This is one of the posibble explanations why static effort can be a reason for the development of Cumulative Trauma Disorders (CTDs).

Also other postures and actions were found where a large muscular load occurred. The muscular load during other subtasks also consisted of dynamic peak values and periods with static effort. The most strenuous worktasks for the workers besides manipulating the hoist, were refilling the boxes above shoulder level (Figure 2) and cleaning the batteries.

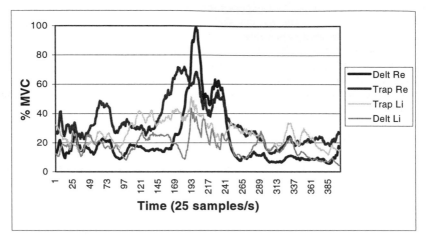

Figure 2. Normalised muscular activity of the left (Trap Li) and right (Trap Re) trapezius and deltoid (Delt Li and Delt Re) muscle while the worker refills the boxes above shoulder level

The muscular activity of the lower back muscles was not measured in this study. However, Hoozemans *et al.* (1998) demonstrated that pushing and pulling is often associated with low back pain. It is therefore recommended in future research to also register the surface EMG-activity of the Erector Spinae bilaterally.

Conclusions

It can be concluded that the muscular load of workers while operating the hoists could be quantified. It was clear from the EMG analysis that using material handling devices does not always decrease muscular activity. Periods with static and dynamic effort could be separated and a clear image on strenuous postures and actions could be obtained.

Several structural, organisational and behavioral recommendations were given. The most important organisational and structural recommendations that were given to the company were adjusting the working heights and restricting the reaching distances. These interventions will mainly reduce the static efforts above shoulder level. Also redesigning the placing of the hoists (i.e. replacing the pillar or reducing the lever of the hoist) and job rotation are some other adaptations on the structural-organisational plane which can reduce the musculoskeletal strain of the neck-shoulder area of the workers. A recent study of Kuijers *et al.* (1999) at a refuse collecting department have indicated that job rotation reduces workers' physical workload. These authors also concluded that job rotation makes the job less monotonous and therefore might increase job satisfaction and motivation of the employees. Besides the organisational and structural recommendations, several behavioral recommendations were given in our study. First of all, we gave the workers the advice to handle the hoist with both hands so the load will be distributed over both arms. When it was not possible to handle the hoist with both hands, the advice was given to alternate between the left and right side. Secondly, during manipulating the hoist the workers have to learn to also replace their feet. This will possibly decrease the load on the neck-shoulder muscles.

Also power- and flexiblity-training can be given which will increase the workers load carrying capacity. It stands clear that there is a relationship between maximal force and the later onset of musculoskeletal disorders. It is important to mention that such adaptations can only be succesful when practical training sessions are organised. Moreover the workers need to be motivated to change their behavior. This is simply manageable by introducing participatory ergonomics.

ACKNOWLEDGEMENTS

We wish to thank the medical department and employees of N.V. Duracell batteries, Aarschot for their contribution to this study.

REFERENCES

Hermans, V., Hautekiet, M., Spaepen AJ., Cobbaut, L. and De Clerq, J. 1999, Influence of material devices on the physical load during the end assembly of cars, *International Journal of Industrial Ergonomics*, **24**, 657-664

Hermans, V., Spaepen, A.J., and Wouters, M. 1999, Relation between differences in electromyographic adaptations during static contractions and the muscle function, *Journal of Electromyography and Kinesiology*, **9**, 253-261.

Hoozemans, MJ., van der Beeck, AJ., Frings-Dresen, MH., van Dijck, FJ. and van der Woude, LH. 1998, Pushing and pulling in relation to musculoskeletal disorders : a review of risk factors. *Ergonomics*, **41(6)**, 757-781

Karhu, O., Harkonen, R., Sorvalli, P. and Vepsalainen, P. 1998 Observing Working Postures in Industry: Examples of OWAS applications, *Applied Ergonomics*, **11(1)**, 13-18.

Kuijers, PP., Visser, B. and Kemper, HC. 1999, Job rotation as a factor in reducing physical workload at a refuse collecting department. *Ergonomics*, **42 (9)**, 1167-1178.

Masuda, K., Masuda, T., Sadoyama, T., Inaki, M. and Katsuta, S. 1999, Changes in surface EMG parameters during static and dynamic fatiguing contractions, *Journal of Electromyography and Kinesiology*, **9**, 39-46.

Mc Atamney, L. and Corlett, N. 1993, RULA: a survey method for the investigation of work-related upper limb disorders, *Applied Ergonomics*, **24(2)**, 91-99.

Spaepen, A.J., Baumann, W. and Meas, H. 1987, Relation between mechanical load and EMG-activity of selcted muscles of the trunk under isometric conditions. In : Bergmann, G., Kölbel, R. and Rohlmann, A. (Eds.) *Biomechanics : Basic and Applied Research* , (Martinus Nijhoff Publishers, Dordrecht), 473-478.

Zipp, P., 1992 Recommendations for the standardization of lead positions in surface emg. *European Journal of Applied Physiology*, **50**, 41-54.

AN ERGONOMIC EVALUATION OF FABRIC SLINGS USED DURING THE HOISTING OF PATIENTS

Laura Norton[1], Christine M Haslegrave[1], Sue Hignett[2],

*[1] Institute for Occupational Ergonomics, University of Nottingham,
University Park, Nottingham, NG7 2RD, UK.
Laura.Norton@nottingham.ac.uk
[2] Nottingham City Hospital NHS Trust, Nottingham, NG5 1PB, UK*

Hoist systems using fabric slings are a familiar tool for the manual handling of patients in hospitals. This study examined the sling as an interface, assessing its safety, usability, comfort and dignity against ergonomics principles of design. Information and opinions were collected through discussions with key stakeholders: patients, nurses, a sling designer and back care advisors. The multifaceted study design included: task and hazard scenario analysis of hoisting, semi-structured interviews with patients and observation of expert and novice nurses. Verbal protocol analysis was used to explore the decision making process for nurses selecting slings from a range of sizes and then fitting them. The findings question several aspects of the usability of current slings. Further analysis is needed to identify key anthropometric dimensions for assessing the fit of a sling.

Introduction

To move freely and transfer the weight of our bodies from one position to another is generally taken for granted. The impact of a disability and the need from a second or third person to achieve the goal of movement increase the complexity of the task. Adding equipment into the equation gives a further opportunity for incompatibility of method, fit or technique between those now involved in the task. However, the need for equipment is well recognised and encouraged (Health and Safety Executive, 1992) and the hazards of patient handling without equipment are also well documented and no longer tolerated (Bell, 1979, Owen, 1987, Hignett, 1996). Manual handling equipment is a necessity and mechanical hoists, figure 1, are now commonly found in the hospital environment. A low level of utilisation is still a problem and the research within this field has revealed the causes to be complex, including work environment, staff opinions, equipment design, availability and patient's opinion (Moody et al, 1996, McGuire & Moody, 1996, Conneeley, 1998 and Bell, 1984). A literature review of the area revealed that most work had concentrated on evaluating current hoist designs in relation to the nurse and tasks required of them (Bell, 1984, Hignett, 1998, Moody, et al 1996, Zhuang, et al 2000) but with minimal attention on the sling to attach the patient to the hoist, (Medical Devices Agency, 1994).

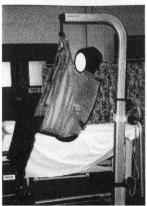

Figure 1. Hoist system involving fabric sling

The sling is involved in the two main interfaces of the hoisting system, being the link between the patient and the hoist, and incompatibility of either of these would be detrimental to the system and the task of hoisting. The operator's ability to assess and apply the sling to both hoist and patient are critical. However, there does not appear to be any direct research related to the usability of the sling.

Reluctance to use hoists has specifically been linked to a nurse's uncertainty in attaching the sling (Bell, 1984, Moody et al, 1996). The patient's experience of hoisting is recognised as being influenced by the nurse's assessment and subsequent choice and application of the sling, with descriptions of "fear", "insecurity", "degrading" and "uncomfortable" all recorded as the patient's perspective towards hoists, with the sling being the most frequent reason for rejecting a hoist (Conneeley, 1998). The decision making process when applying a sling requires skill in assessing the task; and patient's needs, combined with a broad knowledge of the equipment available and the potential mismatches (Bell, 1984, Love, 1996). However, there is little information which addresses the critical factors in this assessment, or whether the design and instructions for slings aid the nurse with the decision process. The principles of ergonomics in product design are to design in order to aid the interaction of products within their intended environments, tasks and users (Norris & Wilson, 1997). The need for sling design to be assessed with these goals in mind and to consider all users is paramount if the critical features that determine comfort, safety, dignity and usability are to be identified and ultimately incorporated to improve user acceptance.

Study Design

A multifaceted study design was used to investigate a U-style sling, figure 1, a) as an interface and b) its compatibility to manual-handling tasks, including the potential hazards and misuse of slings. The goals were to 1) understand the decision-making process and physical design of the sling and 2) establish the sources of error critical to safety, comfort and dignity of the patient during manual handling tasks. This aimed to assist in the development of guidelines to aid the nurse in correct use of the sling and manufacturers in the design of an interface compatible to both the patient and nurse.

The methods used to achieve these goals are summarised in figure 2.

Design Criteria	Interface	Compatibility to Tasks
Discussions with stakeholders Expert Evaluation Anthropometric analysis	Expert Evaluation Anthropometry Task Analysis Observation Patient Interviews	Discussions with stakeholders Task Analysis Observation Patient Interviews

Figure 2. Methods used to investigate ergonomic issues related to the sling as an interface

Expert evaluation
Expert evaluation involved two stages: consideration of the relevant standards and an evaluation by an ergonomist.
Two standards relevant to hoist slings were found:
1) Specification for mobile, manually operated patient-lifting devices (BSI BS 5827: 1979)
2) Hoists for the transfer of disabled persons – requirements and test methods (BSI BS EN ISO 10535: 1998)
The factors relating to hoist slings were developed into a checklist tool and included areas regarding the information on the sling, instructions on use and safety features within the design of the sling. Other literature and standards were also studied, including those related to equipment with similar functions, of lifting and suspending a person, to compare the nature of the information available to these equipment designers (British Coal, 1990, BSI, 1998).
An expert ergonomic evaluation checklist was the second stage and developed based on the literature review to provide a list of the general requirements of a fabric hoist sling. The features identified were grouped into seven main headings; material, size, method of application to patient and hoist, fit to patient in suspension, intended function and labelling. Each sling was evaluated by means of the checklist for its ability to provide a suitable interface with respect to ergonomics principles.

Anthropometric Analysis
The fit of the sling to the body was studied and the dimensions most critical to providing the patient with a safe, comfortable and dignified experience were identified and related to anthropometric dimensions for the UK population.

Task Analysis
An hierarchical task analysis was carried out on the basis of information obtained from an interview with a Back Care Advisor (BCA) and from the manufacturer's video. This provided a basic understanding of the task and typical user in a hospital environment. Three BCA's provided information on the nurse as a user, based on their knowledge of providing manual handling training. There were eight main stages identified in the task of attaching the sling to patient and hoist. Typical hazards identified by BCA's were recorded and used for a more detailed task analysis. Hazard scenarios developed

included: wrong sling size, incorrect placement of straps (sling back and leg straps) and inappropriate method of application of sling to patient or hoist.

Observation

Observation of nurses applying the sling was undertaken. The nurses were required to complete two bed-to-wheelchair transfers with the slings in a mock up of a hospital setting. The tasks were videoed and analysed using an event-recording tool (developed from the hierarchical task analysis). This allowed analysis of each component of the task and the recording of the number of correct stages and any errors performed. Verbal protocol analysis was used to elicit information about the decision making process during each stage of the task. The groups of nurses included both experts and novice users of hoists to explore differences in user behaviour.

A standard 'patient' was represented by, an able bodied 29 year old female. On calculation of more than ten relevant anthropometric dimensions, she was estimated to fall at the 53rd percentile for most dimensions. From an assessment by one BCA and reference to the manufacturers recommendations, based on weight, a medium sling was considered to be appropriate for her.

Patient Interviews

The opinions of patients' with hoisting experience were essential to fully understand the sling as an interface. Patients from a neurological rehabilitation unit were interviewed. The ten recruited were recent users of the hoist slings being evaluated, able to communicate and without cognitive impairment. A semi-structured interview was used with prompts to explore the patient's opinion on the safety, comfort and dignity of fabric slings. Their individual feelings on being hoisted in fabric slings were also explored, in particular how the approach of the nurse influenced them, their perception on the decision making process taken and the level of training and effort which they felt was required of the nurse to apply the sling. The questions were based on the findings from the literature review of previous studies. (Bell, 1984, Conneeley, 1998, McGuire & Moody, 1996, Moody et al, 1996, McGuire et al, 1996). Notes were taken during the interview and transcripts made within two hours. Themes were identified within the responses and the patients' views were summarised

Summary of Findings

The slings performed well against the current standards for patient hoists. However, the standards relating to mountaineering equipment (BSI BS EN 12277: 1998) recommend providing the user with details on sizing of the equipment and how to obtain an optimum fit in the form of a pictogram. This is lacking from the hoist standards but would be of value in view of the hazards highlighted from the task analysis and observations made in this study. The maximum weight and the size category of the sling are required on patient hoisting equipment but there is no recommendation that a method of sizing or fitting should be included on the sling. Weight is the only decision aid suggested as a sizing tool in the instructions (Medical Devices Agency, 1994) and is used by manufacturers. However, as this gives no advice on the patient's proportions or compatibility with the sling chosen, it only provides a guide and does not provide any real information to the decision maker as to which sling will most adequately fit a particular patient, since other body dimensions are not closely correlated with weight.

The sling sizes available appeared to cater for the weight of a population (UK) between the 5th female and 95th male percentile but, as already mentioned, weight is not the only dimension relevant to the fit of a patient. The multiple dimensions considered critical to the fit of a sling included; the circumference of the base of the sling, the circumference of the leg strap and the width and length of the back support. These dimensions would require the relevant circumferential and functional anthropometric data to assist sling designers in designing for a wide population and to provide guidance on sling sizing. These dimensions were not found to be available in current published data and therefore not available to the designers of this equipment. This makes it difficult for designers, at present, to provide detailed guidance to nurses in identifying a sling size to fit a particular patient. The only guidance on sizing found in instruction books for hoists surveyed during the study illustrated how to recognise a poor fit rather than guide a good fit.

With regard to attaching the sling to the patient (and hoist), the sling studied had some instructions on a label but the sling and label used one colour and this did not draw the user's attention to the information. In general given the lack of detailed instructions, it appears to be assumed that nurses will remember any training provided and training is often cited as being sufficient for correct manual handling of patients (Brewin, 1995, Moody et al, 1996). However use of a hoist may be infrequent after training for many nurses (Hignett, 1998) and a design that relies on training for correct usage risks human error.

From the observational studies, it was found that the decision-making process differed between the experts and novice users. The Expert users had an 89% success rate in applying the sling; and often reported that they relied on rules gained from training and previous experiences, these included failures and difficulties. Novices adopted a problem solving approach; but, since they had few prompts from the sling itself, only managed 56% success. Placement of the bottom of the sling on the patient; had been identified as critical to sling safety by sling designer; and BCA as well as during the expert evaluation. However this was noted as being particularly poor in both experts and novices.

The only dimensions considered important in fitting, by both groups, were the width and the height of the back support. These are misleading since they do not take account of the main support surface provided by the bottom of the sling. There appeared no definitive rules adopted for sizing and individual interpretation was evident in the nurse's estimation of the sling size required for the patient.

The patient's gave a slightly different perspective on the decision making process. They thought the choice of sling size was dependent upon their weight and body size but also considered that availability of slings was also a determining factor. They suggested that a nurse's level of experience influenced the efficiency in fitting the sling with one patient commenting that " if there is a wrong way to do it someone will do it".

The comfort of the sling, in the patient's opinion, was related to the use of the correct size and method of application. Overall the patients considered that the slings were safe every time they were suspended but that the nurse's confidence in completing the task influenced their feeling of security.

The potential for error with selecting and fitting slings is evident and consideration of the hazards identified by this study should be considered in future sling design

References

Bell, F. (1979) Patient hoist biomechanics. *Occupational Therapy,* **Jan:** 10-16.

Bell, F. (1984) *Patient–lifting Devices in Hospital. Croom Helm. London.*

Brewin, M.(1995) A qualitative research study into the use of manually operated mobile hoists and the need for tailored 'job aids'. *British Journal of Therapy and Rehabilitation,***2**,(4): 165-173.

British Coal (1990) Guidance on the safe methods of entry, cleaning and maintenance of bunkers. Report No.13: 9-23*Community Ergonomics Action;European Coal and Steel Community.* Luxembourg.

BSI BSEN (ISO 10535: 1998) *Hoists for the transfer of disabled persons-requirements and test methods.* London: British Standards Institution.

BSI (5827: 1979) *Specification for mobile manually, operated patient-lifting devices.* London: British Standards Institution.

BSI EN (12277:1998) *Mountaineering equipment-harness-safety requirements and test methods.* London: British Standards Institution.

Conneeley, A.L. (1998) The impact of the manual handling operations regulations 1992 on the use of hoists in the home: the patient's perspective. *British Journal of Occupational Therapy ,***61**, (1): 17-21

Health & Safety Executive (1992) *Manual Handling: Guidance on regulations. Manual Handling Regulations 1992.* London.

Hignett, S. (1996) Work related back pain in nurses. *Journal of Advanced Nursing ,***23**,: 1238-1246.

Hignett, S. (1998) Ergonomic evaluation of electric mobile hoists. *British Journal of Occupational Therapy,***61**,(11):509-516.

Love, C. (1996) Ergonomic considerations when choosing a hoist and slings. *British Journal of Therapy and Rehabilitation ,***3**,(4): 189-198

Medical Devices Agency (1994) Slings to accompany mobile domestic hoists-An evaluation. Report No A10. London. *Department of Health and Safety/HMSO:1-28*

McGuire, T. and Moody, J. (1996) A study into clients' attitudes towards mechanical aids. *Nursing Standard ,***11**, (5): 35-38.

Moody, J. McGuire, T. Hanson, M and Tiger, F (1996) A study of nurses' attitudes towards mechanical aids. *Nursing Standrd ,***11**, (4):37-42

Norris, B. and Wilson, J. (1997) *Designing safety into products: Making ergonomics evaluation a part of the design process.* Depatment of Trade and Industry. London.

Owen, B.D. (1987) The need for application of ergonomic principles in nursing. *Trends in ergonomics/Human Factors IV:* 831-838. Elsevier Science Publishers B.V. North Holland

Zhuang, Z. Stobbe, T.J. Collins, J.W. Hsiao, H. and Hobbs, G.R (2000) Psychological Assessment of Assistive Devices for Transferring Patients/Residents. *Applied Ergonomics ,***31**,: 35-44.

MANAGING BACK PAIN

PARTICIPATIVE QUALITY TECHNIQUES FOR BACK PAIN MANAGEMENT

S Brown[1], R A Haslam[1] and N Budworth[2]

[1]*Health and Safety Ergonomics Unit, Department of Human Sciences, Loughborough University, Loughborough, Leicestershire, LE11 3TU*

[2]*Health, Safety and Environment Department, NSK-RHP Europe Limited, Mere Way, Ruddington, Nottingham, NG11 6JZ*

Support from people at all levels within an organisation is essential for achieving the changes necessary to reduce back injuries, with active commitment from senior managers especially important. Quality management methods have high credibility among such personnel in the manufacturing industry, providing the rationale for this study. This involved the development and initial trialing of a back injury management package, based on a set of tools known as Plan Do Check Act (PDCA). Preliminary evaluation showed that trained operators were able to make effective use of a tailored version of PDCA. Follow up of SMEs briefed on the package found that several were planning to incorporate material from it into their manual handling training programmes. None, however, was planning to use the back pain management package in full.

Introduction

Despite the reduction of heavy industry over recent years, the prevalence of low back pain in the UK has remained constant, while back-related sickness absence has actually increased (Waddell, 1998; MacDonald *et al*, 2000). In response to this, the Back in Work campaign, a joint initiative of the Department of Health and the Health and Safety Executive, is targeting back pain in the workplace. The study reported here was one of nineteen projects commissioned under the initiative to explore current best practice for reducing back-related sickness absence.

The aim of the project was to develop and evaluate a back injury management package suitable for use in both large and small organisations. The study, which ran for 12 months, concentrated on operators in a distribution warehouse, owned by a large engineering company. Loads handled were mostly pre-packaged boxes of various weights and sizes. Once developed, the back pain management materials were then tested with engineering SME partners on the project. Another member of the project team, Newark Hospital Physiotherapy Department, provided access to 'fast track' physiotherapy treatment for employees in the study group experiencing acute back pain.

Evaluation Data

At the commencement of the study, baseline data were collected using sickness absence figures and a musculoskeletal questionnaire, allowing the effect of changes arising from the project to be assessed. The mean number of days lost per employee through back-related sickness absence increased fourfold from 0.71 days in 1997 to 3.24 days in 1998, reducing slightly to 2.07 in 1999. Among the sickness records, 14% gave insufficient information to identify the reason for absence. At the end of the project the average number of back-related sickness absence days per employee decreased to 0.97 (figure 1).

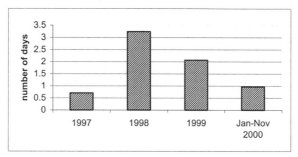

Figure 1. Mean back-related sickness absence days per employee 1997-2000

This pattern was not reflected in the overall sickness absence figures which had risen considerably (figure 2). Back-related sickness absence was 35% of the total sickness absence for the warehouse in 1998, whereas in 2000 it had decreased to 4%. Changes in the sickness absence figures were compared with changes that had taken place in the workplace. A major change occurring in the warehouse in 1997 was the introduction of heavier boxes. This arose due to the closure of a warehouse elsewhere in the UK, with larger, heavier boxes subsequently transferred to the study warehouse for storage and distribution. There was no proportionate increase in staffing.

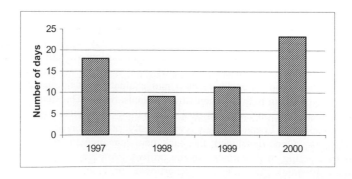

Figure 2. Mean total sickness absence days per employee 1997-2000

A modified version of the Nordic Musculoskeletal Questionnaire, (Kuorinka et al, 1987) was completed by 36 day and night shift operators, 92% of the warehouse workforce, at the start of the project. The data identified the number of workers who had experienced musculoskeletal discomfort, in various parts of the body, in both the preceding seven days and 12 months, and whether the discomfort had prevented them performing their normal daily routines at work or at home.

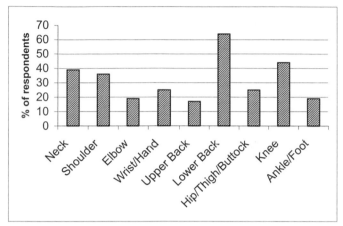

Figure 3. Musculoskeletal discomfort during previous 12 months

Low back pain was the most widely reported problem, figure 3, with 64% of respondents experiencing this in the preceding 12 months. Neck and shoulder problems were identified by more than one third of respondents and almost half, 44%, recorded knee discomfort in the preceding 12 months. The latter finding is significant, as manual handling training has traditionally promoted a 'back straight, knees bent' technique.

Plan Do Check Act

PDCA is a quality management method which uses a problem solving cycle for continuous improvement (Imai, 1986). At each stage of the cycle there are tools allowing the identification and prioritisation of problems and evaluation of changes. A simplified format, using a step by step guide to problem solving (table 1), was developed for use in the warehouse, to identify problems with manual handling postures. This was supported by a checklist for quick and easy monitoring of manual handling postures. A steering group, comprising four operators from the warehouse, was formed at the start of the project, with the group meeting fortnightly for one hour. After receiving training in use of PDCA, the group used the method to analyse manual handling in the warehouse.

The steering group were encouraged to target their ideas for solutions at problems which would be relatively easy to overcome and benefit a large number of operators, giving quick visible changes in the workplace. For example, a storage shelf was lowered to avoid operators having to lift heavy loads above shoulder height. Two major issues identified by the group were access to the storage system and the weight of loads; these issues were not easy or quick to resolve. As an interim measure, it was suggested that heavier boxes should be stored in the most accessible bins, to reduce the amount of reaching and bending required. In the longer term, the packaging department will review the maximum weight of boxes produced.

Table 1. Simplified PDCA problem solving steps

Problem Solving Steps	Technique/Tools
Define the problem	SMART objective
Determine possible causes of the problem	Brainstorming
List the 10 most likely causes in order of importance	Cause and Effect diagram
Calculate the degree of ease of change and its level of impact	Impact assessment
Identify the root causes	5 WHYS
Determine what can be monitored to verify the causes	Checklist
Use the information to determine the three most significant causes	Top 3
Monitor the effects of the changes	Checklist

Back School

A back school was prepared, with the intention of providing the study group with accessible guidance on how to avoid and manage back pain. This gave advice on preventing back pain, both at work and during home and leisure activities, and explained how best to treat any back pain that did occur. The school comprised six one-hour sessions, covering the following topics:

- the structure and function of the spine
- causes of back pain
- self management of back pain
- methods of treating back pain
- stretching exercises
- 24 hour back care
- principles of manual handling

Sessions were designed to be highly interactive, featuring games and quizzes, as a way of unobtrusively repeating key messages. The material incorporated information from the recent guidelines for the management of back pain, promoting activity rather than bed rest and addressing the effects of psychosocial factors (Faculty of Occupational Medicine, 2000). The sessions were expected to run either in work time or as paid overtime, and to be available to all employees in the warehouse, whether or not they had experienced back pain. Attendance was on a voluntary basis.

Fast Track Physiotherapy

Early treatment of some back injuries, within the first six weeks, has been shown to be helpful in reducing pain and is regarded as having credibility with patients. As a part of this project, arrangements were made for employees to be referred by the company occupational health doctor to the 'Acute Back – Fast Track Service' at the local NHS physiotherapy department. Referral could be made if the employee met the medical criteria issued by the hospital, regardless of whether the injury was work related or not.

Discussion

A major part of this project was the attempt to bring a comprehensive package of measures to bear on a warehouse environment with a suspected back pain problem. Evidence regarding the existence of the problem was provided by an analysis of sickness absence data at the start of the project, with absence due to back pain reaching 35% of total sickness absence in the three years prior to the study. Corroboration was provided by a survey of musculoskeletal discomfort, with two thirds of respondents indicating experience of back pain during the previous year. Also, informal assessment by the two ergonomics specialists involved with the project judged the manual handling in the warehouse to be significant.

An interesting finding from the musculoskeletal survey was that approaching half of respondents indicated they experienced discomfort with their knees. The usual manual handling training message 'keep the back straight and bend the knees' aims to move the load by means of the strong leg muscles rather than those of the back. However, people may find the extreme knee position required by a 'back straight, knees bent' posture difficult or impossible to adopt. If this finding is more widespread, then an alternative approach to training, encouraging consideration of manual handling principles, rather than being prescriptive with respect to actual technique, might be more effective.

Although of value, the sickness absence data and findings from the musculoskeletal survey should be treated with caution. While the analysis provided useful indications of the scale and trend of back pain in the warehouse, both may have been influenced by other factors, including psychosocial aspects. A large increase in back-related sickness absence in the warehouse in 1998 followed an increase in volume and weights handled. While the handling of heavier loads may have contributed to an increase in back pain, it is possible that some of the sickness absence at this time may have been associated with pressures arising from work and organisational changes. The vulnerability of the sickness absence data to reporting influences is illustrated by the impressive 50% decline in back-related absence over the project year, at a time when sickness absence in the warehouse for all reasons nearly doubled. The authors suspect that at least some of this 'reduction' is an artefact of the operation of the sickness absence reporting and recording process, rather than entirely due to improvements in manual handling.

Application of the PDCA quality management technique to back pain had mixed success. After training, the warehouse steering group used the method effectively, identifying manual handling problems and possible solutions. Feedback from the SME project partners was that while they found the PDCA approach interesting, they thought it unlikely they would adopt it in full. Quality management techniques, such as PDCA, are often straightforward to use once mastered, but do require an initial commitment to learn. It may be hard for SMEs to justify this against competing priorities, unless already familiar with a method. The SMEs did, however, indicate that some of the material developed in support of PDCA and the back school would be of value for their own manual handling control programmes.

Unfortunately, it proved impossible to run the back school, due to problems with scheduling, particularly identifying a convenient time for operators to attend. Shifts began early in the day, with long hours, and thus overtime was not popular. Shortages of staff in the warehouse arising from sickness absence, holidays and shift patterns, made it difficult for the sessions to be run in work time. Take up of the fast track physiotherapy referral was also limited. Here, there seemed to be a breakdown of referral procedures, together with difficulty accommodating the time required for employees to receive treatment off-site. Also, it is not certain that the research team were wholly successful in persuading both supervisors and operators that treatment for back pain arising from activities outside work was a legitimate and beneficial part of the package.

Conclusions

As might be expected when tackling a complex issue in a real world setting, the project did not proceed entirely according to plan, with difficulties experienced implementing some aspects of the intervention. For this case study, the existence of a back pain problem seemed to be a significant impediment to actually tackling the problem, with obstacles arising from staffing shortages and subsequent low morale among the workforce. However, there were successful outcomes from the study, particularly with respect to the development of high quality assessment and training materials. The judgement of the research team is that while some components of the back pain management package may have had limited success, increased awareness in the study workplace arising from the project is starting to reap benefits over the longer term, as it has focussed line management attention on the economic benefits of effective health and safety management. It is probable in this instance that the benefits from the 'whole' of the intervention will be greater than the sum of its parts.

Acknowledgements

This research was funded by the Department of Health. The authors also wish to thank project partners NSK-RHP, Newark Hospital Physiotherapy Department, and the Engineering Employers Federation. Views expressed are those of the authors.

References

Faculty of Occupational Medicine, 2000, *Occupational Health Guidelines for the Management of Low Back Pain at Work* (Faculty of Occupational Medicine, London)

Imai, M., 1986, *Kaizen*, (McGraw-Hill, London)

Kuorinka, I., Jonsson, B., Kilbom, A., Vinterberg, H., Biering-Sørenson, F., Andersson, G. and Jørgensen, K., 1987, Standardised Nordic questionnaires for the analysis of musculoskeletal symptoms, *Applied Ergonomics,* **18**, 233-237

MacDonald, E. and Haslock, I., 2000, Spinal Disorders. In R.A.F. Cox, F.C. Edwards and K. Palmer (eds.) *Fitness to Work* (Oxford University Press, Oxford), 210-234

Waddell, G., 1998, *The Back Pain Revolution*, (Churchill Livingstone, Edinburgh)

A MULTI-DISCIPLINARY APPROACH TO MANAGING BACK INJURIES IN BRITISH HEALTH CARE WORKERS

Pearce, J.P. [1], Morant, C. [2], Turner, T. [2]

[1] School of Health Professions, University of Southampton, Highfield Southampton SO17 1B, UK.
[2] Occupational Health & Safety Service, Southampton Community Health Services NHS Trust, Moorgreen Hospital, Botley Road, West End Southampton, SO30 3JB, UK.

Despite all the recent initiatives that have tried to reduce the number of back injuries, back pain continues to be a major problem for NHS staff.
This paper will detail through audit how, by providing continuing education, occupational health nurses can utilise current research to effectively monitor injured staff and triage them to appropriate treatments. The cost effectiveness of referral by occupation health nurses to private physiotherapy will be discussed. Consideration will also be given to managing workers who have persistent back problems and explore how a Back School can be used to educate staff into self-management of their problems.

Background

Despite the recent national initiatives (RCN 1996) that have promoted an ergonomic approach to the handling of patients based on risk assessment, there remains a significant problem with staff injuries. Recent estimates suggest that 80,000 nurses receive back injuries annually with 3,600 nurses needing to be medically retired each year because of their injuries (BackCare 2000). These figures have a major impact on the NHS both in terms of treatment costs and decreased productivity.

There are significant financial implications of having staff on sick leave. NHS employees will receive sick pay (full rate) for six months then half pay for a further six months. While they are on sick leave their work still needs to be done. This means that another person, assuming availability, will need to be employed in order for their work to be covered. As the cost of employing a second person is likely to be at least the cost of sick pay, an employer is essentially paying at least double the normal cost to get the job done.

Ongoing changes in staff induce low morale and the perception that the employers do not value their staff and will inevitably mean that the quality of the service provided will diminish. Primarily this is because of the loss of continuity - changing staff, changing approaches, lack of individual patient knowledge. Specific expertise in a given area will be diminished.

The crisis we see in nursing recruitment and retention today means that we desperately need to protect the key resource of staff.. Allowing staff to remain on long term sick leave is not only bad in financial and quality of service terms but also once a person is off work 6 months they have only a 50% chance of returning to their job, after 2 years the chances of ever working again are slight (CSAG 1994). It is however, not only in personnel terms that we need to be concerned - surely there is a moral issue here as well. If our staff are injured going about their duties don't we have a moral obligation to aid their recovery and hence their return to work?

The work presented in this paper was undertaken in a NHS Trust that provides community health care to a large geographical area and employs in excess of four and a half thousand staff across eighty different sites. The services provided include district nursing, elderly care, mental health, child health and learning disabilities.

The occupational health and safety team comprises six occupational health nurses, an occupational health doctor, a health & safety officer, a back care advisor, a staff counsellor and a physiotherapist. The team has the services of the occupational health doctor for only 4 hrs per week. With the medical cover available it was inevitable that the service become nurse led. This in turn means that it is imperative for the nurses to be able to manage not only the usual occupational health issues but also an increasing amount of back and general musculoskeletal injuries.

Despite a comprehensive moving and handling training programme, the provision of handling aids and the redesign of some work areas, staff were still being injured. The service needed to effectively manage back and musculoskeletal injuries. The front line staff who needed to face this challenge were the occupational health nurses. Bearing in mind the breadth of their work and training, if the nurses were going to manage the musculoskeletal problems that came into the department on a daily basis they needed tools and specialist skills to do so.

Methods

As part of a prospective audit this paper will outline four tools that have been utilised and developed in order to manage musculoskeletal and, more especially, back injuries more effectively.

Monitoring the recovery of staff

Monitoring recovery and returning staff to work following injury is complex is often fraught with complications. Physical and psychological risk factors (Kendall et al 1997 & RCGP 1999) need to be considered and decisions often have to be undertaken without the advice of an occupational health doctor. Simply asking the member of staff if they feel well enough to return to work is too subjective. Measuring pain is difficult, measures of disability are more reliable and valid (Waddell 1998). What was needed was a validated, objective tool to measure disability. Greenough & Frasers's (1992) Low Back Outcome Score was designed to assess outcome in patients with low back pain. This is a self administered scale consisting of 13 questions ranging from the amount of rest needed to the patient's ability to undertake activities of daily living.

It was decided that injured staff would be scored at their initial occupational health consultation and during recovery to ascertain that they were indeed recovering. This functional capability score would then be used alongside their subjective opinion to determine return to work.

Triage

Various international authorities (e.g. AHCPR 1994, CSAG 1994, and RCGP 1999) have examined the management of back pain over the last few years. These authorities along with the guidelines recently developed for managing back pain in occupational health (OHG 2000) all consider diagnostic triage to be fundamental to the management of back pain. Diagnostic triage is used to exclude non spinal and serious spinal pathology and to distinguish between a nerve root problem and simple back-ache (Waddell 1998). It was apparent that the team needed an understanding of triage.

We are not expecting the nurses to differentiate between an L1 and an L5 nerve problem but rather to recognise serious spinal pathology or progressive/severe weakness and, in the absence of an occupational health doctor, to make appropriate emergency referrals. Figure 1 illustrates a flow chart approach to determining serious spinal pathology.

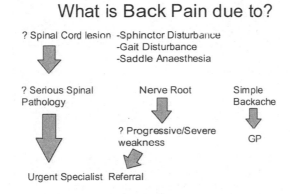

What is Back Pain due to?

? Spinal Cord lesion -Sphincter Disturbance
-Gait Disturbance
-Saddle Anaesthesia

? Serious Spinal   Nerve Root   Simple
Pathology   Backache

? Progressive/Severe
weakness   GP

Urgent Specialist Referral

Figure 1. Flow chart to illustrate diagnostic triage (after CSAG 1994)

Treatment

The ability to monitor and triage injuries is vital however, having gained that information we need to utilise it effectively in order to progress the injured person. Invariably physiotherapy was indicated. A study by Zigenfus et al (2000) looked at 3867 back injured patients and divided them into 3 groups on the basis of their delay in receiving physiotherapy. The study found that initiating early physiotherapy was associated with fewer physician visits, earlier discharge from care, fewer restricted workdays and days away from work. As a team we were acutely aware of the fact that the longer a person is off work the lower the chances of returning to work (CSAG 1994). Accessing physiotherapy through the GP was problematical at this time with waiting lists running at up to 20 weeks. Acute problems being left untreated were becoming chronic. At this time the only option for obtaining physiotherapy was via the person's GP – what we needed was direct, fast access to physiotherapy.

Interestingly across the road from one of our major sites was a high quality, private physiotherapy practice. The concept of referring injured members of staff into the private sector for physiotherapy is not common and to many would be perceived as revolutionary. Before going any further however, the issue had to be taken to the corporate management board. Obviously the argument presented was based on cost:

- Cost – we were able to prove that physiotherapy would be more cost effective as staff would return to work faster.
- No capital costs in terms of equipment
- No accommodation requirements
- Excellent quality - (cutting edge / research based practice)
- Fast access < 48 hours
- Long opening hours – our staff could be seen during evening and weekends thus protecting work time

In 1997 the hospital managers were persuaded of our proposal and agreed to a pilot study. This ran for nine months and was so successful that the service has continued since.

Support – Back Care Programme

Unfortunately not everyone makes a good recovery. The issue of supporting people with chronic long-term problems needed to be addressed. Bearing in mind the bio-psycho-social model of back pain (Waddell 1998) and Kendall et al's (1997) psychological risk factors(yellow flags), the team devised a support programme with input from a clinical psychologist. The programme comprises one session a week for eight weeks and these are run in an informal manner with an opportunity to have a cup of tea and a chat. Each session includes:

- Exercise - Kankaanpaa et al (1999) showed the importance of exercises for chronic low back pain, while Klaber Moffett et al (1999) also proved the benefit of group exercises for chronic low back pain
- Relaxation
- Topic for the week e.g. 24-hr. back care, pain/stress management, ergonomics at work and home, etc.

Results

Over the last two-year period, ending 31 March 2000 two hundred and thirty six staff (236) staff has been treated for back and general musculoskeletal injuries.

We were able to establish that:
27% of injuries were definitely work related
28% of injuries definitely were not work related
45% were of unknown cause

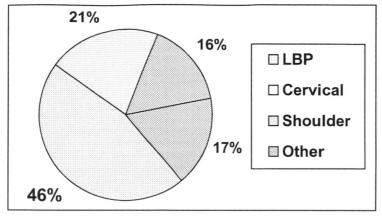

Figure 2. Types of injuries sustained

Figure 2 illustrates the type of injuries sustained by staff. While the low back pain is not surprising the amount of shoulder injuries was not expected.

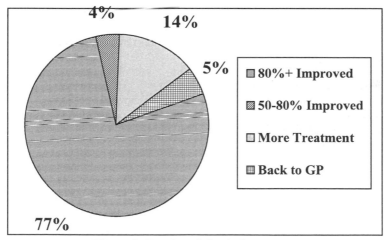

Figure 3. Results of physiotherapy

Figure 3 illustrates that within 5 treatments 77% of staff showed an 80% plus, self reported, improvement in pain, mobility and function. This was done at a cost of £135 maximum per person. The cost of a 'D' grade nurse being unable to work for one week (sick pay + replacement) is £708. So in the majority of injuries the cost of physiotherapy is less than one fifth of the cost of replacing a basic nurse for one week. Fourteen percent needed more than 5 sessions of physiotherapy, while 4% were at least 50% improved and continuing to self manage with exercise(self reported). Four percent needed to be sent back to their GP for either reassessment or specialist referral.

Twenty two percent of staff injured required time off work and of those who had time off 41 % were back to work in 4 weeks.

The majority of participants on the back care programme continue to work but may well be on a managed/staged return to work. Further qualitative research is currently being undertaken by semi structured interview to evaluate this programme. From initial thematic analysis the key issues arising are:
• The support the group offers each other
• The chance to ask questions and explore their problems in a safe, caring environment
• Being valued enough by the trust to be sent on the programme

Conclusion
Despite ergonomic intervention injuries still occur. Equipping the occupation health nurses with tools and resources means injuries can be more effectively managed. This makes good economic sense for the trust as staff return to work faster and has been a real morale boost to staff that the trust is interested in them to the extent where they will pay for their private treatment.

References.

AHCPR 1994, *Management Guidelines for Acute Low Back Pain,* Agency for Healthcare Policy and Research, (US Department for Health and Human Services, Rockville, MD)

BackCare. 1999, Women and Back Pain – some facts and figures. Press release Teddington Middx

CSAG 1994, *Report on Back Pain,* Clinical Standards Advisory Group, (HMSO, London)

Greenough, C. G,. and Fraser, R. D. 1992, .Assessment of outcome in patients with low-back pain, *Spine,* **17**, 1, 36-41.

Kankaanpaa, M., Taimela, S., Airaksinen, O., and Hanninen, O. 1999, The efficacy of active rehabilitation in chronic low back pain', *Spine,* **24**, 10, 1034-1042

Kendall, N. A. S., Linton, S. J. and Main, C. J. 1997, *Guide to Assessing Psychosocial Yellow Flags in Acute Low Back Pain: Risk Factors for Long-Term Disability and Work Loss,* (Accident Rehabilitation and Compensation Insurance Corporation of New Zealand and the National Health Committee, Wellington, New Zealand)

Klaber Moffett, J.., A Torgerson, D., Bellsayer, S., Llywelyn-Phillips, H., Farrin, A., and & Barber, J. 1999, Randomised control trial of exercise for low back pain, *British Medical Journal,* **319**, 297-283

OHG 2000, *Occupational Health Guidelines for the Management of Low Back Pain at Work – Evidence Review,* (Faculty of Occupational Health Medicine, London)

RCGP 1999, *Clinical Guideline for the Management of Acute Low Back Pain,* (Royal College of General Practitioner, London)

Royal College of Nursing 1996, *Manual Handling Assessments in Hospitals and the Community,* (Royal College of Nursing, London)

Waddell, G. 1998, *The Back Pain Revolution* (Churchill Livingstone, London)

Zigenfus, G. C., Yin, J., Giang, G. M., and Fogarty, W. T. 2000, Effectiveness of early physical therapy in the treatment of acute low back musculoskeletal disorders, *Journal of Occupational and Environmental Medicine,* **42**, 1, 35-39

HEALTH AND SAFETY

SAFETY GUARDING OF MANUALLY LOADED, MOBILE WOOD CHIPPERS WITH POWER FEEDS

David Riley[1], Neil Craig[2], Robert Nixon[3], Robert Blogg[3]

[1]Health & Safety Laboratory, Broad Lane, Sheffield, S3 7HQ
[2]Health and Safety Executive, National Agricultural Centre, Stoneleigh, Kenilworth, CV8 2LG
[3]Health and Safety Executive, McLaren Building, Dale End, Birmingham, B4 7NP

A serious accident involving a mobile wood chipper where a young operator sustained an amputated arm prompted an investigation of the circumstances of the accident and the machine design. The use of mobile woodchippers has increased in recent years, and they are now widely used by arboriculturalists and conservation and landscaping contractors to convert unwanted wood and brush into woodchips for mulch. The accident highlighted the potential for the operator to become inadvertently entangled with the fed material while feeding the chipper, and consequently to be drawn into the hopper and towards the feed rollers and blade - the danger zone. The design of the wood chipper, through the guarding and control and safety devices should aim to minimise the possibility of this occurring. The majority of woodchippers on the market are of a similar basic design, but it became apparent that small differences in the guarding and the configuration of the control/safety devices could mean large differences in the degree of protection provided. This paper describes some of the work done in making an appraisal of these machines, and in producing recommendations for design to inform the European Standard drafting process.

Introduction

The use of mobile woodchippers has gradually increased over the last ten years, and they are now widely used by arboriculturalists and conservation and landscaping contractors to convert unwanted wood and brush into woodchips for mulch. They are typically used at roadsides and in parks and gardens by local authorities and arboricultural businesses. It is estimated that there may be around 10000 in the UK. A mobile wood chipper is a trailer mounted machine consisting of a feed chute or hopper into which wood and brush is fed. At the neck of the hopper are either one or two infeed rollers. These grip the material and draw it in against a fast spinning chipping blade(s). The cutting blades are usually powered by a diesel engine with the infeed roller(s) powered hydraulically. A typical pulling force has been calculated as 370 kg force, with quoted feed rates of 46 metres / minute (760mm/sec). The chipped material is blown out of a directable chute, usually into the back of a truck for removal. These machines are capable of chipping wood of around 230 mm in diameter. The operation of the wood chipper is usually controlled by a control bar which also acts as a safety device. The typical design has the bar extending around the sides and top of the front edge of the hopper and pivoting at the base. The action of the feed rollers is controlled by the position of this bar, it having several positions for stop, feed, and reverse functions. The engine is usually controlled by other means as well as an emergency stop switch.

In 1999 there was a serious accident involving a mobile wood chipper where a young operator sustained an amputated arm below the elbow. This prompted an investigation of the circumstances of the accident, the machine design and the relevant safety standards. Since then, a second accident occurred in the UK and others are known to have occurred elsewhere in the world.

The precise details of the accident are unclear, but whilst feeding gorse into the chipper, the worker's right hand or glove became entangled in the gorse and his arm was drawn into the machine through the infeed rollers. As a result his arm was severed below the elbow. It is not possible to tell where he was standing in relation to the hopper while working, but as his arm was drawn into the hopper it was twisted so he was turned face down, consequently he could not reach the control/safety bar. His shoulder was dislocated, presumably as a result of the twisting action and resisting the pull of the machine. The control/safety bar did not work to prevent the worker reaching the rollers while they were active. The key feature was the positioning of the device. The bar was mounted at the front edge of the hopper, pivoting from a point 150 mm up the side and running up the sides and across the top. The bar had three control positions, feed-stop-reverse, the feed position was when the bar was flush with the front edge of the hopper, and stop was when the bar was pushed behind the front edge. As such the device was not capable of being passively actuated as the worker was pulled towards the rollers.

Type A with a control/safety bar present along the bottom and sides of the hopper

Type B with side plates and control bar. Large triangular plates extend from the extension table to the top of the hopper

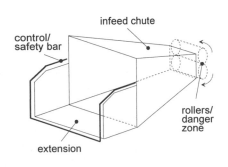

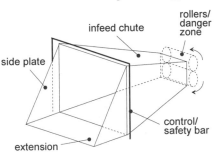

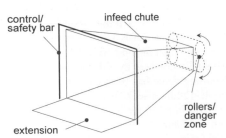

Type C chute with a flat extension table with top and side control/safety bar

Figure 1. Hopper and control/safety bar configuration for the three basic design types

An initiative was started to prevent further use of this type of machine until modifications had been made to the configuration of the control/safety bar. In addition the design of all mobile power fed chippers was to be considered in order to inform the UK input to the prEN. A survey of the machines identified three design types as shown in Figure 1.

Guidance on safe design

Wood chippers are covered by the Supply of Machinery (Safety) Regulations 1992, and at the time of the accident there were two sources of information specific to safe design of wood chippers in the UK. These were an HSE information document and a draft European Standard prEN 13525 (available for public comment April 1999). The documents are similar, and require that an additional safety device is provided since guarding by distance alone is not appropriate (BS EN294:1992). Both require that the control/safety bar should enable the worker to stop or reverse the infeed action by using parts the body other than the hands (e.g. arm, torso, hip etc...). In addition the bar should provide protection over 75% of the vertical projection of the chute and 100% of the horizontal, either at the top or the bottom, depending upon the working height of the chute. If the bottom surface of the chute is above 600 mm then the bar shall extend across the bottom edge, if it is below 600 mm from the ground, it shall extend across the top edge. The 600 mm height criterion is a compromise arrived at by consensus. It is intended to differentiate machines at which the operator might be able to push material in with their foot. There are additional requirements for the action of the control to protect against this.

Entanglement situations

There are a variety of ways in which a person could be pulled into a wood chipper once they were sufficiently entangled in the fed material, and a number of factors influencing how a worker might be drawn into the hopper once entangled, such as: the characteristics and dimensions of the hopper; operating height; feed speed and force; configuration of the control/safety bar in relation to the hopper; the actuation force of the bar; working practices; material being chipped, and; worker size. However, here concerns are centred on the hopper and control/safety bar and the ways in which an entangled worker might interact with them.

In understanding how entanglement scenarios might occur and progress, it is important to consider the environment in which the machines operate and how they are used. The most likely scenarios were considered by starting from the most common working positions and practices. Observation of workers during chipping operations with a Type C machine suggests that workers tend to feed from the side of the hopper and close enough to reach the control bar easily (e.g. to select reverse to free stuck material), while avoiding being struck by material twisting about as it is drawn into the machine. Larger pieces were placed into the feed rollers, shorter pieces were seen to be tossed into the chute and then pushed in with following material to ensure that the rollers engaged on them. When clearing up, workers were observed placing bundles of short branches/brush straight into the hopper from the front.

A hopper and control/safety bar assembly was borrowed from a manufacturer for study. The various potential entanglement scenarios from typical working positions and the resulting movements and body positions were enacted and photographed. Some of these were modelled using the ergonomics Computer Aided Design tool SAMMIE. Some scenarios were associated with particular machine types. Type B design was studied in more detail because of concerns over the effect of the side panels on the function of the safety device.

For all types, the most common case of entanglement is likely to result in the worker moving toward the hopper and activating the safety trip with their arm or other body part, even if they do not operate it deliberately. It is worth bearing in mind that a worker's initial reaction may be to use the free hand to release the entangled one. However, considering the typical working conditions and environment it seems plausible that anyone being pulled toward the hopper while feeding could be caught off-balance and may slip or fall against the front of the hopper/extension. In the worst case they may do this without actuating the safety trip. In these cases, the body of the worker will tend to be pulled in against the bottom corners of the hopper sides. In addition, material will probably be twisting about and taking up much of the space within the hopper. This is likely to impede attempts at activating the safety trip as well as attempts to free an entangled hand. Since the worker may end up lying in the hopper, the safety device should be effective in this situation. Body size data suggests that it should operate within the bottom 200 mm of the hopper vertical sides (5 percentile female waist depth is 171mm, unclothed. PeopleSize2000). For Type B and many Type C machines it would seem unlikely for the bar to be activated in this type of situation. This was felt to be an important shortcoming in the draft standard.

Type B designs have a relatively short hopper depth, but with longer extensions combined with the side plates. The control/safety bar is of similar configuration to the Type C, extending from a pivot point close to the base of the hopper, up the sides and across the top. The control bar is therefore generally only operated by the top section. The main concern is that the side panels may reduce the likelihood of the control/safety bar being tripped passively. Based upon working practices, what were considered to be the most likely entanglement possibilities were as follows:

1. The standing operator's arm is drawn toward the feed roller and the operator is tall enough so that as they move towards the back of the hopper they can move along the outside of the side panel such that the top of the side panel passes under their armpit;

2. The standing worker, as their arm is pulled toward the feed rollers, is drawn against the front edge of the side panel such that their body remains outside the side panel which will lodge under their armpit or otherwise prevent them from moving any further towards the back of the hopper;

3. The worker, as they are drawn towards the feed rollers, is pulled/falls against the hopper or sidepanel.

These assume that the worker is initially standing in a position on the left hand side of the hopper (i.e. with the hopper on their left). Equivalent mirror image scenarios exist for a worker standing at the other side. In scenario 1 the control/safety bar can be passively tripped, but in this situation it is likely that the worker will be able to reach and activate the bar with a free hand before any passive activation occurs. In scenarios 2 and 3, there is no possibility that the control/safety bar will be tripped passively. The only option is for the operator to reach it and apply sufficient force to trip it actively (assuming that they do not have both hands entangled). Also, there is the additional possibility that the worker may become drawn over or around the side panel, especially if they were to struggle or attempt to reach the control/safety bar.

(a). Normal work position

(b). Entanglement scenario 1

(c). Entanglement scenario 2

(d). Alternative scenario 2

(e). Entanglement scenario 3

(f). Alternative scenario 3

Figure 2. Some entanglement scenarios with Type B machines.

It could be that scenarios 1 and 2 would arise from scenario 3 depending if the operator regained their footing. However, if the operator fell more into the mouth of the hopper, rather than with their body remaining outside the side panels, considering the pulling capability of the machines it seems likely that they could be drawn in sufficiently to reach the danger zone. This is more likely when the operator is in front of the machine, as when feeding short material. Opportunity to reach the bar in this case is very limited.

There are two remaining concerns: How people might interact with the hopper/sidepanel if they are drawn against/fall into/onto it, and; What happens if a person is retained by the side panel and their arm is still being pulled by the material they are entangled in?

It is difficult to determine how a person might interact with the side panel in the various situations, but given the magnitude of the infeed roller pull force, they may be dragged into the hopper, or if movement was blocked by the side panel they may be injured by the pull force.

For a worker feeding from the front, the opportunity for passive activation of the safety bar is more limited. Also, some Type B designs do not appear to be as effective in limiting how far the operator can lean into the hopper opening as some Type C designs. This is because there is no obstruction between the side panel tops which can enable operators to reach relatively closer to the danger zone than they might otherwise do with a Type C machine.

Conclusions

The feed speeds and pulling forces that these machines are capable of demand that to minimise the risk of injury, passive tripping of the safety device should occur as early as possible in the entanglement event sequence, and as far from the danger zone as possible.

The purpose of the safety trip mechanism is to prevent a person, if entangled in the feed material, being pulled along with it into the feed rollers. The safety trip mechanism should not necessarily require any conscious effort on behalf of the entangled worker, and it should actuate as a result of the worker being pulled toward the hopper. For this to be the case, it is essential that the actuation of this safety trip mechanism operates before (or as) the worker's body comes into contact with the hopper edge.

There are a variety of hopper and control/safety bar configurations possible, some are more effective than others in terms of the ability to be tripped low on the sides of the hopper and with small actuation forces. While protecting the bottom corner of the hopper is important (because the opportunity for recovery from that position is limited) , the design of the safety trip mechanism should not reduce the level of protection for the more common scenarios.

Based upon information gained from operators, observation and measurement Type A and C designs appear to be more protective since the location of the control/safety bar is at the very front edge of the hopper. An entangled operator is considered have a relatively low likelihood of passively actuating the control/safety bar on Type B machines. This is largely due to the vertical portions of the bar being obstructed by the side panels, and is especially so for an operator working in the area at the front of the hopper. Although it is believed that in a proportion of entanglement scenarios, the side panels may physically prevent the operator from reaching the danger zone, this is itself not without danger since the feed mechanism may still be pulling which may result in serious injury.

Further recommendation arising from this work were for an 'emergency stop' function for the safety trip, to prevent a struggling worker or other person from inadvertently reactivating the infeed rollers, and a specification for the control/safety bar functions such that if freeing an entangled person, the bar does not have to be moved past the feed position in order to reach reverse.

These and other findings regarding the nature of the safety bar and chute design were presented to a meeting of the UK suppliers and Manufacturers in July 1999. At the meeting it was agreed that a new HSE information document detailing revised safety guidelines would be produced, and that from 31 October 2000, manufacturer and suppliers would make available retrofit kits for existing machines and improve the safety guarding on new machines. The contents of the information sheet would form the basis of the UK input to the draft European Standard.

References

prEN 13525 Draft British Standard: Forestry machinery - Wood chippers - Safety. Draft for public comment April 1999. BSI.

BS EN 294: 1992 Safety of machinery - Safety distances to prevent danger zones being reached by the upper limbs. BSI.

PeopleSize 2000 Professional, version 2.00. Friendly Systems Ltd. 1993-1999.

HSE Agricultural Information Sheet No 38. Health & Safety Executive, 2000.

USING FOCUS GROUP DATA TO INFORM DEVELOPMENT OF AN ACCIDENT STUDY METHOD FOR THE CONSTRUCTION INDUSTRY

Sophie Hide, Sarah Hastings, Diane Gyi, Roger Haslam and Alistair Gibb

Health and Safety in Construction Research Group
Health and Safety Ergonomics Unit, Department of Human Sciences,
Loughborough University, Loughborough, Leicestershire LE11 3TU, UK

A major study of accident causality in the construction industry is underway. In order to develop and enrich the study methodology, a series of focus groups with a range of industrial practitioners has been undertaken. Seven focus groups, each including either safety professionals or industrialists involved at client level, senior and site management, and operative grades, were conducted. Participants concentrated on failure occurrence in four discussion areas (1) Project concept, design and procurement, (2) Work organisation and management, (3) Task factors and (4) Individual issues. Among the large quantity of anecdotal material generated a number of themes arose repeatedly. It appears that problems may exist relating to financing and pay; work planning, scheduling and management related issues; difficulties with information transfer; and inconsistencies or inadequacies at a role based level.

Introduction

The construction industry is one of the most hazardous employment sectors in the United Kingdom, with the HSE reporting that construction industry injury rates are among the highest for both fatal and major injuries (HSC, 1999). Existing data and research do not give a conclusive picture of causative factors in construction accidents especially given problems with inadequate data collation, over concentration of interpretation at accident event level (Whittington et al, 1992) and under-reporting (Gyi et al, 1999, HSC, 1999).

Focus groups have been used as a preliminary data collection method to develop the study method for research, which ultimately will entail detailed studies of 100 construction industry accidents. The study method requires careful planning: the construction industry is highly complex by virtue of the development process, and the levels of management and chains of consultation, which are inherent from the initial project concept through to the build process. Whittington et al (1992) identified failure according to three factors: issues relating to headquarter responsibilities, site management, and the injured party or immediate work colleagues. An especially important outcome of this research was the identification of contributing factors that are distal from the site-based issues. More recently, Suraji et al., (2000, in press) has developed a theoretical model, the 'Constraint-response model of accident causation', which enhances further the distal factor contribution that may be incurred throughout the developmental and build process.

Despite the valuable contribution of the Whittington et al and Suraji et al research, there is little up to date material available that has incorporated consultation with industrial practitioners concerning their perceptions of accident causality. This is especially relevant given significant post 1992 legislatory changes (in both general health and safety and construction related aspects), and of the development of initiatives to improve performance, quality and efficiency (Department of the Environment, Transport and the Regions, 1998). Focus groups were selected for this preliminary study as a data collection method that would permit greatest access to practitioner perspectives and viewpoints.

Focus group methodology

Development of discussion materials
HSE classifications of immediate and underlying causes in accidents (HSE,1997) were adopted as main discussion themes. Additionally, a supplementary classification was developed, addressing the early stage concept and development phases, which are unique to the construction industry. Prompts to stimulate discussion were also prepared for each discussion area (but not reproduced here) by allocating key points from the Suraji and Whittington research to the four chosen discussion areas:

- Discussion Area One   - Project concept, design and procurement
- Discussion Area Two   - Work organisation and management
- Discussion Area Three   - Task factors
- Discussion Area Four   - Individual factors

Focus Group Participants
All groups had between 5 and 7 participants and composition is shown in Table 1. Groups One and Three varied from that intended, mostly due to the practicalities of recruiting participants. Minimal changes were made after the pilot study and thus the data from this 'mixed group' was retained for inclusion with the final data set.

Table 1. Focus Group participants

Group	Employment	Target participants
One	Client team	Clients or client representative, Architect, Engineer (Structural / Civil or Mechanical / Electrical), Financial Manager, Project Manager or Design Manager and a Planning Supervisor
Two	Senior managers	From General and specialist contractor firms
Three	Site Managers	representing civil engineering, major building or the residential sectors and from small and large projects
Four	Operatives large site	Tradesmen or general operatives
Five	Operatives small site	
Six	Safety professionals	Industrial safety professionals and Construction Enforcement Officers
Seven	Mixed group (pilot)	A mixed discipline group (trades and professionals)

Focus group procedure and analysis
Each group was scheduled for 90 minutes with time allocations (used flexibly) to ensure that all discussion areas were addressed. The group moderators introduced the prompts prior to each of the discussions, and participants were encouraged to explore beyond the themes introduced. With each prompt, participants were asked to consider "where does failure occur?" and "why do accidents still happen?". All discussions were audio taped and an abridged transcription was made for each focus group. Analysis included an intermediate summarisation of all text into short bullet point statements per group, followed later by comparison and categorisation of all focus group information according to the discussion area headings provided. Findings have been since been reviewed by industrialists as part of the validation process.

Results

Project Concept, Design and Procurement
Participants were often very negative about clients who procure construction work, excepting larger, high-tech organisations such as within the petro-chemical, oil, nuclear and (to a certain extent) retail industries. These companies were often seen as more responsible, although alternatively 'larger' companies were also portrayed as arrogant risk takers, with little interest in the build process. They were also described as ignorant of certain areas of the process, such as their legislative responsibilities under the Construction (Design and Management) Regulations 1994 (CDM), which includes aspects such as the appointment of a Planning Supervisor and ensuring the competency of those that they appoint. Clients were also seen as ignorant of the benefits that can be gained from safety innovations. It was suggested that fear of prosecution acts as their main driver to influence safety considerations.

The desire to maintain a high public profile was reported, yet construction personnel, felt that this resulted in reduced time schedules for the build process, reduced site area, and an increased demand to undertake weekend and night work. The general opinion seemed to be that client pressures and inflexibility causes time and output pressure, while perpetual cost cutting induces a compromise of safe working methods.

The design process itself was the subject of many criticisms, directed towards the use of incorrect or outdated drawings and inadequate provision for safety in the design process. Client team professionals, especially Designers and Architects, were seen as distant from site issues and unaware of or uneducated in their legislative responsibilities under CDM, and of their own impact upon health and safety issues.

Lack of appropriate audit and poor innovation in design were considered as shortcomings. Some saw the advantages of increased use of pre-finished components, which permit greater speed in the build process and compensate for site based skills shortages. These were also seen in a more negative light however; an example, involving timber trusses, revealed that manufacturers are reluctant to alter their designs even when possibilities for improvement have been identified. There are a variety of construction management styles, with participants showing preference for a contractor - client alliance, which encourages practical design reviews.

The procurement process, and especially the selection of contractors, was also criticised. Prominent among points raised was the perceived extreme price competition among contractors in tendering (and hence the apparent advantage of firms who inadequately cost safety), and also in ritualisation that has developed in preparing 'paperwork', such as the pre-tender questionnaire and Health and Safety Plan. These were seen often as time wasting generic materials which offer minimal value to the client in their decision making.

Work Organisation and Management

The quality of Method Statements, procedures and general planning issues occupied a considerable proportion of the discussion time. Firstly, it was thought that although Method Statements may provide a 'task breakdown' (although the task analyses may be inadequately considered from an ergonomics perspective) they do not necessarily provide adequate procedural information. It was suggested that Method Statements are invariably prepared as an office-based exercise, using generic texts and with little consultation among practitioners or understanding of the practicalities of their work. It was reported that operatives do not necessarily see, read or understand the Method Statements, although their non-use was attributed both to habitual practice and a desire to short-cut and make financial gain. Procedural violations were seen as insidious and tolerated, reflecting a wider malaise on site.

Participants described problems with the constant revision of work schedules. Changes to work in progress also contribute to planning problems, arising from modifications in areas such as design, scheduling, as a result of transport and delivery problems, or as a result of weather conditions. The consequences of planning problems were described as trade overlap (and loss of work sequence), work back log, taking short-cuts, and the generation of time pressure – all of which were felt to contribute to risk circumstances.

A number of criticisms were made regarding the move from direct labour towards lengthy chains of sub-contractors. Sub-contractors, and especially those most distal in the chain from Principal Contractors, were seen as distanced from responsibility, often inadequately supervised, and ignorant of and not committed to the common responsibilities of the site.

A range of other comments were offered, again recounting the shortcomings in performance , or circumstances which can have a negative impact upon the performances, of personnel involved in work organisation at site level. Time pressures upon and poor availability of competent foremen were cited especially frequently.

Task factors

The selection of correct tools, materials and equipment received a number of comments, with these appearing to be influenced by availability and work scheduling factors. Although it was generally acknowledged that tools are often good and new to each site, it was suggested that their selection is too cost motivated and that they are not always freely available. There were some concerns about the unknown quality of equipment that is used by sub-contractors, inadequate maintenance and the use of multi-functional equipment.

Participants also discussed the use of personal protective equipment (PPE). It was indicated that availability is plentiful among larger companies, but may be less so or absent among smaller companies. It was indicated that those advocating the use of PPE do not adequately appreciate the considerable loss of mobility and comfort through its wear. PPE seemed to be generally disliked by users and a number of comments were offered suggesting non-use at week-ends (indicating the atmosphere to be more lax).

Participants described different experiences of supervision, with contradictory observations that there is both more and less supervision nowadays. At site level, the efficiency of supervision was seen to deteriorate with a rise in the volume of sub-contractor labour, yet where supervision was regarded as good, sub-contractors would conform to standard. Communication was presented as at times inadequate, both within a same status team, and hierarchically through different grades.

Small jobs, isolated work or short term contracts were seen as those where least forethought is given, and with safety factors more likely to be considered on an ad hoc basis or at an individual level only. It was noted that setting up safely and waiting for arrival of and use of safety equipment can take longer than the job itself and that duration of exposure to a 'risk' influences an individual's choice of safe working methods.

A number of different criticisms about training were mentioned and the first of these concerned the inadequate content and evaluation of site-induction. Likewise, training is often inappropriately seen as a response to all problems. It was indicated that there is a shortage of courses, that training is not provided consistently (absence of manual handling training for labourers for example) and that the training content pays insufficient attention to the development of practical skills. The lack of practical field skills was thought especially important. In this respect, problems were mentioned with one day training courses that provide a certificate of competence, with the recipient expected to display a wide range of skills from a very early stage.

Work load and time pressures, revised work patterns and long hours culture (and consequent disruption to domestic life and lengthy travel time) were reported to be prevalent in the industry. Additionally, there has also been an increase in the introduction of weekend, night and block work by clients – with resulting fatigue considered to compromise safety, decision-making and productivity.

The implications of payment methods upon performance, quality and efficiency were mentioned on numerous occasions. There are reportedly no longer any fixed wages for trades people, as all work is now target or bonus related. Financial expectations are high and exceeding the work target and increasing bonus related pay is considered essential for income and the prime incentive for operatives. Bonus pay may be safety-related, but it seems that most often bonus pay is solely related to task performance.

Individual factors

The discussions indicated that there is a trend towards increased reliance on young and inexperienced employees on sites, raising particular concern about early responsibility and use of dangerous equipment by young workers. Young people were described as more safety conscious and more likely to follow work instructions, but were reported to experience a high accident rate, especially within their first week of appointment. The verification of 'experience' seems especially difficult and it was discussed that there are problems with people with inadequate skills presenting themselves as a skilled trades person. Concern was also voiced about the appointment of trades people from outside the industry and reservations about the transferability of their skills onto site and the verification of competency.

Although experienced workers were described as having fewer accidents, experience was also seen to have a negative side. The range of problems associated with experience was noted as work fatigue, over-familiarity and over-confidence, complacency, omission of or low safety awareness, and difficulties in changing work techniques. A number of comments were made by participants, which revealed the presence of a fatalistic approach to accidents – with luck and chance seen to have a considerable contributory role.

More broadly, varying views and differing perspectives with respect to general health status were noted, with reports of considerable health problems among construction workers. It was reported that light work may be possible for injured employees, but that dismissal is sometimes the alternative. The general impression from participants was that ill health and health-related issues (especially slowly developing health issues) are under-appreciated in the industry and that an increase in the extent of litigious action is anticipated in the future. Limited health assessment occurs reportedly, but is hampered by the high mobility of the workforce.

Conclusion

The focus group discussions have provided a rich source of data with which to supplement existing materials and develop the accident study method. In particular there were repeated references to four specific themes across the discussion areas: financing and pay related issues, work planning scheduling and management aspects, problems with information transfer, and inconsistencies or inadequacies at a role based level. Whilst judgement has not been passed upon whether views presented by group participants are right or wrong (it is possible that in some respects focus group participants may be factually incorrect or hold opinions with which others disagree) the areas where dissatisfaction has been shown will be used to lead the future enquiry.

Acknowledgement

The authors would like to acknowledge the HSE who are funding this research; Dr Roy Duff and Mr Suraji for their contribution with research; Mike Evans of Laing plc, Tony Wheel of Carillion plc and Suzannah Thursfield of the Construction Confederation for their assistance and introduction to participants; and, lastly, to all construction industry practitioners and specialists who also generously contributed their time and knowledge.

References

Department of the Environment, Transport and the Regions. *Rethinking Construction (July 1998)* On-line, internet. Available via< http://www.m4i.org.uk/publications/ rethink/1.htm>

Gyi, D.E., Gibb, A.G.F. and Haslam, R.A. 1999. The quality of accident and health data in the construction industry: interviews with senior managers, *Construction Management and Economics*, **17**, 197 – 204.

Health and Safety Commission, 1999. *Health and Safety Statistics 1998/99*. HSE Books: Sudbury.

Health and Safety Executive, 1997. *Successful Health and Safety Management, HSG65*. HSE Books: Sudbury.

Suraji, A., Duff, A.R. and Peckitt, S.J. 2000. Development of a causal model of construction accident causation (in press - submitted to ASCE *Journal of Construction Engineering & Management*).

Whittington, C., Livingston, A. and Lucas, D.A. 1992. *Research into Management, Organisational and Human Factors in the Construction Industry*. HSE Contract Research Report 45/1992. HMSO: London.

ACTIONS OF OLDER PEOPLE AFFECT THEIR RISK OF FALLING ON STAIRS

R A Haslam, L D Hill, P A Howarth, K Brooke-Wavell and J E Sloane

Health & Safety Ergonomics Unit
Department of Human Sciences, Loughborough University
Loughborough, Leicestershire, LE11 3TU
R.A.Haslam@lboro.ac.uk

An interview and environment survey was undertaken to investigate how older people keep and use their stairs, with regard to stair safety. Visits were made to 157 older people, aged between 65-96 years, in their own homes. Information was collected under three headings (1) behaviour involved in direct use of stairs, (2) decisions and actions which change the stair environment and (3) behaviour affecting individual capability. The findings indicate that behaviour-based risk factors for falling on stairs are widespread among older people, while only 13% of those interviewed were able to remember ever receiving advice about stair safety. Simple, inexpensive measures exist which individuals can take to reduce their risk of falling on stairs. Primary care providers have an important contribution to make in encouraging implementation of these.

Introduction

Older people falling on stairs in the home is a serious problem, resulting in up to 1000 deaths and 57,000 hospital A&E attendances each year (DTI, 2000). In 22,000 of these incidents, casualties suffer a fracture, concussion, or otherwise require admission to hospital for more than a day. The cost to the health services and wider community of caring for these patients is substantial. Also, falls have serious psychological and social consequences for the individual, affecting mobility, confidence and quality of life. Although the personal and environmental factors involved in falls on stairs are well known, the influence of behaviour has received much less attention from researchers (Connell and Wolf, 1997). The aim of this investigation was to improve understanding of how older people keep and use their stairs, considering the implications for stair safety.

Behaviour and falls

A series of preliminary focus groups, conducted at the commencement of the project, generated an initial model demonstrating how the behaviour of older people might contribute to risk of falling on stairs (Hill *et al*, 2000). The model suggests that increased risk of falling on stairs arises from the way individuals interact with stairs directly, as a

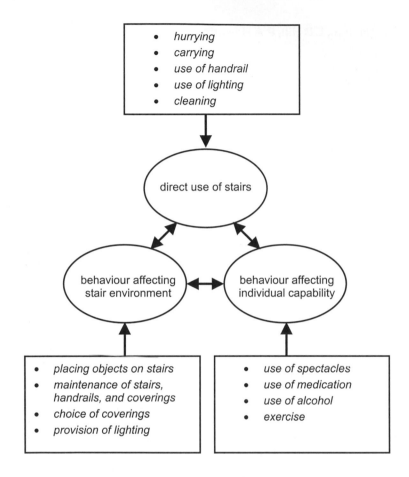

Figure 1. Behaviour-based risk factors for older people falling on stairs

consequence of actions which modify the stair environment, or through behaviour affecting individual capability, figure 1. This framework underpinned the main element of the research, a home interview and stair environment survey.

Interview and stair environment survey

An interview and stair environment survey was undertaken to validate the findings from the focus groups and to quantify different risks. Semi-structured interviews were conducted with 157 community dwelling individuals, aged 65-96 years, in their own homes. The interviews collected qualitative and quantitative information on the behaviour of respondents on and around their stairs, awareness of safety factors, and history of falling. During each home visit, lasting approximately 2 hours, details were recorded and photographs taken as to the design and repair of stair coverings, number and condition of handrails, objects on and around the stairs, and provision of lighting. Measures were also taken of health status and vision.

Participants were recruited on a quota basis, according to age and gender, using estimated UK population figures from the Office for National Statistics. Likewise, dwellings were selected based on national estimates of housing stock, with respect to age and type of dwelling.

Results and discussion

Direct use of stairs
A majority of participants (63%) reported hurrying on stairs on occasions, despite 89% acknowledging this as increasing the likelihood of falling. The most common reasons given for hurrying were to respond to someone at the door, answer the telephone, retrieve items left upstairs or to use an upstairs toilet.

Although 92% of interviewees recognised that carrying items up and down stairs could be hazardous, 29% said they would still attempt to carry something that might cause them difficulty. Among the items interviewees reported carrying on stairs were laundry, vacuum cleaners and walking sticks.

Adequate lighting is important for safety on stairs. Low day time illuminance (<50 lux) was found in 61% of the survey households. Almost a quarter (23%) of interviewees living in these homes said that they do not switch on stair lights during the day. Interestingly, nearly one fifth of participants (18%) indicated that they do not switch on lights when using the stairs during the night. Reasons given for this included that sufficient light was already present, eg from street lighting outside, or not wanting to disturb a sleeping partner, or to make it easier for the person to get back to sleep.

Cleaning stairs appears to be a struggle for many older people, with 58% of interviewees identifying difficulties with this activity. It was reported that problems arise due to a combination of difficult access (eg landing windows) or the need to use heavy equipment, such as a vacuum cleaner.

Behaviour affecting stair environment
Although many participants (64%) recognised that objects on stairs may be a hazard, items were found on the stairs in 29% of households, see figure 2, for example. Objects on stairs may either be permanent, for example furniture, or placed there on a temporary basis. Permanent items were more common in smaller dwellings (p<0.001), presumably because space is at a premium. Discussions with interviewees revealed that temporary items are often left on stairs for taking up or down later.

While most participants (86%) reported their stair covering to be in reasonable condition, in 29% of cases the stair covering was judged by the researcher to be in need of replacement or repair. It seems likely that either the interviewees had a lower threshold for what constitutes reasonable condition, or else they had not noticed wear and tear that may have happened gradually over many years.

Although not quantified, many examples were seen during the home visits where the carpet colour and pattern on stairs makes it difficult to distinguish edges of steps, figure 3. Generally, coverings light in colour and non-patterned make steps more visible, while those that are heavily patterned may cause problems. There was low awareness that this might be an issue among survey participants.

Figure 2. Objects placed on stairs   **Figure 3. Carpet design can camouflage step edges**

Behaviour affecting individual capability

Among study participants, 82% were taking at least one prescribed medication daily, with 16% reporting that their medication makes them feel drowsy, dizzy or affects their vision. Over a third of participants (38%) reported drinking alcohol when taking prescribed medications. Almost all of the sample (99%) had spectacles, with 16% indicating these cause visual difficulties on stairs, as a result of distortion from bifocals, for example. Although not prompted during the interview, a number of participants (10%) specifically mentioned using the stairs as a form of daily exercise. Although exercise is considered beneficial in reducing falls among older people (Health Education Authority, 1999), advice encouraging use of stairs for exercise needs to consider the increased risk this might create for some individuals of falling.

Conclusions

Falls on stairs typically involve an interaction of events (Templer, 1992), illustrated by the following accounts from interviewees:

> *"I was carrying the washing basket downstairs and feeling unwell. I had a lot on my mind and slipped and fell all the way down to the bottom."*

"I got into bed, I'd had a drink and realised that I hadn't got any meat out for Sunday, rushed downstairs and slipped. The shopping trolley was at the bottom, I wouldn't have hurt myself if it wasn't for the trolley ... I cracked my ribs. But that was my own fault, I'd had a drink, and I had nothing on me feet."

Prevention depends on eliminating or reducing as many different risks as possible. Although it may be difficult to improve fixed features of the environment, such as steep or awkward stairs, there is scope to address behaviour-based risk factors. Encouragement comes from randomised trials, which have shown interventions combining medical assessment, home safety advice and exercise programmes to reduce falls (Close *et al*, 1999; Steinberg *et al*, 2000).

The results of this study indicate that occurrence of behaviour-based risk factors for falling on stairs is widespread. While many participants were able to recognise and appreciate risk factors once prompted, many had not previously given much thought to stair safety. A notable finding is that only 13% of the sample recalled ever having received advice about stair safety. The UK government 'Avoiding Slips, Trips and Broken Hips' campaign is a welcome initiative in this respect.

Acknowledgements

This research was funded by the Department of Trade and Industry (DTI). The 'Avoiding Slips, Trips and Broken Hips' falls prevention (for older people) campaign is run by DTI in association with Health Promotion England. An information pack 'Step up to Safety' is available from DTI Publications Order Line on 0870 1502 500.

References

Close J., Ellis M., Hooper R., Glucksman E., Jackson S. and Swift C. 1999, Prevention of falls in the elderly trial (PROFET): a randomised controlled trial, *The Lancet*, **353**, 93-97

Connell B.R. and Wolf S.L. 1997, Environmental and behavioral circumstances associated with falls at home among healthy elderly individuals, *Archives of Physical Medicine and Rehabilitation*, **78**, 179-186

Department of Trade and Industry (DTI), 2000, *Home accident surveillance system including leisure activities, 22nd annual report 1998 data* (Department of Trade and Industry, London)

Health Education Authority (HEA), 1999, *Physical activity and the prevention and management of falls and accidents among older people* (Health Education Authority, London)

Hill L.D., Haslam R.A., Howarth P.A., Brooke-Wavell K. and Sloane J.E., 2000, *Safety of older people on stairs: behavioural factors* (Department of Trade and Industry, London), DTI reference 00/788

Steinberg M., Cartwright C., Peel N. and Williams G. 2000, A sustainable programme to prevent falls and near falls in community dwelling older people: results of a randomised trial, *Journal of Epidemiology & Community Health*, **54**, 227-232

Templer J. 1992, *The staircase: studies of hazards, falls and safer design* (MIT Press, Massachusetts)

PERFORMANCE MODELLING IN THE DOMAIN OF EMERGENCY MANAGEMENT

BECKY HILL AND JOHN LONG

Ergonomics & HCI Unit,
University College London, 26 Bedford Way
London, WC1H OAP, UK
b.hill@ucl.ac.uk, j.long@ucl.ac.uk

The Emergency Management Combined Response System (EMCRS) is the co-ordination system which plans and controls, agencies, such as fire and police, when they respond to major emergencies. An initial combined agency (CA) model of the EMCRS describes its overall 'actual' performance, as concerns its plans. This model supports the diagnosis of co-ordination problems between agencies. However, at present, it fails to take individual agency plans into account. This paper describes an attempt to improve diagnosis of co-ordination problems by decomposing the EMCRS into its parts and modelling each individually, with respect to their plans. The CA model is then used, with individual agency models, to re-interpret overall EMCRS performance. It is concluded that EMCRS decomposition supports more accurate diagnosis of the effect of the co-ordination problems on its overall performance.

Introduction

This paper presents work intended to develop a model to diagnose the ineffective planning and control performance of the EMCRS, as occasioned by co-ordination problems between the Police, Fire, and Ambulance emergency service agencies. Section 1 describes the EMCRS and the initial CA model thereof. The need is identified to decompose the system into its parts and to model each individually. Section 2 outlines the single agency model for the fire agency, with respect to an example co-ordination problem. Section 3 relates the CA model to the single agency model. This relationship provides a more accurate expression of EMCRS overall planning and control performance, and so more accurate diagnosis of the effect of the co-ordination problems. Section 4 presents a summary of the work and future plans.

EMCRS and the Initial CA model

The EMCRS is composed of agencies required for a disaster response, and has a three level planning and control structure. The EMCRS has objectives (plans), common to all agencies, whatever their individual responsibilities. These objectives are (in descending priority): to save life; to prevent escalation of the disaster; to relieve suffering; to safeguard the environment; to protect property; to facilitate criminal investigation and judicial, public, technical or other inquiries; and to restore normality as soon as possible (Home Office, 1994). The EMCRS was set up to support better co-ordination between agencies responding to disasters, such as explosions, aircrashes etc.. The individual agencies relate their own plans by means of those of the EMCRS, to interact effectively with each other. Each agency plan specifies a set of functions, for example: fire service - rescuing trapped casualties; preventing escalation of the disaster; etc..

An initial CA model of EMCRS planning and control has been developed elsewhere (Hill and Long, 1996) from the Planning and Control for Multiple task work (PCMT) framework (Smith *et al* 1997), and data from an EMCRS training exercise. This model distinguishes the interactive worksystem (here the EMCRS, comprising one or more users and computers/devices/equipment), from its domain of application, constituting its work (that is stabilising a disaster). The effectiveness (performance) with which work is carried out, is a function of: the quality of the task (i.e. whether the goals have been achieved), and the resource costs to the worksystem (i.e. the effort etc. required to achieve the work goals) (Dowell and Long, 1989). Overall EMCRS performance thus expresses whether goals, e.g. preventing escalation of the disaster, have been achieved.

The CA model has been used to describe tasks carried out by the combined response system, in terms of the planning, control, perception and execution behaviours and the transformations these behaviours perform in the domain see Figure 1. (A complete description of the behaviours and the domain object transformations for the EMCRS tasks can be found elsewherc (Hill and Long, 1995)). These descriptions have been used to identify planning and control co-ordination problems between agencies, by identifying 'conflict' behaviours. These conflicts constitute the 'co-ordination problem'. A behaviour conflict may reduce overall EMCRS performance by either hindering goal achievement, for example, reducing life saving capabilities, and/or by rendering resource costs un-acceptable, for example, requiring extra personnel, that are not available. However, defining accurately effects of these problems is difficult, due to trade-offs between individual agencies' performances. For example, the police service wish to preserve the disaster site as a 'crime scene' (vandalism is suspected), and to catch the criminals, and so require the fire service not to trample the site; and the fire service, who slow the rescue of casualties, and are less effective in fire prevention, if they do not trample the site. The CA model describes the 'actual' overall CA performance with respect to EMCRS common objectives. The co-ordination problems identified thus do not take account of the performance trade-offs between agencies. For the above example, the CA model describes an overall EMCRS performance deficit which derives from the police and fire services. Trampling by the fire service reduces the chances of the vandals being caught. Carrying out minimal trampling reduces rescue of casualties and control of the fire. These overall deficits derive from the common objectives, i.e. to save life (casualties not rescued); prevent escalation of disaster; (fire not controlled); and to facilitate criminal investigation (vandals not caught).

However, each agency has its own disaster plan. These plans describe agency functions/tasks and their priorities. To express accurately overall EMCRS performance, account must be taken of these plans. For example, the fire service plan states: 'Investigation work will not take precedence over the necessity to rescue casualties, fight fires, or the protection of lives and property from fire or further deterioration. Every effort must be made by the Fire and Rescue Commander to preserve the scene intact.' (Chief and Assistant Chief Fire Officers' Association, 1994). Thus, the fire service should keep their trampling to a minimum, to preserve other fire service behaviours. Thus, for this co-ordination problem, there is no fire service performance deficit, as the actual performance effected by minimal trampling, is equal to the planned performance, which allows effects of minimal trampling. Thus, although there is still a police service performance deficit (minimal trampling still reduces vandal apprehension), the overall EMCRS performance deficit is less than was identified by the CA model.

There is a need, therefore, to decompose the EMCRS into its parts and each agency to be modelled individually with respect to its plans. These single agency (SA) models describe planned individual agency performance with respect to CA actual performance, that is, overall EMCRS actual performance. The CA model can then be re-interpreted with the help of the SA models, to diagnose more effectively overall EMCRS performance, as concerns planning and control co-ordination problems.

Single Agency Models

SA models have been developed for the fire, police and ambulance agencies. They describe 'planned' performance of an individual agency with respect to the CA 'actual' performance. The data are a subset of the CA data. Six training exercise behaviour conflicts have been identified as co-ordination problems. The models describe only planned performance, as it relates to behaviour conflicts and their effects. Only one behaviour conflict is described here, involving fire and police and ambulance services, due to space limitations.

Behaviour Conflict: Trampling

The police declare the site a 'crime scene', because there is some suggestion of vandalism. The fire service and the ambulance service are thus expected not to trample what might be evidence. The CA model describes 'actual' behaviours of the fire and ambulance services as minimal trampling, without affecting their other behaviours (and so the associated performance). For the fire service performance is: expeditious rescue of casualties, and effective fire containment; and for the ambulance service performance is: expeditious casualty access and subsequent transferral to hospital. The fire service SA model shows that actual performance (P_A) is equal to planned performance (P_p), as their plan specifies that minimal trampling may be carried out, but should not hinder rescue or other fire service behaviours. Such behaviours are not hindered, hence $P_A = P_p$. In contrast, for the ambulance service $P_A < P_P$. Their plan does not mention trampling, so minimal trampling reduces their performance to less expeditious casualty access and transferral to hospital.

Relating the Models

The initial CA model describes CA P_A < CA Pp, as a performance deficit is identified for both police and fire services; and police and ambulance services. The overall performance deficit can be re-interpreted using the SA models as follows:

Trampling (fire service)
If SA1 P_A = Pp and SA2 P_A < Pp,

then CA P_A < CA (SA) Pp
and (CA P_A < CA (SA) Pp) < (CA P_A < CA Pp)

If the fire service (SA1) actual performance equals its planned performance (their trampling behaviours not reducing their effective fire containment and casualty rescue), but the police service (SA2) actual performance is less than its planned performance (their crime scene preservation behaviours hindered by fire service trampling, which reduce their effective vandal apprehension), then CA actual performance (CA P_A) is less than the CA planned performance, given SA planned performance (CA (SA) Pp). However, the performance deficit (CA P_A < CA (SA) Pp) from the CA (SA) model is less than the performance deficit (CA P_A < CA Pp) of the CA model (the CA (SA) model identifies only the police service as having a performance deficit). Thus, the overall EMCRS performance deficit is less than originally identified by the CA model above.

Trampling (ambulance service)
If SA1 P_A < Pp and SA2 P_A < Pp,

then CA P_A < CA (SA) Pp
and (CA P_A < CA (SA) Pp) = (CA P_A < CA Pp)

If the ambulance service (SA1) actual performance is less than its planned performance (their trampling behaviours reduce expeditious casualty access and recovery), and the police service (SA2) actual performance is less than its planned performance (their crime scene preservation behaviours hindered by ambulance service trampling, reducing their effective vandal apprehension), then CA actual performance (CA P_A) is less than the CA planned performance, given SA performance (CA (SA) Pp). The performance deficit (CA P_A < CA (SA) Pp) of the CA (SA) model is equal to the performance deficit (CA P_A < CA Pp) of the CA model. A deficit occurs for both the police and the ambulance services. Thus, the overall EMCRS performance deficit, identified by the CA model, in this instance, is accurate. However, the ambulance service SA model has identified that trampling behaviour is not specified in their plan. Were trampling behaviour to be specified, as in the fire service plan, then their performance would not be in deficit.

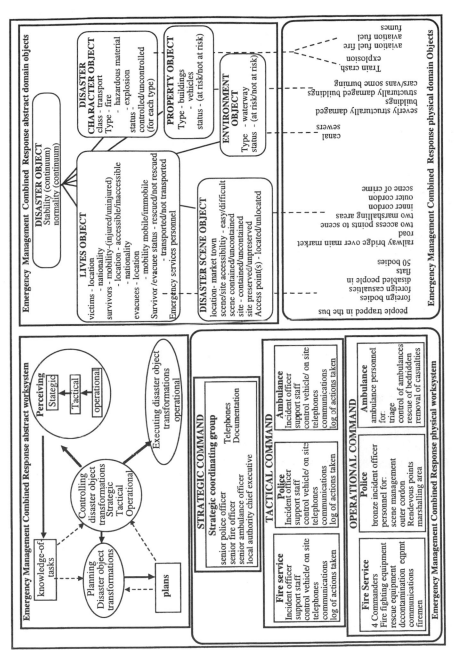

Figure 1. Emergency Management Combined Agency (CA) model

Summary and Future Work

This paper has described an attempt to improve the diagnosis of overall EMCRS performance by decomposing the system into its individual parts (agencies) and modelling each, with respect to their agency plans. It is concluded that EMCRS decomposition supports more accurate diagnosis of the effect of the co-ordination problems on overall EMCRS performance. Future work will acquire additional data to validate further this modelling and decomposition technique.

References

Dowell, J. and Long, J.B., 1989. Towards a conception for an engineering discipline of human factors. *Ergonomics 32,* 1513-1536.

Hill, B., and Long, J.B., A preliminary model of the planning and control of the combined response to disaster. *Proceedings of the Eight European conference on cognitive ergonomics (ECCE8) Granada Spain 1996*

Hill, B., and Long, J.B., (1996) A Preliminary Model of the Planning and Control of the Combined Response to Disaster. *Proceedings of the Eight European conference on cognitive ergonomics (ECCE8)* Granada Spain

Smith, M.W., Hill, B., Long, J.B. and Whitefield, A.D., (1997) A design-oriented framework for modelling the planning and control of multiple task work in Secretarial Office Administration. *Behaviour and Information technology, vol. 16, no 3,* 161-183.

Home office (1994) Dealing with Disaster 2nd edition. HMSO, London.

Chief and Assistant Chief Fire Officers' Association (1994) Fire Service Major Incident Emergency Procedures Manual. CACFOA Services Ltd. UK.

Acknowledgement

This work is part funded by the Home Office Emergency Planning Research group under the EPSRC CASE scheme

A STAGE SPECIFIC APPROACH TO IMPROVING OCCUPATIONAL HEALTH AND SAFETY

Cheryl Haslam[1] and Roger Haslam[2]

Department of Health Studies[1]
Brunel University, Osterley Campus, Isleworth, Middlesex, TW7 5DU.
Health and Safety Ergonomics Unit[2]
Department of Human Sciences, Loughborough University,
Loughborough, Leicestershire, LE11 3TU, UK.
cheryl.haslam@brunel.ac.uk
R.A.Haslam@lboro.ac.uk

This paper explores the potential application of an influential model of health related behaviour (Prochaska and DiClemente's stage of change model) to interventions designed to improve occupational health and safety. According to the model, people progress through a series of stages when modifying their health related behaviour. The model also suggests that a person's stage of change determines their receptiveness to different forms of health education messages. Given the similarities between community health promotion and health and safety interventions in the workplace, this paper considers the possible applications of the model to occupational health and safety. It concludes that health and safety interventions which are matched to the recipient's stage of change are more likely to facilitate the implementation of safer working practices.

Introduction

It can be argued that improving health and safety in the workplace has considerable similarities with health promotion in the community. In occupational settings, behaviour at both the individual and organisational level is influenced by a set of underlying attitudes and beliefs (e.g. Bentley and Haslam, 1998; 2001). Theoretical models of health related behaviour could therefore have much to offer occupational health and safety.

DeJoy (1996) reviewed theoretical models of health behaviour to consider their applicability to workplace self-protective behaviour. DeJoy notes that despite the link between these models and self-protective behaviour at work, scant attention has been paid to how these models might be applied to the workplace. Based on his review of health belief models, he proposes a framework that conceptualises self-protective behaviour as consisting of four stages: hazard appraisal, decision-making, initiation and adherence. Within these stages, five constructs are identified as being of primary or secondary importance: threat-related beliefs, response efficacy (beliefs about the consequences of preventative action), self-efficacy, facilitating conditions and safety climate. DeJoy highlights the need to target interventions to each of the four stages and to consider the environmental or situational factors in promoting self-protective behaviour in an occupational setting.

A staged approach to ergonomics interventions has also been suggested by Urlings *et al.* (1990). They took a model based on social psychology and communication theory concerned with changing attitudes and behaviour through information. This model was developed originally in the area of smoking cessation (De Vries and Kok, 1986; Kok *et al*, 1987). The model comprises 6 stages: giving attention to the information, understanding the information, changing attitudes, changing intentions, changing behaviour and maintenance of new behaviour.

Urlings *et al.* (1990) argue that to implement improvements in health and safety practices one needs to change the attitudes within the organisation. In order to do this, the ergonomist needs to ascertain at what stage managers and employees currently reside. Urlings *et al.* describe the methods of analysing stage as: interviews, questionnaires and direct or indirect (photographs and videotape) observations of the workplace. By ascertaining the stage at which managers and employees are positioned, information can be provided to facilitate progress to further stages and thus implement improvements. This method was demonstrated with a case study in the Dutch furniture industry where managers were encouraged to implement sit-stand seating in upholstering workplaces.

Hence, there would appear to be at least some awareness that health and safety interventions in the workplace could usefully extrapolate from theoretical models of health related behaviour.

Stages of change

Prochaska and DiClemente's (1982) stage of change model has attracted great interest among researchers involved in the study of health related behaviour such as smoking, drinking, exercise and diet. Central to the model is the notion that people attempting to improve health related behaviour progress through predictable stages.

The model assumes that behaviour change is a dynamic process involving five distinct stages. These stages are: precontemplation (not considering changing one's behaviour), contemplation (thinking about changing), preparation (making definite plans to change), action (here the individual has changed their behaviour) and finally maintenance (working to prevent relapse and consolidate the gains made) or relapse. See Figure 1.

Prochaska *et al.* (1994) state that a person's stage determines their receptiveness to different forms of health education. People in the precontemplative stage are more influenced by graphic information about health risks, whereas skills training or practical advice is more appropriate for those in later stages (those who have already decided to change). Proponents of this model argue that interventions need to 'place' recipients in terms of their stage and target health information accordingly.

Risk Perception

The applicability of the stage of change model has been demonstrated in a variety of health related areas, with self-reported stage of change shown to relate to actual behaviour in the direction that would be predicted by the model. A recent study (Haslam and Draper, 2000) has taken this one step further by exploring risk perception and stage of change. These authors found that pregnant women's stage of change in relation to smoking was strongly associated with their beliefs about the risks of maternal smoking.

Women further along the cycle of change were more convinced about the health risks of smoking during pregnancy.   This adds further support to the argument that interventions need to be stage specific. Precontemplative individuals tend to be less convinced of the health risks compared to those in the contemplative, preparation and action stages.   This suggests that precontemplative individuals do require graphic information to convince them of the health risks.   Those in the contemplative and preparation stages show higher levels of conviction about the health risks so, rather than health information, these individuals need practical advice to help them change their behaviour.

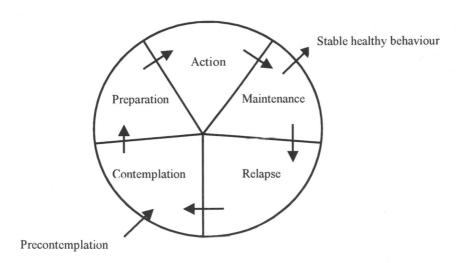

Figure 1. Stages of Change

Implications for interventions

The stage of change model has intuitive appeal and takes account of the dynamic nature of attitude and behaviour change.   The model emphasises the importance of acknowledging people's intentions when designing, implementing and evaluating health interventions.

An important practical advantage is that the model is relatively simple to apply.   It should be possible to assess a person's stage of change in relation to a specific health related behaviour from their responses to a few simple closed questions.

Implications for health and safety interventions

It can be argued that the stage of change model has wide application in workplace settings at both the managerial and employee level. According to the model, people in different stages require different forms of information (graphic information about health risk versus skills training information).

Workers and managers already considering changing working practices should be provided with information to help them implement safer working practices (which may involve learning new skills). Workers and managers in the precontemplative stage (not considering changing working practices) would require a two-phase approach. Firstly they would require information about the possible consequences of current working practices for health and safety. The aim here would be to facilitate movement into contemplation and preparation. When this transition has been achieved, these individuals would require skills training information to move them to the action stage, where they would then start to implement improved health and safety measures. See Figure 2.

Example applications

Two examples are given to illustrate how the model might inform interventions. It is fundamental to the approach that stage of change can be assessed using just a small number of questions. It should be borne in mind that the actual questions asked would need to be assessed for validity and reliability.

A possible application at the individual employee level could be in the context of manual handling.

An employee's stage of change might be determined by asking questions such as:
(a) are you concerned about injuring your back? and (b) would you be interested in learning different lifting techniques? Those answering 'no' to both questions would be designated as precontemplative and could be targeted with information about the risks of injury associated with incorrect lifting. Those answering 'yes' to (a) but 'no' to (b) would be contemplative. While those answering 'yes' to both questions would be at the preparation stage. Employees in the contemplative and preparation stages could be instructed in safer lifting techniques. Part of the skills training would involve answering employees' queries about changing working practice (such as concerns about the impact on work performance). In this way, any potential barriers to change could be dealt with.

An important objective for research in this area would be to determine any differences in responding between those in the contemplation and preparation stages. It may well be the case that health and safety interventions need to be further tailored for recipients in the stages of contemplation and preparation.

When employees have changed their habits and adopted safer working practices, they should be monitored to establish that these practices have been maintained. Where relapse occurs, steps should be taken to move employees back through the cycle to the action stage.

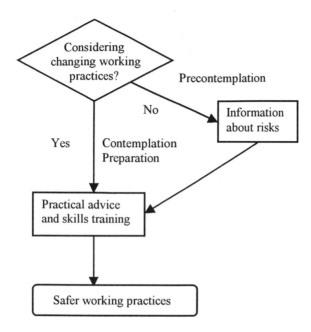

Figure 2. Staged approach to improving occupational health and safety

An application at the organisational level might include managers' attitudes toward stress among the workforce.

Managers might be asked: (a) are you concerned about stress among your employees? and (b) are you planning to adopt practices aimed at reducing stress in the workforce? Managers responding 'no' to both questions could be targeted with information about the health risks associated with employee stress and the long-term impact on organisational performance and profit. Managers answering 'yes' to (a) but 'no' to (b) would be contemplative, while those answering 'yes' to both questions would be at the preparation stage. Managers at both of these stages could be encouraged to make changes and given practical advice regarding the implementation of new working practices.

As with the case of interventions at the individual level, it would be important to determine any differences in response from organisations at the contemplation and preparation stages. If differences were reliably found then interventions could be tailored accordingly.

Again when action has been taken, the organisation should be monitored to assess either maintenance or relapse and appropriate steps taken where relapse occurs.

Conclusions

Interventions aimed at improving health and safety in the workplace are likely to be most effective if they recognise both individual and organisational readiness to change and target information appropriately. The stage of change model offers scope for researchers and practitioners to develop, implement and evaluate health and safety information that is more closely matched to user requirements and therefore more likely to facilitate safer working practices.

References

Bentley, T.A. and Haslam, R.A. 1998, Slip, trip and fall accidents occurring during the delivery of mail, *Ergonomics*, **41**, 1859-1872.

Bentley, T.A. and Haslam, R.A. 2001, A comparison of safety practices used by managers of high and low accident postal delivery offices, *Safety Science*, **37**, 19-37.

DeJoy, D. M. 1996, Theoretical models of health behaviour and workplace self-protective behaviour, *Journal of Safety Research*, **27**, 61-72.

Haslam, C. and Draper, E.S, 2000, Stage of change is associated with assessment of the health risks of maternal smoking among pregnant women, *Social Science and Medicine* **51**, 1189-1196.

Kok, G.J., Wilke, H.A.M. and Meertens. R.W. 1987, *Information and Change* (Wolters Noordhoff, Groningen).

Prochaska, J.O. and DiClemente, C,C. 1982, Transtheoretical therapy: toward a more integrative model of change. *Psychotherapy Theory, Research and Practice*, **19**, 276-288.

Prochaska, J.O., Norcross, J.C. and DiClemente, C.C. 1994, *Changing for good* (William Morrow & Co. Inc, New York).

Vries, H. De and Kok, G.J. 1986, From determinants of smoking behaviour to the implications for a preventative programme, *Health Education Research*, **1**, 85-94.

Urlings, I.J.M., Nijboer, I.D. and Dul, J. 1990, A method for changing the attitudes and behaviour of management and employees to stimulate the implementation of ergonomic improvement, *Ergonomics*, **33**, 629-637.

OCCUPATIONAL HAZARDS

COMBINED EFFECTS OF OCCUPATIONAL HEALTH HAZARDS

Andy Smith[1] and Colin MacKay[2]

[1]Director, Centre for Occupational and Health Psychology
Cardiff University
63 Park Place
Cardiff, CF10 3AS
Smithap@cardiff.ac.uk

[2]Health and Safety Executive
Magdalen House
Stanley Precinct
Bootle
Merseyside L20 3QZ

In many jobs workers are exposed to a combination of potential hazards. Both research and legislation usually considers factors in isolation and we currently have little knowledge of the impact of combinations of factors on health and performance efficiency. Our current research is addressing the issue of combined effects of occupational health hazards and aims to increase our knowledge of the topic by reviewing the literature, conducting appropriate secondary analyses of existing databases, collecting new data on accidents at work, and carrying out laboratory studies of the effects of combinations of potential stressors on physiological functioning, performance efficiency and subjective reports of mood.

Background

There has been previous research on a large number of workplace hazards. These include those arising from the psychosocial environment as well as those due to working hours and physical agents. For the most part the nature and effects of these are considered in isolation. This is not often representative of the real-life situation where employees are likely to be exposed to multiple hazards (e.g. noise, shiftwork, organic solvents). There is limited information on the combined effects of these hazards on health and performance efficiency.

Indeed, there have not even been any systematic reviews of the existing literature, no attempt to produce a coherent framework for studying these factors, and a dearth of studies using multi-methods to investigate the topic.

The first aim of the present project is to review the information currently available on combined effects. Following this a variety of methods are being used to increase information about combined effects. Achievement of these two objectives is an essential first requirement for advancement of the area. Once this has been done it will be possible to determine whether the guidelines suggested on the basis of studying hazards in isolation apply to combinations of hazards. The information will also allow development of a better conceptual framework for studying the combinations of factors. This will lead to more precise modeling of effects and identification of mechanisms underlying them. Once this has been achieved it will be possible to use both preventative and therapeutic means to reduce the extent of the problems. This will be translated into benefits in terms of reduced risk to health, improvement in quality of working life and increased productivity and safety.

Examples of previous work in the area

Epidemiological studies have for a long time shown that combinations of hazards are present. However, the usual approach has been to focus on a single factor and to treat others as potential confounders. For example, it has long been acknowledged that jobs with high noise levels also have other features which may lead to accidents (e.g. potentially dangerous machinery). However, instead of looking at the combinations of these factors studies of accidents have tried to identify the variance associated with individual hazards.

Interactions between stressors have been studied in the laboratory for a long time. Indeed, the first systematic review appears in Broadbent's (1971) book "Decision and Stress". This shows that certain factors have largely independent effects whereas others interact. Broadbent interpreted this in terms of a two-level arousal theory, with some variables influencing the lower level and others altering the function of the upper control mechanism. The studies also showed that it was difficult to predict the combined effect of two factors from their individual effects. For example, noise and sleep loss both increase momentary lapses of attention but in combination they cancel one another out (sleep loss reducing arousal, noise increasing it).

The early laboratory studies were very artificial but led to the realisation that combinations of factors needed to be considered in more realistic designs. For example, Monk and Folkard (1985) argue that "such influences (environmental factors), like those of other stressors such as noise and fumes, must be regarded as an integral part of the shiftworker - performance question". Some areas have been studied in detail because the combination is the defining characteristic of a particular workplace problem (e.g. noise and vibration - see Griffin, 1992). However, even within a particular domain there is still a tendency to consider factors in isolation.

One of the most widely studied combinations has been noise and nightwork. These are both large scale issues and it has been estimated that in the EEC 20-30 million

Table 1. Features of the working environment that should be covered in a study of occupational stress

Organisational culture: e.g.	• Poor task environment
	• Lack of definition of objectives
	• Poor problem solving environment
	• Poor communication
	• Non-supportive culture
Role in organisation: e.g.	• Role ambiguity
	• Role conflict
	• High responsibility for people
Career development: e.g.	• Career uncertainty
	• Career stagnation
	• Poor status
	• Poor pay
	• Job insecurity and redundancy
	• Low social value of work
Decision latitude/control: e.g.	• Low participation in decision making
	• Lack of control over work
	• Little decision making in work
Interpersonal relationships at work: e.g.	• Social or physical isolation
	• Poor relationships with superiors
	• Interpersonal conflict
	• Lack of social support
Home/work interface: e.g.	• Conflicting demands of work and home
	• Low social or practical support at home
	• Dual career problems
Content of job/task design: e.g.	• Ill defined work
	• High uncertainty
	• Lack of variety
	• Meaningless work
	• Under-utilisation of skill
	• Continual exposure to client groups
Workload: e.g.	• Lack of control over pacing
	• Work overload or underload
	• High levels of pacing or time pressure
Work schedule: e.g.	• Shift work
	• Inflexible work schedule
	• Unpredictable working hours
	• Long or unsociable hours
Physical environment: e.g.	• High level of noise
	• Poor heating/ventilation

people are exposed to levels of noise equivalent to continuous noise exceeding 80dBA, and that about 20% of the workforce are involved in some kind of shiftwork (Smith, 1990). Results from studies of industrial accidents and absenteeism suggest that noise and nightwork have independent effects (Cohen, 1973). Similar effects have been obtained in laboratory studies of acute effects of the two factors (Smith & Miles, 1985; 1986; 1987a, 1987b). In addition to physical agents and working hours a number of other features of jobs have been shown to be important (see Cox, 1990). These are summarised in Table 1. It is clearly desirable, therefore, to look at the impact of combinations of these factors on health and efficiency.

Combined effects can also be considered in terms of individual differences. Individual differences may take the form of stable characteristics or more temporary changes in state. The transactional approach to stress (Lazarus & Folkman, 1984) argues that the influence of any stressor will be moderated or mediated by a person's psychosocial resources. Such an approach clearly has great potential for the investigation of workplace hazards. Recent research has also shown that when a person has minor illness, such as a cold, they are more susceptible to factors like exposure to noise (Smith et al., 1993) or prolonged work. This suggests that what is considered safe for the healthy worker may be inappropriate for a person with even a minor illness.

Overall, it is clearly important to have further information on combinations of hazards. The present project has the following specific objectives.

Objectives

(1)   To review the current literature on the combined effects of occupational hazards on health and performance efficiency.
(2)   To examine, using subjective reports of physical and mental health, the impact of combinations of workplace factors on health.
(3)   To investigate, using the 'after-effect' technique, the impact on combinations of workplace factors on performance efficiency.
(4)   To investigate the impact of combinations of workplace factors on work-related accidents.
(5)   To conduct laboratory investigations on the combined effects of different workplace stressors on performance efficiency and well-being.
(6)   To investigate the effects of combinations of occupational stressors on salivary cortisol and other physiological indicators of stress.

Methods

(1)   *Literature review*
This covers the existing published literature on the effects of combinations of workplace hazards on health and performance efficiency.

(2)   *Summary of the effects of combinations of workplace hazards on subjective health*
This has involved analysis of a database from a random sample of 4,000 of the working population who were given a questionnaire describing features of their job (e.g. exposure to physical agents; working hours; demands; social support at work) and also provided information on their current physical and mental health status. Information on demographic characteristics, health-related behaviours and stress outside of the workplace was also collected.

(3)   *Workplace factors and performance efficiency*
Each volunteer carries out a series of performance tasks before starting work and after finishing work. The difference between these two measures has been shown to be related to the workload (Broadbent, 1979). Each person carries out the procedure for several days and records characteristics of each working day (in terms of exposure to physical agents; working hours; demands, etc). Comparisons can then be made both between and within individuals to provide information about the impact of combinations of factors on performance efficiency.

(4)   *Combinations of factors and work-related accidents*
Individuals involved in work-related accidents can be identified from the records of the Accident and Emergency Department. They are then asked to complete questionnaires describing the combinations of factors they are exposed to at work. As in the previous sections, information on moderators and mediators (and non-workplace confounders) is also collected. Comparisons with colleagues doing a similar job who have not had an accident, and those attending the emergency unit for non-work related accidents and injury allow us to assess the impact of combinations of workplace factors on accidents at work.

(5)   *Laboratory investigation of the combined effects of different workplace stressors on performance efficiency.*
These studies will based on previous studies of noise and nightwork and of minor illnesses and noise. Individuals will be categorised according to level of normal exposure and current occupational stress level. They will then be exposed to the noise and/or nightwork condition and performance, mood and physiological functioning assessed to test predictions from the Adaptive Cost Model (Glass and Singer, 1973).

(6)   *Physiological investigations of combinations of occupational health hazards.*
There are now a number of physiological measures which are considered good indicators of stress (e.g. salivary cortisol). These are being measured in workers exposed to a range of occupational factors and associations between the questionnaire measures of combinations of exposure and the physiological assays examined.

Preliminary results using these methodologies are presented in the subsequent papers.

Acknowledgement

The research described in this article has been funded by the Health & Safety Executive.

References

Broadbent,D.E. 1971, *Decision and Stress*. London: Academic Press

Broadbent,D.E. 1979, Is a fatigue test possible ? *Ergonomics*, **22**, 1277-90

Cohen,A. 1973, Industrial noise, medical absence and accident record data in exposed workers. In: W.D.Ward (ed), Proceedings of the second international conference on noise as a public health problem. U.S. environmental protection agency

Cox, T. 1990, The recognition and measurement of stress: Conceptual and methodological issues. In E.N.Corlett and J.Wilson (eds.), *Evaluation of Human Work*. (Taylor & Francis, London)

Glass, D.C. & Singer,J.E. 1972, *Urban Stress: Experiments on Noise and Social Stressors.* London: Academic Press

Griffin, M.J. 1972, Vibration. In: A.P.Smith & D.M.Jones (eds), *Handbook of Human Performance, Vol.1, The Physical Environment.* London: Academic Press. pp. 55-78

Lazarus, R.S & Folkman, S. 1984, *Stress, appraisal and coping.* Springer Publications, New York

Monk,T.H. & Folkard,S. 1985, Shiftwork and performance. In: S.Folkard & T.H.Monk (eds), *Hours of Work.* Chichester: Wiley. pp. 239-252

Smith, A. P. 1990, An experimental investigation of the combined effects of noise and nightwork on human function. In: *Noise as a Public Health Problem, Vol.5, New Advances in Noise Research, Part II,* (eds) B. Berglund & T. Lindvall. Stockholm: Swedish Council for Building Research, 255 - 271

Smith, A. P. & Miles, C. 1985, The combined effects of noise and nightwork on human function. In: D. Oborne (ed.), *Contemporary Ergonomics 1985*. London: (Taylor & Francis), 33 - 41

Smith, A. P. & Miles, C. 1986, The combined effects of nightwork and noise on human function. In: M. Haider, M. Koller & R. Cervinka (eds), *Studies in Industrial and Organizational Psychology 3: Night and Shiftwork: Long term effects and their prevention.* Frankfurt: Peter Lang, 331 - 338

Smith, A. P. & Miles, C. 1987, The combined effects of occupational health hazards: An experimental investigation of the effects of noise, nightwork and meals. *International Archives of Occupational and Environmental Health,* **59**, 83 - 89

Smith, A. P. & Miles, C. 1987, Sex differences in the effects of noise and nightwork on performance. *Work and Stress, 1,* 333 - 339

Smith, A.P., Thomas, M., Brockman, P. 1993, Noise, respiratory virus infections and performance. *Proceedings of 6th International Congress on noise as a public health problem.* Actes Inrets 34, Vol 2, 311-314

A REVIEW OF THE LITERATURE ON THE COMBINED EFFECTS OF NOISE AND WORKING HOURS.

Benjamin Wellens

Centre for Occupational and Health Psychology
Cardiff University
63 Park Place
Cardiff CF10 3AS
wellensbt@cf.ac.uk

There has been considerable research on the effects of noise and working hours on health and performance. In many situations it is the combination of these factors that may be critical, yet there have been few previous reviews of this topic. An initial review of the area is reported here and the following conclusions made. First, there is a small literature on combined effects and the findings are often contradictory. Secondly, the area is complex and comparisons across studies difficult due to different areas covered by the topics of noise and working hours. The literature does suggest that many of the effects of noise and working hours are independent although some studies of physiological function show that the combined effects cannot be predicted from the individual effects of noise and working hours.

Introduction

The individual effects of noise and working hours
The individual effects of occupational noise and working hours on health outcomes, both objectively and subjectively measured, are well studied and the general trends are relatively undisputed. The auditory and non-auditory effects of noise have both been reviewed extensively elsewhere (HSE, 1998, Butler et al., 1999). The effects are wide ranging and have generally been researched in both experimental and occupational environments. The most recent review (Butler et al., 1999) concluded that *'in conjunction with other agents, noise can produce additive, cancelling or interactive effects. However, the complexity of exposure variables makes it difficult to identify consistent and stable response patterns'.*

Organisation of working hours can take any of a number of forms, such as permanent day work, permanent night work and three-shift work. A recent meta-analytic review of the effects of hours of work on health (Sparks et al., 1997) found small significant effects; there were significant positive correlations between overall health symptoms, physiological and psychological health symptoms, and hours of work. Health problems associated with the organisations of working hours are generally related to shift work, particularly night work. In conclusion, Sparks et al. (1997) found problems drawing solid conclusions due to *'the obvious shortage of studies investigating health and working hours'*.

The combined effects of noise and working hours
A considerable problem encountered when such research is attempted is the difficulty in isolating the effects of the occupational stressors in question and the reliable assignment of causation. Certain studies concentrate on one stressor and 'other stress factors', for example, night work, shift work, or noise exposure and other variables. An exhaustive database search of the usual suspects, PsychLIT, Medline, BIDS, ASSIA, Ergonomics Abstracts and New Scientist was made. The Internet was searched for relevant sites with the keywords: 'noise at work', 'occupational noise', 'shift work', 'night work' and 'working hours'. Reviewing the combined effects of noise and working hours literature did not uncover an extensive body of research, but what was found is reviewed below.

Summary of the auditory effects of noise and working hours

Auditory effects are all those effects on health and well-being caused by effects on the hearing organ and effects that are due to the masking of auditory information (i.e. communication problems). The auditory effects of noise are well reported, with set guidelines for recommended levels of noise exposure to avoid acute and chronic effects. Temporary or permanent deafness and tinitus are two likely effects of continued exposure to loud noise. No auditory effects of working hours are known.

Hearing loss and masking effects
There was very little information found on the combined effects on hearing loss. The findings of one study (Holzmuller et al., 1990) suggested that those working three-shift systems had retained better hearing than those working single-shifts systems. These findings suggest that there may be some type of interaction between working hours and noise effects however the paucity of research hinders drawing any conclusions.

Noise can cause communication problems in workplace by drowning out speech sounds. There is also the possibility of masking the sounds of alarms that may be important in the workplace. There is no known research on the auditory masking effects of noise and working hours in combination, other than those attributable to noise only.

Summary of the non-auditory effects of noise and working hours

Non-auditory effects of noise can be defined as *'all those effects on health and well-being...with the exclusion of effects on the hearing organ and the effects which are due to the masking of auditory information'* (Smith and Broadbent, 1992). Such effects include performance effects, physiological responses and health outcomes, annoyance and sleep disturbance. There is also a considerable body of evidence for the non-auditory effects of noise as have been previously outlined (Smith and Broadbent, 1992,

Butler et al, 1999). For the purpose of this review, all the effects of the organisation of working hours will be viewed as non-auditory.

Cardiovascular effects
It is thought that acute noise exposure may influence cardiovascular functions and levels of catecholamines. There is some evidence for increased cardiovascular disease, coronary heart disease and myocardial infarction in those working longer hours, however such research is rare and contradicted by other studies. In combination, work noise annoyance and stressful working conditions were found to result in an increase in blood pressure, concomitantly with heart rate increase (Petiot et al., 1991). These effects were more obvious in men than in women. Another study confirmed the trend. Both systolic and diastolic blood pressures were significantly higher when work noise annoyance and night work conditions were combined than in the work noise annoyance alone condition (Lercher, Hortnagl and Kofler, 1993).

Other physiological and biochemical effects and health outcomes
The individual effects of noise on physiological and biochemical variables are due to stimulation of the reticular activating system that has an influence on cardiovascular function, increased catecholamines and cortisol. The physiological and biochemical effects of working hours are well described and it is likely that there are substantial effects on circadian rhythm variation and fatigue. Catecholamine excretion and electrodermal activity was studied under the combined effects of shift work and noise (Ottmann et al., 1989). Independent effects of shift work and noise were demonstrated. However, the study also found an interaction between type of shift and noise, noise leads to an increase of adrenaline excretion for night workers and to a decrease of excretion for day workers. Catecholamine excretion rates, autonomic reactions, reaction times and ratings of subjective alertness showed changes typical for night-shift work and there was evidence of some interaction effects (Boucsein and Ottmann, 1996).

Effects on sleep and fatigue
It is generally agreed that noise may disrupt sleep in a number of ways. Noise is also thought to have fatigue effects and it is generally agreed that noise during the day influences sleep at night. Long working hours can lead to poor sleep, and night work and shift work can cause serious disruption to the body's circadian rhythms that, although reversible, can lead to long term health problems, thought to be related to accumulative sleep debt. A study looking at the effects of shift work and work environment on sleep found that disturbances of sleep were associated with: rotating shift work, external environmental noise and the physical working environment (Abebe and Fantalum, 1999).

Shift work and noise are seen to induce independent effects (Rutenfranz et al., 1989), explained in terms of 'activation' for shift work and in terms of 'tension' for noise. Therefore, the combination of both factors is partly additive and partly subtractive, as night work and noise negatively affect day sleep.

Psychosocial and psychological effects
Annoyance, with the possibility of the development of psychopathology, is generally associated with noise. However such annoyance is related to the individual's characteristics, and is not a direct effect of noise. Generally, studies have found it

problematic to accurately contribute causation to the effects of working hours, although there is some suggestion that long working hours and non-day work were the main stressors leading to poor mental health. Noise annoyance in combination with other stress factors can affect health. Time stress and self-reported noise annoyance were the predominant factors related to general health, neuro-vegetative disorders and ear, nose and throat diseases (Jansen and Schwarze, 1988). Due to the nature of the study, conclusions cannot be drawn on whether this is due to separate independent effects of noise and time stress or to a combination. Shift workers reported that they perceived noise exposure at both home and work to be more unbearable than day workers did. This greater sensitivity to noise also proves to be related to more health problems, suggesting a relationship between the psychological and physical effect (Koller et al, 1988). The authors suggest that noise sensitive shift workers may be more disturbed by noise whilst trying to sleep during the day, or maybe there is synergism between the stressor of night work and occupational noise exposure, which is moderated by their sensitivity to noise.

Noise, working hours and performance
The individual effects of noise on performance are varied dependent on factors such as length of exposure and dB (a) level. Performance effects of working hours are varied, including time of day effects, fatigue, sleep inertia and sleep loss due to night and shift work. Despite some detailed and thorough studies no combined effects of great significance are apparent. On the whole, the effects of noise and working hours (specifically in this research, night work was studied) were found to be independent. Measures of introversion, neuroticism, anxiety and morningness were included. The only measure of personality that interacted with both noise and night work was introversion. Such effects were only present for a few aspects of performance. Effects of noise and night work on performance are seen to be generally selective and independent. Performance studies have also demonstrated some sex differences: female subjects showed a large night time impairment on manual dexterity, but performed working memory tasks better at night. Male subjects showed a small impairment on both tasks (Smith, 1989, Smith, 1990).

Noise, working hours and accidents
It has long been argued that noise may have a direct causal relationship with accidents. Fatigue caused by long working hours, or other organisation of work hours, has been cited in a number of the most severe man-made disasters of the last 20 years; Chernobyl, the Exxon Valdez oil spill, Three-Mile Island, Bhopal and the decision-making process involved in the Space Shuttle Challenger Accident. It has been suggested that a circadian rhythm of accidents exists and that factors, such as noise, may modify this rhythm (Wojtczak-Jaroszowa and Jarosz, 1987). This rhythm is governed by the interrelationship between a number of endogenous and exogenous rhythmic factors. Such factors largely consist of both human factors (fatigue or circadian fluctuation of biological function) and multiple hygienic and social components that either promote or suppress human performance (including noise or high temperature). Rates of accidents were found to depend on the particular combinations of these factors, which vary around the clock.

A counteracting effect of noise on the sleepiness caused by night work has been suggested (Akerstedt and Landstrom, 1998). The study concluded that night work caused severe sleepiness, impaired performance and increased accident risk. In reviewing potential practical countermeasures to these risks, noise was highlighted as one of a number of promising techniques to counteract sleepiness. This noise was controlled and therefore would likely be experienced very differently to the noise experienced in a noisy workplace.

Conclusions

The most important thing to make clear is the small literature on the combined effects of noise and working hours; all other conclusions drawn should be in seen in the light of the huge difficulties in making representative statements on the basis of such sparse information, where what findings there are, generally contradict one another.

This is a hugely complex area, as reflected in the substantial problem concerning the use of the terms 'working hours' and 'noise'. Noise exposure can either be acute or chronic, it can be either perceived (and as such may well better reflect noise annoyance than noise experienced) or be objectively measured (in controlled environments so that subjects are in either high or low noise groups), and from any one (or more) of a huge number of sources (machinery, other people, vehicles). Working hours could refer to many aspects of the work time organisation, be it night work or shift work (of a temporary or permanent nature), long work shifts or irregular work. Therefore any conclusions drawn must be made in light of the precise research scenarios in which the findings were made.

Regarding future research, the literature review can help construct predictions for effects that may be found. The literature suggests that effects found will be largely independent whether the outcome measures are subjective or objective health measures, performance or accident data. However, combined effects may be seen in certain physiological and biochemical variables such as blood pressure and catecholamine excretion. Small but significant performance effects may also be apparent.

Acknowledgement

This review was conducted as part of a project on the combined effects of occupational health hazards. The project is funded by the Health & Safety Executive.

References

Abebe, Y. and Fantalum, M. 1999. Shift Work and Sleep Disorder among Textile Mill Workers in Bahir Dar, Northwest Ethiopia. *East African Medical Journal*, **76** (7), 407 – 410

Akerstedt, T. and Landstrom, U. 1998. Work Place Countermeasures in Night Shift Fatigue. *International Journal of Industrial Ergonomics*, **21** (3 - 4), 167 – 178

Boucsein, W. and Ottmann, W. 1996. *Psychophysiological Stress Effects from the Combination of Night-Shift Work and Noise.* Biological Psychology, **42** (3), 301 – 322

Butler, Graveling, Pilkington and Boyle 1999. Non-Auditory Effects of Noise at Work: A Critical Review of the Literature Post 1988. *Prepared by the Institute of Occupational Medicine for the Health and Safety Executive*

Health and Safety Executive. 1998. Reducing Noise at Work: Guidance on the Noise at Work Regulations 1989. *HSE Books, Sudbury, Sussex*

Holzmuller, M., Seibt, A., Jakubowski, A. and Friedrichsen, G. 1990. Studies of the Combined Effects of Shift Work and Noise on Permanent Hearing Loss. *Z. Gesamte Hyg.,* **36** (9), 501 – 502

Jansen, G. and Schwarze, S. 1988. *The Relative Importance of Noise within the Scope of Combined Stress Factors at Workplaces. In: Manninen, O. (Editor).* Recent Advances in Researches on the Combined Effects of Environmental Factors. *The International Society of Complex Environmental Studies – ISCES,* 349 - 364

Koller, M., Kundi, M., Haider, M. and Cervinka, R. 1988. *The Combined Effects of Noise and Night Work: Noise Sensitive versus Noise Insensitive Workers. In: Manninen, O. (Editor).* Recent Advances in Researches on the Combined Effects of Environmental Factors. *The International Society of Complex Environmental Studies – ISCES,* 381 - 396

Lercher, P., Hortnagl, J. and Kofler, W.W. 1993. Work Noise Annoyance and Blood Pressure: Combined Effects with Stressful Working Conditions. *International Archives of Occupational and Environmental Health,* **65** (1), 23 – 28

Ottmann, W., Rutenfranz, J., Neidhart, B. and Boucsein, W. 1989. Combined Effects of Shift Work and Noise on Catecholamine Excretion and Electrodermal Activity. *Ergonomia,* **12** (1), 41 – 50

Petiot, J.C., Parrot, J., Lobreau, J.P., Smolik, H.J. and Guilland, J.C. 1991. Combined Effects of Noise and Shift Work Schedule in Experimental Setting. II. Cardiovascular and Rectal Temperature Data. *Archives of Complex Environmental Studies,* **3** (1 – 2), 13 – 24

Rutenfranz, J., Bolt, H.M., Ottmann, W. and Neidhart, B. 1989. Combined effects of Shift Work and Environmental Hazards (Heat, Noise, Toxic Agents). *Arhiv Za Higijenu Rada I Toksikologiju,* **40** (3), 257 – 276

Smith, A.P. 1989. A Review of the Effects of Noise on Human Performance. *Scandinavian Journal of Psychology,* **30**, 185 – 206

Smith, A.P. 1990. An Experimental Investigation of the Combined Effects of Noise and Night Work on Human Function. *In: Berglund, B. and Lindvall, T. (Editors).* Noise as a Public Health Problem. Volume 5. New Advances in Noise Research Part II, 255 - 271. *Swedish Council for Building Research, Stockholm*

Smith, A.P. and Broadbent, D.E. 1992. Non-Auditory Effects of Noise at Work: A Review of the Literature. *London: HMSO*

Sparks, K., Cooper, G., Fried, Y. and Shirom, A. 1997. The Effects of Hours of Work on Health: A Meta-Analytic Review. *Journal of Occupational and Organizational Psychology,* **70**, 391 – 408

Wojtczak-Jaroszowa, J. and Jarosz, D. 1987. Chronohygienic and chronosocial aspects of industrial accidents. *Progress in Clinical and Biological Research,* 227 (B), 415 – 426

THE INFLUENCE OF KARASEK DIMENSIONS OF JOB STRAIN ON HEALTH

Stephanie Sivell

Centre for Occupational and Health Psychology
Cardiff University
63 Park Place
Cardiff CF10 3AS
sivells@cf.ac.uk

The Karasek Job Strain model has been widely used to examine the relationship between job characteristics and health. The present article describes different versions of the model and evaluates the extent to which they predict occupational health. Overall, the literature shows evidence of the independent effects of job demand, decision latitude and work social support. There is also evidence demonstrating that both control and social support can buffer against the negative effects of job demand. There is little evidence on whether the model can predict differences in objective performance and safety outcomes and it is important to provide information on this in future research.

Overview

The Job Demand-Control Model (Karasek, 1979), hereafter referred to as the Job Strain Model, is the most widely utilised model of the relationship between work and health. The model proposes that psychological strain is a product of the combination of the work situation an individual is exposed to and the amount of freedom available to make decisions at work. As job demands increase, job control decreases. When the levels of job demands and job control are high the model hypothesises that the job is 'active' and new behaviour patterns develop both in and outside the workplace. When a job is 'passive' activity levels decrease as do general problem solving activities. If the worker is unable to initiate any action or decision making then the "unreleased energy" may result in mental strain. Therefore, it is the limitations on decision-making rather than the strain of decision-making itself that can be problematic, rendering job strain as a risk for high-status and low-status jobs alike.

'Work Social Support' was added as a third dimension leading to an adaptation to the Job Demand-Control-Support model (Johnson and Hall, 1988). The concept of 'iso-strain' was introduced, where demands are high, control is low and social support is also low. The body of research that exists on the relationship between the Job (Iso-)

Strain Model and health outcomes have generally utilised two related yet different hypotheses: the Strain Hypothesis and the Buffer Hypothesis. The Strain Hypothesis states that employees working in a high strain job experience the lowest well-being. The Buffer Hypothesis postulates that control (and social support) can moderate the negative effects of high demand.

Physical Health

Strain hypothesis
Cardiovascular related health is one particular area of physical health that has been extensively researched. Jobs that are characterised by high job strain have been found to be associated with cardiovascular risk factors (Pieper et al, 1989) and more specifically an increased risk of cardiovascular disease (e.g., Theorell and Karasek, 1998). There is evidence in the Japanese literature that the Job Strain Model is associated with increased levels of blood pressure and serum lipids in the Japanese working population (Kawakami and Haratani, 1999). There is however a substantial body of literature that does not support this (e.g., Hanke and Dudek, 1997). The effects are not always found to be similar in magnitude or in direction for all characteristics, and sex differences are also sometimes shown (Hellerstedt and Jeffery, 1997).

Research carried out on musculo-skeletal health has reported that job strain characterises the psychosocial work environment of certain professions (i.e. musicians, video terminal display operators) and influences musculo-skeletal discomfort (e.g., Fjellman-Wiklund and Sundelin, 1998). Associations between reproductive problems and job strain have also been reported. Both preeclampsia (Klonoff-Cohen et al, 1996) and abortion (Fenster et al., 1995) were found to be related to high job strain in working women. Low job control has been reported in the literature to be a risk factor associated with sickness absence (Bodeker, 2000). In addition to job demands and job control, effort-reward imbalances have been linked to an increased risk of long spells of sickness absence for women.

Buffer Hypothesis
There is not much evidence to support the buffering effects of job control, or social support, on physical health outcomes. An exception is blood pressure, where changes have been reported as being effected by interactions between job demands and control (Chapman et al., 1990).

Mental Health

Strain hypothesis
High job demands have been reported to be associated with an increased risk of psychiatric disorder (Stansfeld et al., 1999), psychological distress (Bourbonnais et al, 1996) and poor mental health status (Yang et al., 1997). Perrewe (1986) found that job demands negatively effect satisfaction and positively effect psychological anxiety.

Buffer hypothesis
There is some evidence that suggests that control over work does have a stress-buffering role (e.g., Kawakami and Haratini, 1999) and low job control and high job

demands can negatively effect psychological distress (e.g., Barnett and Brennan, 1997). Perrewe and Ganster (1989) found partial support for the Job Strain Model in that high control lessened the impact of work overload on anxiety.

Health-Related Behaviours

Strain hypothesis
There is also a substantial literature available on the associations between job demands and job control with health related behaviours although the types of behaviours tend to be limited to alcohol consumption and smoking habits. Low job control has been found to be associated with alcohol dependence in women (Stansfeld et al, 1999). Job control and job demands have been associated with drinking problems in Japanese workers (Kawakami and Haratini, 1999). Low job control, low job demands and low workplace social support have been related to later alcoholism in blue-collar men with 'passive' jobs being associated with an increased risk of a psychiatric diagnosis of alcoholism (Hemmingsson and Lundberg, 1998). Bromet et al (1988) have also found the specific combination of high job demands and low job control to be important in predicting the occurence of alcohol problems.

Hellerstedt and Jeffery (1997) found job demands to be positively associated with smoking, smoking intensity and even high fat intake in men. They have also been reported to be positively associated with Body Mass Index (BMI) and smoking intensity in women. It was also found that high strain male smokers smoked more than other smokers and high strain women had a higher BMI than other women. High job strain has also been linked to increased drug use (Storr, Trinkoff and Anthony, 1999).

Buffer hypothesis
Women who smoke have been found to experience low job control and high job demands (Jonsson et al., 1999). There is also some evidence, although very limited, to suggest that job demands and job control may influence an individual's level of exercise. Men who experience high levels of job control and high levels of social support at work have been found to be more physically active during leisure time (Jonsson et al, 1999). Job control has been reported to be positively associated with exercise in both sexes (Hellerstedt and Jeffery, 1997).

Performance and accidents

Research evidence on the relationship between job demands and job control and performance or accidents at work is extremely limited. However, there is some evidence to suggest that different levels of job demands in combination with job control can have an effect on performance (Sargent and Terry, 1998).

Conceptual issues

The issue of conceptualising job control has been raised in criticism of the model. It has been suggested that this conceptualisation is vital to discriminating between studies that support the model and those that do not. DeCroon et al (2000) argued that job control has been poorly conceptualised. Regression analysis was reported to reveal a significant job demands by job control interaction effect as well as a significant main effect of the two independent variables on psychosomatic health complaints (DeCroon et al, 2000). Similarly, only aspects of job control that correspond to specific demands of a given job moderate the impact of high demands on well-being. Stressors such as job insecurity and broader aspects of job control are also largely ignored by the model (Van der Doef and Maes, 1999). Such factors could be vital in building a more accurate picture of the effects of job demands and control on an individual's physical health.

Warr (1990) found no evidence for an interaction between job control and job demands. In this instance it was argued that job control and job demands each predict different aspects of well-being, depression-enthusiasm and anxiety-contentment respectively. Jimmieson and Terry (1998) also found only limited support for job control and task information acting as a buffer to the negative effects of task demands on adjustment.

The incorporation of social support as a buffer has produced more consistent findings (e.g., Schaubroeck and Fink, 1998), although it is important to note that social support can not only have a positive effect on job strain, but also a negative one, depending on the type of support received (Johnson and Hall, 1988). However, as with much of the research on the Job Strain model, there is also a considerable body of data that suggest that social support does not have a buffering effect against the worst consequences of job demands (e.g., Glaser et al, 1999).

Conclusions

It is difficult to make representative statements about the literature due to the sheer breadth and depth of it. This is a complex area with a vast array of potential methodological issues. Findings have been consistently contentious with some studies supporting the model but others producing conflicting results. Evidence has also accumulated showing that addition of the dimension of social support has improved the model, and that work social support can act as a buffer against job strain.

In conclusion, it has been shown that that the effects of demand, control and social support will exhibit large independent effects on the array of outcome measures. It has also been the case that for the majority of outcome measures, the combination of high demand and low control (and low social support) will often be most deleterious to health. There have also been other interaction effects, particularly involving the buffering effects of social support against the combination of high demand and low control. To a lesser extent some buffering effects of high control against high demand may also be found. The almost non-existent research on performance measures and accident rates means that further new research is required on this important topic.

Acknowledgement

This review was conducted as part of a project on the combined effects of occupational health hazards. The project is funded by the Health & Safety Executive.

References

Barnett, R.C. and Brennan, R.T. 1997. Change in Job Conditions, Change in Psychological Distress, and Gender: A Longitudinal Study of Dual-Earner Couples. *Journal of Organizational Behavior*, **18**, 253 – 274

Bodeker, W. 2000. The Influence of Work Related Stressors on Disease Specific Sickness Absence. *Sozial Und Praventivmedizin*, **45** (1), 25 – 34

Bourbonnais, R., Brisson, C. Moisan, J. and Vezina, M. 1996. Job Strain and Psychological Distress in White-Collar Workers. *Scandinavian Journal of Work, Environment and Health,* **22**, 139 – 145

Bromet, E.J., Dew, M.A., Parkinson, M.A. and Schulberg, H.C. 1988. Predictive Effects of Occupational and Marital Stress on the Mental Health of the Male Workforce. *Journal of Organizational Behavior,* **9**, 1 – 13

Chapman, A., Mandryk, J.A., Frommer, M.S., Edye, B.V., and Ferguson, D.A. 1990. Chronic Perceived Work Stress and Blood Pressure Among Australian Government Employees. *Scandinavian Journal of Work and Environmental Health*, **16**, 258 – 69

DeCroon, A., Van Der Beek, A.J., Blonk, R.W.B. and Frings Dressen, M.H.W. 2000. Job Stress and Psychosomatic Health Complaints among Dutch Truck Drivers: A Re-evaluation of Karasek's Interactive Job Demand-Control Model. *Stress Medicine,* **16**, 101 – 107

Fenster, L. Schaefer, C., Mathur, A., Hiatt, R.A., Pieper, C., Hubbard, A.E., Von Behren, J. and Swan, S.H. 1995. Psychological Stress in the Workplace and Spontaneous Abortion. *American Journal of Epidemiology*, **142**, 1176 – 1183

Fjellman-Wiklund, A. and Sundelin, G. 1998. Musculoskeletal Discomfort of Music Teachers: An Eight-Year Perspective and Psychosocial Work Factors. *International Journal of Occupational and Environmental Health*, **4**, 89 – 98

Glaser, D.N., Tatum, B.C., Nebeker, D.M., Sorenson, R.C. and Aiello, J.R. 1999. Workload and Social Support: Effects on Performance and Stress. *Human Performance*, **12** (2), 155 – 176

Hanke, W. and Dudek, B. 1997. The Effect of Stress in the Workplace on the Risk of Ischemic Heart Diseases: The Role of Epidemiological Studies. *Medical Practitioner.*, **48**, 675 – 686

Hellerstedt, W.L. and Jeffery, R.W. 1997. The Association of Job Strain and Health Behaviours in Men and Women. *International Journal of Epidemiology*, **26** (3), 575 – 583

Hemmingsson, T. and Lundberg, I. 1998. Work Control, Work Demands, and Work Social Support in Relation to Alcoholism Among Young Men. *Alcoholism: Clinical and Experimental Research*, **22** (4), 921 – 927

Jimmieson, N.L. and Terry, D.J. 1998. An Experimental Study of the Effects of Work Stress, Work Control, and Task Information on Adjustment. *Applied Psychology: An International Review*, **47** (3), 343 – 369

Johnson, J.V. and Hall, E.M. 1988. Job strain, work place social support, and cardiovascular disease: a cross-sectional study of a random sample of the Swedish working population. *American Journal of Public Health*, **78**, 1336 – 1342

Jonsson, D., Rosengren, A., Dotevall, A., Lappas, G. and Wilhelmsen, L. 1999. Job Control, Job Demands and Social Support at Work in Relation to

Cardiovascular Risk Factors in MONICA 1995, Goteborg. *Journal of Cardiovascular Risk*, **6** (6), 379 – 385

Karasek, R.A. 1979. Job Demands, Job Decision Latitude, and Mental Strain: Implications for Job Redesign. *Administrative Science Quarterly*, **24**, 285 – 311

Kawakami, N. and Haratani, T. 1999. Epidemiology of Job Stress and Health in Japan: Review of Current Evidence and Future Direction. *Industrial Health*, **37** (2), 174 – 186

Klonoff-Cohen, H.S., Cross, J.L. and Pieper, C.F. 1996. Job Stress and Preeclampsia. *Epidemiology*, **7**, 245 – 249

Perrewe, P.L. 1986. Locus of Control and Activity Level as Moderators in the Quantitative Job Demands*Satisfaction / Psychological Anxiety Relationship: An Experimental Analysis. *Journal of Applied Social Psychology*, **16** (7), 620 – 632

Perrewe, P.L. and Ganster, D.C. 1989. The Impact of Job Demands and Behavioral Control on Experienced Job Stress. *Journal of Organizational Behavior,* **10** (3), 223 - 229

Pieper, C., La Croix, A.Z. and Karasek, R.A. 1989. The Relation of Psychosocial Dimensions of Work with Coronary Heart Disease Risk Factors: A Meta-Analysis of Five United States Databases. *American Journal of Epidemiology*, **129**, 483 – 494

Sargent, L.D. and Terry, D.J. 1998. The Effects of Work Control and Job Demands on Employee Adjustment and Work Performance. *Journal of Occupational and Organizational Psychology*, **71** (3), 219 – 236

Schaubroeck, J. and Fink, L.S. 1998. Facilitating and Inhibiting Effects of Job Control and Social Support on Stress Outcomes and Role Behaviour: A Contingency Model. *Journal of Organizational Behavior*, **19** (2), 167 – 195

Stansfeld, S.A., Fuhrer, R., Shipley, M.J. and Marmot, M.G. 1999. Work Characteristics Predict Psychiatric Disorder: Prospective Results from the Whitehall II Study. *Occupational and Environmental Medicine*, **56** (5), 302 – 307

Storr, C.L., Trinkoff, A.M. and Anthony, J.C. 1999. Job Strain and Non-Medical Drug Use. *Drug and Alcohol Dependence*, **55**, 45 – 51

Theorell, T. and Karasek, R.A. 1998. Current Issues Relating to Psychosocial Job Strain and Cardiovascular Disease Research. *Journal of Occupational and Health Psychology*, **1**, 9 – 26

Van Der Doef, M. and Maes, S. 1999. The Job Demand-Control (Support) Model and Psychological Well-Being: A Review of 20 Years of Empirical Research. *Work and Stress*, **13** (2), 87 – 114

Warr, P.B. 1990. Decision Latitude, Job Demands, and Employee Well-Being. *Work and Stress*, **4**, 285 – 294

Yang, M.J., Ho, C.K., Su, Y.C. and Yang, M.S. 1997. Job Strain, Social Support and Mental Health: A Study on the Male Heavy Manufacturing Workers. *Kao Hsiung Ko Hsueh Tsa Chih*, **13**, 332 – 341

THE COMBINED EFFECTS OF OCCUPATIONAL FACTORS ON SUBJECTIVE REPORTS OF HEALTH

Andy Smith, Carolyn Brice, Stephanie Sivell and Benjamin Wellens

Centre for Occupational and Health Psychology
Cardiff University
63 Park Place
Cardiff, CF10 3AS
Smithap@cardiff.ac.uk

The present paper reports results from a secondary analysis of data from the Bristol Stress and Health at Work Study. In the first, analysis combinations of factors reflecting the physical environment at work (e.g. perceived noise) and working hours (e.g. nightwork) were considered. In the second set of analyses combinations of job demand, decision latitude and social support at work were examined. In the final analyses a measure was derived which represented the total number of negative factors present at work. The first two sets of analyses largely revealed main effects of the factors, with some evidence of additive effects when variables were considered in combination. The analysis of the total negative factors score showed a clear dose response between the number of negative factors present and negative health outcomes.

Background

Reviews of the literature on combined effects of occupational health hazards have revealed a limited knowledge base. Databases do exist that allow this topic to be investigated and the present article is concerned with a secondary analysis of a community survey providing subjective reports of exposure and health outcomes.

The Bristol Stress and Health at Work Study
This study is described in detail in Smith et al. (2000a, b). Data were collected from a random community sample and the analyses reported here are based on those who were working (over 4,000 respondents). In addition, a sub-sample repeated the questionnaire 12 months later and analyses of change scores could be made using this data set. The questionnaires measure aspects of work such as the physical environment, working hours, job demands, decision latitude and support at work. Health outcomes such as perceived stress, mental health (the General Health Questionnaire, Goldberg and Williams,1988; Hospital Anxiety and Depression Scale, Zigmond and Snaith, 1983) and physical health problems (both chronic and acute) were also measured. Health-related

physical health problems (both chronic and acute) were also measured. Health-related behaviours such as smoking and alcohol consumption were also recorded. Previous analyses have shown that demographic and occupational factors (e.g. full versus part-time work, salary) are associated with stress at work and these were co-varied in the present analyses.

The first set of analyses examined combinations of factors relating to the physical environment and working hours.

Combined effects of the physical environment and working hours.

These analyses considered perceptions of (a) exposure to noise at a level that disturbs concentration (b) noise that produces a ringing in the ears or temporary feeling of deafness (c) exposure to fumes or dusts, and (d) handling potentially harmful substances or materials. The working hours variables were (a) nightwork (b) shiftwork and (c) number of hours worked per week. The analyses compared those groups with no exposure, those exposed to single factors and those who were exposed to a combination of the environmental and temporal factors being considered.

The results of the analyses showed a similar pattern for all the combinations of variables. This pattern of results is illustrated by considering the noise disturbing concentration variable and nightwork.

Noise and night-work
The analyses were carried out using a median split on noise and night-work exposure with responses of 'often' and 'sometimes' assigned to the exposure group and responses of 'seldom' & 'never' assigned to the not exposed group. From the analyses conducted to investigate the combined effects of noise and night-work three main patterns of results were identified, namely effects of noise, effects of night-work and the combination of exposure to these occupational factors. Examples of these effects are shown in Table 1. Noise exposure was found to be most influential factor on various aspects of health with the effects being largely negative. Those exposed to noise reported poorer physical health on measures of quality of health over the last 12 months. This same pattern of effects was also found for measures of mental health with those exposed to noise reporting greater anxiety and depression as measured by the Hospital Anxiety Depression scale and having higher scores on the General Health Questionnaire. Poorer health-related behaviours were also reported for this noise group.
Exposure to night-work was also found to have some effects on reported health but the effects were limited (see Table 1).

When the two factors were examined in combination there were few instances where the combined effects differed from those of the individual factors (see Table 1). For the majority of health measures analysed the combined effects of noise and night-work are found to be the same as effects found for the noise only group. Exceptions to this can be seen in Table 1 for measures of stress at work and alcohol units consumed per week where the combined effects of exposure to noise and night-work were found to have greater negative impact on health.

Many of the individual effects of noise and night-work found in the present analyses have previously been documented in the literature, confirming that the present results may generalise to a range of occupations.

Table 1. Effects of exposure to noise and / or night-work on measures of stress and health (scores are means, s.e.s. in parentheses; high scores = greater stress and more negative health outcomes)

Variable	Both noise & night-work (N=319)	Night-work only (N=716)	Noise only (N=597)	Neither noise or night-work (N=2186)	F
12 month health	2.23 (0.07)	1.98 (0.04)	2.22 (0.05)	1.97 (0.03)	[noise] p<0.0001
* Stress at work	3.19 (0.08)	2.80 (0.05)	2.99 (0.05)	2.69 (0.03)	[night-work] p<0.0001; [noise] p<0.0001
Anxiety	7.86 (0.32)	7.03 (0.21)	7.77 (0.23)	6.52 (0.12)	[noise] p<0.0001
Depression	4.43 (0.26)	3.78 (0.17)	4.42 (0.18)	3.42 (0.10)	[noise] p<0.0001
* units of alcohol per week	12.73 (0.99)	11.18 (0.64)	11.20 (0.71)	10.87 (0.39)	[night-work] p⁻0.07 [noise] p=0.06

*= Only outcome measures for which the combined effects of exposure to noise and night-work are found to have greater negative impact on health compared to noise or night-work alone.

Karasek scales

The Karasek scales of 'decision latitude', 'job demand' and 'work social support' were examined alone and in combinations. A median split on each of these three scales was conducted and analyses of covariance carried out to compare the eight groups formed by the combination of high / low sub-groups derived from the scales.

Results

In the majority of analyses there were significant main effects of job demand, control and work social support. There were, however, few interactions between these measures (see Table 2). A notable exception was found for the analysis of stress at work ratings, which showed a significant interaction between demand and decision latitude. If one orders the groups from no negative features (low demand, high social support, high control) to one negative feature (low support or high demand or low control) to two negative features through to the high job strain condition (high demand, low control, low social support) one finds a linear relationship with most health outcomes. This suggests that one way of examining combined effects is to sum the number of negative features in an individual's job or work environment. This was carried out in the next set of analyses.

Table 2. Effects of job strain (job demand, decision latitude and work social support) as measured by the Karasek model on measures of stress and health (scores are the means, s.e.s. in parentheses; high scores = greater stress and more negative health outcomes)

Variable	12 month health	Stress at work	Anxiety	Depression	Total units of alcohol per week
Low JD High WSS High DL (N=502)	1.83 (0.05)	2.45 (0.06)	5.37 (0.23)	2.63 (0.19)	9.84 (0.79)
Low JD Low WSS High DL (N=401)	2.00 (0.06)	2.65 (0.06)	6.54 (0.27)	3.33 (0.22)	11.78 (0.91)
Low JD High WSS Low DL (N=590)	1.92 (0.05)	2.39 (0.05)	5.96 (0.23)	3.04 (0.19)	10.98 (0.74)
High JD High WSS High DL (N=459	1.92 (0.05)	3.01 (0.06)	6.95 (0.26)	3.24 (0.21)	11.59 (0.84)
Low JD Low WSS Low DL (N=557	2.05 (0.05)	2.61 (0.05)	6.70 (0.24)	4.22 (0.19)	10.87 (0.76)
High JD Low WSS High DL (N=424)	2.19 (0.06)	3.14 (0.06)	8.4 (0.27)	4.58 (0.22)	11.73 (0.89)
High JD High WSS Low DL (N=262)	2.12 (0.07	3.13 (0.08)	14.89 (0.34)	3.87 (0.27)	12.66 (1.11)
High JD Low WSS Low DL (N=467)	2.30 (0.05)	3.31 (0.06)	8.78 (0.25)	5.21 (0.2)	10.88 (0.79)

KEY: JD = Job Demand, WSS = Work Social Support, DL = Decision Latitude

Total number of negative factors score

A total score for occupational factors was calculated. This measure was grouped into quartiles and these groups compared. Overall a linear pattern of effects was observed for the majority of factors (see Table 3). For the most part those in the highest quartile were found to report poorer health, for both physical and mental health and health-related behaviours. A linear pattern of effects was found for a measure of quality of health over the last 12 months. It was found that those in the top quartile reported a poorer health score compared to the other groups examined. Those in the upper quartile were also found to report a greater number of reoccurring health problems over a 12 month period, more symptoms over the 14 day period prior to completing the questionnaire and to have the least amount of sleep on average per night. The same linear pattern of effects was reported for measures of mental health, more specifically those in the upper quartile reported greater levels of both work and life stress and higher levels of anxiety and depression as measured by the Hospital Anxiety Depression Scale compared to the lower quartiles. However, a non-linear pattern of effects was reported for the General Health Questionnaire total score in which the third quartile reported the highest and therefore the worst health score. Those in the top quartile also reported a greater number of unhealthy behaviours including smoking a greater number of manufactured cigarettes per day and performing the least amount of moderate and vigorous exercise. However, this group were found to consume significantly less units of alcohol per week compared to the lower quartile groups, with the third quartile reporting the greatest consumption of alcohol.

Table 3. Effects of a total negative occupational factors score on measures of stress and health (scores are the means, s.e.s in parentheses; high scores = greater stress and more negative health outcomes)

Variable	VLOW (N=936)	LOW (N=913)	HIGH (N=847)	VHIGH (N=832)	F
12 month health	1.84 (0.04)	1.97 (0.04)	2.09 (0.04)	2.22 (0.04)	F (3,3483) = 35.63, p<0.0001
Stress at work	2.44 (0.04)	2.73 (0.04)	2.90 (0.04)	3.16 (0.04)	F (3,3485) = 94.06, p<0.0001
Anxiety	5.50 (0.18)	6.88 (0.18)	7.26 (0.19)	8.17 (0.19)	F (3,3448) = 71.64, p<0.0001
Depression	2.64 (0.14)	3.42 (0.14)	3.88 (0.15)	4.95 (0.15)	F (3,3405) = 84.86, p<0.0001
* units of alcohol per week	10.64 (0.59)	10.74 (0.59)	12.16 (0.60)	7.19 (0.59)	F (3,2392) = 2.85, p<0.05

* = outcome measure for which a linear pattern of effects was not found

Conclusions

Analyses of the noise and night-work data showed that noise had a negative effect, night-work had fewer effects and that there was only slight evidence for the combined effects of noise and night-work to be greater than noise alone. A similar pattern of main effects but lack of interactions was observed in the analyses of the other variables from the physical environment and working hours categories. The results of the individual factors confirm effects previously reported in the literature. The greater effects of the physical environment variables may reflect the fact that these are based on subjective perceptions rather than objective measurement (working hours may, in contrast, be more objective). Further research is now required examining associations between objectively measured combinations of variables to determine whether a similar profile of results is observed.

Analyses of the Karasek scores showed largely independent effects of demand, control, and social support. The lack of interactions has been noted in other studies although the present findings do not preclude the possibility of such effects. Linear relationships were found between the number of negative factors in the workplace and reported health. This implies that the best method of predicting the combined effects is to add up the number of negative features reported.

For the total negative occupational factors score a linear pattern of effects was seen for the majority of health measures examined. A similar profile was obtained when change scores were used in the analyses. Although it is difficult to determine the causal mechanisms involved in such associations the strong dose response effect and the presence of the association in both cross-sectional and longitudinal analyses suggests that an intervention study reducing negative features of the workplace is justified.

The present analyses have shown a clear profile of the associations between combined effects of occupational factors and reported health. The next step must be to determine whether a similar pattern is observed when objective outcomes (e.g. accidents, sick leave, performance efficiency) are considered.

Acknowledgement

The research described in this article was supported by the Health & Safety Executive.

References

Goldberg, D. and Williams, P. 1988, *A user's guide to the General Health Questionnaire.* Windsor : NFER Nelson

Smith,A., Johal,S.S., Wadsworth,E., Davey Smith and Peters,T 2000a, The Scale of Occupational Health: the Bristol Stress and Health at Work Study. *HSE Books. Report 265/2000*

Smith, A., Brice, C., Collins, A., Matthews, V. and McNamara, R. 2000b, The scale of occupational stress: a further analysis of the impact of demographic factors and type of job. *HSE Contract Research   Report 311/2000. HSE Books*

Zigmond, A.S. and Snaith, R.P. 1983, *The Hospital Anxiety and Depression Scale HAD).* Acta Psychiatrica Scandinavia, 67, 361-370

COMBINED EFFECTS OF OCCUPATIONAL FACTORS ON OBJECTIVE MEASURES OF PERFORMANCE AND HEALTH

Andy Smith, Carolyn Brice, Stephanie Sivell, Emma Wadsworth
and Ben Wellens

Centre for Occupational and Health Psychology
Cardiff University
63 Park Place
Cardiff, CF10 3AS
Smithap@cardiff.ac.uk

The aim of the research reported here was to examine the effects of combinations of occupational factors on objective measures of performance and health. A secondary analysis of the Bristol Stress and Health at Work data base showed that accident rates were greater when combinations of noise and nightwork were present than when the factors were considered in isolation. Indeed, accidents increased linearly with the number of negative occupational factors reported. Analyses of performance in the laboratory also showed that combinations of factors were associated with greater impairments. Further research using objective indicators is being conducted and these methodolgies are briefly described.

Background

Our previous analyses of the combined effects of occupational factors have been based on subjective reports. It is essential to determine whether a similar or different profile is observed when objective indicators of health and safety are considered. This was initially done by conducting a secondary analysis of the Bristol Stress and Health at Work database (Smith et al., 2000). In this study measures of accidents at work were recorded from the sample who repeated the questionnaire 12 months after the original one (N = 1854). A small cohort study was also conducted (N = 188) and this involved performance of laboratory tasks known to be sensitive to occupational factors (e.g. sustained attention tasks; the stroop colour-word task – a measure of resistance to distraction).

Accidents at work

Three types of analyses were conducted. In the first, combinations of factors reflecting the physical environment at work (e.g. perceived noise) and working hours (e.g. nightwork) were considered. In the second set of analyses combinations of job demand, decision latitude and social support at work (the Karasek job strain dimensions) were examined. In the final analyses a measure was derived which represented the total number of negative factors present at work. The job strain analysis showed no significant effects of any of the dimensions on accident rate. In contrast to this the analysis of the noise and nightwork groups showed that those who were exposed to both had a higher accident rate than those exposed to only a single factor (see Table 1). Similarly, the analysis of the number of negative factors showed that accidents increased linearly with the number of negative features reported (see Table 2). These results should, of course, be treated with caution due to the low incidence of accidents.

Table 1. Effects of noise and nightwork on accidents at work
(mean number in last 12 months, s.e.s. in parentheses

Both noise and night-work	Night-work only	Noise	Neither
0.11	0.04	0.04	0.04
(0.03)	(0.01)	(0.01)	(0.01)
N= 140	282	341	1065

Noise: p = 0.0001
Nightwork: p = 0.004
Noise x nightwork: p = 0.02

Table 2. Number of negative factors present and accidents at work
(mean number in last 12 months, s.e.s. in parentheses)

	Number of negative factors		
Very low	Low	High	Very high
0.02	0.03	0.05	0.06
(0.01)	(0.01)	(0.01)	(0.01)
N= 491	461	435	439

$F (3,1822) = 3.88$ p <0.01

The cohort study

A small cohort visited the laboratory and carried out three performance tasks known to be sensitive to occupational factors (the repeated digits sustained attention task; the stroop colour-word task; and a variable fore-period simple reaction time task). Identical analyses to those described in the previous section were conducted.

In the analysis of noise and nightwork it was again found that those exposed to the combination of factors had the most impaired performance in terms of speed of response to targets occurring at unknown times (see Table 3). The analyses of the number of negative factors also showed that those reporting the highest number of negative features at work were the most susceptible to disruption (see Table 4). Performance of the simple reaction time task was not significantly effected by either single or combined factors. These results demonstrate a number of important points. First, the volunteers were tested outside of working hours and the results show that the nature of a person's job has an impact even after the person is no longer in the workplace. Secondly, tasks shown to be sensitive to laboratory manipulations of stressors are also useful in studying the longer term effects of the nature of the workplace. Finally, effects are selective and not observed in all tasks. This contrasts with the view obtained with subjective reports where a more global profile is observed. It should, of course, be noted that the present results are based on a small sample and further replication and extension is needed. However, it does appear that assessing performance outside of working hours can be a useful way of investigating combined effects and our ongoing studies using this approach are briefly outlined in a subsequent section.

Table 3. Effects of noise and night-work on reaction times in a sustained attention task (scores are the mean rts in msecs, s.e.s. in parentheses)

Both noise and nightwork	Nightwork only	Noise only	Neither
605	542	567	577
(25)	(14)	(12)	(10)
N= 24	26	47	85

Noise x nightwork: F (1,178) = 5.57 p<0.05

Table 4. Number of negative factors present and stroop interference (scores are the means in seconds, s.e.s. in parentheses)

Very low	Low	High	Very high
81.3	81.4	82.6	91.9
(3.4)	(2.7)	(2.2)	(2.6)
N= 43	52	45	51

F (3,187) = 4.12 p<0.01

Ongoing Studies

Having conducted the reanalysis of the Bristol database, further studies were commenced to look at performance measures, physiological correlates of stress and accident data. This was done in two on-going long-term studies. The first investigative study was carried out to increase the information currently available on the impact of combined effects of occupational hazards / workplace factors on health and safety at work, as signified by attendance of the Emergency Unit at Cardiff University Hospital with work-related injuries. The second study was carried out to examine how the demands of the normal working day impact upon cognitive performance and mood. One methodology that has been applied when assessing performance efficiency is to take a measure of mood and performance effects at the beginning and at the end of the working day - the difference between these two measures has been shown to be related to the workload (it provides a measure of how much the day has taken out of you, Broadbent, 1979). The aim of the current investigation was to compare both between, and within, individuals to provide information about the impact of combinations of workplace factors on performance efficiency and levels of the stress hormone cortisol. Details of the two studies are given below.

An investigation of the impact of combinations of workplace factors on work-related accidents

Data collection at the Accident and Emergency Unit
During first contact with the triage nurse at the Cardiff Accident and Emergency Unit, patients were informed about the study and asked whether they would be prepared to participate in the investigation. Only patients who were in full or part time paid employment were selected for the study. An initial short questionnaire consisted of 12 items that offer a basic picture of their recent experiences at work. Any participants willing to take part in further studies were then sent a mail shot consisting of 3 questionnaires:
1 .A short (5-page) booklet profiling events and work characteristics 24 hours prior to the accident. This data provided information about the acute events that preceded attendance of the Emergency Unit, recorded at the time of the accident.
2. A more detailed questionnaire assessing the combinations of factors the person was exposed to at work (including physical agents, working hours, job demands and work social support). Also included in this booklet were questions providing information on the moderators and mediators of the stress / health relationship and potential non-workplace confounders such as demographic characteristics, health-related behaviours and stress outside of the workplace. This questionnaire provided a retrospective account of the work place factors and other life factors preceding the accident that could be compared to the acute accounts collected previously.
3. A second of these questionnaires was sent to each volunteer also. This was to be completed by a colleague employed in a similar role to them at their place of work, who had not attended the Emergency Unit in the previous six months. If a suitable colleague was not found or was not willing to complete the questionnaires then they were asked to return the

questionnaire blank. These questionnaires were completed at home and returned to the Centre in the freepost envelopes provided.

A retrospective postal survey

A retrospective study of work-related accidents in the general population was also initiated. This retrospective study sampled 1000 individuals who were involved in work-related accidents and 1000 individuals who attended the Emergency Unit with non-work related injuries during the last six months. The sample was identified from the records of the Emergency Unit. A series of questionnaires were sent out, analogous to the set sent to participants recruited whilst at the Emergency Unit. Those individuals with work-related injuries were sent the questionnaire for a colleague who was doing a similar job to them at the time of the accident, but those who attended with non-work injuries were not. All posted questionnaires were accompanied by a covering letter explaining the purpose of the study, that participation was completely voluntary and that they were under no obligation to complete the questionnaire(s), and that all information obtained was confidential .

An investigation of the impact of combinations of workplace factors on mood and performance efficiency and levels of cortisol

This investigation involved volunteers carrying out a series of computerised performance tasks before starting work and after finishing work. Each volunteer carried out this procedure on two occasions at the beginning and end of the working week. Characteristics of their working day throughout the week were recorded. Questionnaires providing information as to specific features of their job (e.g. exposure to physical agents, working hours, job demands and social support at work), their current physical and mental health status, demographic characteristics, health-related behaviours and stress outside the workplace were also administered. On each test day mood was recorded before and after starting work as were objective measures of performance, blood pressure and salivary cortisol. Male and female volunteers of various occupations and backgrounds were recruited by opportunistic sampling – from a number of sources, from the staff at Cardiff University and the general public via posters, newspaper advertisements and volunteers from previous studies.

1. Familiarisation. During this session volunteers were asked to read an information sheet providing details of the study and to sign a consent form. This familiarisation session also involved completing a battery of computer tasks (mood, simple reaction time, focused attention and categoric search) to ensure that the volunteer had a full understanding of each of the measures. Full-length test versions were used for familiarisation. At this session volunteers were also familiarised with other measures that were used during the study (e.g. measurement of blood pressure and providing a saliva sample).Volunteers were informed that they needed to attend the Centre on a further two days. The first of these days was at the beginning of the working week and the second was on the final day of the working week. On each of these days volunteers were tested before and after work at a pre-arranged time, although leaving enough time to get to or from work comfortably. On completion of the familiarisation session volunteers were provided with a series of questionnaires that assessed a more detailed profile of stress, physical and mental health and were asked to return the completed booklets at a test session. Volunteers were also provided with a set of salivettes, labelled with the day and time for collection of each saliva sample together with a set of instructions reminding them of how the saliva should be collected and stored.

2. *The first test session.* On arrival at the Centre volunteers were asked a series of questions to ensure that they adhered to the test day requirements. The test day restrictions were as follows: volunteers must have been healthy for a period of seven consecutive days prior to the test session, taken no medication for 24 hours prior to each test session, consumed a maximum of 4 units of alcohol on the evening before each test day, abstained from alcohol and performance of strenuous exercise on each test day, and abstained from consumption of caffeine and smoking for two hours prior to each test session. If the volunteer met these criteria then they were shown to the test room and rested for a few minutes prior to the test session beginning. Volunteers were asked about their general health and of any medications they were taking. Then they completed a questionnaire detailing sleeping and eating practices over the previous 24 hours. Blood pressure was then recorded (systolic, diastolic and pulse rate) using the Omron automatic blood pressure machine set to an inflation of 170. On completion of these tasks participants completed the battery of performance tasks, identical to that practiced at the familiarisation session measuring various aspects of performance (simple reaction time, focused attention, categoric search and mood) known to be sensitive to changes in state.

Visit 3: End of the working week. The procedure for the test sessions on the final day of the working week was identical to that outlined for the beginning of the working week. Additional measures taken during this study were as follows: on each day of the working week volunteers completed a questionnaire assessing characteristics of their working day in terms of exposure to physical agents, working hours, demands etc. Saliva samples were also collected via salivettes to measure salivary cortisol. These cortisol saliva samples were recorded at the following times, (1) Before breakfast on the first day of the working week, (2) Just before bedtime on the first day of the working week, (3) Before breakfast on the last day of the working week and (4) Just before bedtime on the last day of the working week.

Overall, these new studies will extend our knowledge of the combined effects of occupational factors and it will be of major interest to see whether the results confirm the associations with accidents and performance efficiency that were seen in the Bristol data.

Acknowledgement

The research described in this article was supported by the Health & Safety Executive.

References

Broadbent,D.E. 1979, Is a fatigue test possible ? *Ergonomics*, **22**, 1277-90

Smith,A., Johal,S.S., Wadsworth,E., Davey Smith and Peters,T (2000a). The Scale of Occupational Health: the Bristol Stress and Health at Work Study. *HSE Books. Report 265/2000*

ALARMS AND WARNINGS

THE DESIGN OF ALERTS FOR FUTURE PLATFORM MANAGEMENT SYSTEMS

Michael A. Tainsh, F.Erg.S., Eur.Erg.

Centre for Human Sciences, Defence Evaluation and Research Agency,
Farnborough, Hants GU14 0LX, UK

There is a requirement for improved user performance with warnings and alarms within the Platform Management Systems (PMSs) on current ships. A programme has been conducted for MoD/Ship Support Agency to investigate the optimal design characteristics of all alerts for future PMS. The past approach to alarms and warnings was considered 'equipment centred' while the approach taken here is 'user centred'. The requirements for future major Royal Navy platforms were considered: the likely functionality of their PMS workstations, the means of on-line control, and concepts for the design of the human-computer interaction. Hypotheses were formulated on potentially desirable categories of information to be associated with alerts. These categories are identified as reference (including function, location and organisation), significance (covering redundancy, timeliness and hazard) and statements of quantified information including encyclopaedic material. Trials results with Subject Matter Experts (SMEs) supported the hypotheses.

Introduction

The control of a ship and its services is carried out within the Platform Management System (PMS). There is sound evidence that there are problems within current Royal Navy (RN) PMSs. The status information, warnings and alarms are displayed, acknowledged and controlled in ways that do not match user characteristics. In fact, the general design concept that appears to have been applied in many cases seems to be more associated with the design of electromechanical dials. In this case, the physical measure from a sensor (status) is available plus additional information on whether it is slightly different from a desired range (a warning) or substantially different from that range (an alarm). This is an 'equipment centred' approach to displaying information and takes little account of the users' characteristics. In the past, this may have been the only information that could have been provided in non-computer based systems. However, in current and future systems this can be judged a very limited approach.

The Defence Evaluation and Research Agency (DERA) Centre for Human Sciences (CHS) was approached by the MoD/Ship Support Agency (SSA) to study the handling of alerts in future RN ships to support procurement programmes in line with current practice (References 1, 2 and 3). It was agreed that the study should employ a number of techniques, including:

(a) Top-down analysis of known assumptions or system characteristics in order to derive likely human factors issues;

(b) Observations of lessons learned in the past on similar projects and general concepts of best practice;

(c) Exercises in early-prototyping, trials and the use of Subject Matter Experts (SMEs).

In this case, all three were used to investigate design options and mitigate risks. The work described here addresses the issues associated with future ship PMSs and aimed for entirely novel design solutions.

Definition

The initial high level issue stems from the established set of categories for describing alerting information: status, warning and alarm. In this study, a provisional definition of an alert was adopted: the mechanism by which a user is informed of a change of information. The established set was seen as 'equipment centred' based on what electro-mechanical technology could offer. This definition of alerts is both arbitrary and general. An alert is defined in terms of the mechanism for informing the user about a change in information. However, it was believed that this definition would ensure that there would be a broad opportunity for a wide range of possible design solutions to arise from the investigation. It was hoped that this would be achieved by adopting the general definition that makes no *a priori* distinctions between the categories of alerting information that might be provided to a user.

Aim of the Investigation

The aim was to investigate the categories of information (a taxonomy) that users would prefer to be associated with alerts, in distinction to the set that appear to be assumed in all cases: status, warning and alarm.

Study Approach and Findings

The study proceeded in five stages:

(a) Consideration was given to the set of possible RN platforms currently under investigation for future procurement, and their operations. The main concepts to be taken into account here include the consideration of 'Whole Life Costs' which has substantial implications for the introduction of automation and the need for greater integration of all ship systems.

(b) A description was obtained of the possible functionality of a future PMS workstation including the equipment and plant that was likely to be associated with it, and the user organisation. It is envisaged that jobs/roles and tasks may be radically changed as new possibilities for automation are introduced. These will have profound consequences for work organisation, complementing and the layouts of the ships. Improved communications mean that the internal arrangement of spaces is likely to be quite different in the future from any ships in current operation.

(c) An understanding was gained of the alternative means of on-line control of the plant, equipment and services likely to be available to the PMS, including automation and support. The control of the PMS is likely to depend on more sensors and more processed information. In particular, the use of automation to reduce the need for the user to execute manual control functions was seen as central to the consideration of the users' roles.

(d) A concept of the Human-Computer Interaction (HCI) for a future PMS was developed. The implications for the HCI are profound. There may be a central workstation but equally the user may act from a remote position.

(e) The generation of implications of the HCI concept for the provision of alerts. It was inferred that the scope of alert information should cover three main categories:

(i) Reference information. This covers the functions carried out by the equipment and plant, the organisation of the ship's company and in particular the team associated with the PMS, and the spatial layout of the ship and the plant and equipment within it.

(ii) Significance. Priorities must be indicated associated with hazards, availability of equipment and plant (redundancy) and the temporal factors associated with a failure or other undesirable state.

(iii) Statements of quantified information. This was considered to include information associated with the state of the equipment and plant, history including trends and allied encyclopaedic information.

Implications of the Study for Trials

The findings from the study led to the hypothesis that future users may wish to have alert information that covered this set of categories. The questions then are:

(i) Precisely what information might the users want associated with each category;

(ii) how may they wish to have this information presented to them.

The emphasis in the trial was intended to focus on the first of these questions.

Trial Procedures

A scenario for equipment and plant design and configuration and operation was constructed based on possible options for a new aircraft carrier. This was chosen because work is still in its very early phases and none of the Subject Matter Experts (SMEs) involved in the trials would have any well-formulated views about existing or possible design solutions.

Three sets of trial conditions were constructed with ten SMEs (two were used to pilot the techniques with the remaining eight used to generate the results). The starting point was a set of thirteen tasks that could be taken as representative of a day in the life of the Marine Engineering Officers of the Watch and Platform Engineering Officers. The former group is typically highly experienced Chief Petty Officers while the latter are typically senior officers.

The tasks were:
- Platform Engineering Officer of the Watch (PEOOW) comes on duty
- Automatic plant reconfiguration
- Galley fire
- PEOOW watch handover
- Planning of maintenance routine
- Go to State One
- Go to flying stations
- PMS automatically reconfigures to match load requirements
- Failure of weapons lift
- Failure of gas turbine
- Go to action stations/readiness state
- Fire in gas turbine compartments
- Failure in heating, ventilation and air conditioning

Three sets of trial conditions were employed. The first used all thirteen tasks with relatively unstructured displays that lacked detail. The second used only one task with a more structured display based on conventional designs and rather more detail. The third set of conditions used two tasks that were supported by display material based on possible future design solutions with detailed alerting information. There was an increasing degree of structure and detail in the display material from the first to the third. This was intended to provide the greatest possible opportunity for eliciting preferences from the SMEs on their design requirements that would not be unduly influenced by the hypothesis, and the presence of the trials team. The objective in each of the conditions was to discuss the users' information/alert requirements without unduly influencing the outcome of the dialogue. Examples of the trials material are given in Figures 1 and 2.

The trials were conducted within the Human Factors Laboratory at DERA/CHS, Portsdown. Each SME was interviewed separately.

Results

The conversations with the SMEs were analysed for content. They provided evidence of wide range of preferences. In particular, the SMEs requested the supplementary information that could be provided through the addition of reference information on plant and equipment availability. The SMEs were particularly keen to have information on the significance of an alert. This included summary information on the availability of plant and equipment and the provision of information on the priority of the alert in terms of the time required before it had to be actioned, and hazard considerations. This result alone implies a set of display and control designs that could enhance user performance and it is believed it could improve the safety of the system.

Figure 1: Example display from first set of trials

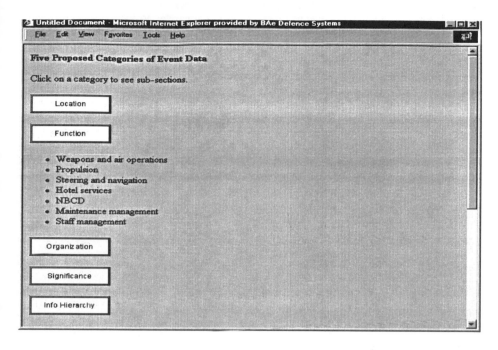

Figure 2: Example Display from third set of trials

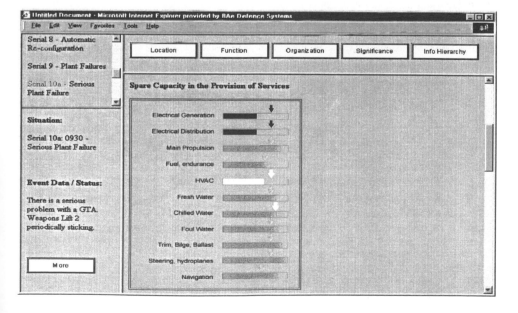

SMEs were interested in having referential information on function, location and organisation. Similarly they wanted quantified information from all sources and other encyclopaedic information provided in a well-organised hierarchy. In general, the taxonomy developed from the top-down analysis was supported.

The user centred approach based on a combination of top-down analysis and user trials has resulted in a concept for alerts that is both novel and in line with SMEs' preferences for future ships. This supports the definition of an alert that was stated earlier. It is more general than the current usage involving alarms and warnings. This study suggests that a new categorisation of alert information, starting from the trials results found here, could usefully be developed to enhance design studies for future PMS.

Conclusions

This investigation has shown that a top-down approach to the design of alerts in combination with trials can point the way to novel design solutions for the provision of information/alerts to users in operational systems. This appears as highly beneficial from the point of view of superseding established 'equipment centred' techniques associated with the provision of status, warning and alarm information and indicates a 'user centred' way forward for meeting the HF requirements for the PMSs of future ships. Additional studies based on the provision of new automation derived from these concepts have already lead to the conclusion that such novel designs could help reduce manpower requirements while maintaining safety levels.

References

1. Chief of Defence Procurement Instructions for Project Management CDPI/TECH/330 Issue 1.0 (Jan 1998) Managing Human Factors Integration

2. Defence Procurement Management Guide DPMG/TECH/330 Issue 1.0 (June 1998) Managing Human Factors Integration

3. ISO 9241: Ergonomic Requirements for Office Work with Visual Display Terminals

Acknowledgements

I acknowledge the valuable contribution of BAESYSTEMS (David Carr, Brian Sherwood-Jones, Tom McClean and John McFarlane) and Mr R Bishop of MoD/SSA to this paper. This work represents the views of the author and should not be taken as representing the views of either MoD or DERA.

BETTER ALARM HANDLING
A PRACTICAL APPLICATION OF HUMAN FACTORS

John Wilkinson & Peter Mullins

HM Specialist Inspectors (Human Factors)
Health and Safety Executive, Human Factors Team
Hazardous Installations Directorate
St Annes House, Bootle, Merseyside L20 3RA
john.HID.wilkinson@hse.gsi.gov.uk
peter.mullins@hse.gsi.gov.uk

This paper provides a practical example of the way the Health and Safety Executive (HSE) is applying human factors in the chemical industry. Major hazard industries, including the chemical industry, place considerable reliance for the safe control of their processes on control room operators who have to respond appropriately to alarms. Past major accidents (one is described) have highlighted the importance of a well-designed and managed alarm system. This paper explains the background to the HSE's strategy on alarm handling by reference to key publications and work on alarm systems and alarm handling. It introduces the human factors approach to alarm handling - essentially a participatory ergonomics and user-centred design and review approach.

Introduction

While alarms and warnings are always likely to be important, the consequences of the failure of an alarm system in managing a major hazard process safely in our sector, the chemical industry, can be disastrous, affecting workers and the public alike. It is also important to remember that alarm systems need to be seen in their full context. For example the inherent safety of the design, appropriate redundancy and diversity of design, management of change, and the correct allocation of function between automatic systems and the operator(s) should also be considered.

We aim here to discuss our current work on:
- Learning the lessons from previous accidents, incidents and near-misses;
- Developing standards for the design and operation of alarm systems and for establishing the competency of designers, installers and operators;
- Providing advice and guidance for alarm system users and designers.
- Promoting the identification and assessment of continuing HF problems in alarm systems and their solutions;

HSE Alarms Strategy

HSE shifted to a more focused approach to alarm handling following the major explosion and fires at the Texaco oil refinery in Milford Haven, South Wales, in 1994. The current HSE strategy is derived from the Texaco investigation report recommendations (HSE 1997) as modified by more recent research and local project work. The following sections describe the development of the HSE alarms strategy.

Texaco refinery incident 1994

The 1994 explosion and fires at the Texaco Milford Haven refinery injured twenty-six people and caused damage of around £48 million and significant production loss. This incident triggered HSE's more focused interest in this issue. This example also provides the simplest way to describe typical alarm handling problems (note: a more detailed overview will be given in the accompanying presentation and is available from the authors on request - and see (HSE 1997)). Key factors that emerged from the HSE's investigation (HSE 1997) were:

- There were too many alarms and they were poorly prioritised
- The control room displays did not help the operators to understand what was happening.
- There had been inadequate training for dealing with a stressful and sustained plant upset.

In the last 11 minutes before the explosion the two operators had to recognise, acknowledge and act on 275 alarms, representing well over 10% of the total alarms configured in the system for that unit.

Key problems contributing to the incident included: alarm floods; too many standing alarms; control displays and alarms which did not aid operatives; no clear process overview to help diagnosis; alarms which presented faster than they could be responded to; 87% of the 2040 configured alarms were displayed as "high" priority, despite many being informative only; safety critical alarms not clearly distinguished.

The other key lesson was that the management of the alarm system cannot be successfully dealt with in isolation from the overall SMS context. In other words this is not just a technical issue and both users and designers/installers need to keep this in mind. The SMS failures identified after the Texaco incident included:

- deficiencies in the plant modification procedure
- an inadequate instrument maintenance system
- inadequate training and competence of operators
- a lack of clear guidance on managing unplanned events and on when to initiate emergency plant shutdown
- a lack of clear authority to initiate shutdown.

Ultimately, plant safety should not depend on an operator response to an alarm where the consequences of failure can lead to a major incident.. In industry the chances of an operator failing to act in such circumstances can be very much higher than you may think. For example, in two recent LPG releases from road tanker hose couplings during loading/unloading (HSE investigations - unpublished), the operators present failed to use the emergency stop buttons to cut off the LPG supply to the leak point. There were a variety of contributory reasons in each case (not least of which were the psychological and physical effects of being enveloped in a vapour cloud of unignited LPG!) but these incidents nevertheless show some real world effects of an emergency or upset situation on human reliability which a designer or risk assessor may not always take into account.

Reliance on people or automatic systems for safety-critical functions needs to be properly allocated at the design stage, and then assured in the same way that any other system - such as quality - is.

HSE Research

HSE subsequently commissioned some broad initial research (Bransby & Jenkinson 1999). The report concluded that the alarm system problems identified as a result of the Texaco investigation were widespread in the chemical industry and that those problems could be solved or prevented.

Hse subsequently ran a local project in the North East in1999. This arose out of the Bransby and Jenkinson report (Bransby and Jenkinson 1999). Its aim was to form a clearer HSE view about how well companies are equipped to deal with human factors in alarm handling and to establish a baseline for further work.

The main findings (taken from the HSE unpublished report) were:

- Reliance on a distributed control system (DCS) was closely linked to a proliferation of 'alarms' which were in fact mostly status indicators.
- HAZOP (Hazard and Operability) studies generally added extra alarms, again often not true alarms.
- Corporate standards are available but varied in quality - the lack of clear standards and benchmarks was confirmed, as was the lack of HF expertise.
- The sites visited had analysed alarm rates and spurious alarms but they had not established performance standards suitable for monitoring progress.
- Simulators were reported - and confirmed - as bringing significant benefits and savings in identifying and remedying some potential operator reliability problems.
- Overall the project showed that, for critical safety-related activity, a good risk assessment was the key starting point.

Amongst the issues arising from the project it was recognised that there were re-engineering difficulties on existing plants (the 'I wouldn't start from here' syndrome) and associated inspection difficulties for HSE. Put simply, implementing remedial action on existing systems required what was in effect the carrying out of the missing design work on the alarm systems after the event ie the detailed analysis, categorisation, justification and recording of all the alarms on the system - not an easy task and very resource intensive.

Setting the standard - the EEMUA Guide

Following the publication of the initial research (Bransby and Jenkinson 1999) and the HSE local project, this was used as the starting point and basis for guidance produced by The Engineering Equipment and Materials Users Association - EEMUA - (EEMUA 1999). HSE regards the guide as being the nearest thing to a standard currently available (no doubt one will emerge from Europe in the future - in the meantime a British Standard based on the EEMUA Guide is being contemplated). The guide was written primarily for engineers and engineering managers. Its purpose is 'to stimulate discussion and encourage industry to develop its principles to meet specific safety applications'. The Guide provides generic all-industry technical and practical guidance. Human factors is the key element which drives and links the guidance. The Guide was written to be goal-setting, and sets some tough (but achievable) performance targets for designers to aim at.

Enter the Human Factors Team

The human factors team was formed in September 1999 and consisted originally of the authors - both experienced field inspectors - led by Dr Debbie Lucas, Principal Pyschologist, who has considerable practical human factors experience both within and outside HSE. Our main aim is to promote and develop the human factors approach to drive continuous improvement in health and safety in the onshore chemical industry. This includes:

- Developing an internal training program aimed at establishing and assuring the necessary competence in our inspectors.
- Providing hands-on inspection and investigation support to the field inspectors on key human factor topics.
- Developing appropriate targeted guidance on key topics.
- Progressing a targeted research program.

In the past human factors development has sometimes been hampered by a research-led focus which has not always delivered a practical product or tool to the user or designer. With this in mind, and following the results of the HSE research and local project, and other user needs communicated to us via the Chemical Industry Forum (a tripartite consultative body consisting of HSE, employers and employees' representatives), we set out to produce simple, free, easily-communicated, and practical guidance for inspectors, employers and employees on alarm handling, providing a link between the research, higher-level more detailed guidance, and the users. The revised HSG 48 (HSE 1999) had, in the interim, been produced (led by Debbie), so generic all-industry guidance on human factors was now available to support this.

'Better Alarm Handling'

The 'Better Alarm Handling' free information sheet promotes a continuous improvement appropoach - there is no simple 'one-off' solution. The guidance sheet was based around the EEMUA guide. Matthew Bransby (sadly, only shortly before his untimely death), who was closely involved in producing the EEMUA guide, commented in detail on the information sheet. A simple human factors - and participatory ergonomics - approach is taken.

The guidance provides a simple 3-stage approach, based on that taken in HSG 48: find out if you have a problem; decide what to do and take action; manage and check what you have done (note: Rather than describe the information sheet in detail here, interested readers are referred to our web site (see References - HSE 2000) where it is freely available).

The results so far, and the early feedback, are very encouraging. We have distributed in excess of 10,000 copies of the information sheet in around 6 months by dint of careful targeting, persistence, conference and seminar presentations, articles, and good advertising and marketing support from our in-house publishers. One of the most pleasing responses has been from actual control room operators who often immediately recognise the situations and problems described.

Setting the standard - competency for designers and users

In addition to the Guide, EEMUA have recently appointed a training contractor to provide training for key persons involved in the use and design of alarm systems. The training (now a 2-day course) is steadily growing in terms of take-up and provides a mix of theory and practical work designed to equip attendees with the information, awareness and skills necessary to tackle alarm system issues. The course is firmly based on the EEMUA Guide and incorporates the 'Better Alarm handling' information sheet (see below).

A competency framework, the development of which was commissioned by the HSE, has also now been launched by the Institute of Electrical Engineers and the British Computer Society (IEE 2000) to provide a level playing field on competency for safety-related system practitioners and other specialists involved with alarm system design, modification and maintenance.

Finally, the human factors team will be carrying out formal training as part of a briefing pack approach to alarm handling for inspectors eg the packs will include suitable question sets with appropriate benchmarks and standards to judge the answers against. There will also be continuing hands-on inspection by the team to promote confidence and to facilitate wider 'cascade' practical training by inspectors in turn. We are also working on developing our understanding of the role of the simulator in operator training.

HSE Alarms Strategy - Inspection and Enforcement

So, in summary, following the development process described above, HSE's current strategy on alarm handling in the chemical industry sector is to: raise awareness and provide information; to promote consideration of the issue within the whole 'life cycle' of the system, and a user-centred approach to design and review; and to promote consideration of the key role of the operator in the system. HSE will consider enforcement where either no fundamental review has been done and/or safety critical alarms have not been identified; and where there is high reliance on operator to react or respond to alarms to prevent a major accident

The profile of this issue is now also being steadily raised through inspection - having successfully raised industry expectations we now need to make sure that these are met when inspectors call. Most recently, for example, Peter and I have carried out detailed inspections at two major oil refineries where significant and widespread problems were found. Leading by example in this way facilitates the growth of site inspector confidence and competence together with planned and more formal training.

General advice for users and designers

HSE expects to see:
- A policy that recognises human factors in alarm handling as a management issue.
- A logical process in train which has assessed/is assessing the current situation.
- A sensible action programme to deal with issues found.
- For COMAH a rigorous demonstration that HF has been addressed adequately where operator response to alarms is claimed as defence against major accidents.

Users should evaluate, prioritise and modify existing alarm systems, taking account of the degree of risk to target their efforts. They should ensure new designs meet EEMUA standards and take into account human limitations. The alarm system should be managed as an integral part of the SMS, and as part of a continuous improvement programme.

Remember also that, no matter how well designed, no alarm system can operate effectively if the workloading and staffing levels do not take account of all foreseeable conditions (from normal through upset, shutdown and start-up, to emergency) and if operators are not competent or if their needs have not been considered in the new design or modification. Companies also need to consider shift lengths and patterns, and fatigue factors - otherwise there may be no response when one is most needed.

HSE expects designers to follow the EEMUA Guide principles with the SMS/Safety Report context also being raised and considered as part of the overall solution for a proposed new or modified alarm system. Remember that the needs of installers and commissioning engineers will not be the same as those of the final system practitioners eg designers, installers, maintainers.

Summary

Dealing with alarm systems is not just about technical specification and engineering - though this is clearly important too - it is part of the overall SMS, a continuous improvement culture and good change management. The role of the operator must be considered throughout the whole life cycle. Human factors is not like a coat of paint which can be added later in the design process. - but neither is it rocket science.

HSE considers alarm handling to be a continuing major safety issue. There is no room for complacency. Despite previous major accidents in which poor alarm handling played a key role, incidents are still occurring involving alarm systems and there are still significant problems with alarms systems on many major hazard sites. Training, competency and user support are still key areas and users and designers need to become better aware of each others' requirements. However, solutions are available by original design or by modification and practical guidance is also available. In general terms HSE expects companies to review their existing alarm systems from a usability and reliability viewpoint and to seek continuous improvement.

The future

In addition to our developing training program, ongoing inspection, and publicity via seminars, conferences and articles, we are aware that the Chemical Industry Forum (CIA) is currently producing its own guidance for members (pitched at an intermediate level between the EEMUA Guide and 'Better Alarm Handling'.

We are also now looking closely at the vexed problem of ensuring that users needs are properly spelt out in user specifications and that, contractually, these survive the often complex - and often multi-handed - commissioning and installation process.

Technical solutions to some key areas of difficulty are also being developed - diagnostic packages are now available (or being developed) by several major suppliers and manufacturers - these will allow users to accurately identify priorities from eg alarm flood data and to tackle these effectively and efficiently.

References

Bransby and Jenkinson 1999, *The management of alarm systems* (HSE Books), HSE Contract Research Report 166/1998 Bransby Automation Ltd & Tekton Engineering
EEMUA 1999, *Alarm Systems, a guide to design, management and procurement* (The Engineering Equipment and Materials Users Association), EEMUA Pubn No191
HSE 1997, *The explosion and fires at the Texaco Refinery, Milford Haven, 24 July 1994: A report of the investigation by the Health and Safety Executive into (etc.)* (HSE Books),
HSE 1999, *Reducing error and influencing behaviour*, HSG48 (HSE Books)
HSE 2000, *Better Alarm Handling*, HSE Information Sheet, Chemical Sheet 6 (available from HSE Books and from the HSE's web site at www.hse.gov.uk/pubns/chis6.pdf)
IEE 2000, *Competency guidelines for safety-related system practitioners,* Institute of Electrical Engineers, 2000

HABITUATION EFFECTS IN VISUAL WARNINGS

Paula Thorley, Elizabeth Hellier and Judy Edworthy

Department of Psychology, University of Plymouth,
Drake Circus, Plymouth, Devon, PL4 8AA, UK
pthorley@plymouth.ac.uk

Research into visual warnings has found that their efficacy may be affected by variables such as signal word, colour, and the influence of others. The literature, however, has revealed little about what happens to warning perception and behaviour over time. The current investigation observed and recorded non-compliant behaviour of a target participant group who experienced repeated exposure to a visual warning where a hazard was not actually present. Two further signs were posted to attempt to dishabituate behaviour. The findings indicated effects of habituation and social influence. The implications of the findings and recommendations for further research are discussed.

Introduction

Although there is much recent research in the area of warnings, there is none that addresses the issue as to what happens to warning perception and compliant behaviour over time. Habituation to warnings is an area largely unexplored in the literature, and is the topic of this paper. Thorpe (1963, as cited in Petrinovich, 1973) defined habituation as "The relatively permanent waning of a response as a result of repeated stimulation which is not followed by any kind of reinforcement" (p.141). This lack of reinforcement could be an issue with visual warnings in particular, firstly because they may not be relevant all of the time, and secondly if the hazard does not become apparent then the warning sign is not reinforced. The memory theory of habituation (Sokolov, 1963, cited in Schwartz and Robbins, 1995) proposed that "...perceived stimuli are passed through a 'comparator' that checks to see if a representation of the event already exists in memory. If such a memory is present, the stimulus is afforded no further processing and no responding occurs" (p.50). Wagner (1976, cited in Schwartz and Robbins, 1995) furthered Sokolov's memory theory of habituation and suggested that knowledge is held in either short-term memory (STM) or long-term memory (LTM). If information is held in STM it is in an 'active' state and results in little or no response to a stimulus, whereas information held in LTM is in an 'inactive' state and the presence of a stimulus will provoke a response.

It was hypothesised that following repeated exposure to a visual warning without reinforcement, behavioural compliance would decrease as a function of time. Moreover, changing the warning to another that differed in appearance, but not in intensity, would dishabituate the existing response and increase compliant behaviour.

The investigation observed door use behaviour and the effects of the presence of a warning sign on that behaviour. The observation was conducted in a lecture theatre at the University of Plymouth. The venue was chosen for reasons that included, a) two sets of double doors that could be easily observed, b) labelling one set of doors 'defective' and asking participants to use the other set was indicative of a moderate cost condition, and c) the target participant group were scheduled to attend the lecture theatre three times per week, therefore experiencing repeated exposure to the stimulus.

Method

Participants
Potentially[*] 220 stage one psychology undergraduates at the University of Plymouth were observed for this part of the study. As they were being observed in an area where strangers would normally view them, there was no requirement for a briefing/debriefing session.

Materials
Three signs, A, B, and C, were created by the experimenter to depict two different levels of hazard; medium (signs A & B) and high (sign C). All signs were designed using CorelDRAW™ 8 and the choice of colours and signal words used were based on the perceived hazard word/colour trade-off proposed by Braun and Silver (1995). Signs A and B (medium level hazard) each comprised of a signal word, ('STOP' or 'CAUTION'), the hazard ('DEFECTIVE DOORS'), instructions as to what to do to avoid the hazard ('PLEASE USE OTHER DOORS') and an icon (an arrow) indicating the direction in which the alternative doors were located. The word 'STOP' was coloured red and the word 'CAUTION' coloured orange. Both were designed using 135-point Verdana font. The hazard and the instructions were designed in black using 60-point and 50-point Verdana font respectively. Both signs A and B were produced in landscape orientation using an inkjet printer on to white paper, 210mm x 297mm (A4). Sign C (high level hazard) also comprised of a signal word, ('DANGER') and the same hazard warning and instructions as signs A and B, but did *not* include an arrow indicating in which direction the alternative doors were located. The words were printed over an icon of a black exclamation mark surrounded in yellow, outlined by a black triangle (created in CorelDRAW™8). The word 'DANGER' was coloured red and was designed using 152-point Verdana font. The hazard and instructions were also coloured red, outlined in grey, and both were designed using 135-point Verdana font. Sign C was also produced in landscape orientation using an inkjet printer on to paper 210mm x 297mm, but was then enlarged by 141% using a colour photocopier.

All of the signs were laminated for both durability, and to be of a similar format to the other signs present around the university.

---

[*] Approximately 220 students were enrolled on the course at the time observations were recorded. However, it was presumed that a variable subset of students would attend each lecture.

Design and Procedure

An observational study was conducted over a period of eight sessions. Using event sampling techniques, door use behaviour was observed. To obtain a baseline figure, the initial recording session logged the number of participants using the target set of doors as both an entrance to and exit from the lecture theatre when there was no sign present. Sign A was then posted on the aforementioned doors both inside and outside the lecture theatre itself. Subsequent recording sessions logged all non-compliant door-use behaviour, and the number of people in attendance at the lecture was also recorded approximately five minutes after the lecture had begun.

The target behaviour was logged as either non-compliant alone or non-compliant in groups of two or more, in order to control for the effects of social influence (Wogalter *et al*, 1989).

When it appeared that non-compliant behaviour had consistently returned toward baseline after five recording sessions, sign A was replaced with sign B and the same behaviour logging procedure was employed. At the next session, sign B was replaced with sign C, again using the same observational and recording techniques. The doors were not locked at any point during the observation sessions.

Results

The frequencies of recorded non-compliant behaviour can be found illustrated in Figure 1.

An overall chi-square analysis of frequency revealed a significant difference across all conditions, $\chi^2 (7, n = 1160) = 168.38$, $p < 0.001$.

Session by session chi-square analysis comparing non-compliant and compliant behaviour is summarised in Table 1. Critical conditions were identified as being session 1 (no sign), session 2 (first showing of sign A), session 6 (last showing of sign A), session 7 (sign B) and session 8 (sign C).

Significant differences were revealed between all critical sessions; between 1 and 2, $\chi^2 (1, n = 403) = 47.43$, $p<.001$; between sessions 2 and 6, $\chi^2 (1, n = 289) = 61.46$, $p<.001$; between sessions 6 and 7, $\chi^2 (1, n = 226) = 19.1$, $p<.001$; and between sessions 7 and 8, $\chi^2 (1, n = 216) = 10.04$, $p<.01$.

Session by session chi-square analysis comparing compliance alone and compliance in groups revealed few significant differences between sessions.

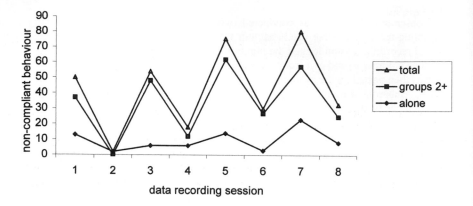

Figure 1. Frequency of non-compliant behaviour recorded over the eight sessions.
1= no sign condition, 2-6 = sign A, 7 = sign B, 8 = sign C

Table 1. Session by session Chi-square analysis for observed compliant and
non-compliant behaviour (df = 1)

Session	2	3	4	5	6	7	8
1	χ^2=47.4 n=403 *	χ^2=1.2 n=395	χ^2=.45 n=.503	χ^2=17.2 n=380 *	χ^2=2.1 n=302	χ^2=47.5 n=340 *	χ^2= .68 n=325
2	-	χ^2=59.2 n=382 *	χ^2=34.9 n=283 *	χ^2=101.7 n=367 *	χ^2=61.5 n=289 *	χ^2=151.3 n=327 *	χ^2=78.3 n=279 *
3	-	-	χ^2= 2.2 n=275	χ^2=9.1 n=359 **	χ^2=.28 n=281	χ^2=33.5 n=319 *	χ^2=2.9 n=271
4	-	-	-	χ^2=14.2 n=260 *	χ^2=3.1 n=182	χ^2=35.7 n=220 *	χ^2=7.3 n=172 **
5	-	-	-	-	χ^2=3.8 n=266	χ^2=8.8 n=304 **	χ^2=.55 n=256
6	-	-	-	-	-	χ^2=19.1 n=226 *	χ^2=1.1 n=178
7	-	-	-	-	-	-	χ^2=10.04 n=216 **

* p<.001   ** p<.01

Discussion

The results indicated that compliant behaviour did decrease as a function of time following repeated exposure to a visual warning without reinforcement, therefore supporting the hypothesis that participants would habituate to visual warnings. Changing the warning in appearance, but not intensity, may have had a dishabituating effect as there was a significant difference in behaviour between signs A and B. However, this was not in the expected direction. Instead of decreasing, non-compliant behaviour increased when the sign was changed from A to B. Compliant behaviour increased however following exposure to sign C, but was not statistically significant from either the last session of sign A, or the no sign condition. Compliance was significantly greater than for Sign B.
This effect is probably due to increasing the overall intensity of the warning according to available data on perceived urgency and arousal strength (e.g. Braun & Silver, 1995).

The findings support a number of the nominated characteristics of habituation (Thompson and Spencer, 1966, as cited in Petrinovich, 1973). Namely, if a stimulus elicits a response, repeated application of the stimulus decreases response strength and habituation of a response to a stimulus exhibits stimulus generalisation. The literature has not specified how the novel stimulus should be designed in order to dishabituate a response, and the results from the current research have not allowed assumptions to be made about novelty.

The observed occurrences of non-compliant behaviour, especially compliance alone frequencies, could be interpreted within the memory theory of habituation framework. For example, when sign A was initially posted, participants may have perceived the stimuli and processed it through a 'comparator'. If there was no representation of it in an active state in short-term memory (STM), a response may have been provoked, as any information about warning signs and broken doors would be held in an inactive state in long-term memory (LTM). The subsequent exposures may then have been encoded in an active state in STM and therefore afforded no further processing, resulting in a lowered response. This could also explain why compliant behaviour increased between sessions 5 and 6 for sign A. An intermediate lecture had been cancelled leaving a greater time between exposures, consequently reverting the information to an inactive state, therefore evoking a response. This is however a tentative suggestion as it is not evident from either the memory theory of habituation or this data for how long information is represented in STM.

Familiarity of the surroundings may also have affected compliance. Godfrey *et al.* (1983, cited in Hyde & Hellier, 1997) and Wogalter *et al.* (1995) reported that product familiarity affects behavioural compliance, and this could presumably be extended to include situational factors. The participants had attended a number of lectures by the time they were observed and this could possibly have prompted non-compliant behaviour, as could familiarity to the warning itself following repeated exposure. Again, further research would be recommended. It would be interesting to explore situational variables, comparing unfamiliar and familiar surroundings within controlled conditions, as well as the issue as to just how prominent and salient a warning would have to be to encourage behavioural compliance under such circumstances.

References

Braun, C.C, & Silver, N.C. (1995). The interaction of signal word and colour on warning labels: differences in perceived hazard and behavioural compliance. *Ergonomics, 38,* 2207-2220

Hyde, C., & Hellier, E. (1997). How do warning labels and product familiarity influence hazard perceptions? In *Contemporary Ergonomics*: Taylor & Francis

Petrinovich, L. (1973). A species-meaningful analysis of habituation. In H.V.S.Peeke & M.J. Herz (Eds), *Habituation I. Behavioural Studies.* London: Academic Press

Schwartz, B., & Robbins, S.J. (1995). *Psychology of learning and behaviour. Fourth Edition.* London: Norton

Wogalter, M.S., Allison, S.T., & McKenna, N.A. (1989). Effects of cost and social influence on warning compliance. *Human Factors, 31,* 133-140

Wogalter, M.S., Barlow, T., & Murphy, S.A. (1995). Compliance to owner's manual warnings: Influence of familiarity and the placement of a supplemental directive. *Ergonomics, 38,* 1081-1091

HCl

YOUNG PEOPLES' VIEWS AND PERCEPTIONS OF INFORMATION AND COMMUNICATIONS TECHNOLOGY

Martin Maguire

HUSAT Research Institute
Loughborough University
The Elms, Elms Grove
Loughborough, Leics LE11 1RG
m.c.maguire@lboro.ac.uk

A survey was carried among a sample of children and young people (between the ages of 5 and 17) on different issues relating to computers, IT and the Internet. The aim was to identify their views and perceptions of current technology and on new ideas for technology products. The survey seemed to show that while young members of society are very accepting of new technology, they are not over impressed by it and can see its limitations. The paper also identifies some differences in perspective between younger and older children from the survey

Introduction

Children represent the next generation of information and communications technology (ICT) users. There is a general feeling that young people have a natural affinity for computers and wish to spend as much time as possible using them. A survey was carried out among a sample of young users between the ages of 5 and 17 to elicit their understanding and views on different kinds of new technology such as computers, email, the Internet and mobile phones. This paper reports the results of the survey, which provides some insights into how children view technology, the practical problems they envisage with certain new developments, and the benefits and drawbacks of using ICT compared with traditional means of communication.

Sample description

The survey was distributed to a sample of 87 children, including pupils from a local primary school in Leicestershire, together with the sons and daughters of friends, relatives and colleagues. The age distribution was as follows:
- Infants (5 - 7 years old) - 48 responses
- Juniors (9 - 11 years old) - 18 responses
- Secondary (12 - 17 years old) - 21 responses

It was found that 80 (92%) of the 87 children had access to, and used, a computer at home. This indicates that the sample were reasonably well experienced with computer technology beyond school, putting them in a good position to think about and give their opinions on it.

Computer usage

Respondents having a computer at home were asked what activities they performed on them. The results are shown in Table 1, split by age group:

Table 1. Activities performed on computer at home (%)

	Infant	Junior	Secondary
Games	86%	89%	70%
Internet	19%	39%	55%
Email	2%	17%	20%
Educational/homework	14%	39%	85%
Writing	7%	22%	10%
Art	17%	22%	0%
Scanning/publishing	0%	0%	10%
Music/video/CDs	2%	0%	10%

It can be seen that while a wide range of activities are performed, playing games is an important activity for all groups! However for the junior and secondary groups, internet and email use becomes significant. Also as the demands of homework increase, so does the use of the home computer to support it.

Spending birthday money

The respondents were asked how they would prefer to spend £10 received for their birthday – either buying via their computer and the internet, or going to a real shop.

Table 2. Spending £10 birthday money

Ownership	Frequency	Percentage (%)
Buy from a real shop	63	73%
Buy online	23	27%
Total	86	100%

Interestingly, as Table 2 shows, the large majority stated that they preferred to spend their birthday money in a real shop. Analysis of the data showed that infants were the most accepting of online shopping (40%), while juniors were less accepting (17%) and secondary children were least accepting (5%).

The main reasons given for preferring a real shop were:
- Because it's got more choice.
- You can see what you are buying.
- There is no guarantee (with the internet) that you'll get the object ordered.
- Because I want to see how much it is.

- Because there would be no p&p and I won't need to wait for it to come.
- Because you can go with friends, look and try before you buy and it is nice to talk to real cashiers.
- Because we get tyanj (change) to buy ice-cream.

Reasons for preferring the internet were:
- It's cheaper and easy to find (things).
- Because its more interesting and fun.
- So you don't have to travel and rush.
- In shops you have to walk around and stay for a long time.
- It's less busy and "you won't get bashed".
- Internet, if I had a little help - but (would do) the rest on my own.

Interest in new TV-based activities

Respondents were asked: if they had a new TV at home, would they like to perform novel activities such (i) as watching new TV programmes from satellite, (ii) sending messages to friends, and (iii) buying things through the TV. Table 3 below, shows that watching new programmes and sending email were relatively popular. It was felt that satellite TV offered "new programmes and series, which you can't get on normal TV" e.g. cartoon network, music channels, films, and provide something interesting to watch at any time. Other reasons were: "because of the gossip at school" and "you see programmes sooner than on normal TV". However one 10 year old was concerned that the channels might be fuzzy, while an 11 year old thought that they might become a "couch potato". Several 5 to 7 years stated that they thought they had plenty of choice already. Reasons for sending messages via TV to friends included: the phone being engaged, it being cheaper than the phone, and easier than the computer which takes a while to set up. Buying items via the TV was much less popular and several respondents lacked confidence in it. One 10 year old felt that "it might never arrive, and you might have to leave a credit card number". Another thought that they "would probably make a mess of it". Several young people from 9 and upwards stated that they enjoyed going to the shops to buy items.

Table 3. Preference for specific TV-based activities

Activity	Yes	No
Watch new programmes from satellite TV	65 (75%)	18 (21%)
Send messages to friends	63 (72%)	21 (24%)
Buy things	36 (41%)	46 (53%)

Working with computers

Another area of the survey was working with computers. Respondents were asked if they would enjoy working with computers, for a lot of the time, as part of their future jobs. Table 4 shows that most children did not wish to do this. This was particularly true of primary school children (5 to 11) of whom only 35% wanted to spend much time working with computers. Reasons given were that they would be too busy doing their job (builder, astronaut, singer, model, vet, etc.) or that it was boring, noisy, slow, and might give them headaches and hurt their eyes. A higher proportion of secondary children (48%) stated that they would like or expect to use computers substantially in their future jobs. Reasons given were: "because it would make work easier", "I can do my designs on a computer" (fashion designer), and "I would need to as a games tester". However other older children felt that computers were unpredictable and frustrating. They could see how they would fit into their jobs, e.g. to record patients notes (as a doctor), or lap times (as a racing driver) but few wanted to use them just for the pleasure in doing so. As one student, who wanted to be a chemical physicist or natural scientist, stated: "I expect that I will need to use one, but I would like to have variety in my work".

Table 4. Spend a lot of time using computer in job?

Response	Frequency	Percentage (%)
Yes	33	38%
No	51	59%
Don't know	3	3%
Total	87	100%

Email versus letters and cards

Respondents were asked if they had sent or received email before. It was found that 57 (66%) had done so. The breakdown in terms of age group were: infants - 52%, juniors - 72%, and secondary - 90%.

They were then asked several questions about whether they would want to send computer messages (e.g. email) or traditional letters and cards for different situations: i.e. writing to a friend, sending a greeting to a grandparent on their birthday, receiving greetings on their own birthday. The results are shown in Table 5 below.

Table 5. Preferred means of writing to a friend

Response	Prefer letter/card	Prefer computer message/email
Writing to a friend	34 (39%)	47 (54%)
Greeting to grandparent on their birthday	83 (95%)	4 (5%)
Receiving greeting on own birthday	76 (87%)	10 (12%)

There was a clear preference for sending email to friends rather than letters. Further analysis showed that the infant group had a stronger preference for letters (48%) compared to junior and secondary (28%). Infants saw the process as easier and more flexible, allowing them to draw and colour pictures within their letters. Other children saw email as quicker and easier, although letters were more satisfying and pleasant. One 12 year old stated that "E-mail is good for a quick message but I'd still send birthday cards or postcards in the post".

A very high proportion of the respondents preferred to send a card on their grandmother's or grandfather's birthday. They thought that cards were nicer, more pleasant, personal and satisfying (as well as the fact some of their grandparents had no computer!). One 6 year old thought his grandparents would like to see his handwriting. Reasons given for preferring to send a computer message were: "I like what I can do on my computer" and "I think cards are nice to have but my grandpa is very up in the world of computers".

A similar picture is revealed when asked about receiving birthday cards. Most respondents preferred to receive cards than emails. Cards were seen as pleasurable ("I like pretty things - 5 year old respondent), traditional, and also containing money! Those who preferred computer messages thought they were modern and 'neat'. Interestingly though, one 11 year old thought that computer messages were more personal than cards.

Mobile versus standard telephones

It was found that a large majority (78%) had used a mobile telephone before - one child was not sure. Respondents were then asked if they would prefer to use a mobile telephone or ordinary (fixed) telephone. Table 6 below shows a clear preference for using a mobile rather than a fixed phone. Respondents liked the fact that they were portable, could be used anywhere, offered additional facilities (e.g. text messaging and 'snakes' game) and would be useful in an emergency e.g. "if your car broke down". However a significant minority still preferred standard telephones because they were thought to be cheaper and safer, with several respondents of junior and secondary age mentioning radiation hazards associated with mobile phones. Interestingly one 6 year old preferred an "ordinary phone because you can carry it round". (They appeared to regard mobile phones as now standard.)

Table 6. Preference for mobile or ordinary telephone

Response	Frequency	Percentage (%)
Prefer to use a mobile	53	62%
Prefer to use an ordinary phone	31	36%
Don't mind/No difference	2	2%
Total	86	100%

Response to innovative technology

Children's reactions to new technology ideas are of interest as possible indicators of the success of future products. The respondents were asked whether they liked the following set of new features and products related to the telephone such as a telephone in a wristwatch. seeing the other person you are talking to (a videophone), buying via a mobile phone and playing games on a telephone.

As shown in Table 7, the idea of a telephone in a wristwatch was quite well received, particularly among junior age children of whom 83% were in favour. They thought it 'brilliant', 'cool', 'wicked' and 'handy'. 76% of the secondary children liked the idea although only 65% of infants favoured it. Some of this latter group thought the phone would stop the watch going round or would "cut into your wrist".

The idea of a video phone was also popular although a greater proportion of infants (83%) and juniors (84%) favoured it compared to secondary school students (76%). Concerns over personal appearance over the phone (of themselves and the other person on the phone) was the main disadvantage for the older group. Contrasting comments from two 7 year olds were: "Yes because it is polite to look at someone when you are talking to them", and "No because you would be seeing your friend all the time and you would be fed up".

Playing games on a telephone was generally thought a good idea (especially for infants if it contained Pokemon, Shac, Snakes or Toy Story 2). However some older children thought it would be too small, unreliable or could be stolen.

Telephone purchasing was less appealing since you could not see the products well and they might not be in stock - although you could buy items on the move.

Table 7. Responses to the following ideas

Activity	Positive	Negative
Telephone built into wristwatch	62 (71%)	20 (23%)
Seeing person when telephoning them	73 (84%)	10 (11%)
Buying items via a mobile phone	40 (46%)	41 (47%)
Playing games on telephone	68 (78%)	14 (16%)

Electronic versus paper books

Finally the children were asked whether they would preferred to read a paper book or an electronic one via a computer screen. It was found that 63 (77%) preferred paper books, indicating that they will continue to be used in the future. This was especially true of the secondary children of whom 90% preferred paper over electronic books.

Conclusion

The survey showed that children and young people could see the cons as well as the pros of ICT. Most would not prefer to purchase over the net, would not want to spend long periods of time at work on a computer and still prefer paper books. However they were stimulated by new ideas such as playing games on the phone, videophones and telephone wrist watches. They liked email but did not see them replacing traditional letter writing or sending cards which were seen as more personal. These results are somewhat encouraging and show that children in the future will keep computers, IT and the Internet, well in perspective. It will not dominate their lives but they will be inspired by new ways of employing it.

Acknowledgement

The author would like to thank the many people who distributed questionnaires to sons, daughters and school classes on his behalf, and to the HUSAT support staff who helped compile the data for analysis.

OLDER ADULTS AND INTERNET TECHNOLOGY

Mary Sheard[1], Jan Noyes[1] and Tim Perfect[2]

[1] *University of Bristol, Department of Experimental Psychology,*
8, Woodland Road, Bristol, BS8 1TN, UK
[2] *University of Plymouth, Department of Psychology,*
Drake Circus, Plymouth, Devon, PL4 8AA, UK

Every year older adults are forming an increasingly larger proportion of the population of the Western world. Further, the developed world is becoming more heavily reliant on technology for basic everyday tasks. It is therefore important to examine older adults' use of the Internet in order to assess usability issues. Two versions of a questionnaire were posted on the Internet, differing in one design feature, response style, menus vs. radio buttons. Participants performance in filling in the questionnaire was compared for both age and response style. Effects of age on performance were found. It was also found that radio buttons have benefits over pull down menus, and should be used wherever possible in the design of interfaces for older adults.

Introduction

The Internet is the world's largest public access information database. It consists of millions of pages, theoretically accessible to all. In practice, however, many sites are not well designed and cannot be accessed easily and effectively by those who wish to use them. Various sectors of society can be expected to encounter difficulties; the elderly, novice computer users, the colour blind and those with disabilities to name but a few examples. The cultural stereotype, particularly of disabled and older people is that they do not wish to use the Internet, or are not capable of so-doing. This is not the case, but it is important that interface design takes into account their needs in order that they are not excluded.

This study compares computer, Internet and email use across people of different ages. Its' main aim, however, is to look at a design issue in form design. Forms are a part of web capability that allows a site owner to receive information from site users. The user may be asked to respond to a multiple-choice format, or to write answers or comments in a text box. They then click a button, which automatically sends their completed form to an email account. This is a new means of data collection for research, but one that is increasing in popularity and has a number of advantages (Pettit, 1999). Once the questionnaire has been programmed and advertised, data is delivered directly to the experimenters' email account in a form that can be copied and pasted into a spreadsheet. As such, costs of both stationery and working hours can be kept to a minimum.

In the study reported here a questionnaire about computer, Internet and email usage was posted on a website in two different form styles. The first, the menu condition, used pull-down menus and lists. Here the phrase 'select one' is visible in a box, and the user must click on a down arrow at one side of the box to reveal the list of responses, then highlight one or more using the mouse. This has the advantage of taking up little screen space, but list items are not always visible, and thus speculative exploration of options is more complicated. The second, radio, condition used radio buttons (round tick boxes) next to each option in a list for the respondent to highlight their chosen response. This has the advantage that the full list of options is visible on the screen without having to click on anything, but in the case of long lists this may result in the system user needing to scroll to move through a document. These two conditions may be compared to different menu styles in applications, where items are visible on the screen constantly, or are pulled down from a list heading at the top of the display.

Little research has been carried out to compare pull down and traditional menu types, with much work focussing on other issues in menu design, such as the optimum breadth and depth of menu structures (eg. Jacko and Salvendy, 1996), the optimum number of items per menu (eg. Boren *et al*, 1997) and conceptual layout in relation to cognitive organisation (eg. Coll *et al*, 1993). However, Carey *et al* (1996) compared pull down and traditional menu systems. These are direct equivalents of the response structures used in the questionnaire in this study, namely the menu and radio conditions. They found that overall, experienced users completed menu search tasks faster than novice users, regardless of the menu style used. The traditional menus elicited fewer errors than the pull down menus from both experienced and novice users, but there was no time difference for task completion between the two menu types.

It is expected that older adults will be less experienced with computers than younger adults. Further, there are implications from Carey *et al*'s work for this study. First, that the menu condition will prove more difficult than the radio condition as rated by a series of usability ratings. Second, that there will not be a difference in time taken to complete the task based on response style, but that there will be a difference in time taken in an age comparison, with older adults taking longer relative to younger adults.

This study provides a simple comparison of pull down menus and radio buttons, about which there is currently little research. It is not complicated by embedded menu levels, complex conceptual requirements, or ambiguous terminology or jargon. Part one of the questionnaire used pull down lists or radio buttons to offer responses to a series of multiple choice questions. Part two consisted of usability measures relating to part one, and a timer. The findings will have important implications with regard to the design of interfaces for use by older adults and novice computer users.

Method

Design

A between participants design was used, with participants being randomly allocated to one of two questionnaire conditions. Both questionnaires were identical, except in the way in which participants selected their responses. In the menu condition, pull down menu lists were used. In the radio condition, radio buttons or tick boxes next to each possible response were highlighted.

Participants

Participants were 237 Internet users, mostly recruited through advertisements on online bulletin boards, in particular at sites aimed at older adults, and also through an email mailing list. One hundred and twenty one of these responded to the menu condition questionnaire, and 116 to the radio questionnaire. Their ages ranged from teens to eighties. In the menu condition, 48 were aged over 61, 36 between 51 and 60 and 37 were aged 50 or under. In the radio condition, 54 were aged over 61, 25 between 51 and 60 and 37 were 50 and under. The majority were from the UK or Australasia, with a minority from elsewhere in Europe, or the USA.

The questionnaire

The questionnaire was located on one page of a web site. The first section consisted of questions about computer, Internet and email experience, and answers were given using either menus or radio buttons. The responses to this section constitute the experience measures. The second section included 10 questions on the usability of completing the first section, rated on a seven point likert type scale from strongly agree to strongly disagree, and a timer.

Procedure

On reaching the introductory web site for the questionnaire, participants clicked on a 'continue' button, which randomly sent them to one of the two questionnaire conditions. Although initially this was set up to give a 50:50 split, it was changed partway through to have a 2:1 bias towards the menus condition to balance the number of questionnaires received from each condition. Participants answered the questions by highlighting their response option using the mouse. At the end of the questionnaire was a timer, from which participants copied down the time they had taken to complete the form in a box. A small number of participants were unable to fill in this part, if their browser was not Java enabled. On clicking on the 'submit' button at the bottom of the questionnaire, the data was transmitted to the experimenter, provided the form had been completed correctly, ie. that none of the compulsory questions had been omitted. If questions had been omitted, a screen was displayed listing them, and the respondent requested to go back and complete those before submitting again. Thus only fully and correctly completed questionnaires were sent to the experimenter.

Results

Effect of age on computer experience

Seven of the eight measures of computer experience used here showed significant age effects; years of computer use ($F_2=9.51$, $P \leqslant 0.01$); frequency of computer use ($F_2=4.37$, $P \leqslant 0.05$); computer experience ($F_2=9.41$, $P \leqslant 0.01$); Internet experience ($F_2=9.03$, $P \leqslant 0.01$); email experience ($F_2=12.64$, $P \leqslant 0.01$); number of emails sent per week ($F_2=25.22$, $P \leqslant 0.01$); number of people in email contact with ($F_2=10.69$, $P \leqslant 0.01$). In all of these, with the exception of frequency of computer use, younger people are more experienced than older people.

Effect of experimental condition and participant age on performance
A multivariate ANOVA was used here. It gave only one significant effect of condition on performance, where the amount of scrolling required by the questionnaire was considered to be a more negative feature of the radio than of the menu format (F_1=5.97, P≤0.05). The remaining nine usability measures and the timer did not differ significantly between conditions.

A number of age effects were found: ease of response choice (F_2=5.65, P≤0.01); ease of reading the text (F_2=7.74, P≤0.01); ease of understanding the instructions (F_2=3.05, P≤0.05); enjoyment (F_2=17.73, P=≤0.01); clarity of layout (F_2=7.29, P≤0.01); reasonable amount of scrolling needed (F_2=3.42, P≤0.05); time taken (F_2=5.83, P≤0.01). In all of these with the exception of time taken, older adults' ratings were more positive (in the direction of ease, as opposed to difficulty, of use) than younger adults'. Older adults took significantly longer than younger adults to complete the questionnaire. No significant effects of the interaction between age and condition were found.

This analysis was repeated with the sample split by age and by condition. In the menu condition there were significant effects of age on ease of response choice (F_2=7.14, P≤0.01); ease of reading the text (F_2=3.39, P≤0.05); enjoyment (F_2=8.75, P≤0.01); clarity of layout (F_2=4.91, P≤0.01); reasonable amount of scrolling needed (F_2=3.39, P≤0.05). In the radio condition there were significant effects of age on ease of reading the text (F_2=4.80, P≤0.01); ease of understanding the instructions (F_2=3.54, P≤0.05); enjoyment (F_2=9.13, P≤0.01); time taken (F_2=4.22, P≤0.05). Again, in all of these, younger respondents' usability ratings were more negative than those of older adults. In the radio condition, the time taken was longer for older than for younger adults.

There were no significant condition effects for the young or middle age conditions. There was one significant effect of condition for the older adults group, reasonable amount of scrolling needed (F_1=4.32, P≤0.05), where the menu condition got more positive responses than the radio condition.

Discussion

The data collected constituted demographic features of the respondents, computer experience ratings and usability measures for the two questionnaire types. It was found that older adults are less experienced with computers than younger adults. Also that older adults took longer to complete the questionnaire than younger adults, although the older adults gave more positive usability ratings. In addition there is a distinct advantage of the radio condition over the menu condition.

Older adults were shown to have significantly less computer, Internet and email experience than the younger respondents. This is consistent with cultural stereotypes and experimenter expectations. Many young people (below retirement age) use computers at work, have little choice in whether or not to use them, and must use them for many hours each day, thus accumulating experience. Older adults may only be using computers if they have an interest, and may have less experience due to financial constraints, ie. that they have to purchase a computer rather than having access to one at work, or restricted accessibility, possibly using computers outside their homes, eg. in public facilities.

Older adults, and also less experienced computer users, took longer than younger people to complete the questionnaire. The usability ratings also showed age effects. Here, however, older adults rated the questionnaire more positively in both the menu and radio conditions than the younger adults. It was predicted that the usability measures would indicate that older adults had experienced more difficulties than younger adults, but this does not appear to be the case. It is possible that older adults, working more slowly than the younger respondents, did have less usability difficulties. An alternative explanation is that the subjective usability ratings are rated by each respondent relative to their expectations. Older adults may have initially been more apprehensive of the challenge of completing a form online, with their positive ratings indicating that it was not as hard as anticipated. Sheard *et al* (2000) found that although older adults performance in a library database search task was poorer than that of younger adults, they rated their enjoyment of the task more positively than the younger group. The usability ratings in the current study may have been influenced by differing levels of enjoyment experienced by the two groups.

The usability ratings of the two response styles, menu and radio, did not indicate any major preference for one or the other, except that the menu condition came out more favourably on the amount of scrolling respondents had to do. This does not emphasise a distinct superiority of one or other style. Overall there were no differences in the time taken to complete the questionnaire for the two conditions, a finding confirmed by Carey *et al* (1996) who found time differences only in comparison of experienced and novice users. However, they did find a difference between the conditions, where more errors were made in the pull-down than in the traditional style of menus. The current study did not explicitly include a measure of errors. However, the web site was set up so that if the form was incorrectly completed the data would not be sent, although no record was kept by the system of the number of failures. Early on in the study it became apparent that the data being received consisted of approximately 20-25% more radio forms than menu forms. Hence the random link generator was reprogrammed to give a bias towards the menus condition in order to keep the number of responses per condition even. This was initially thought to be a coincidental bias in the randomisation, but it later seemed unlikely that this was the case. Several times during data collection a 50:50 split between the conditions was attained by giving the random link a 2:1 bias in favour of menus, and each time once the link was modified back to a 50:50 split the data set again gradually developed a bias with more radio than menu forms being received. This, although not formally measured, provides an indication that there were higher error rates in the pull down menu condition than in the radio (traditional) condition, comparable to that identified by Carey *et al*.

The usability measures did not show a preference for the radio condition consistent with this observation. There are a number of possible reasons for this. First, that people who had difficulties with the form, and may thus have rated its' usability less favourably, would frequently not have been able to successfully submit their data. Second, there is the possibility that the usability questions used here were simply not sensitive to the particular problems encountered by respondents.

In conclusion, this research has indicated that radio buttons provide a means of responding to online questionnaires that is preferable to using pull down menus. This confirms the findings of Carey *et al,* who found a preference for traditional rather than pull down menu styles. Although this finding appears to be robust here, in system design the suitability of one menu design over another must be dependent on more than just usability. Traditional menus or radio button lists take up much more screen space than pull down styles. In many interfaces this will be impractical, particularly when a lot of information must be available on screen. However, if everything else is equal, it is suggested that radio buttons should be used in interface design and especially for older adults and novice computer users. Further, this work has demonstrated the use of the Internet as a viable means of data collection.

Acknowledgements

This research was supported by the Research into Ageing and AgeNet joint studentship number 187s.

References

Boren R.W., Moor, W.C. and Anderson-Rowland M.R. 1996, Optimal number of choices per menu for naïve users of a computer answered telephone system, *Computers and Industrial Engineering*, **32**, 363-370

Carey J.M., Mizzi P.J. and Lindstrom L.C. 1996, Pull-down versus traditional menu types: An empirical comparison, *Behaviour and Information Technology*, **15**, 84-95

Coll J.H., Coll R. and Nandavar R. 1993, Attending to cognitive organisation in the design of computer menus: A two experiment study, *Journal of the American Society for Information Science*, **44**, 393-397

Jacko J.A. and Salvendy G. 1996, Hierarchical menu design: breadth, depth and task complexity, *Perceptual and Motor Skills*, **82**, 1187-1201

Pettit F.A. 1999, Exploring the use of the World Wide Web as a psychology data collection tool, *Computers in Human Behaviour,* **15**, 67-71

Sheard M.C.A., Noyes J.M. and Perfect T.J. 2000, Older adults' use of public technology. In P.T. McCabe, M.A. Hanson and S.A. Robertson (ed.s) *Contemporary Ergonomics 2000*, (Taylor and Francis, London), 161-165

COGNITIVE ERGONOMICS

WHICH WAY NOW? APPLICATIONS OF MALLEABLE ATTENTIONAL RESOURCES THEORY

Mark S. Young & Neville A. Stanton

Department of Design
Brunel University, Runnymede Campus
Coopers Hill Lane, Egham
Surrey TW20 0JZ
United Kingdom

Over the past five years, research in our laboratory has been directed towards the development of a new theory of attention. Termed Malleable Attentional Resources Theory (MART), it posits that situations or tasks which induce mental underload can temporarily diminish the resource capacity of the operator. The empirical support for MART, and ramifications for existing theories, are reviewed in this paper. Previous research which could have been explained by MART is also considered. The paper then moves on to practical implications, concentrating on principles of design and how the knowledge uncovered here would impinge upon them. Finally, the paper concludes with thoughts for future research, and proposes that a shift in thinking is necessary for both theoretical and applied researchers in attention and workload.

Introduction

Previous conference papers in this series by the current authors (Young & Stanton, 1997; 1998; 1999) have reviewed literature on attention and mental workload, to synthesise a theory which could explain why extremely low task demands can be detrimental to performance. Studies of automated systems provided an ideal domain within which to explore the theories, as there had been many previous examples comparing the performance of active controllers against passive monitors. The literature review extracted mental underload as a specific problem for task performance, as many researchers had commented on the possible detrimental effects of excessively low mental demands. Whilst there was plenty of concern for the effects of underload, only a few (e.g., Desmond et al., 1998) had offered explanations as to why low demands had an adverse effect on performance.

To answer this problem, we proposed that task demands might exert a short-term influence on attentional capacity, an idea which had not been previously considered. Thus emerged *malleable attentional resources theory* (MART), which posited that the size of attentional resource pools can shrink in line with reductions in demands imposed upon them. This would explain the detrimental effects of mental underload on performance. In the event that resources are needed (e.g., to resume manual control), the operator is unable to devote the necessary attention to the task, because the resources simply no longer exist. This is where MART differs from other explanations of mental underload. A maladaptive mobilisation of effort theory (e.g., Desmond et al., 1998) implies that there is some level of voluntary authority over investment of effort in performance. MART has no such mechanism – reduced capacity is an involuntary and inevitable consequence of reduced demands.

In our experiments (Young & Stanton, 1997; 1998; 1999), the Southampton Driving Simulator was used to test participants under different levels of vehicle automation. Allocation of attention was assessed by recording eye movements, and mental workload (MWL) was measured using a secondary task. These studies revealed that attentional capacity shrinks as task demands fall – there was a direct correlation between MWL and inferred resource size. Furthermore, this correlation held true for skilled and unskilled participants alike. When called upon to take over in an automation failure scenario, though, participants in the lower skill group were ineffective when MWL was at its lowest, whereas more skilled participants could cope regardless of task demands. Clearly, then, this reduction in resource capacity does not impinge upon the automatic performance of skilled operators. Their reliance upon resource-demanding processing routines had become sufficiently low that it did not affect their ability to resume manual control in a critical situation. Only the unskilled participants, who needed their attentional capacity, suffered performance degradation due to mental underload and resource shrinkage.

In sum, then, the experiments reported previously demonstrate good support for MART. Such a link between MWL and attentional capacity has not been demonstrated elsewhere, and forms the core of a new theory of malleable attentional resources. In its final form, MART is presented as follows. The size of attentional resource pools can be increased or decreased according to the level of task demands imposed on the operator. This mechanism can be used to explain the detrimental effects of mental underload on performance. However, these effects can be mitigated by automaticity. Although the skilled operator does suffer from capacity fluctuations with task demands, processes which are essentially resource-free do not show performance decrements in the same way as controlled, resource-dependent processes.

Implications for current theoretical positions

Any reader familiar with capacity theories of attention, and the dual-task paradigm, will realise that these results have huge implications for research in the past, present, and future. Indeed, a lot of prior research in this field could be called into question.

Multiple resources theory (cf. Wickens, 1992), and many studies based upon it, have implicitly assumed that the size of resource pools is invariant across tasks. The conclusions of such studies often hinge upon the assumption that the total demands of primary and secondary tasks equals a constant. For instance, timesharing or multitasking experiments tend to infer that performance decrements are simply indicative of maximal capacity boundaries being exceeded (Brown, 1978). These inferences do not account for the possibility of the capacity limit adjusting to demands. Many such studies using dual- or multiple-task techniques to assess mental workload and performance may have to be reassessed. It may no longer be possible to directly compare different primary tasks against each other using the same secondary task. Although an increase in secondary task responses would still indicate an easier primary task, this cannot then be extrapolated to make absolute and quantitative deductions about the resource demands of the primary task. By virtue of the fact that the addition of primary and secondary task demands no longer equals a constant, the whole dual-task methodology is thrown into turmoil.

A further implication concerns the traditional views of demand-performance relationships. Fixed capacity models assume that performance remains at ceiling, and is data-limited, as long as demands remain within the attentional capacity of the operator (Norman & Bobrow, 1975; Stokes et al., 1990). Performance only begins to decline as the task demands approach the maximum resource availability. This is the very essence of the dual-task approach. Because two tasks can vary in objective difficulty, yet remain within the total capacity of the operator, overt performance differences will not be observed. A secondary task can assess remaining capacity once the primary task has taken its toll, and can therefore differentiate between such levels of difficulty. However, MART predicts that instead, performance is largely resource-limited for the full range of task demands. This would explain why some researchers have found an inverted-U relation between task demands and performance. At low levels of demand, attentional capacity is reduced, artificially limiting the performance ceiling. If task demands exceed the maximum capacity of the operator, performance degrades. Only at medium levels of demand are resources (and hence performance) optimised.

Applications of malleable attentional resources theory

The following paragraphs detail how MART could be used to explain the results of previous research, and the predictions it would make for future research. Later, the practical implications of MART will be considered, along with suggestions for the design of future automated systems.

Buck et al. (1994) used a 'par hypothesis' to explain compensatory behaviour in their study, and Zeitlin (1995) found similar changes in performance in response to task demands. However, none of these papers implicated reduced attentional capacity or mental underload as the cause of performance changes. MART predicts that capacity shrinks to meet task demands, thus making a task easier does not necessarily improve performance. Under such circumstances, performance would appear to be homeostatic.

Of particular relevance to future automated systems are predictions of performance decrements in extreme situations of mental underload. Typically, in tasks involving automation, the theory would predict impaired performance either during or immediately following low MWL episodes. Monitoring performance has been demonstrated to deteriorate if the task is automated (Wickens & Kessel, 1981). Furthermore, manual task performance has been shown to suffer following automated control, whether this is due to expected changes in function allocation (Scallen et al., 1995), or in response to automation failure (Desmond et al., 1998; Stanton et al., 1997).

So, some general predictions about performance under varying levels of mental workload can be made. The idea of an optimal level of MWL (Hancock & Caird, 1993) is clearly supported, with performance suffering if demands are either too low (underload) or too high (overload). Starting with underload conditions, MART predicts that gradual increases in demands would facilitate performance. Such facilitation is particularly evident if suddenly required to assume additional tasks (or resume control of an automated system). The operator who had been working under higher demands (and therefore increased attentional capacity) will cope better with an emergency situation than the underloaded operator. Indeed, this is probably the single most important prediction of malleable attentional resources theory. If resources have shrunk in response to reduced task demands, a sudden increase in demand – even if it is within the ordinary capacity of the operator – cannot be tolerated.

Malleable attentional resources theory and the design of automated systems

The primary aim of all research in ergonomics is to improve the design of systems, and thereby optimise the performance output of that system. To be of any worth, though, recommendations must be based on solid theoretical grounds. The field has often neglected to draw on established theories of cognitive psychology in its approaches, which has limited the progress of research (Bainbridge, 1991). The final substantive section of this paper addresses this point, relating the theoretical bases derived throughout this work to the design of future systems.

Probably the most relevant design argument to the present concerns is the issue of workload optimisation. Many authors have expressed opinions about how automation can lead to underload, but also how the complexity of new systems can result in overload (Wiener & Curry, 1980). To combat this, the concept of adaptive systems has gathered momentum (Hancock et al., 1996), whereby intelligent devices adapt to task circumstances and change their level of intervention to regulate mental workload. However, apart from the significant practical problems of how and when to switch control from human to machine, the nature of adaptive systems can cause rapid control oscillations which can disrupt performance (Scallen et al., 1995). If workload becomes too high, the automation takes over, which reduces workload, manual control is resumed, workload increases again, and so forth, ad nauseam. Therefore, some advocate a preference for support systems rather than automatic control, as this fosters human strengths while compensating for weaknesses. This latter position certainly fits well with the ideas in this paper.

One aspect which has not been covered in any of these discussions about design, yet constitutes a major element of the MART model, is the effect of skill. Hancock et al. (1996) stated that designers should consider all potential users, regardless of skill. Indeed, our experiments show that it is the unskilled who are particularly vulnerable to the effects of underload. This makes it vitally important to support these drivers rather than try to replace them. Ideally, technological support systems should act like a driving instructor in the passenger seat - subtle enough so as not to cause interference, but accessible enough so as to provide assistance when needed.

Conclusions

Malleable attentional resources theory has been presented with considerable theoretical and empirical support. It will have a significant impact on the field of ergonomics, as researchers in attention and mental workload will have to shift their thinking from fixed capacity models. Although it does not entirely replace resource theories, it offers a new perspective which has important theoretical and practical ramifications.

Not least among these is the understanding of mental workload, and consequent predictions about performance. If task demands are low (e.g., due to the use of automation), then the level of resources required will also be low. However, the consequences for total capacity mean that it is not possible to infer increased spare capacity in direct proportion to the demands. Therefore, although the absolute level of resources required does decrease, the reduction is not so pronounced when taken as a relative proportion of total capacity. This may be part of the reason why vigilance monitoring tasks have recently been thought of as imposing high mental workload (Warm et al., 1996). Subjective perceptions of load could be more sensitive to the relative demands on resources, rather than absolute demands.

Future research should be aimed at finding a definitive threshold for mental underload. That is, at what point of decreasing demands does attentional capacity begin to shrink? Moreover, a time-decay curve needs to be established, to determine how soon and at what rate resources degrade. The results of the present experiments indicate that 10 minutes is more than sufficient for resource shrinkage to occur, and there is some evidence that the process may begin between 30 and 60 seconds (Scallen et al., 1995). Furthermore, it is not sufficient to know how quickly resources degrade, but also how long it takes capacity to return to maximum when demands increase again. It has been demonstrated that a sudden increase in demands is difficult to cope with for some operators. Quite how sudden such an increase can be without causing performance decrements, though, remains to be seen.

Finally, there is an applied angle which will become increasingly salient in the near future. Given the rapid pace of technological developments, and the trend to automate wherever it is possible, a likely scenario for future vehicles involves a deskilling of all drivers, with total reliance on automation. Consider also that, until technical reliability and innovation reaches a point where the concept of a driver is redundant, a human will always be involved in the system to take over in case of trouble. Now, this presents the situation where automation is the norm, and most drivers will not have the opportunity to develop automatic skills. It will be remembered that unskilled drivers did not do especially well at recovering from automation failure, and demonstrated a particular vulnerability to the effects of malleable resources. If future vehicles do evolve in this way, a huge investment in driver training and retraining will be necessary.

These issues are applicable to all systems in which automation is a possibility. The keepers of future technologies should realise the value of human input, and instead of trying to design humans out of systems, they should integrate the human fully and nurture their abilities. A complete knowledge of the relative abilities of human and machine, and the way in which they work together, is invaluable in this endeavour, for which the theory presented in this paper may make a useful contribution. Neither component of the system is infallible, but by exploiting the strengths of each, the system as a whole really can be greater than the sum of its parts. A deeper knowledge of how attention, skill, and automation interact can help future designers to know the human, and to know what situations will favour performance. Malleable attentional resources theory could well provide one of the signposts in that direction.

References

Bainbridge, L. 1991, The "cognitive" in cognitive ergonomics, *Le Travail Humain*, **54**(4), 337-343.

Brown, I. D. 1978, Dual task methods of assessing work-load, *Ergonomics*, **21**, 221-224.

Buck, J. R., Payne, D. R., & Barany, J. W. 1994, Human performance in actuating switches during tracking, *International Journal of Aviation Psychology*, **4**(2), 119-139.

Desmond, P. A., Hancock, P. A., & Monette, J. L. 1998, Fatigue and automation-induced impairments in simulated driving performance, *Transportation Research Record*, **1628**, 8-14.

Hancock, P. A., & Caird, J. K. 1993, Experimental evaluation of a model of mental workload, *Human Factors*, **35**, 413-429.

Hancock, P. A., Parasuraman, R., & Byrne, E. A. 1996, Driver-centred issues in advanced automation for motor vehicles. In R. Parasuraman & M. Mouloua (eds.) *Automation and human performance: Theory and applications*, (Erlbaum, Mahwah), 337-364.

Norman, D. A., & Bobrow, D. G. 1975, On data-limited and resources-limited processes, *Cognitive Psychology*, **7**, 44-64.

Scallen, S. F., Hancock, P. A., & Duley, J. A. 1995, Pilot performance and preference for short cycles of automation in adaptive function allocation, *Applied Ergonomics*, **26**(6), 397-403.

Stanton, N. A., Young, M., & McCaulder, B. 1997, Drive-by-wire: The case of driver workload and reclaiming control with adaptive cruise control, *Safety Science*, **27**(2/3), 149-159.

Stokes, A. F., Wickens, C. D. & Kite, K. 1990, *Display technology: human factors concepts*, (Society of Automotive Engineers Inc., Warrendale).

Warm, J. S., Dember, W. N., & Hancock, P. A. 1996, Vigilance and workload in automated systems. In R. Parasuraman & M. Mouloua (eds.) *Automation and human performance: Theory and applications,* (Erlbaum, Mahwah), 183-200.

Wickens, C. D. 1992, *Engineering psychology and human performance,* Second edition, (Harper Collins, New York).

Wickens, C. D., & Kessel, C. J. 1981, Failure detection in dynamic systems. In J. Rasmussen & W. B. Rouse (eds.) *Human detection and diagnisis of system failures*, (Plenum Press, New York), 155-169.

Wiener, E. L., & Curry, R. E. 1980, Flight-deck automation: promises and problems, *Ergonomics*, **23**(10), 995-1011.

Young, M. & Stanton, N. 1997, Taking the load off: Investigating the effects of vehicle automation on driver mental workload. In S. Robertson (ed.) *Contemporary Ergonomics 1997*, (Taylor & Francis, London), 98-103.

Young, M. & Stanton, N. 1998, What's skill got to do with it? Vehicle automation and driver mental workload. In M. A. Hanson (ed.) *Contemporary Ergonomics 1998*, (Taylor & Francis, London), 436-440.

Young, M. S. & Stanton, N. A. 1999, Miles away: A new explanation for the effects of automation on performance. In M. A. Hanson, E. J. Lovesey, & S. A. Robertson (eds.) *Contemporary Ergonomics 1999*, (Taylor & Francis, London), 73-77.

Zeitlin, L. R. 1995, Estimates of driver mental workload: A long-term field trial of two subsidiary tasks, *Human Factors*, **37**(3), 611-621.

Workload Assessment in Railway Control: Developing an Underload Scale

Nikki Bristol & Sarah Nichols

Institute for Occupational Ergonomics
University of Nottingham
University Park
Nottingham
NG7 2RD, UK

The new changes planned for the railway control working environments are likely to impact substantially on various human factors issues. It is essential that both valid and reliable measures of these factors are developed to enable effective auditing. However, the nature of signal control tasks presently vary from one signal box to the next due to the gradual implementation of automation. This raises various difficulties for Ergonomists trying to find measures that are suitable for both manually operated signal boxes and those that are partly automated.

This paper specifically addresses the problems encountered when developing mental workload measures. Due to the very noticeable differences in signal controller workload in the different types of signal box, alternative subjective ratings scales that are sensitive to changes in work underload are developed for use in the partly automated box.

Introduction

Background

With the gradual move towards centralisation of railway control operations, it is a primary concern for those involved to make sure that new control centres are designed to support both safe and effective working conditions. It is hoped that by bringing all the different railway control operators under one roof, both the co-ordination and communication between them will be greatly improved. However, such a move will also make a considerable difference to the nature of operators' tasks, resulting in changes such as greater distances from areas under control, more automation and larger geographical areas controlled by a single operator.

The work described in this paper forms part of a larger project - the development of a Railway Ergonomics Control Assessment Package (RECAP, Cordiner et al 2000) - that examines the human factors issues associated with the changes being implemented in railway control. As a result of some of these issues, studies to measure signal controllers' workload levels have been carried out in current railway signal boxes (the RELOAD project). The aim of this work was to gain a better understanding of how to construct and combine tasks in the new centres in order that an optimal level of workload is experienced -

one that is neither too high nor too low. Initial studies of controller workload (Cordiner et al 2000, Nichols et al 2000) involved the development of appropriate subjective workload scales and observable indicators of operator workload. The work described in this paper follows on from these studies, and examines the issue of operator underload in detail.

Rationale for development of underload scale

In initial RELOAD studies, two types of signal box were visited; an NX panel signal box under manual control and an IECC (Integrated Electrical Control Centre) under a mixture of automated and manual control. The subjective measures of workload were initially developed in the NX panel signal box only and comprised a combination of the AFFTC (Air Force Flight Test Centre) single workload rating scale and three adapted rating scales from the NASA TLX - amount of thinking and planning needed, time available for tasks and the amount of physical fatigue (Nichols et al 2000). These subjective measures were applied each ten minutes during an observation period and the relationship between subjective ratings and observable indicators of workload, such as the number of trains on the track or the number of routes set, was examined.

The ratings on these scales in the NX panel signal box appeared to reflect changes in the observed workload of the controller - fluctuating throughout the course of the analysis period (see Figure 1a). However, the same set of scales were also used on a visit to an IECC and, as can be seen in Figure 1b, only the AFFTC general workload scale showed any fluctuation. The other scales appeared to be insensitive to changes in any type of workload experienced during the analysis period of one hour. Considering the much greater amount of monitoring behaviour required of the IECC controllers these findings were not altogether surprising and, after discussion with the signal controllers themselves, it was decided that new scales more sensitive to 'underload' should be developed.

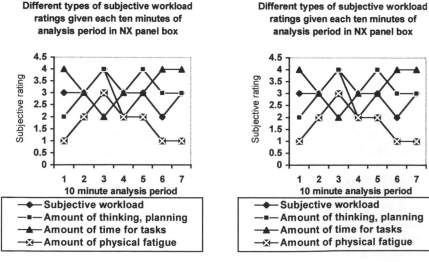

Figure 1a – NX panel controller ratings

Figure 1b – IECC controller ratings

Development of underload rating scales

The three NASA-TLX scales were originally chosen in order to obtain more information about the general AFFTC workload rating given - whether it was due to time restrictions, fatigue or mental effort. New underload scales would need to serve the same purpose, providing additional information about the general workload ratings given.

There are fewer studies to be found on work underload than on workload in the literature. Nonetheless, a study by Braby et al (1993) revealed a method of adjective generation and the formulation of a checklist to measure subjective underload. Although the checklist developed by Braby et al was considered too lengthy and too intrusive to administer safely during the signal controller's shift, a selection of adjectives describing how the controller felt during periods of low activity, combined with a process of paired comparison to refine the selection, was thought to be appropriate for the current project.

Scale development

Site details & materials

The process was carried out in a West Coastal IECC where a male signal controller, who had been working for 4 years at the site, agreed to participate. To prepare for the visit, a list of adjectives was compiled - largely from the study carried out by Braby, but also generated by the author. Eight palm sized pieces of card were cut out ready to use in the paired comparison process, along with an eight point grid for recording the results.

Procedure

The controller was presented with a list of adjectives (see Table 1) and was asked to select eight adjectives either from the list or of his own that would best describe how he felt during periods of low activity during a shift.

Under-stimulated	Unchallenged	Easily distracted
Free from concentration	Daydreaming	Bored
Having difficulties concentrating	Free to think of other things	Needing other activity or stimulation
Alert	Anxious	Nervous
Pressured	Disinterested	Apprehensive

Table 1: List of possible adjectives

The eight chosen adjectives were then written onto different palm-sized cards and also along the sides of an eight point grid. The cards were then presented in pairs and the controller was asked to pick the adjective in each pair that described better how he felt during periods of low activity. Once the cards had been presented in all of the possible combinations and the results noted in the grid, the two preferred adjectives were calculated.

Results

The signaller chose the following eight adjectives partly from the list provided but also from his own suggestions: Unchallenged, free to think of other things, disinterested, bored, alert, easily distracted, having difficulties concentrating and needing other activity or stimulation. Through the process of paired comparison, the signaller eliminated 6 of these, leaving the two preferred adjectives as 'Freeness to think of other things' and 'Disinterest'.

Underload measurement trial

Materials
For the trial of the scale - during one of the controller's shifts in the IECC - , a sheet with two blank 7 point scales and a copy of the AFFTC scale were required. In addition, a recording sheet was needed on which both the time of each rating, the ratings of the workload and underload scales and the frequency of various important tasks or behaviours could be recorded.

Procedure
The two adjectives were written above two 7 point rating scales - scaling between 'not at all' and 'extremely'. Along with the AFFTC general workload rating scale, the two scales were administered at 10 minute intervals during the signaller's next shift. The frequency of various activities were counted during the few minutes around each rating as well as the number of manually controlled trains on the VDU screen – simultaneous to each rating. A video recording was made of the entire shift for later analysis. Any variation in the underload ratings, given in accordance with observed activity level variance, would be taken to mean that the scales were sensitive to perceived changes in workload or underload - at least for this controller.

Results
Unlike with the NASA-TLX scales used previously, the ratings given for the two new underload scales varied throughout the shift; in the opposite direction to the workload ratings. (See figure 2).

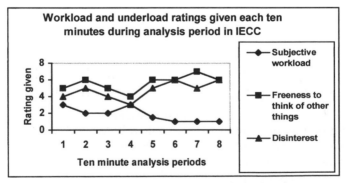

Figure 2: IECC ratings with new underload scales

In fact, comparisons of ratings given with the observed workload showed that only the number of freight trains on the VDU screen and the number of internal phonecalls had any consistent effect on the workload and underload ratings. An increase in either lead to an increase in workload ratings and a decrease in underload ratings - as one might expect:

Workload and underload ratings given when frequency of internal phonecalls made varied

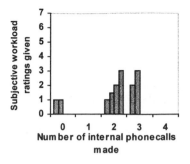

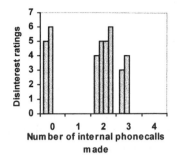

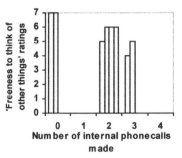

The workload ratings increased and the underload ratings decreased as the numbers of incoming calls increased, suggesting that incoming phonecalls play a substantial role in determining the signallers workload level.

Workload and underload rating given when the number of freight trains on the screen varied

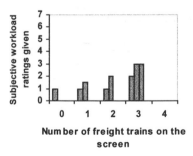

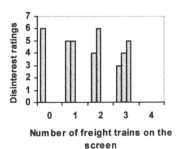

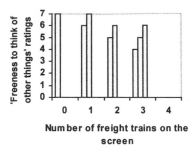

Number of freight trains on the screen

There appears also to be a substantial increase and decrease in ratings as the number of freight or manually controlled trains in the area of control increases. It is likely that this has a significant influence on the signaller's workload level. Frequency of other types of phonecalls showed similar effects on the ratings. However, none of the other behaviours - such as route-setting or simplifier usage – showed any influence on the perceived workload or underload.

Discussion of findings

These results agree with the signal controllers' own opinions about which activities have the biggest influence on their workload or underload levels. Also, the retrospective protocols, previously carried out in the same IECC, revealed that most of the planning and memory work was triggered by either the presence of freight trains on the VDU screen (manually controlled trains) or queries from other operators about the system status.

The two new underload scale ratings suggest that both motivational (disinterest) and attentional (freeness to think of other things) factors contribute to the decreased general workload ratings given. However, there is yet much more work to be done in terms of validifying the scales. After all, the scale is based on the preferences of a single signal controller and it may be that other IECC controllers do not relate to these scales in the same way. Also, on this occasion, the underload ratings varied almost identically in the inverse direction to workload ratings and it may have been that the signaller was simply basing the underload ratings on the workload rating given first. Random ordering of the scale administration may reveal whether or not this is the case. Such developments are currently being pursued by N. Bristol (PhD thesis).

References

Braby C.D., Harris D., Muir H.C. (1993); A psychophysiological approach to the assessment of work underload; Ergonomics, 36 (9), pp1035-1042

Cordiner, L.A., Nichols, S. & Wilson, J.R. (2000) Development of a Railway Ergonomics Control Assessment Package (RECAP). In *Proceedings of IEA 2000*, San Diego, July 30th – August 4th, 2000.

Nichols S., Bristol N., Wilson J. (2000); Workload Assessment in Railway Control; Proceedings of Engineering Psychology and Cognitive Ergonomics Conference, Edinburgh, November 2000.

PSYCHOLOGICAL STRESS IN THE WORKPLACE: COGNITIVE ERGONOMICS OR COGNITIVE THERAPY?

Neil Morris   and   Bianca Raabe

University of Wolverhampton   Manchester Metropolitan University
School of Health Sciences   Crewe and Alsager Faculty
Psychology Department   Humanities and Social Sciences
Bankfield House   Hassall Road
Wolverhampton WV1 4QL   Alsager ST7 2HL
UK   UK

Broadbent's (1971) classic account of work stressors in *Decision and Stress* reviewed the psychological effects of environmental stressors in the workplace. Such cognitive ergonomics clearly places the onus on stress reduction on the employer. Thirty years later a wider range of stressors are acknowledged and there is an emphasis on coping with stressors rather than environmental manipulation to remove them. Many employers now offer cognitive or transactional psychotherapy to aid in coping and resolving stress related problems. Some workers have suggested that the provision of such services, whose efficacy is limited, may be employed as a defensive strategy in the face of possible litigation. We discuss the possibility that one outcome of this would be the attachment of 'blame' for being stressed to the litigant. Cognitive therapy embraces the notion that stress has a cognitive appraisal component and this could be used to bolster this argument.

Introduction

It is now thirty years since the publication of Broadbent's (1971) classic work on human performance *'Decision and Stress'*. In this work Broadbent considers a range of physical environment stressors – heat, glare, noise – and some relating closely to behavioural practices – sleep loss, fatigue and alcohol intake. What is striking, thirty years later, is the change in emphasis with respect to the role of the individual in stressful situations. Broadbent comments on the effects of the above stressors on a range of task performances noting that "there do seem to have been adequate precautions against *contamination effects* of morale and suggestion" (p.411, italics added). This perspective implies a powerfully nomethetic approach with stresses applied to the worker. Idiosyncrasies of individual workers, for example, morale effects, suggestion, are not part of the stress model. By contrast, one could argue that now a distinctly more idiographic emphasis is applied to dealing with stress in the workplace. Different stressors are identified and some form of counselling/therapy is frequently offered. We have moved away from modifying the work environment to modifying the behaviour of the worker. Cognitive ergonomics has given way, since Broadbent's review, to cognitive therapy and this may have serious implications for cognitive ergonomics in the early 21st century.

The rise of counselling in the workplace

Wickens (1984) suggested that "Prior to the birth of human factors, or ergonomics, in World War II, emphasis had been placed on 'designing' the human to fit the machine. That is, the emphasis was on training" (p. 4-5). This is in stark contrast to the tenets of cognitive ergonomics. Singleton (1989) argues that work environments should be designed to be compatible with the physical characteristics and limitations of potential employees and cognitive ergonomics is concerned with work design features that may tax the cognitive capacities of the workforce. There is clearly no conflict between adapting the workplace to the capacities of employees and the provision of training. Wickens observations refer to conditions in which ergonomic considerations were largely disregarded provided an operative could be trained without prohibitive costs and engineering considerations predominated. Such a philosophy allowed companies to implement technologies without serious concern for the welfare of their employees. Furthermore the provision of 'adequate' training could allow 'failure' to be ascribed to the worker rather than the technology. The benefits that have accrued from ergonomic audits have largely negated this largely technology centred approach. However there is some evidence that this more 'humanitarian' approach to work practices, particularly the provision of counselling services for stressed employees, may have some regressive effects.

It has been estimated that in the UK more than 360 million working days per annum are lost through illness and that more than 50% of these losses are attributable to stress related problems (Cooper and Cartwright, 1996). Thus it is not surprising that counselling services for stressed workers have expanded massively in recent years (Carroll and Walton, 1997) and it has been widely argued that therapy, especially cognitive therapy, may be very efficacious for dealing with emotional disorders (Beck, 1989). However counsellors/therapists have pointed out that the support offered in the workplace, or in relation to workplace problems, may be quite different to that offered to alleviate stress in other aspects of life (Carroll, 1997). Such counselling may be very brief or necessarily involve group sessions and the options for stress reduction in the workplace may be considerably narrower than in other areas of life because the stressed individual may not have the power, within the work organisation, to make therapeutic changes.

Reynolds and Briner (1996) comment that Stress Management Training (SMT) is becoming increasingly popular as a counselling service. SMT tends to employ cognitive behavioural therapy combining relaxation training with stress inoculation and cognitive re-appraisal techniques (Meichenbaum, 1985). This does not usually entail any major organizational change and Reynolds and Briner (1996) opine that "Interventions which attempt to change organizational or job characteristics are uncommon, commercially unattractive and very often reported in the context of organizational or socio-technical changes rather than as planned interventions with a specific focus on stress reduction" (p. 144).

Cognitive therapy and coping

Why is cognitive therapy so popular in SMT? It should be noted that other techniques, for example transactional psychotherapy, are also used and many of the same arguments apply but cognitive therapy has a number of characteristics that make it an obvious choice for stress management. These have been succinctly outlined by one of its major advocates. "(C)ognitive therapy consists of all the approaches that alleviate psychological distress through the medium of correcting faulty conceptions and self signals...By correcting erroneous beliefs, we can damp down or alter excessive, inappropriate emotional reactions" (Beck, 1989, p.214). If one works with a definition of stress similar to that outlined by Pratt and Barling (1988), then stress results from the interplay between the individual and his or her environment and involves the stressed individual *appraising* this interaction as being challenging and leading to doubts about their ability to cope with this challenge. Cognitive therapy aims, amongst other things, to help the individual not simply to re-appraise the workplace as over-challenging in some way but to lead the individual into insight into their own role in 'creating' the stressful situation by making them aware of their behaviours that contribute to the stressfulness of the situation. From here it is possible to move on to coping with the situation.

In the wider domain of the individuals complete lifestyle cognitive therapy may be used to challenge beliefs about inferiority etc., and require changes in lifestyle, including perhaps employment changes. Therapy within organizations is unlikely to include advice on changing jobs. Rather workplace based therapy may have as its central tenet the need to keep the individual in the organization, and at work, Such a tenet is in principle laudable but restricts the scope of cognitive therapy to helping the individual survive the current *status quo*. Cognitive therapy, in the workplace, has become a 'downstream' approach to stress – 'unstressed' workers are rarely counselled on how to avoid stress (Bull, 1997). Egan and Cowan (1989) distinguish between two modes of counselling – upstream and downstream help. They argue that it is better to intervene upstream, by modifying work practices etc. than to 'wait' downstream to rescue from 'drowning' those floundering because they are stressed. Helping downstream, they argue, creates a seriously dysfunctional work situation because those aided downstream are returned to the stressful work situation, upstream, to face the same stressors again. A downstream approach, implicit in cognitive therapy – *if one cannot change any causal factors in the organization* – offers coping as the main option.

Dewe, Cox, and Ferguson (1996) define coping as "the cognitions and behaviours, adopted by the individual following the recognition of a stressful encounter, that are in some way designed to deal with that encounter or its consequences" (p. 117). Such coping is not helpful *per se*. Many coping strategies, for example, avoidance, excessive alcohol consumption, smoking etc. may be harmful. Cognitive therapy is unlikely to lead intentionally to any of these options but they highlight that coping comes with costs. There is a difference between ceasing to be stressed because the source of stress has gone away and perceiving a stress reduction because one has changed ones interpretation of the meaning of

a situation. Cognitive therapy
change their work practices
appraising the nature of the
systems embrace denial
but is no substitute f

Work stress ?

There are a nu
suggested by Lowma.
identifies under-commit.
personality disorders. All o   .ow
this can be anything but a down.   .i and
Cowan (1989) arising, unless the w   genesis
of the problem (which may be the case   s life are
responsible for the problem). However the.   is likely to
be a causal factor and practices could be chang   role and from
this perspective role overload can be seen as a sou   . Ganster, 1989),
as can role underload (Cooper and Willis, 1989). Fu.   e is ambiguous or
one has conflicting responsibilities (Nicholson and Goh, .   s may be a source of
stress. Such role strains may be a factor in mediating the seve   ie impact of major life
events (Cassidy, 1999). Other consequences of occupational s. ss may include coronary
heart disease, rheumatic arthritis, ulcers, some allergies, headache, depression and anxiety
(Ivancevich and Matteson, 1980). Cooper and Marshall (1976) stress the behavioural
manifestations – smoking, poor motivation, low self-esteem, absenteeism, lowered
productivity and increased staff turnover.

Blame and legal consequences

Work stress can clearly have serious consequences for health and this has legal implications.
In the USA there have been successful attempts to sue employers for creating physical
and/or psychologically stressful work environments (Matteson and Ivancevich, 1987) and
this has recently occurred in the UK also. Indeed Reynolds and Briner (1996) have
speculated that the provision of SMT may be an employer defensive response to this. Thus
employers may feel obliged to 'blame' the employee for poor job performance to avoid legal
culpability (Lowman, 1993). In the USA trade union members expressed more support for
changing the environment and work practices while management favoured counselling
employees (Neale, et al., 1982) suggesting an unwillingness by workers to accept 'blame'
for stress. On a more positive note Cooper and Cartwright (1996) argue that "activities
aimed solely at individuals reactions to stressful circumstances, and not also targeted at
modify the circumstances themselves, will not be sufficient to avoid the negative legal
ramifications" (p.90). If Cooper and Cartwright are correct then there is a role for cognitive
ergonomics in a domain dominated, recently, by cognitive therapy.

Conclusions

Cognitive therapy in the workplace places the emphasis on changing the individuals cognitions about the workplace. Such redefinition implies coping with stressors not reducing or removing external stressors (unless this involves challenging bullying in the workplace or similar interpersonal problems). The individual, not organisational procedures, are required to change. Implicit in this is that the individual is to blame for their emotional distress (although the counsellor is unlikely to believe this). A company offering such therapy may claim that it acknow

psychological adapt;

domain may well ma

When employe

anything is done th

occupational couns

possibility that the

Lowman, 1993). H

factors/occupational

insight into job desi

Perhaps collaboratio

   Elkin and Rose

- redesign the
- increase the
- analyse wor
- provide sup
- make sure e
- share rewar

The above do not r

in the Post Office (

was reduced after (

Nevertheless Cum

"counselling being

regimes" (p.310). I

workplace to not b

can still see a need

[Handwritten note: An unwillingness to accept 'blame' for stress. Coping with rather than not reducing or removing external stressors.]

References

Beck, A.T. (1989) *Cognitive therapy and the emotional disorders.* London: Penguin.

Broadbent, D.E. (1971) *Decision and stress.* New York: Academic Press.

Bull, A. (1997) Models of counselling in organizations. In M. Carroll and M. Walton (eds.) *Handbook of counselling in organizations.* London: Sage. 29-42.

Carroll, M. (1997) Counselling in organizations: An overview. In M. Carroll and M. Walton (eds.) *Handbook of counselling in organizations.* London: Sage. 8-28.

Carroll, M. and Walton, M. (1997 – eds.) *Handbook of counselling in organizations.* London: Sage.

Cassidy, T. (1999) *Stress, cognition and health.* London: Routledge.

Cooper, C.L. and Cartwright, S. (1996) Stress management intervention in the workplace: Stress counselling and stress audits. In S. Palmer and W. Dryden (eds.) *Stress management and counselling.* London: Cassell. 89-98.

Cooper, C.L. and Marshall, J. (1976) Occupational sources of stress: A review of the literature relating to coronary heart disease and mental ill health. *Journal of Occupational Psychology,* **49,** 11-28.

Cooper, C.L., Sadri, G., Allison, T., and Reynolds, P. (1990) Stress counselling in the Post Office. *Counselling Psychology Quarterly,* **3,** 3-11.

Cooper, C.L. and Willis, G.I. (1989) Popular musicians under pressure. *Psychology of Music,* **17,** 22-36.

Cummins, A.M. and Hoggett, P. (1995) Counselling in the enterprise culture. *British Journal of Guidance and Counselling,* **23,** 301-312.

Dew, P., Cox, T., and Ferguson, E. (1996) Individual strategies for coping with stress at work: A review. In S. Palmer and W. Dryden (eds.) *Stress management and counselling.* London: Cassell. 115-128.

Egan, G. and Cowan, M. (1979) *People in systems.* Monterey: Brooks/Cole.

Elkin, A.J. and Rosch, P.J. (1990) Promoting mental health at work. *Occupational medicine state of art review,* **5,** 739-754.

Ivancevich, J.M. and Matteson, M.T. (1980) *Stress and work: A managerial perspective.* Glenview: Scots Foresman.

Lowman, R.M. (1993) *Counselling and psychotherapy of work dysfunctions.* Washington: American Psychological Association.

Matteson, M.T. and Ivancevich, J.M. (1987) *Counselling work stress: Effective human resource and management strategies.* London: Jossey-Bass.

Meichenbaum, D. (1985) *Stress innoulation training.* New York: Pergamon.

Neale, M.S., Singer, J., Schwartz, G.E., and Swartz, J. (1982) Conflicting perspectives on stress reduction in occupational settings: A systems approach to their resolution. *Report to NIOSH on PO no. 82-1058 Cincinnati, OH.*

Nicholson, P.J. and Goh, S.C. (1983) The relationship of organization structure and interpersonal attitudes to role conflict and ambiguity in different work environments. *Academy of management journal,* **26,** 148-155.

Perrewe, P.L. and Ganster, D.C. (1989) The impact of job demands and behaviour control on experienced job stress. *Journal of Organizational Behaviour,* **10,** 213-229.

Pratt, L.I. and Barling, J. (1988) Differing between daily events, acute and chronic stressors: A framework and its implications. In J.J. Hurrell. L.R. Murphy, S.L. Sauter and C.L. Cooper (eds) *Occupational stress: Issues and developments in research.* London: Taylor and Francis.

Reynolds, S. and Briner, R. B. (1996) Stress management at work: With whom, for whom and to what ends? In S. Palmer and W. Dryden (eds.) *Stress management and counselling.* London: Cassell. 141-158.

Singleton, W.T. (1989) *The mind at work.* Cambridge: Cambridge University Press.

Wickens, D.C. (1984) *Engineering psychology and human performance.* Columbus OH: Charles Merrill Publishers.

COGNITIVE ERGONOMICS: SOME LESSONS LEARNED (SOME REMAINING)

John Long

Ergonomics & HCI Unit, University College London
26 Bedford Way, London WC1H 0AP
j.long@ucl.ac.uk

The past, present and future of Cognitive Ergonomics are reviewed. The past, and in particular its shortcomings, are described in terms of an earlier characterisation (Long, 1987). The present is described in terms of the lessons assumed to have been learned, since that publication. The future is characterised in terms of lessons, which still remain to be learned.

Introduction

Cognitive Ergonomics (CE) emerged in the late 1970s on the back of the 'personal' computer and the 'naïve' user. The contrast was with the 'mainframe' computer and the 'professional' user. The emergence was associated with developments in Cognitive Psychology, Linguistics and Artificial Intelligence (Card, Moran and Newell, 1983). 'Cognitive' Ergonomics was in contrast to 'Physical' or 'Traditional' Ergonomics (Long, 1987). According to Long, "The advent of the computer, together with changes in Psychology, has given rise to a new form of Ergonomics termed 'Cognitive Ergonomics'". This paper considers the past, present and future of CE. The past, and in particular its shortcomings, is described in terms of this earlier characterisation (Long, 1987). The present describes CE in terms of the lessons learned, since that publication. The future characterises the lessons remaining to be learned (Dowell and Long, 1989).

According to Long (1987), "The most general definition of Cognitive Ergonomics is the application of Cognitive Psychology to work ... to achieve the optimisation (between people and their work) ... with respect to well-being and productivity". He goes on to argue: "Cognitive Ergonomics, then, is a configuration relating work to science. In this context, its aim can be defined as the increase in compatibility between the user's representations and those of the machine". Long further claims that CE shows "how the system designer can be provided with information which will help improve cognitive compatibility" and supports "the practical aim of helping designers to produce better systems, and the use of theoretical structures and empirical methods to achieve this end".

Here, this characterisation is considered to be the past, with respect to which present lessons are assumed to have been learned. Lessons still remaining, if learned, would constitute one possible future for CE.

This paper is structured, following the discipline framework proposed by Long and Dowell (1989), as the "acquisition and validation by research of knowledge supporting practices, which solve the general problem, having a particular scope". When applied to CE for present purposes, the framework produces a structure with the following sections: Particular Scope; General Problem; Practices; Knowledge and Research. Each section successively addresses CE past, present, and future.

Particular Scope of Cognitive Ergonomics

Long (1987) references three different particular scopes for CE. First, with reference to the discipline, the scope is people, their work, well-being and productivity ... "the emphasis being on the person doing the work and the manner in which it is carried out, rather than on the technology or on the environment". Second, with reference to humans interacting with computers, as a new technology-driven phenomena, the scope is: "agent(s); instrument; functions; entities; and location." Last, with reference to difficulties experienced by users of computers, the scope is: domain of application; task; and computer interface. However, the scopes are not equally complete. Neither do they exhibit any coherence, since they are not explicitly related. Two main lessons have been learned.

Lessons Learned 1: CE needs to have a single particular scope. Multiple particular scopes, as proposed by Long, need specific justification, since each scope would require a separate mapping to CE knowledge and practices, so militating against coherence and parsimony. Acceptance of the need for a single, particular scope is general (Hollnagel, 1998).

Lessons Learned 2: CE research needs to have the particular scope of: users; technology; work; and performance. CE has always included users and technology in its particular scope. However, except in process control, work has not always been considered separately from the users' behaviours necessary to carry it out (see GOMS Card et al, 1983). There is recognition of the need to include work and performance in the CE scope (Vicente, 1998). Two main lessons remain to be learned.

Lesson Remaining 1: CE needs a particular scope, which relates users and technology together as an interactive worksystem. Performance of work, involving users and technology, cannot ultimately be a function of either alone, but only of both together. Hence, the need for CE to conceptualise them jointly. There is currently some acceptance of this requirement, for example, in the concept of 'joint cognitive systems' (Hollnagel, 1998). All too often, however, performance is only expressed as 'human performance', usually in the form of errors (Reason, 1998).

Lesson Remaining 2: CE research needs a particular scope, which distinguishes performance as task quality and as interactive worksystem costs. Task quality expresses how well a task is performed by the worksystem and derives from an analysis of the work domain. Worksystem costs express the workload incurred (by the users and the technology) in performing the task that well, and derive from an analysis of the worksystem. The distinction between task quality and worksystem costs is one required by CE design practices and supported by CE design knowledge. CE designing requires a balance between (desired) task quality and (acceptable) worksystem costs.

General Discipline Problem of Cognitive Ergonomics

According to Long (1987), the general problem of CE is the provision of advice to designers "to produce better systems ... more usable systems". In addition: "the system designer can be provided with information ... to inform designers ..." etc. The shortcoming here is the restriction of the general problem to advising designers, rather than to design itself. Unless the latter is better specified, it is unclear how the advice to, and the information for, designers can be appropriately acquired and formulated. Two main lessons have been learned.

Lessons Learned 3: CE needs its general discipline problem to be design. The issue is not simply one of providing advice to designers, but the need for CE design knowledge to support specified CE design practices. CE research aims to acquire such design knowledge. The view of CE as a design discipline is generally accepted (Vicente, 1998).

Lessons Learned 4: CE needs to support design practices having the scope of: users; technology; work; and performance. In other words, CE design practices need to include users, technology, work and performance and not just one or the other of these concepts. It is not just a matter, for example, of CE research supporting the design of the interface, without reference to the work being carried out or the performance of the worksystem. Two main lessons remain to be learned.

Lesson Remaining 3: CE needs a general problem, which expresses design as design problems and design solutions. It is not sufficient to express design in terms of the CE particular scope of: users; technology; work; and performance. For the sake of completeness, the CE general problem needs to embody both the start-point of design, the design problem, and the end-point of design, the design solution. How otherwise would design start and stop?

Lesson Remaining 4: CE needs a general problem, which expresses design problems and design solutions, in terms of the interactive worksystem and performance. It is not just a question of expressing design problems and design solutions, actual performance needs to relate to the current worksystem and desired performance needs to relate to some future to-be-designed (target) system.

Practices of Cognitive Ergonomics

According to Long (1987), Cognitive Science (CS) acquires knowledge that CE applies. CS has two primary practices, those of analysis, to produce an 'acquisition representation', and generalisation, to produce the knowledge itself, as required to describe and explain mental phenomena. CE also has two primary practices, those of the particularisation of the CS knowledge to produce an 'applications representation', such as guidelines, etc., and synthesis, as required to optimise "the relationship between people and their work with respect to well-being and productivity". The practices of CS and CE are thus different, the latter applying the knowledge acquired by the former. Two main lessons have been learned.

Lesson Learned 5: CE needs to support design practices, which include design, evaluation and iteration. According to Long (1987), the practice of particularisation transforms CS knowledge into guidelines, checklists, etc., that is, it does not itself contribute directly to design, only indirectly. The practice of synthesis, in contrast, contributes to design directly. However, synthesis here identifies no additional CE practices other than the notion of design itself. CE practices, then, need to be specified more completely, at least as design, evaluation and iteration. This view of CE practices is largely accepted (Reason, 1998).

Lessons Learned 6: CE needs to support design practices, which address: users; technology; work; and performance. CE design practices need to support the particular scope of design. In general, design practices are assumed to address the CE research scope. Two main lessons remain to be learned.

Lesson Remaining 5: CE needs to support practices, which diagnose design problems and prescribe design solutions. It is not sufficient to distinguish overall practice as 'design' or 'synthesis', or indeed the differing CE practices of design, evaluation and iteration. CE practices need to relate specifically to design problems and design solutions. Thus, a CE practice is needed to construct design problems, that is to diagnose and a CE practice is needed to construct design solutions, that is, to prescribe.

Lesson Remaining 6: CE needs to support practices, which diagnose performance as the design problem and prescribe an interactive worksystem as the design solution. As established earlier, the design problem, in which actual performance is less than desired performance, is associated with the current worksystem. The design solution, in which the actual performance equals the desired performance, is associated with the (future) to-be-designed target system. Both expressions of performance and specifications of the worksystem are, thus, different.

Knowledge of Cognitive Ergonomics

According to Long (1987), there are two types of knowledge – science and non-science (that is, "experiential knowledge, both craft and personal"). Scientific knowledge is particularised into guidelines and checklists to advise designers, such that they produce more usable systems. These forms of advice are considered to be a CE 'applications representation' of the scientific knowledge. Two main lessons have been learned.

Lesson Learned 7: CE needs to acquire and validate knowledge, which supports practices of design, evaluation and iteration. It is unclear why Long's 'application representation' is not a third type of knowledge, along with science and non-science, since it is self-evidently neither of these types. If it is a type of knowledge, then it is not simply 'advice to designers'. Last, design knowledge needs to be related directly to more completely specified design practices, such that these practices are supported (and not only generally advised).

Lesson Learned 8: CE needs to acquire and validate knowledge, which references its particular scope of: users; technology; work; and performance. In other words, CE design knowledge needs to reference what is designed, but not simply as a description of the interface, or the user's behaviours or some other individual aspect of the particular scope. If the latter includes some notion of performance, then guidelines need to reference such performance. For consensus on this lesson learned, see Vicente (1998). Two main lessons remain to be learned.

Lesson Remaining 7: CE needs to acquire and validate knowledge, which supports design practices of diagnosing design problems and prescribing design solutions. It is not sufficient for CE knowledge to support design indirectly, nor to support directly only general practices. The knowledge must address directly the specific design practices (that is, of diagnosis and prescription).

Lesson Remaining 8: CE needs to acquire and validate knowledge, which supports the diagnosis of performance, as the design problem and the prescription of the interactive worksystem, as the design solution. Again, this remaining lesson emphasises the specific relationship required between CE knowledge and CE practices, as well as between these constituents and the CE general problem and particular scope. Consider next the discipline of CE.

Discipline for CE

According to Long (1987), "the most general definition of Cognitive Ergonomics is thus the application of Cognitive Psychology to work". In addition, "Cognitive Ergonomics is a configuration relating work to science". Last, "Its aim is the increase of compatibility between the user's representations and those of the machine". CE particularises scientific knowledge into application representations, which are then synthesised to optimise "the relationship between people and their work ... with respect to well-being and productivity". CE, then, is conceived as an applied science discipline.

Lesson Learned 9: CE needs to be a design discipline. As such, CE needs to be essentially an engineering, rather than a science, discipline. Engineering, here, may be informal (craft) or formal and may be related to science (although not as an applied science). CE research thus acquires and validates engineering CE design knowledge (Hollnagel, 1998).

Lesson Learned 10: CE, as a design discipline, needs to acquire and validate design knowledge to support design practices of design, evaluation and iteration, solving the general problem, having the scope of: users; technology; work; and performance. In other words, CE needs to relate to all the requirements of a discipline (see earlier). This lesson learned essentially brings together all those requirements. Two main lessons remain to be learned.

Lesson Remaining 9: CE needs to acquire and validate its design knowledge as design principles, and design practices by solving design problems. In other words, having the constituent elements of a discipline is not enough. The elements need to be validated to ensure the status of the knowledge, as formal design principles, with respect to the practices etc. Validation needs to be formal (rather than informal) to make CE a formal engineering design discipline (rather than a craft one). In this way, CE design knowledge would offer a better guarantee than current knowledge, such as guidelines.

Lesson Remaining 10: CE, as an engineering design discipline, needs to acquire and validate design knowledge to support design practices of diagnosing design problems and prescribing design solutions, solving the general problem having the general scope of an interactive worksystem (users and technology) with a desired performance, expressed as task quality and worksystem costs. The remaining lesson emphasises the intimate relations, that is, the coherence of the CE discipline characterisation. Such coherence is notably missing in Long (1987).

Additional Issues

As concerns the relationship between CE and Ergonomics, no attempt has been made here systematically to relate the two. However, if CE is assumed to be the most recent sub-set of Ergonomics, then they are likely to have lessons remaining in common.

As concerns the relationship between CE and Human-Computer Interaction (HCI), no attempt has been made here to distinguish these potentially different, but actually similar disciplines. Although the origins and the professional pragmatics of the two disciplines may be very different, the two terms are interchangeable. As concerns the relationship between CE and Cognitive Engineering, the relationship is, if anything, closer than between CE and HCI. The same comments hold here, then.

As concerns the relationship between CE and Engineering, the former never operates in a vacuum. Design is not implementation. CE, then, is usually associated with some form of engineering, typically Software Engineering. If CE is conceived as conjoint with some form of engineering, then lessons learned and remaining are likely to be similar.

Summary and Conclusion

This paper has attempted to ground CE within a CE discipline framework. CE can thus be expressed: completely; coherently; and in a manner which is fit-for-purpose. The paper reviews the past (Long, 1987), the present, as lessons learned, and the future, as lessons remaining. In conclusion, the major CE challenge in the new millennium will be the enhanced guarantee offered by CE design knowledge to support CE design practices. The need is for CE research to acquire and validate knowledge to support the transition of CE, from an essentially craft (engineering) discipline based on informal knowledge, supporting 'trial-and-error' practices, to an essentially theoretical (engineering) discipline, based on formal knowledge supporting 'specify-then-implement' practices. The present paper is intended as a (modest) contribution to the achievement of that transition by encouraging consensus, concerning the discipline of CE.

References

Card, S., Moran, T. and Newell, A. (1983). *The Psychology of Human-Computer Interaction*. New Jersey: Erlbaum.

Dowell, J. and Long, J. (1989). Towards a conception for an engineering discipline of human factors. *Ergonomics*, 32, 1513-1535.

Hollnagel, E. (1998). Comments on 'Conception of the cognitive engineering design problem' by John Dowell and John Long. *Ergonomics*, 41, 160-162.

Long, J. (1987). Cognitive Ergonomics and Human Computer Interaction. In *Psychology at Work*, P. Warr (Ed). England: Penguin.

Long, J. (1998). Specifying Relations between Research and the Practice of Solving Applied Problems: An Illustration from the Planning and Control of Multiple Task Work in Medical Reception. *In* D.Gopher and A. Koriat (eds), *Attention and Performance XVII*. Cambridge, MA: MIT Press, 259-284.

Long, J. and Dowell, J. (1989). Conceptions of the discipline of HCI: craft, applied science, and engineering. In A. Sutcliffe and L. Macaulay (eds), People and Computers V. Cambridge, UK: Cambridge University Press.

Reason, J. (1998). Broadening the cognitive engineering horizons: more engineering, less cognition and no philosophy of science, please. *Ergonomics*, 41, 150-152.

Vicente, K. (1998). An evolutionary perspective on the growth of cognitive engineering: the Riso genotype. *Ergonomics*, 41, 156-159.

SYSTEM DESIGN

COTS EQUIPMENT IN ADVANCED MILITARY SYSTEMS – A USABILITY PERSPECTIVE

Iain S. MacLeod & Karen P. Lane

Aerosystems International
West Hendford
Yeovil
Somerset BA20 2AL
iain.macleod@aeroint.com

Driven by military politics, new technologies can be adopted to provide more sophisticated and effective systems. Further, through the use of Commercial Off The Shelf (COTS) systems, it is considered that the required sophistication can be provided at a lower cost and in shorter time scales. However, the more advanced and complex the host system, the greater the problems with COTS adaptation. These problems are mainly associated with the many diverse new system issues introduced by COTS components and can result in the development of systems with poor usability and sub-optimal performance. The implications from a usability perspective are discussed in this paper.

Introduction – System Complexity and Costs

In an effort to improve mission effectiveness, modern military advanced avionics systems are becoming increasingly complex through the use of new technologies. A trade-off exists between the use of minimum personnel, and the successful operation of the systems. Since Military budgets in the Western world are shrinking, the politics of military system procurement are emphasising the need to use Commercial Off The Shelf (COTS) systems, where possible, with promises of savings in terms of cost and system development time. COTS equipment refers to a product, such as an item, subsystem, or system, traded to the general public in the course of normal business operations at costs based on current market prices. Spiralling costs have been particularly evident where avionics hardware and software were specially designed to meet military requirements. The bespoke specification of military equipment was an attempt to improve system reliability and utility, thereby decreasing through-life costs. However, such specialised development is costly when considered in terms of both the large amounts of money, and time, needed to design and develop such systems. Moreover, the time needed to bring such systems into operation meant that they were often seen as 'out-of-date' before they became fully operational.

With the vastly improved reliability of many commercial applications, the justification for developing costly applications under purely military specifications has diminished. Military procurement requires systems with greater flexibility but at as low a cost as is effectively possible. However, achieving large savings through use of COTS systems, carries the cost of system acceptances by being largely bound by that marketplace's realities and rules. COTS equipment is produced to meet a specific commercial market requirement. Thus, the combining of such components, to meet a complex advanced military system specification, may fail to satisfy the system requirements dictate of the "whole being greater than the sum of the parts". While there are many outlined benefits in using equipment originally developed for purely commercial purposes, there are also many drawbacks. One particular drawback of using COTS equipment can be a loss of 'fitness-for-purpose' or usability. As the operation of systems increasingly becomes more of a cognitive than a physical task, a lack of design consistency will cause quandaries in operator understanding and their ease of use of a system. Thus, system requirements for cognition, a 'cognitive' functionality, should also be considered in the logical requirement specification of systems (MacLeod, 1998). Inconsistency within a system will affect the ease of operating the system and decrease its overall utility.

Why consider COTS usability?

Human Factors and Test Engineering offer methods of evaluating the mandated integration of COTS products, and the evaluation of COTS products as stand-alone tools. There are many tried and tested methods, which can be used to analyse dimensions of system usability in a particular COTS design. The performance of such analyses must result in well-founded recommendations to system designers. Recommendations to system designers should cover as a minimum:

➢ The required system functionality to support system operation;
➢ Indications on the system engineered support necessary for operator task performance;
➢ Suggested system operating sequences;
➢ Indications on how to decrease the system performance criticality of particular operator activities;
➢ Advice on how to maintain consistency within the Human Computer Interface (HCI) design;
➢ Guidance on the appropriateness of the various COTS HCI forms available.

Recommendations must be placed within the constraints of the intended system usage and its operating environment. However, the aim for overall usability in the design must be derived through an understanding of the intended operating environment for the system and an explicit anticipation of future operator(s) tasks. In particular, the adopted HCI form for the system should be primarily dictated by user needs, and good ergonomics practice, and not by the best form available from a selection of often equally unsuitable but readily available COTS HCIs. Experience within the design and development of many large military avionics systems suggests that good planning and commitment to the promotion of system usability by companies is rare. Rubin (1994) described the most frequent reasons for poor system usability as being due to:

➢ A broad user base, with a discrepancy between the user and designer in terms of skills and expectations;
➢ The system components related to usability being developed independently by specialized teams;
➢ The product development emphasis being on the machine or system, rather than on the person who is the ultimate end user;

➤ The design of usable systems being a difficult endeavor, yet many organizations treating it as though it was just common sense.

In addition, the authors' experience with several major UK organisations suggests that a lack of good early proactive systems programme planning is a major reason for inadequate attention to COTS usability issues during design and development. Deficient early proactive planning results in poor reactive planning and the crisis management of many unexpected issues as they occur. In such a regime, early concerns tend to focus on the physical aspects of engineering. Software concerns follow later, this resulting in usability activities being then addressed too late in the design and development programme for it to be truly effective.

Usability Issues in the adoption of COTS systems

Although COTS products are already in service as part of avionics-based systems, many of the issues affecting the efficient adoption of COTS have yet to be properly addressed. In the USA there has, over the last decade, been a considerable body of work to develop standards appropriate to the adoption of COTS within military systems. Moreover, in the near future avionics systems predominantly based on COTS will enter military service in the UK, this despite the fact that there is little UK industrial experience in the design, development, or production of such systems. The following sections outline some of the main issues affecting the adoption and usability of COTS components embedded within avionics systems.

Specification and evaluation of COTS products
In traditional development, the principal constraint on a system may be its requirements. A COTS-integrated system, however, is constrained both by the system's requirements, and by the capabilities and constraints of the available COTS components. The traditional sequence of specifying requirements and designing a system cannot be applied in the hope that the performance of the implementation phase will simply be conducted by the purchase of COTS products. To do so would be naïve, unless the COTS product, by some chance, was a perfect fit to the requirements. Instead, we must now do a significant and careful appraisal of the products prior to selection and include product evaluation in both the implementation free, and physical stages, of system design and development.

Most performance failures are due, at least in part, to a lack of consideration of performance issues in requirements specification or during the early stages of system development and integration process. The tradition in systems design and development is to defer consideration of performance until the later testing phase. Further, there is the fatalistic tendency to believe that performance cannot be engineered because the COTS product performance is 'fixed'. The quantity of modification permissible on a purchased COTS product will depend on the nature of the contract and, in the case of large software packages, on whether an operating system is being bought or whether the software comprises both operating and application software. Performance problems, however, may arise from many sources in addition to individual COTS components.

Within complex systems incorporating many COTS products, there are many mediators on system performance. These mediators can include software interfaces, the integration of components, the interactions between components / timing on a network, hardware, or customer designed software supporting the new system. Thus, if possible, it is necessary to engineer additional performance into complex advanced systems using

COTS products. Further, the evaluation activities required are also different from traditional systems activities. The traditional process of evaluation mandated that specified system requirements were met in some form or another. With COTS products, more than with previous systems, it is necessary to evaluate not only whether the functionality of the system meets the required specification, but also what the COTS system can do within the spectrum of the overall host system. Particular attention should be paid to usability in this area.

The trade-offs in the selection of COTS products
Design activities have always included trade-offs. With COTS products, however, the types of trade-offs made are significantly more diverse. For example, regardless of how much the reliability or usability of a component is considered, the reliability of its vendor must also be factored in, both in terms of their staying in business and in offering reasonable product support. Consideration must be also be made of the cost of through life maintenance to the vendor of the product; and the impact of design upgrades and their frequency, on both the product and the other COTS products used in the system.

COTS products as part of a new system or an addition to an existing system
If the COTS product is part of a new system, the problem of selection is different from the problem of adding a COTS package to an already existing system. If the COTS system is a simple one-to-one application package such as a word processor, then selection is a relatively easy matter of choice based on product price, maturity, and compatibility with other word processing packages. If the new system is to consist of many hardware and software COTS packages, then the decision is much more complex, and most of the issues raised in this paper will need to be considered along with many others not mentioned here. It is in this kind of complex new system that consideration of usability is particularly important as good usability is essential if system performance requirements are to be met. However, under current design practices it is most likely to be considered as a secondary problem when compared to the large engineering problems involved in interfacing all the COTS packages into an intended host system.

COTS system updates
Even a simple COTS system can be costly to upgrade both in terms of the length of support that the vendor will give the purchased version of the COTS, but also in terms of the effects on any software packages that a change to the COTS may involve. One recently observed example was the upgrade of an image processing graphics terminal, where the upgrade to the operating system would not support the imaging processing application that was being used. Further, any upgrade to operating systems may also add unwanted functionality and may affect established HCI practices by changing the interface style.

Costs of COTS adaptation
The costs of COTS adaptation will vary with the complexity of the system, and the security and pertinence of the interface standards to which each COTS product subscribes. In a complex system, the costs of a COTS package integration can be vast and can largely negate the expected cost savings in using the COTS product. Further, poor proactive programme planning, and a last minute crisis management approach to the interfacing of the various COTS products, will increase this cost impact. The more reactive the planning, the less time that is available to consider integration issues, and the less likely adequate attention will be applied to the usability issues. With software based

COTS there is usually some scope to change the apparent functionality, possibly by HCI adjustments, provided that this does not appear to seriously redesign the product. With a single or simple COTS system, the emphasis is mainly on tailoring it to requirements.

With more complex integrated systems, the problem emphasis is on interfacing the COTS subsystems with other subsystems. Naturally, there are many levels of consideration required between the tailoring of a simple COTS system and the interfacing associated with high complexity. With larger systems, it should be possible to encompass the COTS systems within an overarching system framework and interface, including a common HCI. However, the development of such a systems framework requires good proactive planning and team commitment to the achievement of good system usability. Further, the development of such a framework may decrease ownership costs over the life of the system, but would impose an increase in the design and development costs when compared to a COTS based system without such an interface. At acceptance, both parties agree to accept the system, normally with noted reservations on the part of the user or their representatives. Thus, acceptance has similarities to the traditional approach to systems design and development but can be achieved with savings on the costs and time needed for traditional system design and development.

Over and under functionality of COTS

COTS products are sensibly procured as being the best fit for purpose from a selection of like products. However, it is possible that the product selected may be under or over functioned when compared to the functionality required. If the functionality is less than required, a possible solution is to purchase another less complex product to address the shortfall. This will only be a viable proposition if the product can easily be integrated with the system as a whole. If the COTS product is over functioned, then it may be possible to ignore the excess functions, provided that the required functions do not depend on the use of the excess functions under the product design. Whatever the issues with unwanted or missing functionality, the existing functionality must be identified and an analysis performed on its related impacts on the host system. In some cases, work will have to be performed to prevent activation of unwanted functionality or to equate mission functions, and costs will be incurred. (Kohl, 1998). In addition, it should be noted that from a usability perspective, an over-functioned HCI can be a source of operator error through evoking unwanted operator behaviours (Langer, 1992).

Security

Though most avionics COTS products are commercial and partly software based, it is very unusual for the customer to obtain the product source code for any software. This not only causes problems related to gaining assurance that the software is secure in its build, but also opens up the product to two possible security threats, namely:

➤ That the product is purchased by other parties who may have access to the source code and, therefore, may be able to find ways of damaging your system;

➤ The source code as delivered may be flawed or contain a routine that could be evoked in the future to damage system effectiveness by making it unusable.

The above points suggest that military contracts should stipulate that the vendor supplies software source code. A possible alternative is that the reputation of the vendor is assessed as being highly capable e.g. being assessed as Level Three or above on the Software Engineering Institute (SEI) Capability Maturity Model (Widmann & Mindlin, 1998).

Maintenance

Maintenance and development of COTS are much the same. The product is sold with guarantees on its quality and life span. Maintenance does not really occur as an activity, but any updates to the product do. However, the buyer has little control over the form and timing of updates, this control normally the remit of the vendor. Frequently, updates become the market standard and vendor support of previous versions can become increasingly expensive, thus forcing the buyer to invest in the latest version. This may present problems in a simple system where certain applications may not be able to work under any changes to a host operating system. However, system updates to a complex system containing many COTS components, will almost certainly require much effort (and cost) to assess the system interfaces, reliability, and produce sufficient evidence to support or maintain the system's certification. Thus, changes incurred in this area may affect not only the usability of the overall system, but may also be detrimental to the retention of the overall avionics system certification.

Conclusions

This paper has presented a brief coverage of some of the issues affecting the adoption and usability of COTS components within the avionics domain. In particular, the potential adverse effects of COTS on system usability and performance have been emphasised. A primary cause of poor usability is an incomplete understanding by both procurers and engineers of the influences of COTS on traditional system design practices. An associated secondary cause is poor planning of the incorporation of COTS into its host system's design and development. Such poor planning normally results in reactive planning to resolve unforeseen engineering problems at the expense of the final quality of the system's ease of use and utility. As a result, the system performance is often suboptimal on acceptance into service. Further, it is suggested that for future advanced aircrew systems, such as complex airborne mission systems, the mandating of a system-wide and effective overarching interfacing system, including a system-wide HCI, would facilitate the achievement of easy to use high utility systems and their certification.

References

Kohl, R.L. (1998), When Requirements are not isomorphic to COTS Functionality: 'Dormant Code' within a COTS product, in E.E.Barker, A. Morrison & K.Toth, (Eds), *Proceedings of the 8th Annual International Symposium of the International Council on Systems Engineering*, July 26th to 30th, p. 75-78.

Langer, E.J. (1992), Matters of Mind: Mindfulness / Mindlessness in Perspective in *Consciousness and Cognition*, 1, p. 289-305.

MacLeod, I.S. (2000), *A Case for the Consideration of System Related Cognitive Function Throughout Design and Development*, Systems Engineering, Vol.3, No.3, Wiley, p. 113-127.

Rechtin, E. and Maier, M.W., (1997), *The Art of Systems Architecting*, CRC: Florida.

Rubin, J. (1994), *Handbook of Usability Testing*, Wiley, New York.

Widmann E.R. & Mindlin W. (1998) Key Features of the "Merged" EIA/IS 731-1 Systems Engineering Capability Model in Proceedings of the *8th International Symposium of the International Council of Systems Engineering*, July 26-30, Vancouver, Canada.

A CASE STUDY OF PROBLEMS IN PROCESS IMPROVEMENT IN A PUBLIC SERVICE

Vicky Malyon[1] and Murray Sinclair[2]

[1]*Human Applications, 139 Ashby Rd, Loughborough*
Leicestershire, LE11 3AD
vicky@humanapps.demon.co.uk

[2]*Dept Human Sciences, Loughborough University, Loughborough*
Leicestershire, LE11 3TU
m.a.sinclair@lboro.ac.uk

The paper outlines an investigation at a local Fire and Rescue Service (F&RS) for one particular process. This formed a case study for the FR&S focussing on process innovation, measurement and management. The F&RS knew that future needs for quality of service meant a change to their current ways of process execution. This paper discusses the process characteristics and contextual problems, and outlines some recommendations to overcome these problems and to deliver an improved future service, based on business process re-engineering. This case study represents a microcosm of the generic problems an ergonomist may face in process improvement in the coming century.

Introduction

'Today there is wide agreement…..that continual improvement in quality, cost, lead time and customer service is possible, realistic and necessary.' (Schonberger 1986). This was said some time ago, and this paper is a current illustration of its truth. Organisational effectiveness in a context of ever-increasing demand for higher quality of service depends on a number of characteristics.

- A continual focus on the development and maintenance of a culture that supports organisational change and growth (Eason 1988)
- Efficient capture and utilisation of knowledge, both from within the organisation and from its environment (Stewart 1997; Siemieniuch and Sinclair 1999)

- Effective communication in both internal and external relationships (Siemieniuch and Sinclair 2000)
- Continually-revised 'current best' business processes that have been designed from a stakeholders' perspective to ensure that processes run smoothly and are accepted by all involved, together with mature processes for their revision. (Humphrey 1989; Nagel and Dove 1992; Nonaka and Takeuchi 1995; Hammer 1996; Kaplan and Norton 1996)
- Process measurement to establish current performance, to carry out benchmarking of processes, to set 'stretch targets' and to identify opportunities for improvements (Prokesh 1997)

And, as so many managers fail to recognise in their actions, organisational effectiveness depends highly on employee motivation and satisfaction, this is improved through their involvement and recognition in decisions and plans, in access to information and to appropriate support and resources necessary to carry out tasks.

These points are illustrated in a case study undertaken in a public-sector organisation, a local Fire and Rescue Service (F&RS), covering a population of about one million people.

The context for the case study

The statutory duties of an F&RS, emanating from a Local Government Act, include that under a 'Best Value' requirement the F&RS must deliver services to clear standards, covering both cost and quality, by the most effective, economic and efficient means available.

An F&RS is a monopoly and has no competition and it should be acknowledged that economic pressures are not the sole motivation for re-examining their structures and procedures.

The 'One Stop Shop' (OSS) is a service based at the Service headquarters where a small number of operators deal with enquiries. Its main function is to record and progress all non-emergency contacts with the Brigade and provide a direct link with internal and external users. However, over time the function of the service has expanded and it now acts as an information resource for the Brigade and carries out a variety of clerical tasks associated with these functions. It has become known as the 'hub of the Service.' As a result the knowledge required of the staff has been forced to expand. OSS members engage in collaborative query refinement with customers. Consequently, it is imperative that they have an in-depth understanding of the fire safety and legal information that the customer requires, since they form the intermediary between the customer, the databases and the rest of the F&RS.

The Case Study

At the time of the study, the performance of the OSS was well above the minimum acceptable standards, as assessed by the government Fire Inspectorate. However, the ever-increasing quality of service requirements, coupled with the change in philosophy raised doubts about its future effectiveness, and it was decided to 're-engineer' the OSS and its processes.

The main aim of the case study was to clarify the OSS processes, based on the aims of a 'Performance plan' recently drawn up for the OSS. The goals of this clarification were efficiency (a measure of how economically the organisation resources are utilised when providing a given level of customer satisfaction (Neely, Gregory et al. 1995)and effectiveness (the extent to which customer requirements are met, (Neely, Gregory et al. 1995)). The approach was based on the principles of 'Process Innovation' or 'Business Process Re-Engineering.'

An ethnomethodological approach was adopted. The design of the investigation was 'emergent' (Lincoln and Gruba 1985) and, although idiographic in nature was intended to harvest nomothetic insights into the importance of process definition and management and the supporting infrastructure. Semi-structured interviews regarding processes, tasks, roles and jobs were held with all members of the OSS, and stakeholders of various ranks. A full ergonomics audit was also carried out within the OSS premises, the workstations and associated facilities. A combined set of recommendations were made; for this paper only those concerned with the main process are reported. This is the Fire Safety Inspection process.

Findings

The established process is outlined below. There is a requirement that all phone calls are answered within 15 seconds, all correspondence within 4 days, and all jobs completed within 14 days.

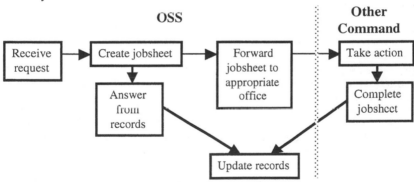

Fig 1. Outline of main OSS process

The main findings of the study are as follows:
- Responsibility and supervision is unclear, particularly after enquiries have left the OSS
- There is a time lag in actions due to distributed locations and shift patterns
- The main IT support for this process is a technocentric office application that is too slow for on-line work as planned and is not available to all Commands. Consequently, paper trails exist in parallel, with distributed versions of files.

- There is a reciprocal relationship with other Commands, with OSS as the process buffer for the customers. OSS have no authority to ensure smooth, efficient operation of the process. There is no audit trail of enquiries or information, which causes repetition and unnecessary 'chasing.'
- Commands do not have a proper understanding of the role of OSS or the enquiry process. This is due to a dislocation between structure and process and an inherent lack of information.
- There is room for improvement in task training, process understanding, and required background knowledge. The F&RS does not yet have in place a strategy for organisational learning or an official training procedure.
- Performance measures for OSS are affected by the actions, or inactions of other groups over which they have no authority. However, such measures are seldom taken or acted upon.
- The office layout and its environment is a constraint to the efficient conduct of tasks within the OSS.

Recommendations

- Re-engineer the main process based on stakeholder analysis, resource requirements, communication channels and its measurement and management. Suggestions were made, though this must be the responsibility of the F&RS.
- Provide better IT support for the process and its users, as a socio-technical system
- Ensure physical environment meets legislation and allows efficient conduct of tasks.
- Define ownership of the process at a suitable level to ensure authority over all parts of the process. Given that the process runs across Commands, a matrix management philosophy may need to be adopted where 'key' players are identified.
- Process performance measures should be chosen carefully; given that the process crosses Commands. These should be measurable at the hand-over interfaces as well as for the entire process.
- Provide training based upon communication with stakeholders requirements. Consideration should be given to the establishment of an organisational learning strategy. This should cover both fire risk issues in the community and process improvement issues.

As result of the case study the first author has developed a model of Process Innovation, Measurement and Management.

REFERENCES

Eason, K. D. (1988). Information technology and organisational change. London, Taylor & Francis.

Hammer, M. (1996). Beyond re-engineering, Harper Collins.

Humphrey, W. S. (1989). Managing the software process. Reading, MA, Addison-Wesley.

Kaplan, R. S. and D. P. Norton (1996). The balanced scorecard, Harvard Business School Press.

Lincoln, Y. S. and E. G. Gruba (1985). Naturalistic enquiry. London, Sage.

Nagel, R. and R. Dove (1992). 21st century manufacturing enterprise strategy. Part 1: an industry -led view. Part 2: infrastructure, Iacocca Institute, LeHigh University.

Neely, A., M. Gregory, et al. (1995). "Performance measurement system design: a literature review and research agenda." International Journal of Operations and Production Management **15**(4): 80-116.

Nonaka, I. and H. Takeuchi (1995). The knowledge creating company. Oxford, Oxford University Press.

Prokesh, S. E. (1997). "Unleashing the power of learning: an interview with BP's John Browne." Harvard Business Review(Sept-Oct): 5-19.

Schonberger, R. J. (1986). World Class Manufacturing: The Lessons of Simplicity Applied. New York, NY, The Free Press.

Sicmieniuch, C. E. and M. A. Sinclair (1999). "Organisational aspects of knowledge lifecycle management in manufacturing." International Journal of Human-Computer Studies **51**: 517-547.

Siemieniuch, C. E. and M. A. Sinclair (2000). "Implications Of The Supply Chain For Role Definitions In Concurrent Engineering." International Journal of Human Factors in Manufacturing **10**(3): 251-272.

Stewart, T. A. (1997). Intellectual capital: the new wealth of companies. New York, Doubleday.

THE PROCESS OWNER - A ROLE TO OVERCOME PROBLEMS OF MANFACTURING COMPLEXITY AND ORGANISATIONAL LEARNING

C.E. Siemieniuch[1] & M.A. Sinclair[2]

[1]*Research School in Ergonomics & Human Factors, Loughborough University, LE11-3TU; c.e.siemieniuch@lboro.ac.uk*
[2]*Dept of Human Sciences, Loughborough University, LE11-3TU m.a.sinclair@lboro.ac.uk*

The paper outlines some implications of complexity as applied to supply chains. The Process Owner role (and its required support) is outlined, as a way of preserving and evolving the organisation's knowledge. The key advantages that emerge from the incorporation of this role are that: capability acquisition processes can be properly managed; work can be moved to the people, rather than the people to the work; audit trails are easier to identify; organisational learning becomes more manageable; and the usual drift towards unsafe operations can be resisted. Some of the disadvantages will also be outlined: the organisational culture must be adapted for the role to work effectively; top management support for the role is critical; the role requires wise people.

Some effects of complexity on manufacturing organisations

Complexity is of interest in business circles for its strategic implications (e.g Kelly and Allison 1999; Rycroft and Kash 1999). A set of characteristics that cause complex behaviour in enterprises have emerged from several studies. Some of these are:
- Many agents, of different kinds (e.g. individuals, teams, business units)
- Lots of parallel connections between agents (e.g. email, phone, fax)
- Effects of an evolving environment (e.g. legislation, e-business, globalisation)
- Effects of co-evolving agents (e.g. competitors, suppliers)

Managing within complexity leads to these conclusions, especially for large firms:
- 'normality' will require unremitting effort (Gregg 1996; Rycroft and Kash 1999);

- the competitive environment will change: information overload; unbundling of products; a move to delivering lifecycle services (Martin 1998)
- there will be a need for a modular, 'business unit' company structure to tolerate internal cannibalisation among these business units (Nadler and Tushman 1999).
- this cannibalisation will create conflict and insecurity; consequently, a conflict-minimising culture (job security, re-skilling, etc.) will be essential. Furthermore, conflict management will be a necessary skill for all middle and top managers
- careful strategic planning will be required, both to predict the occurrence of conflict and to ameliorate its effects. This must include: shared values, culture and goals; organisational design with an emphasis on feedback loops, clarity of policies, processes, roles, responsibilities, authority, resources, rewards, etc.; and shared knowledge management processes.

There are seven bases on which organisational effectiveness depends in order to manage the characteristics discussed above:
- A devolved organisational structure.
- The development and maintenance of a culture that supports organisational change
- IT-based decision tools to business decisions, for (a) projecting the future, and (b) configuring the enterprise's resources to address this future environment..
- A modular, open IT&T infrastructure serving the whole extended enterprise.
- Revised, 'current-best', business processes designed from a stakeholder perspective, using a proven capability acquisition process
- Process measurement, to measure performance, to create benchmarks, to set 'stretch' targets, and to identify opportunities for improvements.
- Efficient capture and use of knowledge over its lifecycle, within the enterprise and from its environment - technological knowledge; organisational knowledge, and network knowledge, inherent in the alliances and relationships that exist.

However, these bases must be used by roles; a new one of interest for complexity and evolving environments is the 'Process Owner' (Hammer 1996).

Process ownership and control

It is the processes which the organisation utilises and the classes of knowledge required for them to work, which comprise the core of the organisation's competence. In view of the issues described above, some companies require that each process within the core competence of the company is owned by a 'Process Owner' centrally in a generic form. The process is then 'licenced' for use by a Process Manager within a business unit, who utilises and manages a tailored version of the generic process as appropriate for the unit's segment of the market and the characteristics of its customers. In the case of manufacturing organisations, the process may be physically instantiated in the form of a factory site with linked machines and offices; in this situation there may be several business units making use of the process, with inevitable scheduling and other internal political problems.

This approach does have some advantages from a complexity viewpoint:
- It permits a longer-term view to be taken of the process, its resource requirements, and its development
- It focuses attention on the key issues for the organisation in its environment, and in effect prioritises the problems that require attention
- In the short-term, work can be brought to the people, rather than the people to the work. For both regional and global companies, this is a significant advantage
- It encapsulates customer-related, supply chain and operational complexity issues within a single business unit, with localised feedback and control
- It encourages senior management to focus their attention on capability acquisition and organisational renewal with less need to 'fire fight' short term problems
- It provides an organisational structure by which to address the teleology of safety

There are some disadvantages as well:
- As discussed earlier, there can be severe problems of rivalry and resource hoarding, leading to dissipation of useful effort
- This can only work well in an organisation or a supply chain which has shared values and secure people.
- A generic process that is deployed into geographically distributed sites may encounter significant cultural and legal differences to which it must be tailored.
- A process whose product serves markets in different regions of the world will have to be tailored to accommodate regional, cultural and legal differences.
- Big companies frequently grow by the use of mergers and take-overs; this can leave 'cultural legacy' problems in the implementation of generic processes.
- The beginning and end of a process may not be on one site; hence the federated control characteristics of supply chains may apply (Siemieniuch and Sinclair 2000)
- Constant management is needed to prevent different instantiations evolving in disparate ways, thus dissipating commonality of process
- The Process Owner's role will be nearly impossible unless an appropriate culture exists in the organisation. The split of ownership from operational responsibility, although it has particular advantages from an organisational learning perspective and from the point of view of quality, internal coherence and economy of resource use, also has drawbacks. These include the differences in goals and the possible differences in response latencies between operations and process change, which provide ample scope for conflict and failures of communication.

The role of the 'Process Owner'

The primary tasks for the Process Owner are:
- to document and proceduralise the process, its controls and resource requirements in a generic form that represent best current practice.
- to ensure that this generic process is maintained as best current practice in its various instantiations in different facilities.
- to authorise variations from the generic process to fit local circumstances in a given instantiation, while ensuring that the safety and integrity of the process is not compromised.   This is a particularly important teleological issue, as Rasmussen (1997, 2000) has cogently argued: 'A closer look at the recent major accidents show that they are not caused by a stochastic coincidence of failures and errors, but by a systemic migration of organisational behaviour towards the boundaries of safe operation.   Major accidents are the side-effect of decisions made by several decision-makers, in different organisations, and at different points in time, all doing their best to be effective locally'.
- to accept and authorise improvements to the process as an ongoing evolutionary strategy, again with regard to process integrity and safety
- to ensure that any proposed changes to the process do not have deleterious effects on other, related processes (and *vice versa*).
- to support the change process by which process improvements are introduced

For this role to be operated effectively, a number of organisational issues must be addressed:
- The Process Owner must have sufficient responsibility and authority to make changes when they are required, and must have sufficient support from senior management to be able to resist special pleading regarding changes.   Secondly, the Process Owner must have in place a mature process for changing the process. It is essential that the Process Owner is seen as a key component in the organisation's learning processes
- For the Process Owner to be responsible for process integrity and safety, he or she must be in a position to know what is happening in the process instantiations wherever they may be and whatever the security that surrounds them, and must have the authority to manage and resource the development and implementations of the process. A clear implication of this requirement is that the processes as implemented are properly metricated for this purpose.
- Identification of appropriate metrics (e.g. to measure process resilience), with the usual reliability, precision, and accuracy problems, and the issues of acceptability, ease of collection and relevance to the organisation and its people

- A critical issue is the identification of the generic process itself, the scope for change, what constitutes a 'change' (and therefore requires formal procedures), and how this will affect related processes. Note that Process Managers have as their primary goal the efficient delivery of a product from the process, and they may introduce deviations from the prescribed process in order to address the problems associated with this goal. These problems may arise from outside the process (e.g. the client demands extra work) or internal pressures (e.g. the formal change process is too slow), or from normal *kaizen* activities.

Finally, we point out that most of the issues discussed above fall well within the boundaries of socio-technical systems theory, and as such, represent a re-arrangement of well-known ideas and concepts; indicating once again that human beings have not changed much in half a century, and that the pioneers in this field were indeed visionary.

References

Gregg, D. (1996). Emerging challenges in business and manufacturing decision support. The Science of Business Process Analysis, ESRC Business Process Resource Centre, University of Warwick, Coventry, UK, ESRC Business Process Resource Centre.

Hammer, M. (1996). Beyond re-engineering, Harper Collins.

Kelly, S. and M. A. Allison (1999). The complexity advantage: how the science of complexity can help your business achieve peak performance, McGraw Hill.

Martin, P. (1998). Time for new trade-offs. Financial Times. London: 42-43.

Nadler, D. A. and M. L. Tushman (1999). "The organisation of the future: strategic imperatives and core competencies for the 21st century." Organisational Dynamics **Summer**.

Rasmussen, J. (1997). "Risk management in a dynamic society: a modelling problem." Safety Science **27**: 183-213.

Rasmussen, J. (2000). "Human factors in a dynamic information society: where are we heading?" Ergonomics **43**(7): 869-879.

Rycroft, R. W. and D. E. Kash (1999). The complexity challenge. London, Pinter.

Siemieniuch, C. E. and M. A. Sinclair (2000). "Implications Of The Supply Chain For Role Definitions In Concurrent Engineering." International Journal of Human Factors in Manufacturing: 10(3), 251-272.

METHODOLOGY

Practical considerations for undertaking ergonomics research in rural sub-Saharan Africa

Marc B McNeill[1] and Dave O'Neill[2]

[1]Accenture
1 Kingsway
London
WC2R 3LT
marc.mcneill@accenture.com

[2]International Development Group
Silsoe Research Institute
Wrest Park
Silsoe
Bedfordshire MK45 4HS

Conducting research in an Industrially Developing Country can present practical challenges that are not encountered in developed countries. Cultural taboos, limited vocabularies and low levels of literacy which can lead to inaccurate translations, logistics and bureaucracy can conspire to make conventional ergonomics working practices and methodologies difficult to implement. In conducting ergonomics analyses in rural agriculture, an adaptive ergonomics approach is often required, with, in the first instance, a core participatory focus, allowing a rapid identification of problems to be made by farmers themselves. Where appropriate, objective ergonomics measuring tools are used to reinforce the findings of the participatory work. This paper draws from experiences of working in the field in rural sub-Saharan Africa and discusses issues that must be considered when conducting ergonomics work in such circumstances.

Introduction

The majority of the population in Industrially Developing Countries work in the agricultural sector, with human effort being the primary source of mechanical power. The multi-disciplinary approach of ergonomics addresses the input of human energy to work, aiming to improve the efficiency of human labour, reduce drudgery and improve safety and health. With the achievement of these goals, increases in productivity and well-being can be expected. Ergonomics research in industrially developing countries (IDCs) is being practised with increasing momentum. With an increasing focus upon sustainable rural livelihoods in Development Programmes, the role of the ergonomist in helping reduce drudgery is becoming more widely recognised. Working as an ergonomist in an IDC however, can present practical challenges that are not encountered in developed (i.e. industrialised) countries. This paper addresses some of the issues encountered when conducting field work in IDCs. Primarily drawing from research in Ghana (McNeill, 1999), and anecdotal evidence and experiences in working with rural communities, practical implications of undertaking ergonomics research in the field are addressed and the implications discussed.

Use of tools

The ergonomist working in an IDC requires methods, tools, techniques and standards to apply. These may be well-established in industrialised countries, however the question of whether they can be used in IDCs has rarely been raised. McNeill (1999) suggested that they will generally remain the same whatever the client group's needs are. In an IDC however the approach may require adapting to the local context whilst maintaining the fundamental principles. In the field, objective tools that simple, reliable, robust, and easy to calibrate and maintain were considered most appropriate and usable (Table 1). Participatory methods proved invaluable for preliminary ergonomics analyses of smallholder agriculture, enabling a rapid identification of problems to be made by the participants.

IDCs are usually characterised by low levels of literacy. In 1995 47% of females and 24% of males aged over 15 years in Ghana were illiterate (World Bank, 1997). The use of written documents such as questionnaires and scales cannot therefore be used. Questions may be presented orally whilst scales may be adapted either by reducing the number of points or using visual prompts. Interpretation of pictures however was often confused and it is important to be aware that photographs, pictograms and diagrams are often culturally bound.

Language

Communicating with farmers in their vernacular often creates problems because of the limited vocabulary. Rarely will the language have equivalent scientific terminology (Barasa, 1987). Concepts used in one particular study often do not transfer unambiguously from one society to another, and in some cases even from urban to rural areas (Bulmer and Warrick, 1983). For example the concept of thermal comfort was not readily assimilated by Akan subjects. Nor did they associate "working hard" with exertion. In *Twi*, the Akan language that is spoken by many people in Ghana, subjective feelings were described with more linguistic description, rather than a single word. When presenting questions on thermal stress, whilst there is a word for 'warm' it is rarely used. Instead a sentence is constructed to express that feeling.

Courtesy

In carrying out research in rural communities it was first important to make formal introductions with the chief or village elders. This could be a time consuming activity, particularly if they had to be summoned from some distance and could involve anything from a declaration of the research 'mission' to performing libations. It has been suggested that contacting local government officials before commencing any project is necessary for success (Mikkelsen, 1995).

Receiving peoples' hospitality can result in a bias when they respond to questions. This is known as courtesy bias where the respondent provides information that s/he feels will please the interviewer (Mitchell, 1983). It was apparent on numerous occasions in

Ghana and has been reported elsewhere in sub-saharan Africa (e.g. Mudamburi, 2000). It is also particularly evident in Asia, where as Wuelker (1983) describes:

"Any attempt at sending out non-Asians to interview Asians would be a fiasco. Asians are far too polite to tell a foreigner anything he might not like to hear, and a European will always receive rose-tinted answers to his queries for fear of offending him."

Table 1 Considerations for tools to be used during fieldwork in IDCs

Criteria	Comments
Cost	• Affordability to recipient, both initial cost (although this may be borne by donor) and running/ operating costs • Customs clearance may be required for bringing expensive tools and equipment into country
Robustness	• Will it withstand the rigors of fieldwork? • Is it dust proof?
Size	• Large, conspicuous equipment may appear threatening to rural communities
Portability	• Transporting to country • Transporting to field
Quality of information	• Validity and reliability of results
Time	• Speed of response • Time to set up tool
Sophistication	• The tools should be matched to the needs of the work. Over-sophisticated tools will probably be inappropriate in rural agricultural field work
Calibration	• Availability of calibration equipment in-country (e.g. gasses)
Compatibility	• Hardware-hardware/ Hardware-software should be matched. Assumptions that software will run on local systems or connecting cables, adapters etc. will be available should not be made
Language	• Software, hardware labelling and operating instructions should be in the local users language
Spare parts	• Availability and cost
Servicing	• Availability of trained service personnel
Operating range	• Climatic conditions
Operating requirements	• Constant electricity supply cannot always be relied on. Generator or power regulator may be required
Country of origin	• Many donors insist on procurement of equipment from donor country

Suspicion towards researcher

Field work often resulted in suspicion. During research into working practices on the farm, several groups of farmers refused to participate because of mistrust of the foreign research worker or his motives; 'what is in it for us?' On more than one occasion field work discussions were disrupted by individuals believing the European researchers had come to sabotage the community. During one participatory rural appraisal (PRA), when talking to women involved in brewing an alcoholic beverage that involved a particularly arduous process, husbands tried to reprimand the women for giving away secrets of the process, claiming the researchers were stealing the recipe for their own ends. Suspicion towards foreign researchers working in the field is well documented in the anthropological and ethnographic literature. For example Dubois (1986) noted how the question of one's integrity as a scholar may come under suspicion, giving an example from India where a European carrying out research by asking questions was suspected of being a spy. Similarly Sudarkasa, (1986) found Yoruba traders in Kumasi, Ghana to be suspicious of her notebook, believing her to be collecting information for the government. In the best of circumstances the field worker will initially be an object of curiosity, however fear and hostility may not be unusual.

Cultural taboos

Superstition plays an important role in many rural African communities. This may include one day of the week being considered 'taboo,' and thus no farm work would be conducted on that day. In Zimbabwe this is called a *chisii day.* Farmers are often reluctant to allow visitors on the farm for fear of bad *juju* (voodoo) being placed on their crops. The work of the ergonomist can thus be hampered by being unable to visit the place of work, particularly if time is limited. In Muslim communities during Ramadan, the month of fasting, when the faithful neither eat, drink nor smoke between sunrise and sunset, their working capacity may be affected and may cause ergonomics observations to be unreliable. Furthermore, during this period they are liable to become weak through the stress of fasting and no extra physical effort, such as participating in research, can be expected of them (Wuelker, 1983).

Time and dates

In many languages time is a fluid concept; for example in *Twi* the words for 'yesterday' and 'tomorrow' are identical. Often the western planning model based upon 'economic time' conflicts with time interpretations which value 'social time' (Mikkelsen, 1995). It was apparent during the research that deadlines and schedules were rarely adhered to and it would be unreasonable to adhere to similar time scales for projects that may be undertaken in industrialised countries.

There is a general ignorance of dates and ages in many communities in IDCs (Blacker and Brass, 1973). In many sub-Saharan African communities the concept of 'dates' is virtually unknown. In the two principal local languages in Gambia (*Mandinka* and *Wollof*) there is no word for date (Blacker and Brass, 1973). Many people who were interviewed during the research in Ghana were unaware of their ages, giving an arbitrary guess rather than an accurate figure. By discussing it with others in the village, using memories of historical events and reference to the age of the eldest a more accurate estimate of age than the farmers first guess was obtained, but a random bias was still possible.

Location of research

Working in a local academic institution presented problems. Lack of funding and resources resulted in low motivation and regular absences of staff. In universities, work can often be hindered by negative perceptions towards researchers from industrialised countries. It is not unusual to encounter the attitude that foreign academics go to developing countries to "exploit local scholars and go home to advance [their] careers at [the local scholars] expense" (Dubois, 1986).

The success of PRA discussions often depended upon location in which they were undertaken. As reported in Mukherjee (1996), shade, space and quiet location were necessary for focused and open discussions amongst eligible participants. Conducting the PRA near the participants' homesteads had the advantage of causing minimum disturbance to their daily routines however it often also resulted in, initially at least, interference from large crowds of curious on-lookers.

Bureaucracy

Administration in many IDCs is characterised by inefficient bureaucracy and officialdom. Obtaining clearance for research can take long periods of time. Similarly if scientific equipment is to be taken to an IDC it is important to ensure that customs clearance will either not be necessary or is arranged before travelling to the IDC. Barasa,(1987) considers bureaucracy to be a major impediment in disseminating agricultural research findings, with results having to go through policy makers; conversion into simpler forms by specialists; then through various levels of extension before they finally reach the farmer. By this stage the information may be out-dated and no longer of use.

Conclusions

Conducting research in an IDC can be a difficult, frustrating and sometimes problematic task. Abbot (1995) suggested that methodology has to mould with the situation, arguing that it is important that methodology "does not remain alien to its setting so that it can satisfy the criteria of 'sound research.'" This is not to suggest that research in IDCs will lack scientific integrity. However, to maintain this integrity when working in rural IDCs, an adaptive ergonomics approach should be adopted with a core participatory focus and objective ergonomics measuring tools used to qualify the findings of the participatory work.

Matters that need to be taken into account include the following:

- Awareness of possible communication problems and need for relevant training of project staff.
- Awareness of local customs, protocol, and cultural taboos, both when working with local communities and when meeting officials.
- Development of mutual confidence to ensure honest answers and avoid suspicion towards researcher. This process cannot be hurried
- Understanding of time and dates as practised by subject population.
- Consideration of logistical problems such as ease of access to communities and reliability of utilities when planning time schedules.
- Consideration of impact on the research of climatic and other physical environmental conditions.
- Availability and dependability of human resources and equipment for research.
- Awareness of bureaucracy and security of adequate funding from the outset.

Acknowledgement

The authors acknowledge sponsorship from DFID (Department for International Development, London) in contributing to the production of this paper. However, the views expressed are those of the authors and not necessarily those of DFID.

References

Abbot, D. 1995, Methodological dilemmas of researching women, *The Journal of Social Studies*, **70**, 97-113

Barasa, D.W. 1987, *Agricultural information in sub-Saharan Africa; problems of participating in AGRIS with particular reference to Kenya*, (M.Sc. Dissertation, Loughborough University)

Blacker J. and Brass, W. 1983, Experience of retrospective demographic vital rates. In Bulmer M and Warrick D.P. (eds.) *Social Research in Developing Countries*, (John Wiley and Sons, Chichester)

Bulmer and Warrick, 1983, Data Collection. In Bulmer M and Warrick D.P. (eds.) *Social Research in Developing Countries*, (John Wiley and Sons, Chichester)

Dubois, C. 1986, Studies in an Indian town. In Golde, P. (ed.) *Women in the field; anthropological experiences, 2nd edition*, (University of California Press, Berkeley)

McNeill, M. B. 1999, *Ergonomics issues and methodologies in industrially developing countries* (PhD. Thesis, Loughborough University)

Mikkelsen, B. 1995, *Methods for development work and research; a guide for practitioners*, (Sage Publications, New Delhi)

Mitchell, R. E. 1983, Survey materials collected in developing countries: sampling, measurements and interviewing obstacles to national and international comparisons. In Bulmer M and Warrick D.P. (eds.) *Social Research in Developing Countries*, (John Wiley and Sons, Chichester)

Mudamburi, B. 2000, Significance of animal traction technology development in the smallholder farming areas of Masvingo and Matabeleland: the animal drawn plough and the donkey harness. In O'Neill, D (ed) *Optimising DAP for cropping (Proceedings of Project Workshop, 21-22 September 2000, Harare)*. Report No. IDG/00/22, (International Development Group, Silsoe Research Institute, Silsoe, UK)

Mukherjee, N. 1996, *Participatory Rural Appraisal and questionnaire survey: comparative field experience and methodological innovations*, (Concept Publishing Company, New Delhi)

Sudarkasa, N 1986, Fieldwork in a Yoruba community. In Golde, P. (ed.) *Women in the field; anthropological experiences, 2nd edition,* (University of California Press, Berkeley)

World Bank, 1997, *World Development Report,* (Oxford University Press, Oxford)

Wuelker, G. 1983, Questionnaires in Asia. In Bulmer M and Warrick D.P. (eds.) *Social Research in Developing Countries*, (John Wiley and Sons, Chichester)

RISK PERCEPTION: LET THE USER SPEAK

Freija H. van Duijne, W.S. Green, H. Kanis

School of Industrial Design Engineering, Delft University of Technology
Jaffalaan 9, 2628BX Delft, the Netherlands
f.h.vanduijne@io.tudelft.nl

The literature on risk perception has been reviewed to gain insight into risk perception in consumer product use. It appeared that most studies are carried out in the psychometric tradition. This approach addresses risk perception as a matter of judgement, instead of an interactive process emerging from the product use situation. This paper proposes to study risk perception on the spot, using observations and in-depth interviews. The findings from a study onto a do-it-yourself task are discussed.

Introduction

Over the last decades, various publications have considered the perception of risk in industries, products, and foods. However, it turns out that most of the reported studies are based on surveys and experiments in which people are asked to rate risks. Examples of studies that have taken into account a more elaborate investigation of the user's perspective on risk perception are scarce.

It is questionable whether a summative, psychometric approach can sufficiently describe risk perception in product use and support the safe design of consumer products. In the product use situation, many influences may shape the user's perception of risk which are not captured by ratings. Design is only supported by elaborate descriptions of actual user activities; summative measures tend to be of little or no use (Kanis and Green, 2000). Therefore, an alternative qualitative approach is proposed that combines in-situ observations of product use with in-depth interviews. This approach has been tried out in the study of a woodworking task. The preliminary results provide an illustrative description of emergent risk perception from the user's perspective.

Methods applied in the risk perception literature

Reviewing the literature, it appeared that most research papers have addressed risk perception by survey studies and experiments in accordance with the psychometric tradition. This quantitative approach uses scaling techniques and multivariate analyses to gain insight into risk perception. The data are derived from self-completion questionnaires in which participants are asked to rate the risks of products and activities. In this section we will briefly describe the core findings of quantitative risk perception studies, which originate from different fields of practice. Additionally, a study of risk perception in product use will be described that is based on an alternative qualitative approach.

For the purpose of improving communication on hazardous industrial activities and technologies, Slovic (1987) introduced a two dimensional representation of risk attitudes and perceptions that intends to make clear people's understanding of risks. In this factor analytic representation, which has been based on survey questionnaire data, perceived risk is composed of "dread risk" and "familiarity with the risk". In the same vein, the influence of risk perception on health and environmental behaviour has been studied to anticipate public beliefs. Rundmo (1999) has analysed and displayed in a path analysis the relationships between perceived risk, concern and behaviour.

In the field of consumer product safety, the assumed relationship between perceived risk and willingness to pay attention to product warnings has stimulated risk perception research. Based on survey questionnaire data, correlations and regression analyses have been used to find the best fitting predictors of product hazard perceptions, which have turned out to be "perceived severity of potential injury" and "likelihood of injury" (e.g. Wogalter *et al*, 1991; Wogalter *et al*, 1993). The accuracy of people's risk perceptions has been tested by comparing risk judgements to accident population statistics on injury frequency data (Lichtenstein *et al*, 1978; Wogalter *et al*, 1993). Comparisons between judgements have also been applied in other studies on product safety. Risk judgements and precautionary intentions on both a traditional and an experimental version of a cigarette lighter have been compared in order to assess the safety benefits of child resistant designs (Viscusi and Cavallo, 1994).

Judging the risks of consumer products has been a part of the experimental task in some experiments on warning design, (e.g. Duffy *et al*, 1993; Hatem and Lehto, 1995). In these studies, risk perception indicators, warning notice and recall have been collected by means of a questionnaire and compared to observed compliance to the product warning, using non-parametric statistics, t-tests and ANOVA. Other studies have addressed the predictability of warning compliance and subsequent safety behaviour in product use by means of a survey instrument that measures personal characteristics (Purswell *et al*, 1986). Observed behaviour has been classified into safe and unsafe categories and related to a list of personal variables in order to predict safe usage. The authors suggest that it is possible to predict safety in behaviour by a questionnaire that assesses risk taking attitudes.

In contrast to the psychometric and experimental approach, Weegels and Kanis (2000) have used a different method to study risk in product use. In their study, risk perception has been an aspect of the on-site investigation of domestic accidents with consumer products. By means of in-depth interviews with the victims and accident re-constructions the authors have demonstrated that the majority of the participants had not noticed the particular risk that contributed to the accident before the event had happened. Several informants had been aware of other risks than the one that contributed to the accident, or they had simply not suspected that anything would happen. Sometimes, the product gave the user a false impression of safety. The findings suggest that risk perception and behavioural adjustment are dependent on the situation and the featural and functional characteristics of the product.

Limitations of experimental and psychometric approaches

The overview of the studies on mainstream risk research yielded remarkably few design recommendations for consumer products. This shortcoming may be related to the impediment that the applied methods present to gaining in-depth accounts of the product users' perspectives on risks. The decisions on the practical feasibility of the concept, the selection of a measurement instrument, and analytic techniques have narrowed down the range of outcomes of a study. This section will illustrate these decisive steps and the consequences for risk research in the psychometric tradition.

Often, risk perception has been operationalised as if it were synonymous with a risk judgement. Products and industrial activities are rated in several risk dimensions listed in a pen and paper questionnaire. As a consequence, these risk judgements are independent of the specific situation in which a person is confronted with a hazard associated with the product or activity from the list. As such, questionnaire findings are based on hypothetical events that often tend to have little in common with real life use situations (c.f. Murphy *et al*, 1998). Surveys explain little of user-product interaction and the risk considerations arising from featural and functional product characteristics and other 'contextuals' in actual usage situations.

A questionnaire format with close-ended Likert scale questions imposes additional limitations to the kind of information that will be obtained. When responses are pre-coded, comments or user (re)actions that do not fit the format are usually left out of consideration. Therefore, questionnaire responses form a very restricted interpretation of people's risk perception. Additional problems arise when questions are interpreted differently by participants or when artificial answers occur (c.f. Bercini, 1992; Foddy, 1998). Social desirability in answering patterns may hide a personalised version of experiences. When social desirability is suspected in a face-to-face interview situation, the interviewer can reveal a personal account by more intense questioning of the subject. Interviewers can interact with the interviewee by rephrasing in order to have the interviewee interpret the question as intended.

In the analytic phase, further restrictions on the interpretation of the findings emerge from the summative approach of quantitative analysis techniques. The outcomes are mostly based on mean scores instead of the personal accounts of individual participants. It has been argued that summative evaluations are often of limited relevance to designers since these do not provide any insight into what happens between user and system. In-depth knowledge of user activities has been shown to offer the required support for adjusting system properties to use situations (Kanis and Green, 2000).

Concluding discussion of the psychometric tradition, it is considered that this approach is only suitable for describing people's opinion on risk. However, a general opinion is not relevant for risk perception in product use. Instead, risk perception research should address the use situation in great detail in order to gain insight into the process of risk perception. Observations and interviews may offer a method of studying risk perception in a naturalistic environment, and to demonstrate the role of featural and functional product characteristics on the emergence of risk perception. Such findings can be applied in product redesign.

Empirical work

The study by Weegels and Kanis (2000) has demonstrated that in-depth interviews and demonstrations of product use can provide informative accounts of the risks perceived by product users. However, these authors have studied only people involved in accidents with consumer products. It is conceivable that risk is perceived differently by people who are not involved in an accident. Therefore, it has been decided to explore risk perception in product use more generally in a common do-it-yourself product use situation.

Seven participants were observed while making a halving joint. The tools they used were a hand saw, a chisel and hammer, an electrical drill, and a screwdriver. Afterwards, in a semi-structured interview they were asked to clarify their user actions and the risks perceived. Then, the analysis of the video-recordings was used to generate questions for a second interview. This interview addressed the participants' perceived risks of tool usage combined with their personal experiences. By referring to observed user actions and previous tool usage, the participants were invited to use these contexts to explain how they perceive and evaluate risks in tool usage.

From the findings, it appears that risk perception is difficult to observe in user actions. This may suggest that the tools used in this study do not raise risk perception to a conscious level. Most user actions are motivated by "ease of use", or people simply do not know why they place their hands here or there. Frequently, safety is at the most a secondary motivator. For instance, participants do not know why they place their free hand on the bench vice during sawing. It is convenient to place the hand somewhere, but preferably at a safe distance from the saw blade. Likewise, it is convenient to put long hair in a ponytail, which also prevents the hair getting stuck in machine parts.

Contrary to the observed tool usage, the personal experiences of the participants do create a basis on which to discuss risk perception. For example, participants indicated that minor injuries are "part of the game". The experience of a scrape or a small cut does not necessarily make a product dangerous in the eyes of the user. The piece of work to be processed may be of far more importance. It should not get damaged and the aim of the task must be achieved. For this purpose people may become inventive in finding solutions. Then, it can happen that one will take some (reasonable) risk.

From the tools in the study, the chisel is considered the most dangerous, because it can hit one's stomach. However, when the tool is directed away from the body, the product is perceived to be safe, as shown in Figure 1. According to the participants, electrical tools are more dangerous than hand tools. Band saws and sanding machines belong to the most hazardous tools; they have fast moving parts, and in case of an emergency, those machines are difficult to stop in time.

Figure 1. Using the chisel safely

To gain insight in people's understanding of the occurrence of accidents with do-it-yourself products, i.e. their cognitions, participants were asked what reasons in their case may cause an accident. They indicated several contributors: lack of knowledge of how to use a tool safely, insufficient caution, and lack of concentration. Overestimation of one's ability has also been singled out as a contributor to accidents. Likewise, trying to get something done with a tool that does not fit that particular task is perceived to be of risk.

Conclusion

The review of the literature has demonstrated that risk perception has been assessed as a if it were a belief or a judgement. This approach ignores the influence of context on the emergence of risk perception. Therefore, the psychometric approach reflected in experiments and survey studies seems of little relevance for understanding risk perception in product use.

An alternative view maintains that in product use, risk perception is situated and locally produced. Thus, insight into risk perception should be derived from the user's experience. This perspective resembles the ethnomethodological framework that in recent years has been embraced in the studies of human-computer interaction (Dourish and Button, 1998). In this tradition, observations of activities in interaction form an empirical and analytic perspective which may be used in the design of technology.

However, of equal interest to ethnomethodological observations is the study of human predispositions. Experiences, and cognitive and antropometric characteristics may directly or indirectly set boundary conditions for the perception of risks. This perspective of the combination of methods will be further explored in the near future.

References

Bercini, D.H. 1992, Pretesting questionnaires in the laboratory: an alternative approach, *Journal of Exposure Analysis and Environmental Epidemiology*, **2**, 241-248

Dourish, P. and Button, G. 1998 On "Technomethodology": Foundational Relationships between Ethnomethodology and System Design, *Human-Computer Interaction*, **13**, (4), 395-432

Duffy, R.R., Kalsher, M.J. and Wogalter, M.S. 1993, The effectiveness of an interactive product warning in a realistic product-use situation, *Proceedings of the Human Factors and Ergonomics Society, 37th Annual meeting*, 935-939

Foddy, W.H. 1998, An empirical evaluation of in-depth probes used to pretest survey questions, *Sociological Methods and Research*, **27**, 103-133

Hatem, A. and Lehto, M. 1995, Effectiveness of glue odour as a warning signal, *Ergonomics*, **38**, (11), 2250-2261

Kanis, H. and Green, W.S 2000, Research for usage oriented design: Quantitative? Qualitative? *Proceedings of the IEA 2000/HFES 2000 Congress*, 925-929

Lichtenstein, S., Slovic, P., Fischhoff, B., Layman, M. and Combs, B. 1978, Judged frequency of lethal events, *Journal of Experimental Psychology. Human Learning*, **4**, 551-578

Murphy, E., Dingwall, R., Greatbatch, D., Parker, S. and Watson, P. 1998, Qualitative research methods in health technology assessment: a review of the literature, *Health Technology Assessment*, **2** (16) University of Nottingham

Purswell, J.L., Schlegel, R.E. and Kejriwal, S.K. 1986, A prediction model for consumer behaviour regarding product safety, *Proceedings of the Human Factors and Ergonomics Society, 30th Annual Meeting*, 1202-1205

Rundmo, T. 1999, Associations between risk perception and safety, *Safety Science*, **24**, (3), 197-209

Slovic, P. 1987, Perception of risk, *Science*, **236**, 280-285

Viscusi, W.K. and Cavallo, G.O. 1994, The effect of product safety regulation on safety precautions, *Risk Analysis*, **14**, (6), 917-930

Weegels, M.F. and Kanis, H. 2000, Risk in consumer product use, *Accident Analysis and Prevention*, **32**, (3), 365-371

Wogalter, M.S., Brelsfort, J.W., Desaulniers, D.R. and Laughery, K.R. 1991, Consumer product warnings: The role of hazard perception, *Journal of Safety Research*, **22**, 71-82

Wogalter, M.S., Brems, D.J. and Martin, E.G. 1993, Risk perception of common consumer products: Judgements of accident frequency and precautionary intent, *Journal of Safety Research*, **24**, 97-106

THE METHODS LAB: DEVELOPING A USABLE COMPENDIUM OF USER RESEARCH METHODS

Alastair S Macdonald[1] and Cherie S Lebbon[2]

[1]*Head of Department, Product Design Engineering*
Glasgow School of Art, 167 Renfrew Street, Glasgow G3 6RQ, Scotland
a.macdonald@gsa.ac.uk

[2]*Research Fellow, Helen Hamlyn Research Centre*
Royal College of Art, Kensington Gore, London SW7 2EU
c.lebbon@rca.ac.uk

This paper discusses the origins and development of the Methods Lab, the objective being to build a definitive resource of user research methods in design. This has resulted in a compendium of user-research methodologies, presented in a 'designer-friendly' format. This format and the content of the Methods Lab are described. Its design is intended to be practical to help designers evaluate and choose the most appropriate user research methods. Although developed primarily for design students, students of user research sciences, and designers and researchers in their first years of professional practice, the Methods Lab will be of interest to students of ergonomics and to ergonomists.

Introduction

The creation of successful designs that suit the greatest number of users, to be 'inclusive', is an ever more challenging task. In recent years, this process has benefited from some convergence among the disciplines of design and its related specialisms. User research methods, which range from near-market to the highly conceptual, from the conventional to the experimental, from the quick and easy to the detailed and exhaustive, have become increasingly vital in understanding users' behaviour and needs, but until recently, there has been no typology to assist designers in the selection of methods. This paper discusses the origins of a typology of user research methods which later become known as the Methods Lab, the objective being to build a definitive resource of user research methods in design.

Development

Earlier work by i~design (Inclusive Design) consortium member IDEO, in particular the report produced by Glasgow graduate Ewan Duncan, which included a user research methods typology, was later taken up by the Helen Hamlyn Research Centre (HHRC) as part of the Presence research programme - one of thirteen EU-funded projects under i3 (the European Network for Intelligent Information Interfaces). Through a series of Presence working groups, known as 'Tea Parties', the format and content of the Lab was developed. The 'user discussion method', an iterative process of discussion, refinement, critique, and evaluation which led to the Methods

Lab in its final form, comprised a process of repeated peer evaluation during the development of the Method Lab.

A mix of designers, researchers and ergonomists, working together with a coordinator and an editor-in-chief, met following a 1997 Presence workshop to consider user research inputs to the programme. By way of preparing for this meeting, initial responses to the IDEO report had been requested. Discussion was focused through two questions - 1) what was the best basis for building and structuring a resource of user research methods and case studies, and 2) what value did user research really add to the design process? Tea party participants were also asked to provide a working definition of what they meant by user research.

Following on from this initial gathering, the aims of the second more in-depth discussion were to determine what kinds of user research information would be most useful to designers and how they might best be communicated. A series of further meetings were held to develop material for the Lab, to focus on developing a publication, either web-based or print, and a decision was made to develop a compact print version of IDEO's original typology of methods. (An interactive web-version turned out to be difficult to produce although a simple text version was also eventually made available on the site).

Discussion and brainstorming produced leads for the development of the Methods Lab section of the Presence web-site. A number of experts in their fields were identified for several of the methods and each was commissioned by the editor to produce a concise 200 word texts describing their method. Several meetings followed at the HHRC to develop and refine the method selection format and iconography. An art-director and designer worked with the editor and coordinator to produce a quick and easy way of rating and comparing the appropriateness of each method. Once there was sufficient material, a scoring template was developed which the coordinator and editor distributed amongst tea party members and others involved in developing the Lab. These same ergonomists, designers and researchers individually scored each method against a number of criteria, e.g. cost and expertise, leading to consensus formats. Important throughout was the idea that designers were evaluating material designed for designers.

The Design

The art-director for the original Methods Lab booklet was Jennet Jessel. Over a twenty year period her experience in the commercial world had involved new product development, packaging, corporate identity and corporate information. Her brief on the 'Design Aid' project completed by the Design for Ability unit at Central St Martins in 1998 was to make statistical data and other information attractive and accessible to designers and manufacturers alike. As part of this process she came up with a set of assumptions about the way designers like to take in information and used this to develop her method for communicating to designers.

Jessel's Assumptions:
- 1 Designers don't read (particularly lots of text)
- 2 Designers don't always listen
- 3 Designers think with pictures
- 4 Designer think in 'blobby' often unrelated thought patterns
- 5 Designers go through a process of translating received information into imagery they can work with
- 6 Designers often have trouble translating visual concepts back into non image-led communication.

- 7 Designers still have a lot of the 'kid' in them, i.e. they like to play, explore and be rewarded particularly by the creation of new 'stuff'
- 8 Designers like 'pretty things'.

She didn't deny that designers need to work with hard facts but it is the fact that they always have a visual agenda for a project that it is crucial to understand. She pointed out that designers prefer 'soft' information preferably in 'bite-sized chunks'. Any information also needs to be presented in a visually attractive way so that it can more readily be translated and absorbed by the designer. This had to be balanced out with the need for text-based information that could be incorporated into reports, proposals, etc.

Her experience from the development of the Design Aid CDRom informed her in the design of the Methods Lab booklet which contains a wealth of hard factually based information. However, by using a range of interesting and appropriate icons she was able to translate this information. The result is a visually rich piece of graphic design which appeals to the designer yet holds all the information that the designer will require in order to select an appropriate user research method. It is also schematic and supported with text, making it equally accessible to other disciplines and the business community. Its format means that it can be used as a ready reckoner of user research methods, easily slipped into a jacket pocket or bricfcase and crucially to the designer, a 'pretty thing'. Its identity is communicated through a distinctive elongated orange format.

Description of the Methods Lab

The Methods Lab comprises several discrete components : the 'Lab': the 'Map', the 'Icons', the 'Finder', and the 'User Research Methods'.

The Method Map
The Methods Map positions each Method listed in the Finder at a point along two axes that reflect designers' concerns. The horizontal axis represents the external reference a method requires. At the left end of the scale, 'Designer centred' projects require no such reference. The right end 'User Centred' projects, tends toward an ideal in which each users' needs would be individually met. The vertical axis depicts design projects concerned with purely visual qualities at the top, ranging to those where functional qualities are predominant at the bottom. It is intended to be a quick way to identify candidate methods for a given aspect of a product. The intention of Malcolm Johnston, one of the design team, and originator of the diamond-shaped Methods Map itself, was to provide a designer-centred tool. He said,"While the map is offered as a prototype with only a small number of methods analysed and placed, its value for locating a method is immediately apparent. It is not a precise tool, but a quick way to identify possible candidates for a given aspect of a project - and also a way to keep in mind the scope of a design project."

The Finder
The Finder comprises a table with the following headings: Typology (methods are grouped in typologies for rapid navigation); Particular Methods (as the Lab is a work in progress, only a selection of methods are described in full and further entries will be added in future); Method Number (these are located in the Methods Map for comparison); Output (these give the main benefit to the design team); and Input (resources required - expertise, time, staff, and costs). These inputs of expertise, time and staff or manpower required, and costs for each of the Methods are represented through 'Icons' with a five level graphics scale indicating minimum to maximum level of input required.

Scenarios | 4

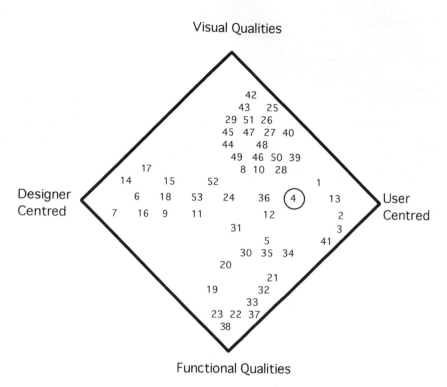

Visual Qualities

Designer Centred

User Centred

Functional Qualities

Figure 1. Scenarios - Methods Map 4. The numbers refer to different Methods, and the 4 is circled to indicate this 'Scenarios' Method and its position within the Map.

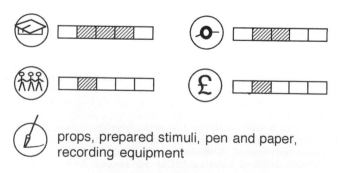

props, prepared stimuli, pen and paper, recording equipment

Figure 2. Scenarios - Input Icons

(Illustrations adapted for purposes of reproduction)

The Methods
Each Method was written by an individual who is globally recognised as an authority in his/her field. They were required to write with maximum concision - each Method in just 200 words. References or web links were also provided for additional information.

User's comments

Student Users
The Methods Lab was evaluated by both student users and professionals. Sixteen fifth year product design engineering students were surveyed individually for their views on the Methods Lab as published in the 'New Media for Older People' text, a second version of the Lab, printed in a different format to the original 'orange' publication. All students, prior to introduction to the 'Lab' had informally used a variety of user research techniques which varied with each individual and with the nature of the product being designed.

All students had positive comments to make regarding the idea, in principle, of the 'Lab' as a tool. A number found the structuring and variety of methods outlined helpful for broadening their understanding of methods available, that it was useful at the start of a project and for those who were unfamiliar generally with the broad spectrum of user methods. One felt that case studies showing worked examples of methods being applied would have been useful, and two felt it would be difficult to know which method was most appropriate before one had begun to research and design a product fully.

The group was then asked to comment on the separate components of the 'Lab': The Map, The Icons, The Finder, and the User Research Methods.

Responses to the 'Map' were mixed. This may have been because of some design students identified more with the designer end of the horizontal axis, and their difficulty in being totally objective. Comments varied from 'quick and simple' to 'a bit confusing', and from 'the numbers in map too close to one another', and 'slightly arbitrary' to 'good for extreme methods, too subjective to differentiate the rest (in the centre)'. Many felt that a better explanation of the meanings of the axes would be helpful. Some thought the use of worked case studies would have been helpful.

The 'Finder' provoked very positive responses from the whole group: 'clear and concise','well structured and laid out', 'easy to follow', 'straightforward', and 'makes perfect sense'. There was an indication that this was one of the most important components and would have benefited from a larger, more legible format - two pages or a fold-out format. One felt some of the Method titles difficult to understand without back-up descriptions, and another that it would be even more useful if it was on the same page as the 'Map' so that all information was able to be viewed at once. These points had been considered by the art director and designer as a larger format had appeared in the originally published version of the Lab, with a fold-out finder which could be viewed in conjunction with the individual methods, each of which contained its own Icons and Map. This confirms the importance of the original design specification and the format of the original published version.

'Icons' had a mixed, though generally positive response. The main issue of contention was the grading system which some students thought approximate, a scale more relative to one's experience rather than being objective. Some thought this might have been improved using a numerical rather than visual value. Responses to the section on 'Methods' again proved positive, the students appreciating the clarity and conciseness of descriptors compiled by the different experts. A few would have preferred the information conveyed through bullet-points.

Most felt each of these components were crucially inter-related and that one would not be useful without the others. However, when asked to rate them, 'Research Methods' was seen as the most useful, followed by 'Finder', then closely by the 'Map' and finally by the 'Icons'.

Professional Users
The Lab was distributed to a number of professional users of research methods, including IDEO, Acquaman, Cambridge Consultants, Telenor, Brunel University, Glasgow School of Art, and P5 Consultants. Responses were very positive, and one interesting fact emerged: although consultants in the field had used methods for a number of years, one comment, typical of many responses was,"there were lots of new things I hadn't come across before".

Conclusions - Context and future plans

The first responses to the Methods Lab have been very positive. It will obviously benefit from further development, user trials and further feedback. The ambition is to list some fifty Methods in total, with each description written by an international authority. There are now plans to build on this work as a basis for a more extended organisation and eventual publication of the output of the HHRC strand of the i~design programme - 'Inclusive Design: strategies, tools and user-research methods'. There are also plans for visual impairment access in the next version. The contents would include: a timeline bringing together key events, legislation, publications and design exemplars to demonstrate the emerging trend towards inclusivity; a rationale for the grouping and organising of the strategies (e.g. 'Transgenerational Design'), tools (e.g. Pirkl's charts of ageing characteristics and design responses), and user-research methods (e.g. 'Immersive Experience'); an overview of relevant inclusive design strategies (theories and concepts), their background, key features, and historical origins, 20+ having been identified to date; a reference listing of available design tools for inclusivity, 30+ having been identified to date; and a mapping of relevant user research methods, 50+ having been identified to date.

References

Aldersey-Williams, H., J. Bound and R. Coleman (eds) 1999. *The Methods Lab: User Research for Design*, Design for Ageing Network, Royal College of Art, London.
Duncan, E. 1994. *User Research* (an internal draft discussion document) (IDEO, San Francisco) - *by kind permission of Bill Moggridge*.
Hofmeester, K., and E. de Charon de Saint Germain (eds) 1999. The Methods Lab: User Research Methods, in *Presence: New Media for Older People*, (Netherlands Design Institute) 118-165
Jessel, J. 1998. www.presenceweb.org (Discussion Forum section) *(Web site no longer linked but document available on request.*

Acknowledgement

To Hugh Aldersey-Williams, editor in chief, John Bound, project coordinator, and Roger Coleman, Co-Director of the HHRC, original editors of The Methods Lab for making their material so freely available and allowing us to borrow from their words.

EDUCATION

THE TEACHING OF ERGONOMICS IN SCHOOLS

Andree Woodcock[1] and Howard Denton[2]

[1]*VIDe Research Centre, School of Art and Design, Coventry University ,Gosford Street , Coventry, A.Woodcock@coventry.ac.uk*
[2]*Department of Design and Technology, Loughborough University, Loughborough, H.G.Denton@lboro.ac.uk*

The relationship between ergonomics and the disciplines it informs has always been tenuous. Woodcock and Galer Flyte (1997) hypothesized that teaching ergonomics in schools would lead to a greater acceptance and willingness to learn and use ergonomics techniques during tertiary education and once in professional practice. This paper discusses the relationship between ergonomics and design, and considers the teaching of ergonomics in secondary schools. Preliminary results of surveys conducted with first year undergraduates to investigate the teaching of ergonomics they received at both 'GCSE' and 'A' level are presented which indicate that most ergonomics education occurred, as expected, in design and technology courses, but was also present, though patchy, in other disciplines.

Introduction

Gaining acceptance and use of ergonomics by designers and other professions has been difficult. This seems somewhat strange, firstly because humans have what appears to be an almost innate capacity for understanding and applying the principles of ergonomics. For example, Australopithicus Prometheus selected pebble tools and made scoops from antelope bones to make tasks easier to perform; in their use of computers for distance co-operation, designers will select those features which enable them to complete the task as effectively and efficiently as possible (Scrivener, Chen and Woodcock, 2000); additionally it is believed that children, even at pre-school level perform user centred design activities in their play and make activities (Woodcock and Galer Flyte, 1998). Secondly, ergonomics has proved to be beneficial when used to inform product and work place design. Thirdly, industry is seeking graduates who are able to design for niche markets.

There have been attempts to improve communication between ergonomics and the disciplines it seeks to inform. Woodcock and Galer Flyte (1998) reviewed these approaches which have included in-house training schemes (Shapiro, 1995), production of designer-friendly literature (Chapanis, 1990), development of techniques which could directly benefit many stages of the design process (Simpson and Mason, 1983) and the integration of ergonomics into other undergraduate courses (Woodcock and Galer Flyte, 1997). These are gradually changing the design and engineering climate. However, these approaches focus on practicing designers or students in tertiary education.

Informal observation has revealed some antagonism towards ergonomics amongst undergraduate engineering and design students. If unchallenged, this attitude might predetermine the use of ergonomics in later careers. For example, Meister (1982) showed that the attitude of senior managers affected the tone of the department, and that one of the ways this could be changed was through the greater integration of ergonomics and the continuing education of managers. If undergraduate engineers and designers are unappreciative of the benefits of ergonomics then we need to consider the reasons for this. This may include multidisciplinary course structure; assessments emphasising the mastery of skills and creativity at the expense of user evaluation. It is hypothesized that the failure to appreciate the value of ergonomics may be partly based on pre-university experiences of ergonomics.

Previous research suggests that pre-school children are cognisant of the need to consider others in their play and make activities and that this empathy continues in primary schools, but is downplayed in secondary school curricular. Secondary schools are largely structured around the examination of individuals. Long tradition of solo endeavour may result in undergraduates who (for the most part) are antipathetic in their views towards user issues. It is argued that this is a lost opportunity and that the discipline of ergonomics has much to offer teachers, pupils and curriculum developers not only as a discipline in its own right, but also as a means of integrating diverse areas of the curriculum and enhancing learning experiences. The rest of the paper discusses investigations undertaken to assess the way in which ergonomics is taught in secondary schools..

Secondary education

A computer based search of the National Curriculum requirements for all subjects at Key Stages 3 (age 11-14) and 4 (age 14–16) showed no specific mention of 'ergonomics', 'anthropometrics' or 'human factors' (National Curriculum, 1999). Currently the Qualifications and Curriculum Authority (QCA) is conducting a major review of National Curriculum requirements and qualifications such as 'A' level. Their guidelines for 'A' level subjects were searched for the terms 'ergonomics', 'human factors' and 'anthropometrics' in the subjects: design and technology, physics, maths, biology, physical education and art. The only hit scored was on 'ergonomics' in 'A' level design and technology guidelines. Here the requirement is for all new 'A' level Design and Technology syllabi to include a section on 'planning and evaluating' and within this is: section c. 'use ICT appropriately for planning and data handling, for example, the use of data base, drawing and publishing and design software. Interpret design data such as properties of materials, ergonomics and nutritional information.' (Midland Examining Group, 1999). 'A' level grade descriptions for Design and Technology is the statement (grade A) that: 'when developing and communicating ideas, take into account functionality, aesthetics, ergonomics, maintainability, quality and user preferences.....'

The overall position, however is clear: a student with 'A' levels in maths or a science may have looked at ergonomics, but it is more likely that they will not. A student with 'A' level 'Design and Technology' should have learned something about ergonomics and should have applied this knowledge to project work. The following investigations were carried out to verify this in terms of undergraduate knowledge and attitudes towards ergonomics.

Investigation 1

The aim of this study was to investigate further the current teaching of ergonomics in schools at GCSE and 'A' level to ascertain the subjects in which ergonomics is being taught, and the methods used to teach it. Such information is necessary for curriculum developers who might wish to co-ordinate activity across subjects, educational product designers who might wish to develop teacher support material and university staff who wish to better understand the prior learning of students. The questionnaire consisted of five main sections:

1. Personal details relating to the age and gender of the respondent, academic course, and willingness to participate in follow up interviews.
2. School details - to enable a follow up survey of selected schools, to discover how ergonomics is taught in these schools.
3. Subjects taken at 'A' level. This also attempted to establish the amount of ergonomics in these subjects (coded into a four point scale from nil to high) and the manner in which that ergonomics was taught (video, reading, lesson, applied in a project etc
4. Subjects taken at GCSE, using the same format at (3) above.
5. Ergonomics questions - to gain an impression of the respondents depth of understanding of ergonomics and perception of the value of ergonomics in their future careers.

Most of the sample were enrolled on engineering courses-Mechanical Engineering, Mechanical Engineering and Manufacturing (MEM), Civil Engineering, Engineering, Science and Technology (EST), and Aeronautical and Automotive Engineering (AA) with a high proportion from Industrial Design/Industrial Design and Education courses, and Product Design / Product Design and Manufacturing (PDM). A third, smaller group consisted of ergonomists, psychologists and human biologists. Over 80% of the respondents were male, of age 18 or 19, with very few mature entrants. This is a fair reflection of the trend within Loughborough University. The following results are based on over 350 responses.

At 'A' level it is only in Design and Technology that ergonomics is taught to any great extent, as is to be expected. However, it is interesting to note that ergonomics was being taught in some schools in other disciplines, most notably IT, sport and geography. Only in a few cases were single methods (e.g. lecture, handouts) used to teach ergonomics. It was shown, not surprisingly, that students who had been exposed to a greater number of teaching methods had a higher level of understanding of ergonomics. These results would seem to indicate that multiple teaching methods appear to be more effective than the single use of any method including project work.

Other trends in the data revealed that younger respondents had slightly more knowledge of ergonomics than older ones, but the sample size for older participants was too small for this to achieve any levels of significance. With regard to the extent to which ergonomics was considered to be important in a later career, 299 thought it was important and 32 that it was not. Of this latter group, 20 of were civil engineers and 9, mechanical engineers. The results also indicated a gender difference, with females having a higher regard of the value of ergonomics to their future careers (mean = 2.11), than male respondents (mean = 1.86).

Only those students who have completed a full GCSE in Design and Technology and particularly those who have taken an 'A' level in Design and Technology can be assumed to have studied ergonomics at school. Individually, students may have covered aspects of ergonomics in independent study or where ergonomic data is used as an illustration of data handling. Older teachers of Design and Technology will not have learned ergonomics as part of their initial training. Students are dependent on these staff having read up on the subject when it entered these syllabi in the 1970's; not all can be relied upon to have done this.

At GCSE level less than half the sample had some remembrance of ergonomics being taught, for example, in technology, art, PE, dance and drama, and the humanities (most notably geography), as well as in the more design related courses. However it formed a very minor element unless developed by the teacher out of personal interest.

Investigation 2

In 2000 a second questionnaire survey was conducted with 66 consumer, product and transport design students to gain further insight into the teaching methods and their level of understanding of ergonomics when they entered the first year of their university studies. The responses from the mainly male (86%) respondents may be summarised as follows.

83% would have liked more ergonomics tuition at school either because it would have made their present course easier (30%) or because they believed it to be an integral part of the design process (50%). Those 17% who did not want ergonomics tuition at school perceived it to be boring and uninteresting, as restricting the creative part of the design a premium was placed on appearance at 'A' level, the use of ergonomics never really changed the end design that had been planned, and it detracted from the more important aspects of the course.

In terms of the teaching of ergonomics, 38% felt that they had been taught well, but a third thought that the teaching had been poor. The remaining third had had no experience of using or being taught ergonomics. As in the previous investigation, teaching was again through multiple teaching methods (11%, 1 method; 15%, 2 methods; 9%, 3 methods and 20% having received ergonomics teaching through multiple methods (4-7 methods)). These included handouts (17%), projects and discussions (both 14%) and reading. Other methods included lectures, video and television, tutorials, past exam papers, practicals and design journals. Topics which were covered included anthropometry (30%), human machine interface (16%) and others including vision, comfort, aesthetics, texture, psychology and social aspects.

When asked to consider the most appropriate method to teach ergonomics, just under a third of the students felt that it should be a more hands on experience, with examples and interactive lectures. 9% felt that if ergonomics was to be taught then it should be considered an important, integral part of design and that this should be reflected in assessment. Under 5% felt that there was scope for ergonomics to be considered earlier than at secondary school.

In terms of the subjects where ergonomics was taught the results confirmed the earlier study. Ergonomics was never taught in languages, classics, maths, humanities or economics. There was considerable amount of variation (possible due to the interest and knowledge of the teacher) in the amount taught in computing, art, IT, 3d design and physics. Most ergonomics was taught in design/technology, art and design, design, design and graphic communication.

In the design related subjects ergonomics was mainly taught through large projects such as transport design (e.g. car, lorry, scooter), chairs (orthopaedic, bench, chair, wheelchair), product (lectern, shrine, sundial, multi-gym). In small projects, design was more varied and product related. In terms of when ergonomics should be used or had been experienced, 34% answered during planning, 32% during design and 20% in evaluation. This implies a naiveté in the sample as to where ergonomics could/should be used during the design cycle.

In the first investigation it had not been possible to clearly ascertain the level of ergonomics knowledge of the undergraduates. Here, when directly asked 17% admitted to having no knowledge at all, 26% claimed little knowledge, 39% some and 17% thought they had a good deal. At the end of the questionnaire students were asked to consider ergonomics aspects of the interior of vehicles in terms of broad categories (e.g. seating, dashboard design), and then to break these down into subcategories (e.g. visibility, adjustability, comfort, reachability). The extent of ergonomics knowledge displayed by some of the respondents was surprisingly high and diverse. This might have been an artefact of the cohort who hoped to specialise in automotive design.

Discussion and future work

The results presented show that ergonomics is being taught in schools in design related disciplines. Clearly some teaching occurs in other subjects where appropriate. A diverse range of teaching methods are employed. The results indicate that where students are exposed to a diversity of educational material, teaching and learning methods they may gain a higher understanding of the discipline.

Some students in the second investigation expressed an antipathy towards ergonomics, and felt it was not part of the design process. There is a slight indication that this may be due to the way in which it is taught in secondary schools. The results show a need to develop student centred, experiential material and to consider the manner in which ergonomics is assessed if it is taught so it is not seen as detracting from the main focus of design.

Future work should address the development of student centred resource packs which will enable students to grasp the main principles of ergonomics in an enjoyable manner and what works well in current practice. From the research we know which schools were rated highly by students in their teaching of ergonomics and these can be targeted for further study.

No evidence emerged in these investigations for the use of ergonomics in cross curricular activities. This might have been due to the manner employed in the data collection, an inability of the students to identify those areas in the curriculum where they had been taught ergonomics by a different name, or it may simply not be happening. For example, with children and schools developing their own web sites, there are opportunities to consider hci issues; to develop cross cultural studies and experimentation through the use of shared resources; in the development of more complex areas such as citizenship. For example a safe driving campaign could consider ways to make passengers, pedestrians and drivers safer, consider vehicle dynamics and the human factors issues surrounding accident causation and what happens during accidents.

As the National Curriculum is flexible in terms of the precise nature of what is taught and how it is taught teachers could use ergonomics as a vehicle for learning in several subjects, for example, within maths, ergonomics could be a subject for statistical data management and interpretation; physical education and art might also raise the subject in various ways. Such educational resources could not be developed by the teachers themselves but possibly could in partnership with ergonomists.In conclusion it would appear from our work that little attention has been given to the teaching of ergonomics to the under 18s and that this area is ripe for the development of innovative, challenging and multidisciplinary teaching methods and resources.

References

Chapanis, A. 1990, To communicate the human factors message you have to know what the message is and how to communicate it, *Communiqué, Human Factors Association of Canada,* 21/2, 1-4

Denton, H.G. and Woodcock, A. 1999, Investigation of the teaching of ergonomics in secondary schools - preliminary results, In Juster N.P. (ed) *The Continuum of Design Education,* (London: Professional Engineering Publishing), 129-138

Meister, D. 1982, Human factors problems and solutions, *Applied Ergonomics,* **13**, 3, 219-223.

Midland Examining Group Design and Technology: Resistant Materials 1999. Page 10.

The National Curriculum 1999, Department for Education and Employment, (http://www.qca.org.uk/) April

Scrivener, S.A.R, Chen, C.D. and Woodcock, A. 2000, Using multimedia mechanism shifts to uncover design communication needs, in S.A.R. Scrivener, L. Ball and A. Woodcock (eds) *Collaborative Design, CoDesigning 2000,* September 13th-15th, Coventry, 349-359

Shapiro, R.G. 1995, How can human factors education meet industry needs, *Ergonomics in Design,* 32.

Simpson, G. C. and Mason, S. 1983, Design aids for designers: An effective role for ergonomics, *Applied Ergonomics,* **14**, 3, 177-83.

Woodcock, A and Galer Flyte, M. 1997, Development of computer based tools to support the use of ergonomics in design education and practice. *Digital Creativity,* **8**, 3 & 4, 113-120

Woodcock, A. and Galer Flyte, M. 1998, Ergonomics: it's never too soon to start, *Product Design Education Conf*erence, University of Glamorgan, 6th-7th July

Woodcock, A. and Galer Flyte, M. 1998, Supporting the integration of ergonomics in an engineering design environment, *Tools and Methods for Concurrent Engineering '98,* 21-23rd April, Manchester, England, 152-168

ARTICULATING RESEARCH RESULTS

[1]Andrée Woodcock, [2]Margaret Galer Flyte, [2]Sarah Garner

[1]*VIDe Research Centre, School of Art and Design, Coventry University, Gosford Street, Coventry, UK. Email: A.Woodcock@coventry.ac.uk*

[2]*Formerly, Department of Human Sciences, Loughborough University, UK*

Ergonomists, and other academics, have frequently been accused of not making their results accessible to those who might benefit from them or wish to apply the information. This, for ergonomics, is especially true for the design professions. This paper describes the way in which a multidisciplinary team of ergonomists and designers worked together to consider the in-car safety and security needs of women drivers and their passengers, and focuses on the manner in which dissemination was undertaken.

Introduction

A consideration of dissemination activities now features widely in research proposals. Indeed dissemination of research results, at whatever level is considered to be an important part of research activity. For example, Archer in his opening address to the CoDesigning 2000 conference exhorted the audience to 'publish promiscuously'.

Unfortunately articulating research results is not something which comes easy to most academics. For example Sabey (1999) "regrettably the failure to follow through good quality research to influence and change policy and practice is not confined to..........It is a widespread failing amongst research organisations, many of which do not consider they have a role or responsibility for encouraging action on their findings,"

Ergonomists have been previously criticised for their inability to write their research in a manner which engages their readers. For example, Meister and Farr (1967) commented that designers do not like long passages of text; Klein and Brezovic (1986) found that designers considered the information too general or the methods used irrelevant to their work and Chapanis (1990), that design applicability in much ergonomics reporting was missing.

The in -car safety and security of women project

This 9 month, DETR funded research project concerned firstly understanding the in-car safety and security needs of women drivers and their passengers in relation to accidents, social and behavioural issues (the main findings of which may be found in Woodcock et al, 2001) and secondly, communicating these issues to an audience of women motorists, stakeholders, researchers, car manufacturers and designers. The communication to different groups was facilitated by the employment of two groups of student designers; namely graphic designers, and automotive designers throughout the lifetime of the project

Developing communication mechanisms with designers

The research team was multidisciplinary, comprising of ergonomists and designers. During the early phases of the work the ergonomists were involved in the development of research methodologies and data analysis. Although this was a potentially 'quiet' time for the designers, they participated in the round table discussions, were invited to take part in data gathering activities such as focus groups, and developed stimulus material. For example, the graphic designers developed a project image and designed driving diaries used to gain insight into driving patterns. The automotive designers developed concept and mood boards to be used during the focus groups.

This strategy engaged the designers in the project and kept them informed about the way in which the work was developing, and allowed them early insight into the results which would inform their work. This engagement is seen as important for multidisciplinary working and was not an experience student designers usually acquired as part of their degree programme.

Communicating research results

The research results arising from this project have important ramifications for both automotive manufacturers and designers, fellow researchers and motorists. This paper forms part of the dissemination programme adopted for dissemination to researchers. The other groups have proved more difficult users of ergonomics information

Communicating to motorists

Two strategies were employed to disseminate the research findings to the general public. The focus groups had established times at which women (in particular) needed more advice or were particularly vulnerable; namely making sure that they were correctly seated, advice for when they had broken down, when they were parking the car and when they were threatened on the road. This information was given to the graphic designers who produced a sequence of posters which could be used in a campaign highlighting these issues (Figure 3).

A second poster was produced which would enable women motorists to perform their own safety audit. Whereas the other series of posters were designed for visual impact this was designed to be read at a close distance either on a notice board or as a leaflet (Figure 4).

In conducting a survey of guidelines to women motorists, we found that the Internet was frequently used to disseminate advice and information especially to younger age groups so we looked at creating a site which would be both entertaining and informative, and which would offer practical advice to young female motorists. Two screens from the web site are shown in Figure 1 depicting what to do in a breakdown on the motorway and how to deal with aggressive drivers. The screens either showed a video, or provided a puzzle to which there was a right or wrong answer.

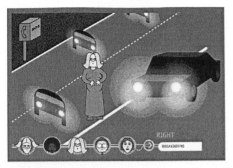 

Figure 1. Examples from the women and cars web site

Communicating to and through car designers

The previous section has focussed almost entirely on what motorists can do to make themselves safer and more secure. This part of the work was fed almost entirely by the results of the qualitative analysis (*i.e.* focus groups, diary, internet and stakeholder analyses).

A second part of the research considered the design of the car, those aspects which made it less safe and secure for women, and which are not suited to their requirements. It is usual practice for such information to be given to designers as a series of reports, guidelines or recommendations. Information presented in such a manner is not usually found to be well received, found likeable or usable by designers (*e.g.* Woodcock and Galer Flyte, 1998).

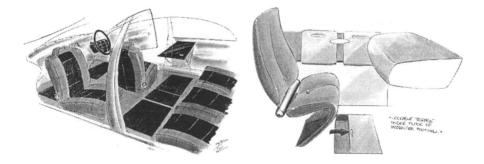

Figure 2. Examples of vehicle interiors and provision of storage areas

As the results from the accident analysis emerged these were relayed to the group in presentations by the researchers, and then discussed more informally. This met with some success. The automotive designers were responsive to the ideas and worked on the development of a car package starting with the anthropometrics of female occupants. Car packages are normally built around the male driver, so this in itself was an innovation. Neither the graphic nor automotive designers were accustomed to working closely with researchers or using emerging data to guide their designs. It is hoped that they found the experience useful as well as challenging.

The concept drawings produced of the interiors and the exteriors (Figure 2) were not solutions in themselves but were used as communication devices which could form a focal point for discussion between the research group. They will also be used to communicate the requirements of women drivers and their passengers in a manner which is more acceptable to the target audience; in this case automotive designers and manufacturers.

Discussion

Dissemination of research results to the different audiences who might derive maximum benefit from them is difficult, but is a consideration which the research community is increasingly being asked to make. This research provided an ideal opportunity to work with designers in exploring the potential for different dissemination activities.

This was in fact a learning experience for all members of the design team, and as such we did suffer some of the problems of multidisciplinary working, with ergonomists having to try to articulate to the designers the key research findings in a manner which the designers could use to inform their creative output. This paper shows that this was possible, and it is believed that both the designers and the ergonomists found it to be a valuable experience.

A further phase of the research will evaluate the success of different forms of dissemination in communicating research findings. Strangely, this type of study is not always undertaken by student designers, so we do not know how effective, likeable of useable the web site was, or whether the poster did raise awareness of car safety and security.

Acknowledgements

Project members: Department of Human Sciences: Sarah Garner and Margaret Galer Flyte: ICE - Ruth Welsh and Jim Lenard; Graphic Design Team, Coventry School of Art and Design - Paul Barrett, Lee Barrett, Richard Clark and Matthew Fish and Automotive Concept Design Team, - Leia Anastasiou, Joonkian Chua and Anna Louise Clough, Advisor Sam Porter; Project Advisors: Anna Humpherson and Mike Bradley, Ford, UK

The work reported herein was carried out under Contract PPAD9/72/36 funded by the Department of Environment, Transport and the Regions. Any views expressed are not necessarily those of the Secretary of State for the Environment, Transport and the Regions

References

Archer, B. 2000, Opening Address, CoDesigning 2000; Coventry, September 2000.

Chapanis, A 1990, The International Ergonomics Association: its first 30 years,*Ergonomics,* 33,3, 275-282

Klein, G. and Brezovic, 1986, Design engineers and the design process; decision categories and human factors literature, *Proceedings of the Human Factors Society of 30th Annual Meeting* (Santa Monica, CA. Human Factors Society), 771-775

Meister, D. and Farr, D.E. 1967, Human factors; engineering blind spot, *ElectroTechnology,* August

Sabey, B. 1999, *Road Safety:Back to the Future,* AA Foundation for Road Safety Research

Woodcock, A. and Galer Fyte , M.D. 1998, Supporting ergonomics in automotive design, *International Journal of Vehicle Design,*19,4, 504-522.

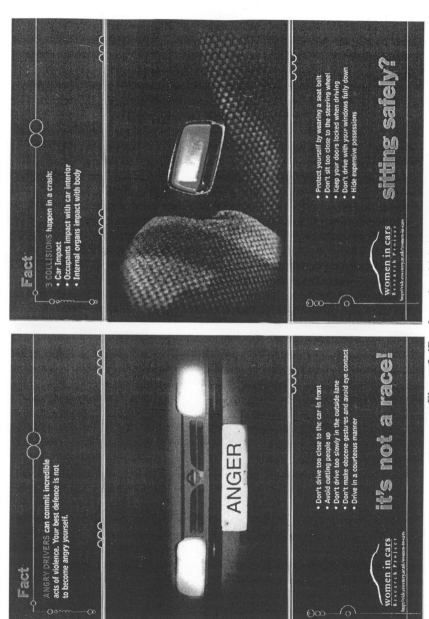

Figure 3. 'Road rage' and 'sitting safely' posters

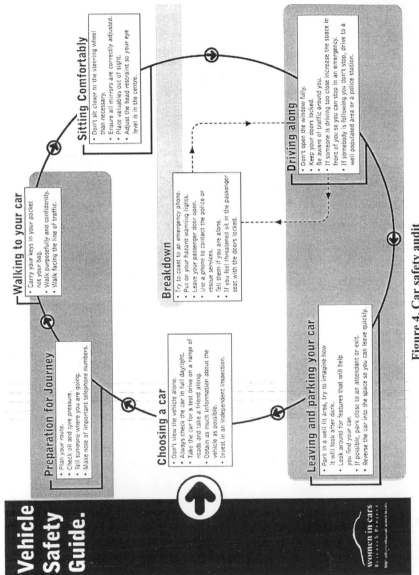

Vehicle Safety Guide.

women in cars
Research Project

Walking to your car
- Carry your keys in your pocket not your bag.
- Walk purposefully and confidently.
- Walk facing the line of traffic.

Sitting Comfortably
- Don't sit closer to the steering wheel than necessary.
- Ensure all mirrors are correctly adjusted.
- Place valuables out of sight.
- Adjust the head restraint so your eye level is in the centre.

Preparation for Journey
- Plan your route.
- Check oil and tyre pressure.
- Tell someone where you are going.
- Make note of important telephone numbers.

Choosing a car
- Don't view the vehicle alone.
- Always check the car in full daylight.
- Take the car for a test drive on a range of roads and take a friend along.
- Obtain as much information about the vehicle as possible.
- Invest in an independent inspection.

Breakdown
- Try to coast to an emergency phone.
- Put on your hazard warning lights.
- Leave your passenger door open.
- Use a phone to contact the police or rescue services.
- Tell them if you are alone.
- If you feel threatened sit in the passenger seat with the doors locked.

Driving along
- Don't open the window fully.
- Keep your doors locked.
- Be aware of traffic around you.
- If someone is driving too close increase the space in front of you so you can stop in an emergency.
- If somebody is following you don't stop, drive to a well populated area or a police station.

Leaving and parking your car
- Park in a well lit area, try to imagine how it will look after dark.
- Look around for features that will help you find your car.
- If possible, park close to an attendant or exit.
- Reverse the car into the space so you can leave quickly.

Figure 4. Car safety audit

ERGONOMICS IN NOTTINGHAMSHIRE: A SURVEY OF KNOWLEDGE, ATTITUDE AND IMPLEMENTATION

Philip D. Bust

COPE
www.cope-ergo.com

Despite references to ergonomics being contained in every part of the current 'six pack' guide to the work regulations it appears that implementation in industry is low. Using a questionnaire developed at The University of Ulster at Coleraine an attempt was made to assess the impact of ergonomics on businesses in the Nottinghamshire area. The study looked at three perspectives 1) knowledge and awareness of ergonomics; 2) attitudes towards ergonomics and 3) the extent to which the principle of ergonomics were put into practice in the workplace. The questionnaire was sent to 700 organisations representing the manufacturing, service and retail industries, with size varying from 10 to over 6500. While attitudes to ergonomics were found to be positive, knowledge and implementation was poor.

Introduction

This study was carried out in order to find how ergonomics was understood and used by businesses in the Nottinghamshire area. An understanding in attitudes towards ergonomics in business is necessary to assist in its promotion. The questionnaire had three main objectives: To find the respondents level of knowledge about ergonomics; opinions of ergonomics and to what extent were ergonomic principles put into practice in their organisations.

Method

The questionnaire used in the study was developed by The University of Ulster at Coleraine for a similar study in Northern Ireland in 1994 and on inspection was considered to be relevant for this work. A sample of 700 businesses was produced using the Dun and Bradstreet business directory of Nottinghamshire and Lincolnshire. It was attempted to include as wide a selection of types of business as possible and only businesses with at least 10 employees were chosen. Otherwise the selection was random. The final 700 consisted of 319 from the service sector, 240 from the manufacturing sector and 141 from the retail sector. The size of company varied from 10 to over 6500 employees. There was a decrease in the number of companies with an increase in size in the sample (see figure 1).

The questionnaires were sent by mail with a covering letter and return stamped addressed envelope. All questionnaires and return envelopes were coded to enable follow up calls to be made one week after issue. When necessary, after the follow up calls, additional questionnaires were issued by e-mail. The covering letter asked that the questionnaire be completed by the person responsible for the company's compliance with the Health and Safety at Work Regulations: 1974 and Management of Health and Safety at Work Regulations: 1992.

A statistical analysis (paired Pearsons comparison) of the returned questionnaires was carried out on responses to the attitude questions (see figure 3). The five point responses to the attitude questions were also adjusted by pairing the two positive and two negative responses together to give broadly positive and broadly negative responses with neutral responses discarded.

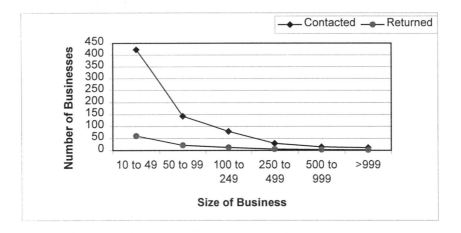

Figure 1 Graph showing Distribution of Business Size in the Sample

Results

One hundred and six of the 700 questionnaires were returned (15.1%) with 49 of the 319 (15.4%) in the service sector; 33 of the 240 (13.75%) in the manufacturing sector and 15 of the 141(10.6%) in the retail sector being returned and completed. The remaining 9 returned were either blank or unopened.

In answer to the first question - Do you have any idea what is meant by the term ergonomics? - there were 29 respondents claiming to have no understanding of the term ergonomics. Only one of these requested further information.

The second question asked – Please provide your definition of ergonomics? There were 67 definitions from those claiming to have knowledge of the term ergonomics.

About two thirds of these were reasonably close definitions while the remainder were mainly either statements on health and safety or a collection of relevant words.

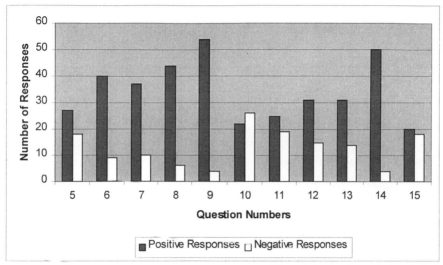

Figure 2 Bar Chart Showing Responses to Attitude Questions

It was found that around 40 of the definitions could be placed into one of four categories. The first of these related to definitions found in 5 dictionaries (three Oxford, the BBC and Longmans) which referred to ergonomics relating to a person's efficiency at work. There were 9 of these responses which also contained the word efficiency or efficient. A second category related to definitions found in 3 dictionaries (Two Collins and the New Penguin) which referred to a scientific study of the relationship between man, equipment and environment at work. There were 13 of these responses and 10 of them also contained the word relationship or inter relationship. The third category related to definitions found in the Chambers dictionary referring to fitting machines and environmental conditions to the person. There were 12 of these and 7 also contained the word match or fit. The final category related to the definition contained in CollinsTodays English dictionary. This referred to reducing effort, making easier or making comfortable. There were 7 of these.

The third question in this study – What if any is the presence of *ergonomics* in your company? - was meant to determine the presence of ergonomic professionals in the respondents' organisations and should have read – What if any is the presence of *ergonomists* in your company? The results can not therefore be compared to those of the first study. The overall response, however, to this question was a 46:5 ratio of low to high presence of ergonomics within the organisations. With question 4 – Do you use outside ergonomists for your needs? - Only 6 out of 68 respondents (8.8%) claimed to use outside ergonomists for their needs.

Responses to the attitude questions (see figure 2), with the exception of question 10 (see table1), were positively skewed towards a favourable attitude to ergonomics. The overall ratio of broadly positive to broadly negative responses to the attitude questions was 381 to 143, i.e 72.7 % of the opinions expressed were broadly favourable to ergonomics. This compares with 79.4% in the original study. The most favourable response (79.4% positive) was to question 9 and the least favourable response (38.2% negative) was to question 10 (see table 1 for list of questions). The intercorrelations of the scores for all 11 statements are shown in figure 3. This shows good interrelationship between the 3 implementation statements together with responses to question 8 for the IT systems. There were also significant relationships between responses to questions 13 and 14 and again with questions 14 and 15.

Discussion

First the good news. As in the original study responses to the attitude questions were positively skewed towards ergonomics. Respondents were agreeing that ergonomics is cost effective (question 6) and that ergonomic issues are essential for health and safety at work (question 14). They also feel that ergonomics is not a fashion trend (question 9) but were less encouraging when deciding whether ergonomically designed products meant better products (question 11).

Table 1. Likert-scale statements used in the questionnaire

5	(Implementation) Ergonomic considerations are taken into account in every aspect of the working environment in this organisation?
6	(Direct attitude) The cost of implementing good ergonomic practices in the workplace probably outweighs the benefits?
7	(Implementation)Ergonomic considerations are always taken into account when buying machinery and/or office equipment for my organisation?
8	(IT Systems)The computer system(s) in my organisation is/are very easy to use?
9	(Indirect attitude) Ergonomics is more of a fashion trend than a necessity?
10	(Indirect attitude) People will always complain about their working environment no matter how good it is?
11	(Direct attitude) Ergonomically designed products will always be more popular than unergonomic products?
12	(IT Systems)Our computers readily accomplish all the tasks that we require of them?
13	(Indirect attitude)You need special training in order to be competent in ergonomics?
14	(Direct attitude) These days, consideration of ergonomic issues is essential for safety and health at work?
15	(Implementation)How many instances have occurred in your organisation during the last 12 months when consideration has been given to ergonomic issues?

In the original study the questionnaire set out to provide data about attitudes to ergonomics and about its implementation. The correlation matrix from the original study

Figure 3. Pearson's product-moment correlations between responses to questions 5 through to 15

	5	6	7	8	9	10	11	12	13	14
6	0.077529									
7	**0.420055** **	-0.19693								
8	**0.285178** *	0.123431	**0.307982** *							
9	0.130458	0.263112 *	-0.0121	-0.09518						
10	0.021153	0.206172	-0.05934	-0.06355	0.171212					
11	0.057166	0.064019	0.128475	0.106966	0.160314	-0.14311				
12	0.115729	-0.01477	0.193797	0.181638	0.057468	-0.0695	-0.16207			
13	-0.11274	0.210836	-0.17569	-0.10111	0.055982	0.175369	-0.19642	-0.07509		
14	0.183108	0.16419	0.134774	0.080187	0.143255	-0.15287	0.08606	0.099515	**0.266419** *	
15	**-0.38948** **	-0.07227	**-0.29121** *	**-0.36915** **	-0.19902	-0.04083	-0.16254	-0.17316	0.074933	**-0.25748** *

n = 68 * p< 0.05; ** p < 0.01; *** p < 0.001   (See table 1 for list of questions) suggested a degree of internal consistency between the items that supposedly pertain to implementation. In this study there was a significant intercorrelation between these items. Neither the original nor this study was able, as hoped, to show high intercorrelations between the direct attitude items. It was also hoped that the indirect attitude questions correlated more with each other than with any of the other items. This did not occur in either study. Where this study was successful and the original not was in finding a relationship between one of the functionality of computer systems statements and the implementation statements. There were significant correlations between responses to Question 8 – The computer system(s) in my organisation is/are very easy to use? and all of the implementation statements.

The not so good news is that of the 67 respondents who claimed to know what the term ergonomics meant there were many definitions that fell short of achieving this. There were also many definitions which were very close to dictionary definitions and also 29 people who had no knowledge at all of ergonomics.

The appearance of key words and style of definitions leads to the conclusion that reference to dictionaries when answering the questionnaire was fairly common. Ten dictionaries were inspected and a definition by W.E. Le Gross Clark was uncovered in the Oxford English Dictionary. 'A man and his machine may be regarded as the functional unit of industry, and the aim of ergonomics is the perfection of this unit so as to promote accuracy and speed of operation, and at the same time to ensure minimum fatigue and thereby maximum efficiency'. It appears that the current dictionary definitions may have borrowed from this over the years without keeping the whole meaning. They talk of efficiency but not how it is achieved or reducing effort by worker without mentioning the benefits. It is no wonder that business organisations lack the correct knowledge about ergonomics if, as it appears, they rely upon dictionaries for this knowledge.

The bad news is that despite the praise for ergonomics and positive response to the implementation statements the presence of ergonomics in the businesses was low and the use of outside ergonomic professionals was also low. This shows that we have a battle to get the message across that the implementation of ergonomics within industry can prove beneficial to organisations to improving efficiency and health and safety. This study was useful in this respect as it enabled the author to contact a large number of people in business and provide them with information about ergonomics and the Ergonomics Society. A further benefit was to develop some of the contacts into working relationships.

Reference list

James D., Grennan S., and Mulhern G. 1994, Ergonomics in Northern Ireland: a survey if knowledge, attitudes and implementation in industry and the public services, *Ergonomics*, **37**, 953-963

Business Directory of Nottinghamshire and Lincolnshire (Dun and Bradstreet, www.dnb.com/uk)

SELLING AND COMMUNICATING ERGONOMICS

IMPLEMENTING AN ERGONOMICS MANAGEMENT PROGRAMME IN THE WATER INDUSTRY

Wendy Morris[1a], John R Wilson[1b] Chris Sharp[2] and Sophie Hide[3]

[1a] Research Assistant, [1b] Director, Institute for Occupational Ergonomics, Dept M3EM, University of Nottingham, University Park, Nottingham, NG7 2RD.
[2] Group Medical Adviser, awg, Anglian House, Ambury Rd, Huntingdon, PE29 3NZ.
[3] Health and Safety Ergonomics Unit, Department of Human Sciences, Loughborough University, Loughborough, Leicestershire LE11 3TU, UK

AWG (formerly Anglian Water) invited the Institute for Occupational Ergonomics to assist them in the development and implementation of an ergonomics management programme. The particular emphasis of this programme of work was to reduce sickness absence due to work related musculoskeletal disorders (WMSD) and injuries. To lay the foundation for such a programme a number of steps were taken. These included; identification of areas where there was a risk to employees of WMSD, prioritisation of work, undertaking ergonomics assessments of two high priority activities, developing and delivering a company specific training course to a multidisciplinary team of employees and implementing a shadow training programme. A systems approach to implementing ergonomics was built on the foundation of the in-house ergonomics resource established.

Introduction

Following the privatisation of local water authorities the water industry has gone through a period of considerable change. The core business of AWG has been essentially to supply fresh water to customers and to treat their waste water. They have, however, a number of other functions such as billing, customer services and engineering development to support the business. The recent acquisition of Morrison plc has also widened the business to an integrated infrastructure management group, with a corresponding increased range of customers. This new market is intensely competitive. AWG has a significant international operation, serving more customers abroad than in the UK. Like many of the utility companies, AWG has undergone a reduction in their operational workforce. Employees of the company perform a wide variety of tasks in offices, call centres, laboratories, treatment centres, outlying pumping stations which are usually unmanned, the public highways and on customers' properties. Although technologies have been developed to support the core water business there is still a considerable requirement for employees to undertake a variety of manual tasks.

The Institute for Occupational Ergonomics (IOE) was invited by the Occupational Health Department of AWG to assist them in the development and implementation of an ergonomics management programme. The focus for this programme of work was to reduce sickness absence due to musculoskeletal problems. The importance of ensuring the well-being of employees was recognised as a key influence on the commercial success of the group and support the group's vision of being the UK number one water and wastewater company in 2002 (achieved in 2000) and a world leader in 2007.

The role of the ergonomist has been to:

- Identify common musculoskeletal problems suffered by employees in the group and which jobs are particularly linked to these problems.
- Undertake initial ergonomics assessments of those jobs which were identified from the previous stage and make recommendations for appropriate intervention.
- Develop and deliver ergonomics training for a core team from AWG which was both 'classroom' and 'field' based (using a system of shadow training).
- Develop a programme of work to build on the foundation of ergonomics knowledge and to support its implementation into work practice within the organisation. The role of the ergonomist changed then to become more of an 'expert' support than the active change agent.

Problem identification

The primary focus of the work was to identify the risks for employees developing work related musculoskeletal disorders (WMSD) and injuries. The first stage of the programme of work was to analyse the company's sickness absence data to identify those tasks which may present a greater risk for the development of musculoskeletal disorders and to allow the prioritisation of particular areas for initial ergonomics assessments. This is an approach which follows the risk assessment strategy laid down in the UK Health and Safety legislation, with which the group must comply (HSE 1999). It is also important that the activity is focused on the most appropriate areas to ensure that the programme of work is efficient and effective (Hanson 1998). The analysis of the sickness absence data identified a number of problems with the system at the time, which we suggest are common to many organisations. The data were collected primarily to support the efficiency of the payroll system and did not appear to have considered how other departments within the organisation may wish to use the data. The standardised categories used by the human resources (HR) department for coding sickness absence had a single category for musculoskeletal injury, although depending on the information given on the sickness absence form certain musculoskeletal disorders may be reported as traumatic injury. One of the recommendations from this early stage of work was that staff in the HR department should receive some training as to how to assign reported sickness absence into the various categories.

Manual sorting of the data was required to identify people who may be at increased risk of developing WMSDs. The manual analysis indicated that there was some over reporting of short periods of absence and under reporting of longer periods of absence. For the company, although the number of days lost would remain the same, the percentage of absence episodes due to musculoskeletal disorders would be over reported. The data were therefore analysed according to both the number of absence periods and the number of days lost per employee, within the different business units of the company. Within those business units identified as having the highest rates of absence per thousand employees, the data were then analysed according to job title of employees and geographical area. Consideration was also given to the pattern of absence during the year and the length of absence. From this work, a number of jobs were identified and then prioritised for ergonomic assessment. The IOE consultants visited a number of sites and facilities in the early stages of the programme to familiarise themselves with the work of AWG and to gain a greater understanding of the influences and interactions that shape the organisation.

Laying the foundation

Ergonomics assessments of two tasks identified on the 'priority list' were undertaken by the IOE consultants. These were formally reported to the company including recommendations to eliminate or reduce risk factors for the development of WMSD where these had been identified. The material from the assessments was also used to tailor a training programme for a team of AWG employees. The training programme was designed to introduce the team to basic ergonomics principles and practice in the workplace. The use of information and terminology that was familiar to the team in their own workplace training programmes was considered important to enable the process of adult learning (Joyce 1999). The training programme consisted of both lectures and practical exercises, where the information presented was then applied to their work situation (van Dijk 1995). Delegates were recruited to the training programme, wherever possible, by personal contact. The team was intended to include all the Occupational Health Advisers working within the company and representatives from as many different departments as possible. The literature suggested that a multidisciplinary approach would facilitate the incorporation of ergonomics into the wider organisation (van Dijk 1995, Joyce 1999). The initial team who took part in the training comprised eleven staff in total from Occupational Health, Safety, Procurement (including a representative from the transport team), Facilities Management, Training and Water Production.

Following the training course a series of shadow assessments were arranged of other AWG tasks that had been identified on the 'priority list' for assessment from the first stage of the programme. The shadow assessments were intended to be a partnership of the IOE consultant and an AWG Ergonomics Core Team member undertaking an ergonomics assessment of a workplace, reviewing the findings and providing a formal report to the company. A field assessment checklist was developed, for the shadow assessments, to guide the AWG team member through the various elements of an ergonomics assessment and included standardised questions to ask the employee in the form of a semi-structured interview. The IOE consultant acted as a reference point for the AWG team member while working through the field assessment checklist. They also undertook the collection of workplace measurements as required and took responsibility for the collection of video data during the assessment. The assessment itself took between 3 and 4 hours. The data collected during the assessment were then reviewed immediately afterwards and the AWG team member given responsibility for part of the analysis of the data. The formal reporting of the assessments was undertaken by the IOE consultant due to work demands and time constraints faced by the AWG team members.

Once all team members had undertaken a shadow assessment, a team meeting was held to review the findings from the assessments and allow an opportunity for team members to feedback their experiences of the shadow assessments, evaluate the programme to date and identify areas that may require further training. It was not possible at this meeting to set an agenda for the future work of the team for a number of reasons, some of which are discussed in the following section.

Difficulties

The aim of the programme of work was to assist the company in developing and implementing an ergonomics management programme to devolve responsibility for ergonomics throughout the company. However the involvement of the IOE coincided with a period of considerable change within the company, involving both the management organisation and the operational systems. Some departments, particularly the Technology and Water Quality business units were in the process of relocating from outlying sites to a central location. It was therefore not possible to recruit potential team members from these departments as other activities were considered of greater importance. This was a potential weakness of the programme as the work of the Technology business unit has a considerable impact on the work environment and equipment that other AWG employees are required to interact with. Steps taken to address this weakness are discussed below. The scope of the changes to be made within the company were far reaching. There were changes to the operational systems, the managerial reporting line and overall employee numbers. The initial analysis of the sickness absence data was more complex than originally anticipated as there had been changes to system by which the data were both collected and reported during the time period under review. It was therefore necessary to review the data and undertake some manual analysis to minimise errors in the comparison of data from different time periods.

The business reorganisation also had an impact on the team of employees that had been trained to form the Ergonomics Core Team. By the end of the period of work the number of core team members had been reduced to eight due to staff changes as part of the reorganisation. Those team members that were still able and willing to continue with this additional role faced a number of conflicting demands on their time at work. As the ergonomics programme was very much in its infancy it had to be prepared to 'take a back seat' while the impact of the reorganisation was accommodated. It was for this reason that the formal reporting of the shadow assessments was fully undertaken by the IOE consultant although the original intention was that the team members would be actively involved in the process.

During the latter part of the project there were also external factors acting on the company that had implications for its organisation and management. AWG was preparing for the Water Industry Regulator's (OFWAT) determination on new limits for the water and sewage prices in England and Wales, which are linked to performance targets for the industry. It was anticipated that this would have a considerable influence on the operations of the group and might pose severe financial constraints on its planned activities. The impact on the group of the review and the business reorganisation also delayed the continuation of the ergonomics management programme into a second stage and it is a credit to those employees that made up the Ergonomics Core Team that the level of interest and enthusiasm was maintained.

Deliverables

Despite the background of change in the organisation it has been possible to slowly build a foundation of ergonomics knowledge and awareness into the organisation. The company had a number of tasks formally assessed and reported on as part of the programme of work. These made recommendations to eliminate or reduce risks for the development of WMSD where these were identified in the reports. These reports were communicated to the appropriate managers for their consideration. One task assessed was that of drivers working within the Regional Tankering Services business, which moves sludge from small sites to the larger wastewater production facilities. There are a number of designs of tanker in use within the unit and two of these designs were assessed in depth. A number of issues related to the design of the tankers as well as aspects of the tanker driver's tasks were identified as presenting risks for the development of WMSD. Subsequent to the delivery of the report to Manager of the business unit, there was a meeting with his management team, the occupational physician, the transport manager from the procurement department and the IOE consultant, to discuss the findings of the report and review a design of tanker which may replace some of the older tankers currently in service. This approach has led to improvements being built into the specification of new tankers, which will improve usability and comfort for the drivers, and reduce musculoskeletal risk. Further evaluation of the new tankers once they are in operational use will be undertaken to further inform the procurement process.

The company also now has an Ergonomics Core Team, who are interested in the application of good ergonomics practice into their workplace and the wider organisation. The interest of the team members has been maintained throughout the period of reorganisation although other demands within the workplace have tended to take precedence and their role as a company team was not well defined. In the second stage of the work with the company however there has been a commitment from Director level to 'release' the team members one day a month to focus on the development of the ergonomics programme.

The training of the team was an important element in developing an in-house resource. However the training of such a team should be viewed as "a means to an end, not the end itself" (Joyce 1999). A summary report of the programme of work delivered to the company, proposed a systems approach to the implementation of ergonomics within the organisation: the training of the core team was identified as one part of a much wider programme of work. A number of recommendations were made which included the following areas of work:

- Improvements to the system for reporting sickness absence. Managing the human resource of the organisation requires effective and appropriate systems for data collection, identification of areas of risk, to design better jobs and to retain staff.
- Develop best practice guidance for tasks that are common to all areas of the organisation. These should be developed by a team of appropriate employee groups (stakeholders) who have a working knowledge of the practical issues surrounding the various tasks.
- Develop ergonomics checklists for departments within the organisation such as Procurement and Technology, to ensure that all user needs are considered in the purchasing of equipment and the development of new designs.

Future directions and conclusion

A foundation for ergonomics within the organisation has been laid as a result of this programme of work. The recommendations will facilitate the incorporation of good ergonomics practice into all areas of company business. The Ergonomics Core Team should be enabled to develop their skills and expertise in the field of ergonomics to support the company's activities. The IOE consultant is continuing to work with the company but the role has changed to be more of an 'expert' support than the active change agent. The Ergonomics Core Team are involved in a number of different projects. One of these is a training session for design engineers to increase their awareness of the need to consider the capabilities and limitations of users of equipment / plant that they design. It is also planned to train a further small group of AWG personnel to increase the size and representation of the Ergonomics Core Team. The work of the Ergonomics Core Team is also being publicised on the group's intranet and through its internal publications which will increase awareness within the organisation of this in-house resource.

The work with AWG is ongoing and although some analysis of the benefits to the organisation is planned there are a number of difficulties with this. These are the changing nature of the organisation, making comparisons of like with like impracticable, the quality of information available on which to base the cost benefit analysis and the time required to allow the impact of changes to take effect. These and other difficulties have been reported by other authors (Mossink 1998, Lischeid & Roy 1999). The future programme of work therefore has been planned to undertake a number of small scale projects which may allow many of the confounding factors to be more fully controlled or accounted for and the cost / benefits of the ergonomics intervention to be evaluated. Such evaluations of the work of the Ergonomics Core Team are important to illustrate to the company that the application of ergonomics principles and practices is a value adding activity and will support the organisation in achieving it's goal of becoming a world leader in its market place.

References

Hansen, M.D. 1998, Employing a team to manage ergonomics. *Risk Management/Insurance Division,* Winter 1998, 9.

HSE, 2000, Management of health and safety at work - Management of Health and Safety at Work Regulations 1999. Approved Code of Practice. HSE Books, Sudbury

Joyce, M. 1999, The Role of Ergonomics Training in Industry. In W. Karwowski & W.S. Marras (eds.) *The Occupational Ergonomics Handbook,* (CRC Press, Boca Raton) 1631-1640.

Lischeid, W.E. and Roy, D. J. 1999, The Cost Benefit of Ergonomics: A Corporate Perspective. In W. Karwowski & W.S. Marras (eds.) *The Occupational Ergonomics Handbook,* (CRC Press, Boca Raton) 1541-1558.

Mossink, J. 1998, The Difficulties with cost-benefit analyses: Good safety and health care at work ensures greater competitiveness. *Janus,* **28,** 17-19.

van Dijk, F. J. H. 1995, From input to outcome: Changes in OHS-education and training *Safety Science* **20,** 165-171.

THE COSTS AND BENEFITS OF OFFICE ERGONOMICS

Sue Mackenzie[1] and Rachel Benedyk[2]

[1] Nestlé UK Ltd., Hayes, Middlesex, UB3 4RF
[2] University College London, 26 Bedford Way, London, WC1H OAP

The aim of this project was to identify the costs and benefits of office ergonomics. The project centres on a computerised office where operators input newspaper advertisements. Since 1992 the company concerned has realised the potential of this work to put staff at risk of developing musculoskeletal disorders. Aware that the ergonomic intervention within the office has been costly though beneficial, the Health and Safety staff were keen to compare the costs and benefits.

The analysis showed that the resulting benefits would equal the cost of the intervention in approximately two years following the implementation. The study also revealed a number of additional relevant benefits, which could not readily be costed or included in a financial analysis.

In addition the study concluded that there are a number of aspects regarding the use of cost benefit analysis methods for ergonomic interventions which require further investigation.

Introduction

Loot is a leading free advertisement paper in the UK. The Free Ads department employs approximately 170 members of staff – the majority of whom are involved in copytaking - inputting adverts received via post, fax, mail, live telephone, recorded telephone messages and e-mail (these staff are referred to as ad takers). This work is repetitive in nature and often involves working to deadlines. Workstations are shared, rather than owned by ad takers, as space in the department is limited. Loot early on recognized the potential for their employees to develop WRULDs (work related upper limb disorders) as a result of this work.

Since 1992 Loot have made efforts to improve the working environment of their offices with prevention of WRULDs in mind. A considerable amount of time, money and effort has gone into these improvements. A consultant Ergonomist assessed the work environment and suggested a number of changes to the work environment. The changes ranged from alterations in the routine of ad takers, changes in work organisation, through to recommendations for lighting and adjustable furniture to suit all workers.

Cost and benefit comparison

Selecting a tool
Simpson and Mason 1995 suggest the identification of core costs arising from health and safety issues to illustrate the benefit of ergonomics. This involves calculating the sickness and disruption costs avoided by the use of ergonomics. The pay-back method (Oxenburgh 1991) goes one stage further than this, calculating the effect of sickness and disruption on the productivity of staff and producing a period in which the cost of the ergonomic changes is repaid. This has the advantage of being a method and term with which company managers will be familiar. Both of these methods concentrates on cost avoided, and do not attempt to cost any benefits gained which are not avoided costs.

Pay-back period
This analysis is based upon data gathered concerning the costs of the work force. These are usually calculated for the current or pre-intervention state and then reworked for the predicted or post-intervention state. The costs can then be compared and cost of the changes included in the calculation. Any improvements related to prevention of injuries will be shown as a reduction in absence time, decreased staff turnover and related indirect costs.

Data selection and collection
The main aim of the Health and Safety staff was to assess the costs associated with the development of WRULDs and compare it with the cost of the changes. After studying the literature, including the core costs proposed by Simpson and Mason (1995) a comprehensive list was developed of possible costs to the company of a staff member needing to leave their job due to WRULDs. Further items were added following staff interviews and some items were discarded as irrelevant in the situation at Loot.

It was estimated that an injured worker would be absent from work on full pay for two weeks. If WRULDs is diagnosed, then Loot would then pay for extended sick leave - six weeks in each case. It is also likely that prior to taking any sick leave the employee would be less productive at work - in the calculation this period is three months. If the medical diagnosis confirms that the injured worker has WRULDs, Loot will offer the injured employee physiotherapy, counseling and retraining.

As few pre-intervention data were available, two scenarios were developed in order to estimate the pay-back period that may apply to Loot:
1) The current situation - using current data on staff turnover, absence and number of staff employed as a base line.
2) The current situation data is reworked to include upper limb injury rates prior to the ergonomic intervention. In 1991-1992 the number of staff leaving their job in Free Ads due to WRULDs was between 2 and 5 per year. The ergonomic changes started after the consultation of an Ergonomist at the end of 1992.

The entire cost calculation comprises of four elements or groups.
Group 1 : the average number of hours for which each employee is productive per year.

Group 2 : the average cost of the wage for each employee per productive hour.
Group 3 : costs associated with losing staff and recruiting new staff.
Group 4 : calculated costs associated with covering or hiring temporary staff to cover for sickness and reduced productivity related to poor equipment or injury caused by the work place.

These are first calculated for scenario 1 and then reworked for scenario 2 so that the difference between the two scenarios can be identified. If the costs are greater in scenario 2, then the difference will indicate the benefit achieved through the reduced injury rate in scenario 1.

Table 1. Pay-back period calculations

	Current situation Scenario 1	Including injury rate from 91- 92 Scenario 2	Benefit value of Scenario 1 over Scenario 2 (Difference)
Group 1 Average productive hours per employee per year	910	907	3 hours/year
Group 2 Average wage cost per employee per productive hour	8.11	8.14	0.03 £/hour Total, £4,859 / year
Group 3 Cost of recruitment	121,817	131,187	9,370 £/year
Group 4 Cost of productivity losses due to injured workers with WRULDs	0	5,374	5,374 £/year
Net benefit. This is the total benefit from groups 2,3, and 4.			19,603 £/year

Table 1 shows that a saving of £19,603 per year is achieved in scenario1 (i.e. with the ergonomic intervention). This is calculated by adding the differences or benefits from group 2,3 and 4 calculations (group 1 calculations are necessary only to feed into the calculations of group 2,3 and 4). This saving would occur for every year that the injuries in scenario 2 are avoided.

When the differences are examined on a per employee basis they seem to be small, e.g. the difference in average wage cost per employee is only an extra three pence per productive hour in scenario 2. When these are worked up for the total workforce of ad takers the differences are more substantial, nearly five thousand pounds per year all together.

The pay-back period is calculated by dividing the cost of the improvements by the net benefit:

Pay-back period = $\dfrac{\text{Cost of improvements}}{\text{Net benefit}}$

The cost of the ergonomic intervention was £43,270. This includes the cost of the Consultant Ergonomist, all the furniture and equipment purchased and any necessary maintenance.

$$\text{Pay-back} = \frac{\text{Cost of improvements}}{\text{Net benefit}} \qquad = \frac{43,270}{19,603} \qquad = 2.2 \text{ years}$$

The result in this case is 2.2 years, i.e. the cost of the ergonomic measures is recouped in two years and 3 months. Beyond this period, actual savings are realised.

During discussions with staff a list was developed of costs to the company of a typical WRULD sufferer. As some of these were not easily identifiable in the pay-back calculations, they were listed and approximate costs were then calculated. This is presented in table 2, note that the costs are shown in the equivalent number of hours of the injured employee. The wage cost for the employee is taken at £7.10 per hour.

Table 2. The cost of one worker developing WRULDs

Cost item	Cost in Employee hours equivalent	Explanation
Normal sick pay	40	2 weeks (part time staff working 20 hours/week
Extended sick pay	120	6 weeks
Cover for sick leave	160	8 weeks
Decreased productivity, due to injury	60	Estimated at 25 % reduced productivity for thre months prior to sick leave
Visits to Doctor/therapist	6	Estimated at 2hrs away from work x 3
Doctors report fees	12	2 letters @ £45 each
Cost of physiotherapy	25	4 sessions @ £45 each
Cost of recruiting new staff member	125	Advertisements, interviews, administration and 25 % lowered productivity for first 3 months du to inexperience
Cost of training new staff	85	Induction and on the job training for 3 months
Discussions with Health and Safety staff or Personnel	8	Estimated @ 1 hr of employees time and 2 othe staff (paid 3.5 times injured staff salary)
Counseling	16	4 sessions @ £30 each
Retraining programme for injured staff member	300	Estimated wages of another staff member training plus injured staff member on full pay
Loss of skill and knowledge		Unable to estimate cost
Impact on other staff, morale, motivation		Unable to estimate cost
Compensation possible plus cost of resulting negative publicity for the company		Unable to estimate cost
Total	957 hours	Employee hours equivalent

Table 3: The cost of ergonomic changes

	Employee hours equivalent	Explanation
Cost of Ergonomic changes	6090 hours	For the same cost as the ergonomic changes (£43,270) a worker could have been employed for 6090 hours (at £7.10)

The cost of the changes can be seen to be equivalent to 6090 hours of the injured employees time (table 3). When this is compared to the total cost of the worker developing WRULDs and leaving their current job (957 hours), it can be seen that the cost of the changes are approximately 6 times the cost of a single worker becoming injured and leaving.

Factors missed in a purely financial analysis

It has not been possible to estimate certain items such as the impact of the injury on other members of staff which may result in decreased productivity and poor morale. This was a point raised by one of the interviewees, who remembered a time in 1992 when five employees were reporting symptoms of WRULDs. The cost to the company of lowered morale at that time was thought by the interviewee to be large, in terms of reduced performance and productivity of the staff in the Free Ads department.

Injuries sustained by individual workers carry costs which must be borne by the individual (e.g. reduced quality of life) and by the state (e.g. medical treatment, sickness benefit) (Cherniack and Warren 1999, Levenstein 1999). Often the extent of suffering and personal cost may be hidden due to a reluctance of workers to report injury. None of the methods reviewed include specific guidelines for including these individual or society costs.

Some staff believed that the ergonomics intervention may have contributed to the image of the company as a caring employer. The attention that the company pays to looking after its staff attracts potential employees to apply for jobs with Loot and current staff are encouraged to stay at Loot as employees due to this culture. A recent survey of staff in similar working environments gave the top five factors for retaining staff as: a caring company culture; team spirit; competitive salary; supportive and effective team leaders (Hills 2000). At Loot the use of ergonomics has been part of a wider health and safety department campaign to ensure the well being of the staff and this health and safety policy has now been integrated into the overall functioning of the business. This has created a good image for the company externally and internally.

There were few negative aspects highlighted by the interviews. Some staff commented on the way in which staff have raised expectations of their employer and come to take the efforts toward preventing WRULDs for granted. One commented that the attitude of 'I would rather have the money' (than have better equipment) existed, and that many were not very appreciative. On this point, Oxenburgh (1991) believes that if workers did not expect much from the employer, if they considered that injuries were part of the job, then the solutions implemented would be superficial and in the long term less viable.

Concluding remarks

It is concluded that there have been many benefits to Loot as a result of the use of ergonomics within the Free Ads department. The monetary benefits were demonstrated using the calculation of the Pay-back period, which showed that the cost of the ergonomic changes would be recouped in approximately 2.2 years, by avoiding the costs associated with WRULDs. Beyond this period actual financial benefits would be realised. In the example of Loot, the use of office ergonomics has proved beneficial, it is suggested that it has played a major role in decreasing injury and that has resulted in large cost avoidance. There were also some benefits revealed, during interviews, which were not costs avoided, such as an increase in morale or improved company image. These benefits are less tangible and difficult to apportion to ergonomic intervention.

Cost benefit analysis for ergonomics is not straightforward. There are few validated methods for carrying out a cost benefit analysis for use by ergonomists and none that give guidance or attempt to include more subjective factors. The analysis also requires much data to be gathered which may be difficult to obtain either because figures are not kept or because of the sensitive nature of some company records. The lack of suitable data means estimates are made which may bring into question the full validity of the analysis.

References

Cherniack, M and Warren, N. 1999, Ambiguities in office related injury: the poverty of present approaches.
Occupational Medicine: State of the Art Reviews, Vol. 14, No. 1, 1 – 15

Hills, F. 2000, Human resources, east meets west.
Connect, June 2000, 56

Levenstein, C. 1999, Economic losses from repetitive strain injuries.
Occupational Medicine: State of the Art Review, Vol. 14, No. 1, 149 – 161

Oxenburgh, M. 1991, *Increasing productivity and profit through health and safety.*
(CCH International, NSW Australia)

Simpson, G. and Mason, S. 1995, Economic analysis in Ergonomics. In Wilson J. and Corlett E.N. (ed.) *Evaluation of Human Work* Second Edition (Taylor and Francis, London), 1017 – 1037

SCALING UP SMALLHOLDER FARMING TECHNOLOGY IN INDIA

Tahseen Jafry

International Development Group, Silsoe Research Institute, Wrest Park,
Silsoe, Bedfordshire, MK45 4HS
tahseen.jafry@bbsrc.ac.uk

Ergonomics interventions have contributed to poverty alleviation through the introduction of alternative equipment and practices to the rural poor in the DFID/Government of India Rainfed Farming Project in eastern India. The equipment has been enthusiastically adopted by project villagers to reduce drudgery, accidents and injuries. The equipment includes mouldboard ploughs, paddy threshers, maize shellers, pulse processing equipment, improved stoves, handcarts (for women) and seed storage bins. Over 3 years, there has been firm evidence that interventions have been adopted and are being utilised successfully.

Against this background, there is the potential for scaling up the successful technologies so that that they reach beyond the villages already benefiting. This paper highlights a possible strategy for a scaling up programme.

Introduction

The UK's Department for International Development (DFID) and the Government of India are co-funding a project on rainfed farming to help poor people move out of poverty. 'The East India Rainfed Farming Project' (EIRFP) focuses on developing a farming systems/holistic approach to the alleviation of poverty. Elements of the farming system being addressed include fish production, irrigation, the role of livestock and seed priming and production. Elements of the holistic approach include ergonomics and social development. The project is located in the eastern plateau of India covering three states - Jharkand (formerly part of Bihar), Orissa and West Bengal.

Ergonomics Assessment in Project Villages
The purpose of the ergonomics activities was to introduce new agricultural tools and working practices, which enhance health, quality of life and productive capacity, especially for rural women. In order to achieve this purpose, an ergonomics needs assessment was conducted. In 1996, a participatory assessment was conducted in 300 villages across Bihar (now Jharkand), Orissa and West Bengal (Jafry 1999). The findings revealed that

- Women do more work than men, despite being physically weaker. Being responsible for crop production, processing, domestic chores and income generation. They are so busy during the day that sometimes they do not have time to stop and eat.
- Most of the farming and income generating work is done using simple, traditional hand tools and implements, which makes the work time-consuming and full of drudgery.
- Due to poor working conditions, many people, especially women, suffer from musculo-skeletal disorders, illnesses and injuries sustained from work e.g backpain from weeding, deep wounds to the hand from harvesting, spine and neck problems from carrying heavy loads such as fuel wood and fodder.

The five most strenuous activities identified by women were carrying loads (fuel wood, crops, marketing goods, manure etc), weeding, threshing, harvesting and transplanting. These findings led to three questions:

i)   For how long can women continue to bear the daily heavy burden of agricultural work?

ii)   What physical inputs (agricultural tools and equipment) are needed to capitalise human capacity and capabilities to improve work outputs?

iii)   What can be done to reduce the social cost of work-related diseases to improve quality of life?

In order to address these questions, activities for promoting safe and improved technologies was developed.

Ergonomics Interventions

A number of technologies, specifically for crop production and processing, were introduced to the project villages. The aims of these technologies were to help people to do agricultural work more productively and with less drudgery.

The top six technologies successfully adopted by project villagers in Jharkand, Orissa and West Bengal were:
- mouldboard ploughs
- pedal-operated paddy threshers (for threshing paddy)
- maize shellers (for shelling maize from cobs)
- sickles (for harvesting paddy)
- Krishak Bandhu pumps (for general irrigation especially during the dry summer months)
- seed storage bins (for storage of rice and other grains)

Five other technologies with potential for success were also identified and are listed below. However, further research and development is required on these technologies before they can be recommended as an appropriate intervention for the intended beneficiaries.
- hand carts for women designed and developed by the EIRFP project and the local agricultural university in Ranchi. Prototype II is now being tested and feedback is being collected.

- knapsack sprayers – although there are many types of sprayers, work is still needed to identify the safest to use from those available in the market.
- winnowers - development work is needed to alter the current design to make them safer to use .
- transplanters – development work is needed to evaluate and produce safe, user friendly transplanters
- wheat threshing machines – although much research has been done on improving design to make threshers safe to use, action is needed to promote manufacture of safe designs in the private sector (Mohan and Patel 1992).

Adaptive research and development are needed to ensure that the technologies are safe to use and are designed taking into account gender differences so that they can be operated by women.

Benefits

Feedback from women and men farmers has been collected by community organisers, who live near to the villages and have been trained to introduce project activities and monitor progress using participatory techniques. The feedback has indicated benefits from using the technology. The benefits recorded include the following:

i) Health - reduced drudgery, less injuries, less absence from work, improved nutrition. For example, less damage to the thumb from using the maize sheller instead of manual shelling.

ii) Income gains - opportunity cost of time saved from using improved technology, income gains from hire schemes, gains from sale of produce. For example, 100kg of paddy can be threshed in 2 hours using the paddy thresher compared with 4 hours if done manually. The time saved is being used to earn income through wage labouring.

iii) Reduced vulnerability - food security, less injuries, ability to pay for medicines. For example, vegetables of nutritive value are grown with the help of irrigation pumps. These vegetables are consumed or sold for income.

iv) Empowerment of women - women involved in training and demonstrations. Women have been actively involved in learning about the new technology. They have better knowledge now and are benefiting from this by being pro-active in making decisions about procuring agricultural equipment.

v) Environment - trees saved from using mould board plough compared with wooden ploughs. With traditional wooden ploughs, one tree is needed to make three plough bases. Each base needs to be replaced, on average, once in a growing season. With the new ploughs, the mould board part of the plough is made from metal thus saving trees.

Scaling-up Strategy

The ergonomics activities at the EIRFP has made progress by contributing to poverty alleviation through the introduction of alternative equipment and practices to the rural poor. These have been enthusiastically adopted in project villages, to reduce drudgery, accidents and injuries, create time for new productive activities and create income from equipment hire schemes.

Assessments, indicating success, have led to the DFID Rural Development Group in India wishing to transfer the interventions to other rural development projects in India i.e to initiate a process of scaling up. Before this can be done, a procedure is needed to identify the elements of the holistic approach and relevant external influences that need to be in place if scaling up is to be successful and sustainable. A compilation of checkpoints was developed through discussions with project staff and is given below. These checkpoints could then form the basis of a strategy for scaling up technology in other rural development programmes in India and to similar projects elsewhere in south Asia.

Checkpoints of Issues for Scaling-Up Technology

i)   Vicious Cycle of Demand and Supply ("Catch 22")
A search for improved tools and implements in the EIRFP project villages revealed that there are no reliable suppliers. Organisations are not willing to manufacture or stock supplies for which they feel there is no demand. However, demand for tools is not generated because farmers are not aware of the alternative or improved technologies that could be made available. It is important to establish demand and supply of tools in a scaling up exercise.

ii)   Extension methods
- demonstrations and training (particularly for women) are essential. There are many mechanisms for this e.g training days at block level, especially on market days to get maximum exposure or training at village level. Village level training may, however, be the best option to encourage women farmers.
- Promotion of technologies via video, radio, leaflets.
- Providing farmer support and collecting regular feedback. A programme for 'scaling-up mechanisation' needs to keep track of the successes, constraints, opportunities and threats. For instance, solving operational problems quickly, Attempts to 'mechanise' India in the past have failed due, inter alia, to a lack of farmer support and feedback mechanisms. A strategy for participative farmer support needs to be developed which would link poor farmers with designers, manufacturers and suppliers.

iii)   Availability
- Identifying procedures to make tools and equipment available locally and nationally is required.
- Encourage the private sector to supply at a local level - this, however, requires evidence of farmer demand for equipment

iv)   Maintenance and repair
- Establishment of agro-service centres to act as sales centres, custom hire service, repair and maintenance centres. Lessons from the past indicate that if there are no effective and affordable repair facilities, equipment will be abandoned and people will return to their traditional methods of production. People will lose enthusiasm and motivation for the improved technology. Thus, is it imperative that maintenance and repair issues are tackled early on in the scaling up process.
- Training of local blacksmiths and, thereby, making use of local skills is essential.

v) *Linkages and Institutional Support*
- It is important to identify procedures that will enable women to network and develop links with agencies for credit, marketing, information and other inputs. Links with the state-run agencies, including agricultural extension and universities as well as the private sector, are necessary.

vi) *Safety*
- Common problems related to the use of agricultural machinery include deaths, limb amputations, eye injuries, cuts/wounds and broken bones (Varghese and Mohan 1990). Therefore safety of equipment is paramount. Although, there exist safe design guidelines for some agricultural equipment, these are rarely (if at all) followed by manufacturers. For other equipment, safe design guidelines do not exist. Procedures must be established to ensure that all equipment is safe to use and only those that are safe to use are promoted and disseminated. Action is needed to ensure that manufacturers do not put cost of production before the safety of their equipment.
- Educating farm workers in the immediate care of injuries is necessary. The development of educational material, promotion of useful local healing treatments and publicising simple and inexpensive first aid administration will help to reduce healing time.

vii) *Ownership*
- The cost of owning improved technology is often not viable for the resource-poor farmer. There are a few options. For example, co-operatives can form to procure tools, promote a tool supply service with the provision of long-term loan schemes for individuals.

Conclusions

If the checkpoints given above were taken into consideration by a development aid project on scaling up and dissemination of improved agricultural machinery, it would lend itself to a number of outcomes. These include:
- Less accidents and injuries sustained from work
- Poor people more knowledgeable about improved technology
- Poor people having access to improved technology and are aware of organisations that could provide help and support in procuring improved technology
- Agricultural work becomes less drudgerous for women
- Women can work more efficiently and productively and therefore have a better quality of life.

According to the World Health Organisation (WHO 1995), ergonomics related and occupational injuries are third among the causes of morbidity on a global scale. This highlights that illnesses induced by work-related and work generated hazards are significant determinants of the state of world health. The project described in this paper, is the first DFID funded project to tackle ergonomics issues in its rural development programme in the context of seeking to improve the lives of poor people and enhance the sustainability of their livelihoods. Ergonomics has contributed towards this by encouraging assessment of the relationship between workers, equipment and the working environment.

References

Jafry, T. (1999) Ergonomics - Attacking One of the Root Causes of Poverty, Human Drudgery in the East India Rainfed Farming Project, Overseas Development Institute, Agricultural Research and Extension Network, Newsletter 40, 17-20

Mohan,.D. and Patel, R. 1992, Design of Safer Agricultural Equipment: Application of Ergonomics and Epidemiology, International Journal of Industrial Ergonomics, 10, 301-309

Varghese, M. and Mohan, D. 1990, Occupational Injuries Among Agricultural Workers in Rural Haryana, India, Journal of Occupational Accidents, 12, 237-244

WHO.1995, Global Strategy on Occupational Health for All, the Way to Health at Work, (World Health Organisation, Geneva).

Acknowledgements

This publication is an output from a project funded by the United Kingdom Department for International Development (DFID) for the benefit of developing countries. The views expressed are not necessarily those of DFID.

USABILITY

EXPERIENTIAL USABILITY STUDY FOR MULTIDISCIPLINARY DESIGN

Mirja Kälviäinen

The Kuopio Academy of Design
The North Savo Polytechnique
PL 98, 70101 KUOPIO, FINLAND
mirja.kalviainen@pspt.fi

This paper will present the usability study of eleven supervision units of industrial environments. The study was conducted to achieve functional and experiential information for the concept design of the interiors and working clothes. The study is combining the questions of usability and meanings associated with the working environment for the purpose of creating an environment where the users' suitable experiential emotions are taken into consideration. This kind of working environment can have contrasting experiential requirements about work and rest, which the study tried to reach by using several study methods including projective techniques. The members of the multidisciplinary design group themselves conducted the study so that they had the possibility to empathise with the product users.

Introduction

The usability study for the design purposes of supervision units in industrial environments has been implemented with multiple methods in eleven different supervision units. The purpose of the study was to increase knowledge about the general usability issues and collect mind images about user experiences and ideals in the supervision units. The design group has the task to produce visual concept design for the interior of the supervision unit and supervision workers' clothing. The main issue of this paper is to introduce the results of the projective methods used in collecting the mind image information as a means for the designers to achieve the right experiential feel for the supervision unit design. This paper was written in the phase where the study visits had been completed and the analysis of the results and the concept design was starting.

The supervision unit cases

The client for the design concepts is a Finnish company called OC-systems, which is a producer of the supervision unit buildings. This company wants to spread its production into a more holistic service for the industry. To achieve this it has gathered a network of other enterprises to be able to produce, in the future, the whole interior, furniture and clothing of the supervision unit as a holistic, multidisciplinary package for their client industries. The collaborative Finnish companies are Ergo furniture, which produces ergonomic office furniture and Varpuke Oy, which is a producer of working clothes.

A multidisciplinary design student group is preparing diploma works in this project. They are two students of furniture and interior design Liisa Korhonen and Jaana Tullila and two students of clothing design Sari Kinnunen and Minna Nevalainen. These designers planned the study together with the OC-systems designer and managing director and with a tutor group of industrial design, interior design, furniture design and clothing design supervisors from the Kuopio Academy of Design. After the plan for the study was made the design students carried out independently the case studies in the supervision units. It was important that the members design group themselves conducted the study so that they had the possibility to empathise with the actual experiential situation of the end users.

The study was conducted in industrial properties mainly in the Eastern-Finland area. Seven industrial and other enterprises were studied. These represented production of newspapers, dairy products, brewery products and timber. Also metal, paper and chemical industries were involved in the study. The variation in the purpose of the supervision units was thus rather wide and covering. The designers saw this as an adequate amount to produce information about the physical environment so that only industrial environment for some completely different purpose and processes might have brought out new aspects. In the study eleven supervision units were documented and studied with thirteen user interviews.

The supervision units were documented by taking photographs and videos from the working process and environment. Also the noise level in the supervision units was measured. Observations were made about the size of the rooms, the placement of objects, clothes used, materials, objects, lightning, ventilation, social action and atmosphere. In addition to this, information that is provided by the local working environment research unit is applied. This research unit has conducted measurements for the supervision unit producer OC-systems that help to define the optimal levels of noise, humidity, draft and temperature for healthy and pleasurable conditions in the supervision unit.

Other information was obtained by interviewing the people working in these environments. The interviews consisted of questions concerning the working conditions, shifts, breaks, the amount of people, the tasks to be preformed, positions used in the tasks, lightning, ventilation, temperature, actions taken during the breaks, social life and eating. The respondents were asked about their use of clothes at the moment and how they used the current furniture and environment. The questions tried also to get the respondents to mention any problems they might have with their working clothing or interior. It seemed difficult for the users to express the problems even when clear faults could be observed.

Especially in the case of the working clothes this was evident. The respondents noticed more problems in the furniture, for example in the office chairs, and they complained about problems such as draft. In general there seemed to be 'this is well enough for a working environment' - attitude.

Besides the spoken questions the interviews consisted also projective tools to obtain material from the mind images and associations connected to the suitable experiences in the working environment. The projective study seemed to provide more information to the visual and image design that the other methods. The information obtained with the projective tools was even in contrast with the information obtained by straight questions.

The projective methods

The projective techniques used consisted of two distinction studies. The first one was about car brands. The respondents were asked to relate the current working clothes with five car brands shown as car pictures: Lada, Porsche, Toyota, Volvo and, Volskwagen. The respondents were asked to make this distinction and association thinking both about the visual image and the technical performance of the clothes in connection to the same features in the car brands.

The car brands seemed surprisingly easy to correlate with the working clothes. The few female respondents in the study seemed to have difficulties with this task as they did not seem to have as strong sense of status and difference about the various car brands as the male respondents had. It was easy for the respondents to put the five car brands into order and think both about the technical and visual aspects of these different brands. For the designers the five different car brands provided information about experiential features and images connected to the working environment.

Lada proved to be the worst in status and it was used for describing a really bad image and technical performance. Volskwagen got the best ranking in combination of both the visual image and technical qualities. Volvo had the best status in technical judgement. Toyota was ranked in the middle with rather good looks and technical qualities. Porsche was ranked in the middle technically but the positive visual image raised it higher. Porsche was, however, seldom chosen to visualise the desirable supervision unit and the designers interpreted that the associated sports and speed were not describing the suitable supervision unit image.

Toyota and Volkswagen were mostly used as the image for the current working clothes. Especially Volkswagen and also Volvo were used as the image for the ideal working clothes. The contradiction in the spoken interviews and in the projective result came out when the respondent did not see anything wrong with their working clothes. Yet they chose another, higher status car, to present their ideal working clothes than the one they connected with the current working clothes. The most striking example of this was the respondent who saw the Lada as the image of his current clothes but still felt that there was nothing wrong with them.

The other means to get knowledge about the users mind images was a set of 44 interior design pictures. They were chosen for this project through a pilot study in the Kuopio Academy of design were design students and staff were asked to write down associations from an even larger set of pictures. The pictures were chosen as ones giving associations about wide variety of interior atmospheres. These pictures were used to get the respondent to choose the most desirable interior for pleasure and then to choose pictures about an ideal supervision unit. Also the most unsuitable environments were chosen. The reasons for choosing certain images were discussed with the respondents.

The interior design pictures seemed to be a bit difficult for the respondents to relate to their own pleasure or work. It seemed that the respondents related too much concrete thinking to the use of the environments. If the picture was a bar or a home the use of the environment influenced the responses. It also proved to be difficult for some of the respondents to analyse from the pictures the causes of the pleasurable feel but it did help when the interviewer promoted the discussion. It proved good to have a wide set of pictures. Different pleasurable features came out of different pictures in the discussions such as colours, scenery or materials.

Even if the range of the pictures was rather wide some of the pictures came out as the most popular. The chosen interiors did not contain too much ornaments or small objects. Neither did they have a clinical, purist atmosphere. The traditional Finnish look was popular with wooden materials, greenery and large windows. Also spacious interiors were appreciated. The interiors chosen for a suitable supervision unit were clear with lots of empty space and free routes for passage. For the supervision unit choices especially the dim and soft lightning was mentioned as the criteria.

The interior pictures shed light into the desirable atmosphere of the interior design. However, for the interior design also the spoken interviews provided good information about the colours and other pleasure related aspects of the environment. In spite of concentrating either on the clothing or on the interiors both of the projective studies are used in the design phase of these different product areas to inform the designers about the experiential feel questions.

Brainstorming and design

After the information had been collected the design team held a brainstorming session, where they produced mind maps stating the important issues, features and problems of the environment. The aim of the brainstorming session was to combine the information obtained by observation, interviews and projective study quickly after the study tour to the supervision units.

The designers produced a big mind map looking at the general usability issues. Usability seemed to concentrate on the physical environment. Issues such as different kinds of office chairs for different types of supervision work could easily improve the usability. In the observation it came out that the users had made their own adjustments and misuse for the products when the environment did not correspond to their needs.

The placement of the various objects, papers and technical equipment needed in the work could be improved by the furniture and interior design. In many cases the breaks and eating pauses were held fitted into the working process and for this reason it was important to have possibilities for cooking or eating at the premises. The lightning and reflections proved to be a problem area causing even headaches and back problems when the users had to move their heads in inconvenient positions to see properly. Often the fluoricent lamps were the cause of this problem.

In the clothing the materials and protection proved to important. Many had protective clothes made for industrial purposes so movements and cover did not cause major problems. There were, however, problems of sweating. Also some work included going in and out of the supervision unit and the different spaces had different temperatures that should have been taken into consideration in the working clothes.

Two smaller mind maps concentrating on the pleasure issues and the image of the supervision units were also produced. In the mind map looking at the pleasure felt from the working environment the social contacts came out as very important factors. This is not of course an issue that a designer can solve alone, but at least it is important to bear in mind that the design should support this social interaction. Pleasure was related also to the meaning the users felt their work had.

The workers had provided themselves some pleasure features into the environment. Some entertainment equipment seemed important such as radio or television. A pornographic calendar was found in every supervision unit. Flowers were also usual. Preferably they should have been real ones but silk flowers were accepted if there were no windows to give the real ones proper light.

Colours came out as important wishes to make the environment less gloomy. Also it seemed important to be able to see out, preferably to the nature outside. This came especially out in the preference for interior pictures with a window and scenery. Also dim and soft light seemed to be preferred. Noise was something the workers had used to put up with but when asked they agreed that measures should be taken against the noise. Temperature caused problems especially when it was different outside and inside the supervision unit. Also the state of ventilation and possible draft were connected to pleasure or displeasure. Smells were stated as annoying and the distraction of them could improve the enjoyment of the environment.

The issues connected to the image delivered through the environment were gathered into a mind map. This image was part of the users contentment and feel that their work was meaningful. The image seemed to be connected closely to the visual elements in the environment. The quality, shape and tidiness of the environment affected into the judgement of the image. In addition to this, there seemed to be some value-based issues such as ethics and ecological considerations that weighed in the judgement. The mental atmosphere in general was considered to be important. The image was also influenced by the company's orientation to the future such as possible drive for renewal in the atmosphere and environment.

In the future the supervision unit producer plans to incorporate intelligent technology into the supervision unit environment. Although this diploma work project does not consist of design suggestions about the use of the intelligent solutions, the studies conducted in this project can be helpful in the future in incorporating user centred technology into the supervision unit environment. Technology could solve problems of achieving pleasurable humidity, temperature and ventilation. Technology could be incorporated into the clothing or into the supervision unit. Also the problems in lightning and adjustability could be solved with the new technology.

From this point the design process continues with more analysis of the case study information. Especially the projective material will be analysed more thoroughly to provide visual information about the right experiential feel for the design concepts. The cars are put into more detailed scales with the help of grids. The most popular interior design pictures will be analysed accurately. Also the documentation of the environment with photographs and videos needs a more thorough analysis. The results of the observations, interviews and the projective distinctions will be compared to point out the differences in the result of the different methods.

The analysis provides the information for the important features in the new concepts of the interior design and clothing. The designers mean to collaborate also during the design phase so that the interior design and the clothing design solutions match to each other. The design phase will also consider the possibilities of the production in the companies involved in this supervision unit project.

The clothing designers aim to work out a function analysis of the clothes and continue designing through that. For them the image seemed to be the most important feel quality to guide their design. The designers aim to create a collection of working clothes from which one sample will be chosen as the concept that can be used for the alterations necessary in different orders of the supervision units. These changes are due to, for example, the colours used in the marketing image of a certain industry.

The interior design will proceed in another manner. The designer will provide a set of guidelines for the interior design of the supervision units. They will also provide an example of how the guidelines can be used and create some module furniture for their example. The guidelines are built to give information about the complexity of the supervision unit interior design. If one point changes it influences how some other point should be organized. The atmosphere and the functionality are the main points in the interior design.

Conclusion

The use of different methods proved to be useful in this study. The different inquiry methods revealed various, even contrasting, information. The projective tools proved to be useful as the source for the ideal and pleasurable feel of the future environments. The observational and question based information was important for the functional features.

This kind of working environment can have contrasting experiential requirements. A previous design work for the exterior of the supervision units in the company OC-systems suggested that the feeling of power is important for the professional image of the workers. On the other hand, the interior of the supervision unit often functions also as a rest room for the industry staff. This use emphasises the possible pleasure aspects of the environment and associations that disconnect the users form the industrial image.

In this project it seemed evident that the supervision unit workers did not see their job as a powerful or special and in the straight questions they did not see much need for improvements in the environment. The meaning of the environment to the status and enjoyment of work came out in the cases where inside the same industry there had been improvements and new supervision unit for other teams. The workers in the older supervision units evidently were jealous if colleagues got a more modern and technically advanced unit. The projective techniques brought out these ideal images of the working environment that did not come out otherwise. These positive environmental experiences are important for increasing the desireability of jobs in the industrial environment.

Rapid design and the online experience: incorporating the human factor into the process

Marc McNeill

[1]*Accenture*
1 Kingsway
London
WC2R 3LT
marc.mcneill@accenture.com

The need for usability in creating online services is well documented. However, there are still many online services that fail to engage the user and are difficult to use. In addressing the need for good design, this paper discusses a rapid and intensive process for the evolution of a value proposition, and its realisation as an online service. Key to the process is the incorporation of timely feedback from the customers themselves and involvement of all key stakeholders. This includes qualitative analysis of the interaction between customers and the prototype service in a dedicated suite. The results of these design evaluation sessions are promptly fed back into the design process, helping to ensure usability, engagement and satisfaction for customers on service release.

Introduction

It is argued that creating an engaging customer experience is a powerful and cost effective way of improving conversion rates and customer retention. Indeed by focusing on purchasing conversion rates ventures can significantly increase revenue figures. Focusing on customer retention as well can increase revenue figures for a high volume site even further. Where competitors are only a click away, ensuring customers enjoy a compelling experience is paramount and critical to the success of interactive services.

The online customer experience is formed from the entire interaction that customers have with an e-business. It is the sum total of the customer touch points, which span from the point at which the customer first becomes aware of the e-business right through to the after sales service long after the transaction has been carried out.

Five key elements have been identified, which collectively provide customers with an engaging online experience spanning from initial awareness through to after sales service (Table 1). However, there are few e-businesses that incorporate all these elements in their on-line experiences. Their failure to do so may be costly, with the benefits of focusing upon the customer experience and the human factor being considerable. In monetary terms, Hurst (1999) suggests that $1 spent on advertising produces $5 in total revenue whereas $1 spent on customer experience improvements may yield $60.

Table 1 Key elements present in a compelling on-line customer experience

Key elements	Description	Example characteristics
Personality	The way in which a site develops a connection with its audience at an emotional level and the means by which it immerses users in an experience that appeals to the user style and preferences	• Editorial style • Community • Interaction Style • Presentation Style
Content	Content is the primary method by which a site delivers its message to users. Refers to the copy and graphics that populate a site	• Clear • Up-to-date • Clear policies • Personalised
Usability	More than just the look and feel of the site, usability encompasses how the user interfaces with the site and how easily they are able to accomplish what they set out to do	• Clear navigation and structure • Consistency • Simple page design • Speaks the users language • Easy to use
Functionality	A functional site is one that offers supportive processes and mechanisms to carry out the users instructions and efficiently guide them through their intended tasks	• Efficient registration and transaction processes • Supporting customer service
Operational excellence	Achieved when an e-commerce proposition delivers in line with customer expectations from the point of logging on to the arrival of goods and beyond	• Robust • Fulfilment • Cross platform

Working in eCommerce presents an increasing demand to work at "e-speed" and rapidly develop applications. The traditional approach to such projects is to allow customer feedback either early on during concept definition or in late-stage usability testing. Site release reflects compromises made in the rush to get to market, compromises which are typically made based on judgment of developers and anecdotal feelings on what customers want/need. Design testing, if done at all, is typically done in the form of formal usability testing towards the end of the project, when major alterations of information architecture or user interface design are costly to implement.

This failure to integrate customer feedback into the critical design issues made throughout the project lifecycle is likely to impact on the ultimate success of the site. There is clearly a need to consider the human factor in eCommerce projects, and involve the end user throughout the design lifecycle. This paper discusses elements of such an approach, while continuing to consider the compromises required during a project (e.g. tight timescales).

Processes

A process for integrating customer feedback and usability into the design lifecycle has been used by the Accenture Interaction Design Group with successful results without impacting deadlines. A number of elements are critical to this process. These include;

- Identifying target users
- Identifying competitors
- Creating information architecture
- Building storyboards

The key to the success of the process has been its iterative nature, involving end users throughout the whole design cycle and into the technical build. This involvement typically occurs in regular design evaluation sessions. These sessions borrow from usability testing, however rather than employing a quantitative, task focussed approach, they adopt a more interactive 'listening' and qualitative approach with active facilitation. The results of these sessions are fed directly into design workshops, allowing decisions to be made quickly, based upon customer research rather than the 'hunches' of the design teams.

Design evaluation sessions

The process for the design evaluation sessions (Figure 1) provides rapid and timely evaluation of the customer experience. Six representatives of a given target customer segment are recruited by an independent field recruitment agency. Nielsen (1993) has found that 85% of usability problems can be uncovered by such a number of users; anecdotal evidence from design evaluation sessions suggests that major issues will be highlighted by even fewer users.

The sessions take place in a dedicated lab facility, with two test rooms; one set up as a lounge, simulating a home environment, the other an office environment. Both rooms allow testing of applications on PCs and mobile devices, whilst the former also includes interactive TV devices and games consoles with Internet connectivity. Each room has two discrete wall mounted cameras and a ceiling camera with desktop and ceiling microphones.

Each representative participates in an hour-long 'depth' session with their behaviour, physical actions and comments being remotely observed' digitally recorded and logged using dedicated software (Noldus™ Observer Pro). A viewing room with a plasma screen allows the development / client team to observe each session.

The approach adopted during the session depends upon the level of maturity of the site being investigated, varying from 'wire frame' prototypes that are task and scenario driven to fully functional sites in which the user is left to explore the product/application, to find out what is of interest. These approaches are discussed below.

Following each session, debriefs are conducted with the facilitator, logger and observers, covering synthesis of observations, analysis of web pages and germination of solutions. Results are presented to a design workshop the following morning, allowing for immediate action. If required, a presentation detailing the results with supporting video clips is also delivered a day and a half after the participant sessions, allowing stakeholders to review the user feedback. This rapid turnaround of results ensures timely user input into the process, and the ability to react to this when making key decisions.

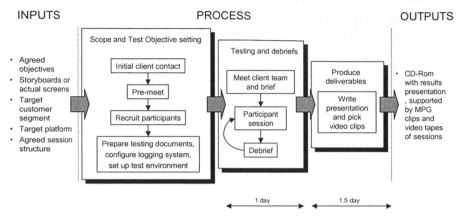

Figure 1 Process for Design Evaluation

Stakeholder involvement

From the outset of the project it is essential to identify all stakeholders (business sponsors) within the company to ensure buy-in to the project and an understanding of what is in and what is out of scope. It is not unusual for different parts of the business to approach the project well into the design and build phase requesting content to be included. Regular meetings are essential, with workshops following design evaluation sessions reviewing the results and steering the project based upon the feedback from the users.

Identify target users

It is essential to have a clear purpose of the offering so it will support customers' interests, needs and goals. Customer insight and an understanding of target market are essential, with conventional demographics, (evaluating market data, customer requirements and developing an audience and customer description) being augmented by an understanding of online user behaviour. On the Internet different people interact in different ways, it is important to consider different interaction styles and ensure that the approach of the web site will appropriately accommodate all target user styles, such as task orientated as opposed to opportunistic, serendipitous browsing.

Competitor analysis

There is a great deal that can be learnt from an eBusiness's competitors; both those competing directly in their market space and investigating how other sectors deliver value using different interaction styles, mechanisms, content and tools. Design evaluation sessions are run with target users interacting with direct competitors' sites. The approach for these sessions is driven by the users, who are invited to browse the site as they would do at home/ work and discuss the experience as they interact. There are no set tasks, the objective being to identify how the site engages the user *overall*, with feedback being qualititative, probing the user for his/her perception of the experience, ease of use and willingness to return.

Task based testing can be successfully used when investigating competitor sites. In particular, when investigating how different tools and mechanisms can be used to support a process. Whilst the content within the tool may be different, much can be learnt on the usability of the presentation method. For example using tools for comparing different products such as tools for comparing electronic products that a company offers may highlight important issues for their implementation and delivery that could be employed in a tool comparing financial products. Similarly, the use of animation for helping customers select clothing for use may be explored for its utility in helping customer select financial products in banking sites.

Information architecture

Information architecture has been defined as the 'art and science of organizing information to help people effectively fulfil their information needs' (Hagedorn, 2000). The process is presented in Figure 2. The organization of the content is undertaken with the co-operation of users in workshops whereby content headings are presented on flash cards and customers are asked to divide and 'chunk' content into appropriate, logical units (Rosenfeld and Morville 1998). The organisational schema are used to inform the process and help develop the structure and architectural blueprint, from which storyboards are developed.

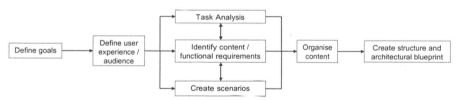

Figure 2 High level process for defining information architecture

Build storyboards

In order to validate the information architecture and to investigate high-level page design, storyboards are mocked up. These are typically crude 'wire frame' illustrations of pages mocked up in a graphical or presentation software package (e.g. Figure 3). They are simple and allow rapid modifications to be made. Storyboards will walk through a particular scenario and are presented to users in design evaluation sessions. Whilst they are task orientated, and linear, users are invited to talk around the page, discussing what their expectations are, both for what they would do next, and around the different elements of the page. Storyboards have the additional advantage of allowing the development team to express their ideas in a simple to understand format to sceptical parties rather than complex functional and technical requirements documents that can be difficult to relate to.

Figure 3 Example of home page for storyboard

Discussion and conclusion

This methodology has been used successfully with Internet, intranet, interactive TV and mobile device applications. Working with numerous clients, by including the human factor in the design process and by involving end users throughout, the result has been a shared vision of a customer-orientated solution. It has also reduced the time to validate a design through the focused and iterative nature of the process, improving usability & increased customer satisfaction. As an example, the immediate benefits of the design evaluation session were demonstrated by a high street grocery retailer with an online presence. The 6 customer sessions focused on searching for products within the site, the registration process, product purchase and the overall customer experience. Issues were identified and fed back to the development team. The findings were used to inform the redesign process. In conjunction with a marketing campaign of hanging boards in stores, leaflets and extra loyalty points for the first purchase, the redesigned site resulted in a sustained increase in sales of 700% and 300% for the sections of the site investigated and an improved conversion rate of 6-7%, well above industry norms.

References

Hagedorn, K. 2000, *The Information Architecture Glossary: A publication of the Argus Center for Information Architecture*, http://Argus-acia.com/

Rosenfeld, L., Morville, P. 1998, *Information Architecture for the World Wide Web* (O'Reilly & Associates, New York)

Nielsen, J. 1993, *Usability Engineering*, (Morgan Kaufmann Publishers Inc., San Francisco)

Hurst, M. 1999, *Holiday '99 E-Commerece: Building the $6 Billion Customer experience Gap*, http://creativegood.com/holiday99/

DESIGN

A survey of the design needs of older and disabled people

Ruth Oliver [1], Diane Gyi [1], Mark Porter [1], Russ Marshall [1], Keith Case [2]

[1] *Department of Design and Technology,*
[2] *Wolfson School of Mechanical and Manufacturing Engineering,*
Loughborough University, Loughborough, Leics., LE11 3TU, UK

The challenges facing design teams with respect to older and physically disabled people are only now beginning to be addressed, largely due to the fact that the population is ageing. In order for designers to consider the needs of these people and design inclusively, it is necessary to understand the requirements and preferences that are experienced in Activities of Daily Life (ADL), as people interact with everyday products, environments, and systems. This paper presents the results of a survey into the needs of older and disabled people today. The results show that, despite advances in technology and design, participants still have difficulty performing the everyday activities that most of us take for granted.

Introduction

In the literature there are a number of studies investigating the problems with Activities of Daily Life (ADL) that older people experience (Weber *et al*, 1989; Ashworth *et al*, 1994; Department of Trade and Industry – DTI, 2000). However, none really address what these people really want to be able to do, or need to be able to do more easily.

The survey detailed in this paper forms part of an Engineering and Physical Sciences Research Council (EPSRC) 3-year project under the EQUAL initiative, aimed at Extending QUAlity Life. This project aims to develop a computer design tool to help designers 'design for all'. The survey detailed here aimed to get a broad range of views from older and disabled people as to what activities cause them problems, and what they would like to be able to do more easily. The results will direct the next phase of the research, the selection of tasks to focus the collection of task-specific data for the computer tool.

Method

A semi-structured questionnaire was used to investigate these issues and provide the interviewer with a frame to work within. Detailed responses to the questions were noted, and a scale was used by the interviewer to facilitate assessing the level of difficulty (Table 1). In this way both quantitative and rich qualitative data were collected.

Table 1. Interviewer's scale (with description) used to assign rating based on participants' responses

Scale point	Description
Easily	No problems performing activity
Some problems	Some difficulty but no need for assistive devices (such as levers, bath chairs, and so on)
Some help	Basic assistive devices needed (e.g. tap levers, stool in bath/ shower)
Considerable help	Assistance from another person and/or complex device needed (e.g. hoist for bath or toilet)
Impossible	Participant unable to perform activity
Not done	Activity not attempted for reasons other than physical difficulty (e.g. lack of interest, necessary equipment not owned)

The interviews were approximately 20 minutes long and mostly conducted face-to-face, although some were by telephone. It was felt that a personal approach would result in higher response rates than postal questionnaires, and remove the possibility of any practical problems for participants completing the questionnaires themselves. This approach also allowed for discussion of interesting points raised, clarification of any misunderstandings, and demonstration of any particular difficulties.

The questionnaire was divided into seven sections, each concentrating on a different area (Table 2). The different sections were designed to be specific and provide focus, rather than asking very broad, general questions. Tasks were selected to be generic in the movements required to complete the task. For example, one question in the kitchen section was 'how do you manage with lifting a small saucepan onto the back hob?'. This asks about a specific activity, but it was felt that the answers would be similar to those for other activities that require lifting and reaching an item of a similar weight at about waist height.

Each section ended with a question asking if there was anything that participants' would really like to be able to do, or do more easily (within the limitations of their disability), if the equipment or environment were designed differently. This information was important in order for the research team to understand what the priorities for older and disabled people are when it comes to design improving quality of life.

It was intended that 50 people be interviewed (25 men and 25 women). The participants were not intended to be representative of the population as a whole, but to provide valuable information as to the problems and needs of people across a wide range of age, abilities and disabilities. Broadly, the breakdown was as follows: ten aged 18-32 years (disabled), ten aged 33-47 years (disabled), ten aged 48-62 years (disabled), ten aged 63 years + (disabled), and ten aged 63 years + (able-bodied). These strata were to ensure a spread of ages within the sample.

Participants were personally recruited from local clubs for disabled or older people and by the handing out of fliers at the Motorbility Roadshow, held at Donnington Park in June 2000. Informed consent was obtained from all participants prior to the interview commencing.

Analysis involved taking counts of the number of participants reporting the same level of difficulty with the same tasks, and also the number of similar comments being made by different participants. The main findings are presented in this paper.

Table 2. Structure of the interview questionnaire

Section	Detail
Personal details	Age, gender, home type, mobility at home/outside, reliance on others, use of stairs, main problems due to disability
Kitchen	Using hob, oven, taps, doing washing-up, washing clothes, reaching high shelves, filling kettle
Bathroom	Using bath, shower, toilet
General household	Opening/closing windows, doors, plugs into/out of sockets
General away from home	Into/out of cars, using buses, trains, cash machines, getting petrol, shopping in large shops
Work (where applicable)	Access, movement within area, reaching
Leisure	Activities undertaken, using leisure centres, garden

Results

Personal details

To date interviews have been conducted with 43 older and disabled people (18 men aged 21-99 years, 25 women aged 21-82 years). This includes younger disabled people, 5 people over 63 years with no specific impairments, and 10 older people with disabilities.

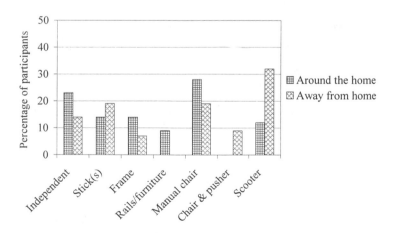

Figure 1. Percentage of participants using various modes of mobility within the home and away from home (n=43)

The results show that, not surprisingly, most people wanted to maintain independence and perform the every day activities that other people take for granted, and all but 2 participants lived at home. Figure 1 shows the number of participants reporting different levels of mobility and any assistance needed to be mobile (for example a stick, or wheeled frame), both in the home and away from home. Seventeen participants were actually in a wheelchair for the majority of the time. Around the home 23 % of all participants were fully independent whilst only 14 % were when away from home. This pattern of reduced independence in mobility is repeated throughout Figure 1.

Bathroom tasks

Figure 2 shows that 42 % of participants found it impossible to use their bath or needed considerable help, for example another person and/or a hoist. Two participants did not use the bath for fear of slipping and falling, and four participants could only use the bath with two handrails fitted. Nine participants had a seat to assist them in using the shower. The three participants needing 'considerable help' to use the toilet needed another person and a hoist to lift them on and off the toilet.

Kitchen tasks

Figure 3 details the kitchen tasks that participants required considerable help with or found impossible. Activities that involved less lifting and reaching resulted in fewer problems (such as washing up small items). Many coping strategies were mentioned, including sitting to do tasks (5 participants), sliding heavy items along surfaces rather than lifting (7 participants), moving heavy items in stages - from worktop to stool, from stool to oven (3 participants), and sitting/kneeling on the floor for low tasks, such as using the oven (2 participants).

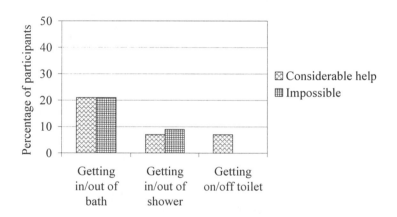

Figure 2. Percentage of participants requiring considerable help or unable or complete bathroom tasks

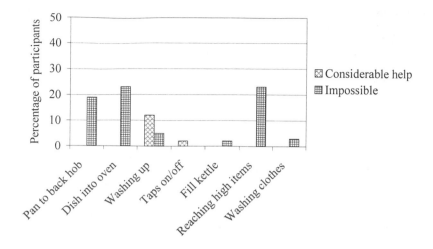

Figure 3. Percentage of participants requiring considerable help or unable or complete kitchen tasks

General household and general away from home
Participants reported that opening and closing windows only caused problems if they were high (14 participants) or too heavy (2 participants). Opening and closing doors caused problems if too narrow for a wheelchair (2 participants).

   With regard to transport, 8 participants reported needing considerable help getting in and out of cars. Buses were inaccessible to scooters (14 participants), whilst lower, 'kneeling' buses were easier for ambulant participants. Only 19 participants had ever tried to use trains, with the remainder saying that they were put off by the difficulties involved due to their disability, or that they had alternative means of transport. Cash machines were often at the wrong height (13 participants). Shopping in large shops was possible for 38 participants, but help was needed with reaching and carrying items (6 participants).

Work and Leisure
Interestingly, 24 participants were of working age (18-65 years of age), but only 7 were actually working. No problems were mentioned by these individuals with the working environment, due to the companies having made any necessary alterations prior to them starting work.

   Participants were involved in a wide range of leisure activities, with a surprisingly high number (20 participants) saying that there was nothing that they wanted to do that they were unable to do. Access caused the most problems, with problems with access to swimming (8 participants), cinema (3 participants), and going on holiday/to other peoples' houses (5 participants).

Contribution of the data to the main study
The results of this survey will be used to inform the collection of detailed, task specific, data from 100 individuals for the next stage of the project. Access issues resulted in the

highest number of activities that people most wanted to be able to do, or do more easily (27 participants), but access is beyond the scope of this investigation, with the focus being more on the design of products and systems.

Kitchen tasks contain many generic actions that are applicable to other ADL: reaching high (impossible for 23 %), bending (impossible for 23 % to use oven, and 3 % to use washing machine) and lifting (impossible for 19 % to lift a pan). Being able to use the oven was also something that 9 participants most wanted to be able to do. It was therefore decided that the detailed data collection phase would involve the lift-bend-reach activities concerned with cooking and using an oven. The activities involved in this task are also suitable for modelling on the computer.

Discussion and conclusions

This study, despite the small sample size, supports the findings of similar studies (such as DTA, 2000) that older and disabled people experience many difficulties in what should be simple ADL. A person unable to cook for themselves, or dress and toilet themselves clearly must rely on the assistance of others. This, along with the fact that the population is ageing (Sandhu, 1997), shows that there is a potential market for products designed with the needs of older and disabled people in mind.

Looking at the overall difficulty participants experienced, only 2 participants could accomplish all the tasks they were questioned about easily or without needing any assistance. At the other extreme, 4 participants needed considerable help or found some tasks impossible. A further 7 participants reported needing considerable help or found some tasks impossible in all but one area of ADL asked about.

People have coping strategies, but the division between managing with a struggle and failure is wide. The number of participants reporting that they manage 'with considerable help' is far fewer than the number who simply do not do tasks that require such a level of assistance. By encouraging designers to consider the needs of such people when designing products, the number of people able to accomplish tasks easily should increase. This would not only make these peoples' lives easier, but we would know that our own future lives would be easier thanks to the forethought of designers today.

References

Ashworth, J.B., Reuben, D.B. & Benton, L.A. 1994, Functional profile of healthy older persons, *Age and Aging*, **23**, 34-39

Department of Trade and Industry Consumer Safety Research. 2000, *A study of the difficulties disabled people have when using everyday consumer products.* DTI, London

Sandhu, J. 1997, Profit by Design. In I. McLaren (ed.) *Proceedings of a conference organised by the British Institute for Design and Disability, London, Oct 1997,* (European Institute for Design and Disability), 12-13

Weber, R., Czaja, S. & Bishu, R. 1989, Activities of daily living of the elders – a task analytic approach. *Perspectives: Proceedings of the Human Factors Society 33rd Annual Meeting, Oct. 16th-20th, Denver, Colorado,* **Volume 1**, 182-186

A REVIEW OF THE METHODOLOGY FOR THE INTEGRATION OF ERGONOMICS INTO THE SUPERMARKET DESIGN PROCESS

Simon Layton, Adam Whitlock, Ruth Chadwick

Human Engineering Limited, Shore House
68 Westbury Hill, Westbury-on-Trym, Bristol, BS9 3AA

This paper presents a practical review of a structured design process that guides the application of ergonomics through the project life cycle. The key stages are: research and technical input, concept generation and review, detailed design, prototyping & user trials, modification, launch and post implementation review. The category of 'Ergonomics Input' features in many of these key stages.

This paper attempts to describe the suitability of this design process from the perspective of logical and appropriate ergonomics integration. This is demonstrated in this paper by describing the life cycle of a number of projects, and assessing the level and impact of the ergonomics support at different stages.

Introduction

The objective of Human Factors Integration (HFI) is to ensure that human factors and ergonomics are applied consistently and appropriately during projects, in order to deliver a safe and efficient product to the customer.

When ergonomics principles are applied to the design process, the user is at the centre of the design process. This involves consideration of the physical and mental (cognitive) characteristics of users that should be applied to the design of an object, component of a system, or environment. Failure to do so may result in the creation of an unsafe and inefficient operation.

The application of human factors can lead to improved:

- efficiency,
- ease of use,
- health and safety,
- comfort, and
- quality of working life

This contributes to fulfilling organisational goals, such as improved productivity, profitability and quality.

The Design Process

The supermarket design department have developed a structured design process that attempts to integrate ergonomics throughout the life cycle of the project. This paper describes the stages where ergonomics support was sought and discusses the impact that this had upon the project.

Figure 1 shows the high-level stages of the design process and the relative amount of ergonomics effort allocated to each stage.

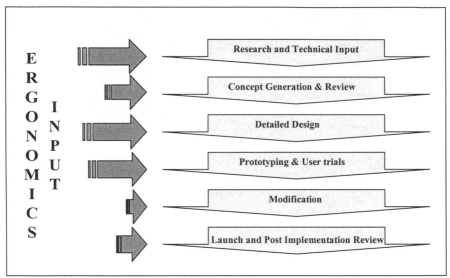

Figure 1. The Design Process schematic

The pictures and drawings within this paper are taken from two projects:
- Pharmacy Re-design
- Customer Services Re-design

The following sections describe the stages in the methodology that was used to integrate ergonomics into several projects, and describe the impact and influence that was achieved.

Research and Technical Input

The initial stage in the project was to perform a review of the existing workplace. This would typically involve the collection of objective and subjective data in order to develop a tabular task analysis (TTA) for all job roles.

The TTA was then used to guide the assessment of a range of human issues, using techniques such as timeline analysis, allocation of functions, and link analysis. The physical workplace was assessed by using established guidelines and ergonomics principles relating to anthropometry and access / egress requirements. In addition, an ergonomic assessment was performed on the equipment, addressing interface usability and other safety, health and performance related issues.

The output from this stage was an 'ergonomics design criteria checklist' which was referenced throughout the project to ensure that the design does not 'wander' from the original guidelines.

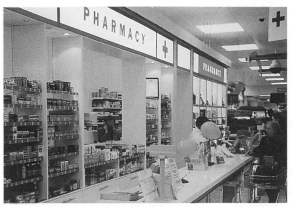

Figure 2. Assessment of the existing workplace

Concept Generation and Review

The design agency (contracted by the supermarket) then produced a number of early concept drawings taking into consideration the ergonomics guidelines generated from the review of the existing workplace and the design specification produced by the supermarket.

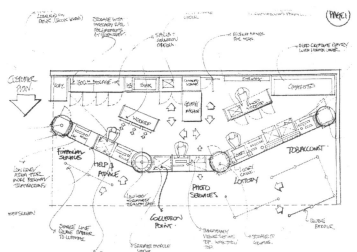

Figure 3. Early concept drawing review

These early concept drawings were then subject to an ergonomic review to ensure that the design was consistent with the ergonomic recommendations and to verify the general concept. A number of design issues / concerns were highlighted and a range of improvement measures were developed and discussed.

Detailed Design

Once the early concept drawings were approved, both from an ergonomics perspective and by the Project Team, the design agency then developed detailed technical drawings (as shown in Figure 4).

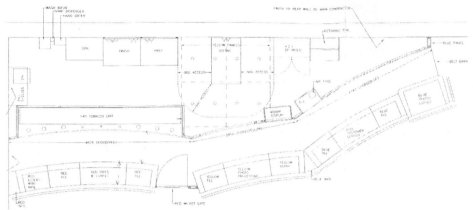

Figure 4. Technical drawing review

These were once again reviewed by the ergonomists and any issues were fed back to the design agency for discussion and changes agreed.

Prototyping and User Trials

A scaled, re-configurable mock-up was then used in support of the next design review meeting to agree the detailed design. At this meeting the ergonomists facilitated a session where a number of scenarios and operations were 'walked-through' in order to identify any design or operational issues. Any changes were agreed by the Project Team before the development of a full size prototype.

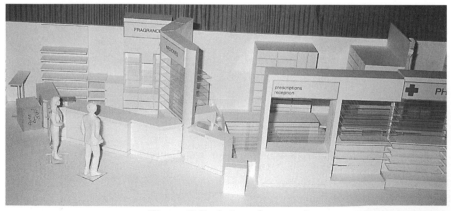

Figure 5. Scaled mock-up review

In addition, the ergonomists checked that the design conformed to the 'ergonomics design criteria checklist' (developed in the initial research stage).

A full size prototype was then developed and set-up in the supermarket's 'shopping clinic'. The 'shopping clinic' is the shell of a store that is used solely for assessment of products and working areas. A user trial was performed using a cross-section of employees (based on age, sex, stature, and level of experience), and a number of task and working scenarios were simulated. Observational data were noted by the ergonomist (and recorded on video) and subjective data were collected via structured discussion and a questionnaire.

The results were delivered in a formal report, and presented and discussed at a prototype review meeting.

The prototype itself could be modified and introduced into a working store (assuming the modifications were not to significant).

Figure 6. Full prototype review

Modification

The ergonomists had minimal input during this stage. Their role was to ensure that the modifications with ergonomics implications are introduced correctly and to ensure that the modifications as a whole comply with the 'ergonomics design criteria checklist'.

Launch and Post Implementation Review

After the modifications were made, the new design was installed into the selected trial store. Once the employees had sufficient time to work with the new design and overcome any initial resistance to change, then the ergonomists visited the store to talk through any usability or operational issues with the employees and to observe working methods and user performance. Once again the design was verified against the 'ergonomics design criteria checklist', and any remaining issues reported back to the Project Team.

Figure 7. Store trial

Review of Methodology

The ergonomics integration methodology has been applied on a range of projects for the supermarket. The ergonomics input received positive reception throughout all projects, which was thought to be due to the combination of a number of factors, including:

- the ergonomists were seen as an integral part of the project team from the start of the project
- the ergonomists were aware of the constraints pertinent to the project
- ergonomics was considered and applied throughout the entire design process, so the likelihood of major re-working or major modifications was reduced
- the ergonomic recommendations were prioritised into 'mandatory', 'highly desirable' and 'desirable'
- the 'ergonomic design criteria checklist' provided a framework to work within
- the users were consulted throughout the project and were represented on the project team

Conclusions

In summary, the methodology for the integration of ergonomics into the supermarket design process has been proven to be very successful, and has guided the development of usable, safe and healthy working environments within the store.

AN ERGONOMICS DECISION SUPPORT SYSTEM FOR DESIGNERS

[1]Andrée Woodcock and Margaret Galer Flyte

[1]VIDe Research Centre, School of Art and Design, Coventry University, Gosford Street, Coventry, UK email: A.Woodcock@coventry.ac.uk

The need to bridge the gap between ergonomics and design had been well documented over the past 30 years. Following this tradition, this paper considers issues surrounding the development, realization, evaluation and iterative development of an ergonomics decision support system to aid designers in their consideration of user issues during concept development. The paper closes by considering what the development of the system has shown about the relationship between ergonomics and design.

Introduction

Ergonomics can have a vast impact on the specification, usefulness and usability of designed artefacts. It can be used to inform all stages of the design process, but has shown to be most effective when employed during concept design where it can have a proactive role in shaping design decisions. When changes are made later in the development cycle they are likely to be reactive and preventative rather than proactive (Simpson and Mason, 1983), are more likely to be superficial, become more costly (Grudin et al, 1987) and may be received with hostility.

It has been postulated that ergonomics is not incorporated effectively and efficiently into design because of, for example, differences in philosophy towards and perception of the user (Pheasant, 1988), linguistic issues (e.g. Haslegrave and Holmes, 1994), presentational difficulties (Allison and Maguire, 1986), material failing to reach the intended audience (Eason and Harker, 1991), organisational issues (Meister, 1982), the perceived usefulness of ergonomics information (Meister and Farr, 1967), the role of ergonomics in the design specification (Morley and Pugh, 1987; Elliott, Wright and Galer Flyte, 1999), the designers frame of reference (Lincoln and Boff, 1988). These issues all contribute to the opaque interface (Woodcock and Galer Flyte, 1995) which hinders effective liaison between ergonomics and design and reduces the amount of formal ergonomics information which might be disseminated to designers. In such cases designers may adopt more informal methods, perhaps based on their own knowledge and experience or that of others in the belief that the product user will adapt to the product (Pheasant, 1988).

A case study of automotive design (Woodcock and Galer Flyte, 1998) confirmed that these factors influenced the utilisation of ergonomics by automotive designers and more importantly the designers and the managers knew that they were not producing the most ergonomically sound vehicle they could and wanted assistance in the employment of ergonomics, especially at the early stages of design. The interviews and questionnaires which formed the case study provided a set of user requirements for the design of an ergonomics decision support system. These requirements included the need to provide a system which would deliver support quickly in a manner compatible with engineering requirements (ie as a matrix), did not require a great deal of time to learn. The system should also allow design decisions to be traced, support joint as required and would not require a change in working methods.

The aim of the research was therefore to develop a computer based ergonomics decision support system which could be used to structure and systematise the consideration of user issues in the concept design in a way which would facilitate their shaping of product requirements. Such a system could be used to draw together both formal information and the more informal knowledge and experience gained by designers in the course of their practice.

Many methods have been developed to enable user centred design such as HUFIT (Allison et al, 1986), INTUIT (Russell, Pettit and Elder, 1992), QFD (Griffin and Hauser, 1993). Of these the philosophy behind HUFIT was chosen to guide the development of the ergonomics decision support system. The next section provides a brief overview of the system.

System design

The ergonomics decision support system was designed to aid designers incorporate ergonomics information and support a user centred approach to design from receipt of the initial design brief through to the specification of the product requirements. No assumptions were made concerning the designers/users working methods. The resulting system was believed to meet most of the initial user requirements. It consists of three main sections:

Clarification of the design brief
The initial stages of design are ill specified (e.g. Pugh, 1990) with design frequently proceeding without all members of the design team having a clear idea of the intended market or product functions. The opening stage of the system provides a series of questions regarding the users and purchasers and the functions of the product which provide an opportunity for early discussion and resolution of misunderstandings. This section was seen as one of the most valuable as it allowed ideas to develop prior to commencing with the design.

Specification of user issues
The ergonomics decision support system was created to support 'design for users'. Designers might have a great deal of information about the potential product users accrued from their own experience, market reports, literature and perhaps formal investigations. These need to be

systematised and brought to bear on product development allowing gaps in knowledge about potential product users to be quickly identified. The system provides an initial and extensive template tree structure of user issues (grouped into user, task, environment and usage characteristics) which forms the basis of discussions and which may be tailored by designers to reflect the characteristics of the user and stakeholder populations.

The specification of the design issues requires the designer to partition the intended product into discrete units (for example, car interior may be broken down into instrument panel, displays, controls, seats). The manner in which the design is segmented is left to the individual designer and will depend on which issues they wish to focus. The specification of the design parameters should be based on an understanding of the user requirements. Naturally this should proceed after some work on the user issues, although no restrictions are imposed on having to complete the user issues first, or in its entirety, before commencing with the product specification.

Development of the Functionality Matrix

The functionality matrix is a 2 dimensional representation of the linkages between (potentially all) user issues and product requirements, with the cell values indicating the level of compatibility between a user issue and product requirement. These values may be entered into the matrix individually or during work on the tree level. The matrix therefore represents the ergonomics specification of the product.

Although the original system, ADECT (Automotive Designers Ergonomics Clarification Toolset) was originally developed to assist automotive design, it was believed that the key user concepts remain constant across products, so the system can, potentially, be generic in nature. This was tested by the evolution of the system into DETECT to support general product design within a large company.

ADECT

ADECT has been described in detail in Woodcock and Galer Flyte (1997). It was evaluated in the context of a final year undergraduate project with 13 students over a period of 8 weeks. The students worked in groups, with unlimited access to the system and Internet resources to develop a Driver Information System for Car 2010. Their usage of the system and their attitudes were measured using SUMI (Software Usability Measurement Instrument), questionnaires, focus groups, video and breakdown analyses.

The attitudes expressed in the focus groups were mostly favourable. However, the SUMI analysis revealed that the software was rated as being just below state of the art. The breakdown analysis of the 1772 minutes of system usage revealed the main areas where problems had arisen. These are shown in Table 1 where breakdowns are characterised in terms of the TUTE framework (task, user, tool, environment) and against the part of the system where they occurred. This revealed that most of the usability problems arose during the editting of the user trees. A number of differences in group behaviour were also noted, in particular in relation to user-user and user-task breakdowns. The usability issues were fed into the next stage of system development.

Table 1. Summary of breakdowns in ADECT

	Part of System						
Breakdown	Strategy	User tree	Product tree	Matrix	General ADECT	Computer	Total
user - tool	10	99	2	12	13	7	143
user - task	10	17	0	0	9	1	37
user - user	10	18	0	0	11	0	39
user - environment	1	1	0	0	7	19	28
tool - task	0	6	0	3	3	0	12
tool - environment	0	3	0	4	3	3	13
task - environment	1	0	0	0	0	0	1
total	31	144	2	19	46	30	272

DETECT

The second, more general purpose system, had a slightly different specification determined by a client organisation. This required the redesign of the matrix, the import and export of data to and from Excel spreadsheets, and the automatic generation of summary reports.

The evaluation considered the way in which the system was used to support a specific design task, in this case the design of a toothbrush. The template was prepared for 8 participants (four designers and four ergonomist) who had to adapt the template and design a toothbrush using the system. Approximately one hour was allowed for training, and two hours for the design of the product. The trials were video recorded. Participants completed a SUMI usability assessment and were interviewed at the end of the session.

The SUMI analysis showed that the system was rated as above state of the art on the overall scale and all subscales. Although not directly comparable to the earlier result, this might be interpreted as a significant improvement. This was also reflected in the semi structured interviews where both the ergonomists and the designers commented favourably on the system.

In terms of the use of the system, all except one participant completed the task in the allotted time, and all except one worked through the system in a systematic and predictable manner (gaining an overview of the user issues before looking at the design ones). All participants thought using the system was valuable, gave them confidence and believed that it would help them to communicate with other members of the design team and clients.

An analysis of the video showed a difference between the way in which the two groups worked (see Figure 1). The designers spent more time considering the user issues (consumer tree) and the product strategy and less time on the construction of the matrix and altering cell values (indicated by 'both' in Figure 1), choosing to spend their allotted time developing and enhancing sketches. For the ergonomists the most fun in the system was the functionality matrix and trying to achieve an optimum design. On the other hand designers spent their time working on the sketches. The semi structured interviews confirmed that the ergonomists had liked the matrix and found it a valuable rapid prototyping tool. The designers developed a sketch as the result of their design session. A number of respondents expressed concerned that the matrix might be incompatible with the outputs of concept design, i.e. the concept sketch and as such might not be used. The implications of this finding for the design of the ergonomics support system are discussed in the following section.

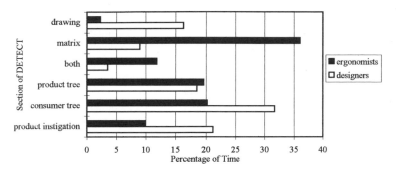

Figure 1. Ergonomists and designers use of DETECT

Integrating ergonomics and design

In terms of the ergonomics decision support system, the software representation enables user issues to be made more explicit and linked to product requirements aiding systematic, focussed discussion of user issues which was valued by the users. The software development process itself as manifest in the functionality of the final system, revealed the complexity of user centred design, the difficulty of mapping user issues on to multiple product features and showing these in a concise, coherent format as an ergonomics specification.

The research has shown that it is possible to use computers to support ergonomics in concept design. The computer based ergonomics decision support system, consisting of a general set of user issues, can be used successfully by individual designers, and small design teams in the early consideration of user issues, and be tailored by them to meet the characteristics of their target population. The recommendations for moving the work forward indicate that further integration of ergonomics into design practice may require consideration of visual representations for the communication of ergonomics requirements between stakeholders in the design process.

References

Allison, G., Catterall, B., Dowd, M., Galer, M., Maguire, M. and B. Taylor, 1992, Human Factors tools for designers of information technology products. In Galer, M.D., Harker, S. and J. Ziegler (eds.), *Methods and Tools in User-Centred Design for Information Technology* (Elsevier Science Publishers B.V :North Holland) 13-27

Allison, G. and Maguire, M. 1986, Some comments on the nature and use of human factors standards for software interface design, *Journal of Information Technology*, **2**, 2, 40-42

Eason, K. D. and Harker, S. D. P. 1991, Human factors contributions to the design process. In B. Shackel and S. J. Richardson (eds.), *Human Factors for Informatics* (CUP)

Elliott, A. C., Wright, I. C., and Galer Flyte, M. D. 1999, Human factors design priorities during new product development; An empirical study. In M. A. Hanson, E.J. Lovesey and S.A. Robertson (eds.), *Contemporary Ergonomics* (Taylor and Francis. London) 369-373.

Griffin, A. and Hauser, J. 1993, *The Voice of the Customer*. Report No 92-106. Marketing Science Institute: Cambridge, Massachusetts

Grudin, J., Erhlich, S. F. and Shriner, R. 1987, Positioning human factors in the user interface development chain, *CHI and GI* (ACM Press. New York), 125-131

Haslegrave, C. M., and Holmes, K. 1994, Integrating ergonomics and engineering in the technical design process, *Applied Ergonomics*, **25**, 4, 211-220.

Lincoln, J. E. and Boff, K. R. 1988, Making behavioural data useful for system design applications: Development of the Engineering Data Compendium, *Proceedings of the Human Factors Society – 32nd Annual Meeting*, 1021-1025

Meister, D. 1982, Human factors problems and solutions, *Applied Ergonomics*, **13**, 3, 219-223

Meister, D. and Farr, D. E. 1967, The utilisation of human factors information by designers, *Human Factors*, February, 71- 87.

Morley, I. E. and Pugh, S. 1987, The organisation of design: An interdisciplinary approach to the study of people, process and contexts, *ICED '87*,210-222

Pheasant, S. T., 1996, *Bodyspace: Anthropometry, Ergonomics and the Design of Work*. (Taylor and Francis. London)

Pugh, S., 1990, *Total Design* (Addison Wesley Publishers Ltd. England)

Russell, F., Pettit, P. and Elder, S. 1992, A computer assisted software engineering support for user-centred design. In Galer, M.D., Harker, S. and Ziegler J.(eds.), op.cit

Simpson, G. C. and Mason, S. 1983, Design aids for designers: An effective role for ergonomics, *Applied Ergonomics*, **14**, 3, 177-183

Woodcock, A. and Galer Flyte, M.D., 1995, The opaque interface - the development of an on-line database of ergonomics information for automotive designers, *HCI'95 Adjunct Proceedings*, 96-104

Woodcock, A. and Galer Flyte, M.D. 1997, ADECT - Automotive Designers Ergonomics Clarification Toolkit, *Contemporary Ergonomics '97*, 123-128

Woodcock, A. and Galer Flyte, M. D. 1998, Supporting ergonomics in automotive design, *International Journal of Vehicle Design*, **19**, 4, 504-522

On observation as inspiration for design

T.Verheijen, H.Kanis, D.Snelders, W.S.Green

School of Industrial Design Engineering, Delft University of Technology
Jaffalaan 9, 2628BX Delft, Netherlands
t.verheijen@io.tudelft.nl

The purpose of this paper is to clarify the concept of needs in relation to discontinuous innovation. Needs are used as a vehicle for gathering consumer information that is considered crucial for product development and is therefore considered important for discontinuous innovation. As an alternative to the technology-driven character of current discontinuous innovation, eliciting consumer needs could be used to implement user-driven innovation. Problems with conceptualisation of needs are presented first. Confusion that might occur in relation to the entire concept of needs in the current literature is illustrated. Next, problems with the alternative approach described in the literature, namely 'latent' or 'unexpressed' needs, are pointed out. Another approach to the conceptualisation of unexpressed needs in relation to discontinuous innovation is advocated, relating needs not only to products but to behaviour as well. Future research will be aimed at studying whether observation can really be used as inspiration for discontinuous innovation.

Introduction

In recent studies in the field of design engineering, it has been observed that many companies realise that their success and survival depends on continued innovation (Veryzer, 1998; Song *et al*,1998; Kotler, 1991). In 'traditional' (or non-innovative) product development, it is assumed that researchers (market researchers, ergonomists et.al.) can obtain information from consumers in such a way that this information can be used to define specifications for the development of product functionalities. Well-known methods, such as Quality Function Deployment (Griffin and Hauser, 1993) or Need/Problem Identification (Kotler, 2000), are based on this premise. However, when this innovation refers to *discontinuous* innovation a problem arises. Here, discontinuous innovation refers to product concepts that are entirely new, that incorporate new technologies and functionalities, are aimed at new markets and, as a result, "... involve dramatic leaps in terms of customer familiarity and use" (Veryzer, 1998, p.137). As O'Conner states (1998, p.151): "In all likelihood, customers will not be able to describe their requirements for a product that opens up entirely new markets and applications." Apparently it is difficult to acquire useful customer information for the development of radically new products (Burns, 2000; Slater and Narver, 1998; Leonard and Rayport,

1997) and this may be the reason why so much discontinuous innovation is technology-driven (e.g. Veryzer, 2000).

One recurring concept described in the literature, which is used as a vehicle for analysing consumer information, is the concept of 'need'. It is suggested that some derivation of what are called latent or unexpressed needs from consumers could help to provide design engineers with the information essential for the development of radically new product concepts (Veryzer, 2000; Burns, 2000; Slater and Narver, 1998, Conner, 1999). The purpose of this paper is to examine what these needs are and how information can be obtained about needs that are latent, unexpressed or inexpressable. The paper explores this issue and offers proposals for methods of observation and analysis on the basis of the idea that observation of people could offer the information about needs that design engineers require in order to develop radically new product concepts. To research the possible role of the user in discontinuous innovation, the project "From observation to innovation" was formulated. It is an explorative study of the possibilities of arriving at innovative products by combining 'technology-driven' innovation with 'usage/behaviour-driven' innovation, in the hope of increasing the success rate of radically innovative products. The objective is to provide design engineers with a structural means of arriving at innovative concepts through analysis of observational data.

A conceptualisation of needs

Whether or not "human needs are ontological facts of life" (Kamenetzky, 1992, p.181), the extensive use of the word, and the concept it entails in various scientific disciplines, presupposes that we have a mutual understanding of its meaning, or of some phenomenon it represents. Various disciplines use the word 'need' for different purposes, and in that respect one can assume that different meanings are intended. Also, the word lends itself easily to confusion. In the literature on the subject one can find that the terms 'need', 'motivation', 'instinct, 'want' and 'desire' are sometimes used as separate concepts and sometimes used as synonyms (e.g. Campbell, 1998; Lea et al., 1987; Rogers, 1995). And extensive use in combination with various adjectives confuses the matter even further, as in 'human needs', 'basic needs', 'general needs', 'implicit needs', 'latent needs', 'intermediate needs', etc.

To arrive at a common understanding of needs in general and latent or unexpressed needs in particular, it is illuminating to examine what the frame of reference of 'needs' is in the literature. Roughly speaking, the term 'needs' is used in two ways. First, it is used to explain some characteristic or intrinsic trait of mankind, associated with individual and social behaviour and motivation, especially found in philosophy, biology and the social sciences literature. This meaning appears to have some biological or sociological component. It is used to describe broad concepts, such as 'mankind has a need for food', 'mankind has a need to procreate', etc., and in this respect the concept of 'need' is often related to the necessities of mankind and of individuals in society, regardless of culture, politics, location, etc. Within this framework, 'needs' are usually presented in classifications or models, with loosely defined categories of needs. Different examples of this grouping of needs can be found in the literature (Maslow, 1970; Kamenetzky, 1992; Doyal and Gough, 1991), which suggests that the authors do think of distinctions between groups of needs. This type of need conceptualisation will be adopted as 'basic needs' in this paper, because of its inevitable character and because it is the most commonly used.

A second way the term 'needs' is used is to establish some connection between a user and an artefact. This is the most prevalent way of using the term in ergonomics, business and design engineering literature (e.g. Griffin and Hauser, 1993; Hauser and Clausing, 1988; Helander and Du, 2000, Karlsson, 1996). In this literature 'needs' are in some way related to product functionalities. Needs are not presented as fixed categories or groups of similar needs but on a much more detailed level, to represent the requirements users have for a particular product, such as 'I need this product to be easy to operate'. The purpose of consumer or user needs is to elicit requirements which in turn can be used to conceptualise product functionalities that can be used to conceive product specifications. Of course in this case, the assumption is that consumers or users can make their needs explicit for the purpose of product development. They will therefore be called 'expressed needs' in the rest of this paper, which is the commonly used phrase, although 'expressable needs' would actually be more appropriate.

The distinction indicated above can also be seen in two roles of ergonomics pointed out by Wilson (2000). According to Wilson, the first role of ergonomics is "... to fundamentally understand purposive interactions between people and artefacts and especially to consider the capabilities, needs, desires and limitations of people in such interactions..... The second role comprises a contribution to the design of interacting systems, maximising the capabilities, minimising the limitations, and trying to satisfy the needs and desires of the human race." (p.12). Seen in this light, ergonomics can deal with both the concept of needs in relation to the human-artefact interaction and the concept of needs in relation to mankind.

From a design engineer's point of view, the concept of needs can be used to conceptualise product functionalities that fulfil (satisfy) a need. Both approaches to the concept of needs establish some sort of relationship between need and satisfaction, and in that sense they are not entirely incompatible. However, these approaches have practical problems in relation to discontinuous innovation. As previously stated, the user-artefact related, expressed needs approach requires some form of detailed elicitation of information from the users of an artefact and it is therefore unsuitable for radically innovative product development (see introduction). On the other hand, the concept of basic needs 'in relation to mankind', seems to offer information that is much too general to be practically useful for the design engineer (Schoormans and De Bont, 1995).

Despite these potential problems, the concept of needs is of interest for discontinuous innovation, because it could provide design engineers with *customer-based* information and inspiration for the development of new product concepts. In the literature the idea of 'latent' or 'unexpressed' needs is brought forward as a possible solution for this problem (Veryzer, 2000; Burns, 2000; Slater and Narver, 1998, Conner, 1999; Leonard and Rayport, 1997).

Identification of unexpressed needs

In the current literature, confusion in relation to the concept of needs called 'latent' or 'unexpressed' seems to continue. Similar adjectives, such as 'unarticulated', 'dormant' or 'future current' needs are used, but sometimes appear to have slightly different meanings. Where the term 'latent' is used it indicates that something is existent or capable of existence, but is not manifesting itself. It gives the impression that something is there, but this has some tricky implications. It would be difficult for a researcher to claim that one has observed a latent need, without some kind of confirmation from the observed person. One way of trying to reduce possible observer bias is by having group sessions to analyse observational data, but this is still doubtful when trying to establish a plausible link between inferred latent needs and the observations. In a list presented by Cross (1989), for example, regarding needs for an electric toothbrush, it is not clear if the needs that are presented are elicited from consumers or constructed by a design engineer (p.78).

The term 'unexpressed', on the other hand, would indicate that something is present in the mind which is not articulated but which *could* be elicited from the consumer. This approach seems less questionable than the so-called 'latent' needs. A distinction can be made in this respect. Obviously, as the term suggests, unexpressed needs are those needs which users *cannot* or *will not* explicitly mention in a study situation. *Will not*, when the needs are considered inappropriate or irrelevant by the user. In other words, when the user consciously makes the decision not to mention a possible need or requirement (e.g. 'secret needs' as mentioned in Kotler, 2000, p. 21). When the user *cannot* express needs, they may not be consciously aware of their presence or at least may have no means of expressing them directly.

In the literature, methods have been described for eliciting these unexpressed needs, regardless of the reluctance or incapability of the consumer. A few typical examples are Empathic Design (Leonard and Rayport, 1997), Lead-user analysis (Von Hippel, 1999) and Delighter clinics (Burns, 2000). The majority of these approaches basically look at three things: problems with products (frustrations, anxieties, doings things wrong, misuse, etc.), the adaption of products (unexpected usage, customer modifications, lead users) and relations to products (triggers of use, interactions with the environment, intangible attributes). A problem with these methods of research in relation to discontinuous innovation can be noted: in the examples presented; the direct involvement of an artefact (product/services) is still imperative for conducting the research. Although this is not surprising, considering that all methods seem to have evolved from the existing practice of eliciting expressed needs, this fixation on relationships between users and a particular existing artefact appears limiting. It still requires a product or product concept to have been developed *prior to* gathering consumer information. Consumer information still does not seem to be used as an inspiration for discontinuous innovation.

Unexpressed needs and observed behaviour

There are some important points to consider, when trying to elicit unexpressed needs. With regard to products, we need an object to fulfil a certain function and thereby satisfy a certain need, but it should be noted that the type of function of the artefact is based on the context in hand (Kroes, 1998). As an example, a baseball bat could be seen as a required object in the game of baseball, or as a weapon for hitting someone over the head. Functions and their related needs are not necessarily the same for each behavioural context. Also, products could lead to changes in behaviour and as a result change what a person may need (Rogers, 1995).

With regard to behaviour, purposeful human action can be seen as situated in (and organised around) the context of particular circumstances (Dourish and Button, 1998). Many actions, also based on habit or routine for example, can perhaps be related to needs as well. To summarise, one should not assume one-on-one relationships between needs and acts of behaviour, needs and product use and needs and product functionalities. It is important to realise this when researching consumer needs.

But where should researchers and design engineers direct their attention when gathering and analysing observational data in order to elicit needs, when all things observed seem relevant? Although it is a pragmatic argument, constraints which direct the focus of the research could be provided by core competences of the business. Market, technological or organisational competences can determine the domain on which design engineers should focus their expertise.

Discussion

There is no reason to assume that observation of behaviour in general, regardless of the involvement of products, could not lead to inspiration for new product concepts. This is not to say that observation of product use should be ignored, but it should be part of the observation of behaviour and, if related to product use, not tied to a single specific type of product. All this inevitably raises the question of what a researcher can learn from manoeuvring through all the relationships between behaviour, function and artefact, thereby obtaining a more complete picture of the situation at hand when trying to elicit needs from consumers. Alternatives to the current approaches of eliciting consumer needs, that lend themselves to this type of research are, for example, ethnography or retrospective interviews with consumers regarding their behaviour. One of the challenges of the research project "From observation to innovation" is to examine alternative approaches to developing radically new product concepts, on the basis of observation of behaviour and its relation to consumer needs, as described above.

References

Burns, A.D., Barrett, R., Evans, S. and Johansson, C., 1999, Delighting Customers through Empathic Design, *Proceedings of the 6th International Product Development Management Conference*, (EAISM, Cambridge)

Campbell, C., 1998, Consumption and the Rhetorics of Need and Want, *Journal of Design History*, **11**, (3), 235-246

Conner, T., 1999, Customer-led and market-oriented: a matter of balance, *Strategic Management Journal*, **20**, 1157-1163

Cross, N., 1989, *Engineering Design Methods*, (John Whiley & Sons, Chichester)

Dourish, P. and Button, G., On "Technomethodology": Foundational Relationships between Ethnomethodology and System Design, Human-Computer Interaction, 13, 395-432

Doyal, L. and Gough, I., 1991, *A Theory of Human Need*, (Guilford Press, New York)

Hauser, J.R. and Clausing, D., 1988, The House of Quality, *Harvard Business Review*, May-June, 63-72

Helander, M.G., and Du, X., 1999, From Kano to Kahnema. A comparison of models to predict customer needs. *Proceedings of the Conference on TQM and Human Factors*, 322-329, Linkoping University, Sweden.

Griffin, A. and Hauser, J.R., 1993, The voice of the customer, *Marketing Science*, **12**, (1), 1-27

Kamenetzky, M., 1992. In Ekius, P. and Max-Neef, M. (eds.), *Real Life Economics: Understanding Wealth Creation*, (Routledge, London)

Karlsson, M., 1996, *User Requirements Elicitation: A Framework for the Study of the Relation between User and Artefact*, Chalmers University of Technology, unpublished thesis

Kotler, P., 2000, *Marketing Management*, Millenium Edition, (Prentice Hall, New Jersey)

Kroes, P., 1998, Technological explanations: the relation between structure and function of technological objects, *Philosophy & Technology*, **3**, (3), 18-34

Lea, S.E.G., Tarpy, R.M. and Webley, P., 1987, *The individual in the economy: a survey of economic psychology*, (Cambridge University Press, Cambridge)

Leonard, D. and Rayport, J.F., 1997, Spark innovation through empathic design, *Harvard Business Review*, November-December, 102-113

Maslow, A.H., 1970, *Motivation and Personality*, Second Edition, (Harper & Row, New York)

O'Conner, G.C., 1998, Market Learning and Radical Innovation: A Cross Case Comparison of Eight Radical Innovation Projects, *Journal of Product Innovation Management*, **15**, 151-166

Rogers, E.M., 1995, *Diffusion of Innovations*, Fourth Edition, (The Free Press, New York)

Schoormans, J. and De Bont, C., 1995, *Consumentenonderderzoek in de produktontwikkeling*, (Lemma BV, Utrecht)

Slater, S.F. and Narver, J.C., 1998, Customer-led and market-oriented: let's not confuse the two, *Strategic Management Journal*, **19**, 1001-1006

Slater, S.F. and Narver, J.C., 1999, Market-oriented is more than being customer-led, *Strategic Management Journal*, **20**, 1165-1168

Song, X.M. and Montoya-Weiss, M.M., 1998, Critical Development Activities for Really New versus Incremental Products, *Journal of Product Innovation Management*, **15**, 124-135

Urban, G.L. and Hauser, J.R., 1993, *Design and marketing of new products*, Second Edition, (Prentice Hall, New Jersey)

Veryzer, Jr. R.W., 1998, Discontinuous Innovation and the New Product Development Process, *Journal of Product Innovation Management*, **15**, 304-321

Veryzer, Jr. R.W., 1998, Key Factors Affecting Customer Evaluations of Discontinuous Products, *Journal of Product Innovation Management*, **15**, 136-150

Von Hippel, E., Thomke, S. and Sonnack, M., 1999, Creating Breakthroughs at 3M, *Harvard Business Review*, September-October, 47-57

Wilson, J.R., 2000, Fundamentals of ergonomics in theory and practice, *Applied Ergonomics*, **31**, (6), 557-567

SCIENTIFIC CREDENTIALS FOR QUALITATIVE AND QUANTITATIVE RESEARCH IN A DESIGN CONTEXT

H. Kanis

School of Industrial Design Engineering, Delft University of Technology, Jaffalaan 9, 2628 BX Delft, the Netherlands, <h.kanis@io.tudelft.nl>

The recent emphasis on context in ergonomics raises the issue of the viability of current quantitative research criteria ('reliability', 'validity') in qualitative research. In discussing such criteria, qualitative circles, though dealing with re-/interactive phenomena, tend to present themselves as subordinate to quantitative traditions. Studies of the literature have revealed various kinds of deficient practices in the application of research criteria in quantitative studies. In this respect they have nothing to offer to qualitative research. It is rather the reverse, especially in a design context.

Introduction

Given that ergonomics is the study of human-system interaction as well as the application of insights from that study to the design of products/systems, the notion of context seems a key-construct in discussing the (future) impact of ergonomics. Focussing on work (workplaces, work practices), Moray (1999) speaks of the design of sociotechnical systems, which should go beyond technical solutions such as better designed interfaces and better training regimes, and should also include organisational, legal, economic and cultural considerations. In addition to the perspective of this broad context in the ergonomics/human factors (E/HF) research area, Moray pleads for openness to and co-operation with other disciplines, particularly ethnography, as an alternative to the traditional research approaches of psychology and engineering.

In a similar vein, Wilson (2000) conceives such factors as financial, technological, legal, organisational and social considerations as the context for ergonomics. At the same time, this author, who does not specifically focus on work conditions in discussing ergonomics, emphasises the importance of understanding how (and why) people think when facing real problems which require perception and action in real environments, i.e. behaviour in the context of everyday settings rather than people's behaviour under carefully controlled laboratory experimentation (eliminating context rather than appreciating it). In this respect, Wilson also points to the significance of ethnographic research for ergonomics.

These references illustrate that the call for context may involve quite different factors:
(*i*)   the inclusion of new areas of interest to reinforce E/HF for future developments, and
(*ii*) the trend towards the observation of real human behaviour emerging under natural circumstances.

These aspects tend not to be of equal importance in directing E/HF research as a means to support the design of consumer products, with the actual usage of these products in terms of usability taken as the main design target.

Context in ergonomics/human factors

New fields of interest (ad i)
Wider societal perspectives (legal, economic) do not seem to be of primary interest for usage centred design. This is also true of organisational considerations. Consumer products tend to be used individually, in private or informal settings, rather than in co-operative work such as in control rooms, with training and supervision involved. This makes the use of consumer products difficult to control, as distinct from operations in an industrial setting, and particularly in the military area (cf. Green and Jordan, 1999). Cultural considerations, however, may have a direct bearing on usage centred design, since products can be seen as the tokens of the 'baggage' of the possessor/user. An example may be the stigma associated with those featural and functional characteristics which are designed to enhance the usability of a product (e.g. large controls, big displays for 'elderly products'), but which may easily seen as exposing and accentuating one's individual deficiencies, or even one's lack of good taste, if the product concerned is considered, in the assumed opinions of 'significant others', to lack appropriate style or currency. The result may be non purchase of products with good usability, or waste of money due to non-usage of such products, which also ends up in the uselessness of designed usability. Note the natural context, which in this case may involve exposure to the general public, as indispensable for the emergence of stigmatisation. This brings about the second aspect (*ii*) of context involved E/HF research.

Natural settings (ad ii)
In usage centred design, the more insight into (prospective) human-product interaction the better. Some of the questions which designers have to face in this field typically emerge from the unpredictability of the future activities of users, i.e. their individual perception, cognition and use actions, including any effort involved. Variation in usage between people, which designers somehow have to cope with, may reflect different ways of perceiving and interpreting such things as featural and functional product characteristics, and also of (other) contextual elements of product usage. Inter-individually, use habits tend to diverge, for instance in relation to distinctive forms of experience. Thus, the interactivity of product usage emerges from the individual conditions of users in natural contexts. It is produced on the spot, and that is where it has to be observed. This approach closely accords with the tenets of ethno-methodology as an empiric and analytic frame, (see Dourish and Button, 1998). In this respect, the plea for context-involved E/HF research in order to support usage oriented design makes perfect sense and, simultaneously, raises the question of what is the current situation for which 'going context' would be the desirable alternative.

Before discussing this question it is appropriate to first consider the observation of human-product interaction in context.

Observing interactivity, emerging reactivity, achieving generality

Interactivity
Methods of studying human-product interaction for the identification of user activities often cannot be anything other than obtrusive, and therefore cannot avoid becoming an inherent part of the interaction they are meant to address. Generally, a method for studying human-product interaction relies at least in part on the same mechanisms which generate the questions to be answered by its application. Thus, interactivity may condition the observability of its constituents, which can then never be adequately defined by any

description of a method. (Note the consequence that methods can never be identified as 'reliable' or 'valid', addressed later.) In this way, the individuality outlined above may boil down to uncertainty about the meaning of people's utterances, for instance in finding out exactly how tasks are understood, or in analysing self-reports about effort experienced in relation to internal references, or perceptions/cognitions related to featural and functional product characteristics, e.g. in terms of use cues.

An inkling of possible consequences at issue here can be gained from studies of questionnaires which show that people being asked the same question may very well answer different questions, (see Foddy (1998) for a recent study in this area). Another interference may occur when standardised interviewing is thwarted by rules of ordinary speech (Houtkoop-Steenstra, 2000).

In the field of E/HF, the uncertainties arising from various sorts of interactivity, although immanent in studies involving the inter-individual variety in human-product interaction, tend not to be paid much attention in observational research. It is notable that conditoning of the observability of human-product interaction due to interactivity is always an issue, in a natural context as well as in controlled, experimental conditions. The study by Houtkoop-Steenstra (op.cit.) shows that it is the very standardisation which may produce biased results in interviewing; this makes the author plead for flexibility in asking questions and discussing answers.

Reactivity

Measurement and observation involving humans, as already indicated, appears to be riddled with interpretative diversity. There are also the possible effects of what may be called the reactiveness of participants in being aware of a research setting. This reactiveness may result in biased observations due to social desirability. The effects of social desirability do not have to be limited to self-reports etc., but can also bias use actions. The obvious remedy is unobtrusive research, i.e. a natural context, which occasionally may be achieved to some extent by distracting the participants. However, elicitation techniques are often at odds with any form of unobtrusiveness. Consider, for example, 'thinking aloud'.

A second manifestation of reactiveness in measurement and observation involving humans is the intra-individual effect of 'carry-over' during the course of a study, when participants become skilled, tired, bored, etc. In the field of E/HF, research papers regularly deal explicitly with the possible effects of carry-over, particularly in discussing so-called test-retest reliabilities (see later discussion).

Generality

So far a few instances have been mentioned which favour natural conditions for the observation of human-product interaction. The most coercive argument for as natural as possible a context involves the generality of observations. The key issue is that interaction is studied for reasons of not knowing, in view of the diversity between users, their predispositions and flexibilities and the unpredictability, particularly on an individual level, of their activities in actual product use, i.e. in everyday settings. The number of possible (combinations of) influences (co-)shaping interaction seems infinite. Once this is recognised, any reduction of the 'situatedness' of everyday usage calls for an explanation when assessing the generality of findings. However, such explanations typically tend to be unavailable, since they presume a deep insight into human-product interaction, the absence of which is the very reason to resort to empirical observations.

Scientific status and 'going context'

For observational studies of human-product interaction, as natural as possible a context seems to be a prerequisite in order to avoid the emergence of artifacts, and to enhance the generality of the findings. The call for E/HF research involving natural contexts suggests that at the present time it is not sufficiently employed. As a remedy ethnography is proposed, by Moray and by Wilson (see above). In discussing these matters both authors raise the issue of scientific credentials in terms of reliability and validity. Wilson points at the negative trade-off in experiments in gaining reliability at the cost of restricting validity, resulting in findings of little practical significance (p.562). Moray maintains that in ethnographic studies validation and reliability require an approach which is very different from that used in traditional experiments (p.861). These statements seem to indicate that current scientific standards are inadequate for the assessment of the scientific status of observations of human-product interaction under natural circumstances. They suggest that the concepts of reliability and validity need at the least revision, if only to eliminate them as factors preserving the type of controlled research that tends to be uninformative for design (Kanis and Green, 2000). Then, in terms of reliability and validity, the issue of what is the current situation for which 'going context' would seem the desirable alternative boils down to two questions:
- what have other disciplines (ethnography, or often qualitative research) to offer, and
- what is the price to be paid for giving up research standards from the quantitative tradition.

Criteria in qualitative research
Several authors struggle with the question of how to deal with research criteria for qualitative research, given the quantitative research tradition in terms of reliability and validity (Silverman, 1997; Murphy *et al*, 1998; Seale, 1999). Here, qualitative research may be conceived as the study of local production of meaning in specific contexts (Seale, p.120), with meanings "shaped and constrained by the circumstances of their production" (Bloor, quoted in Murphy *et al*, p.181). A main source of trouble seems to be the role of contractors, who tend to adhere to current, quantitative criteria when commissioning research grants. The basic difficulty with these criteria is their reference to a stable world, existing 'out there' independently from the (observing) human mind. This positivistic view runs counter to outcomes of qualitative research, in yielding findings on what is happening (ethnography), how people accomplish realities in everyday life (ethnomethodology; Seale, p.39), and why interactions proceed as observed; insight into this why question is indispensable for fully exploiting observations to support design, i.e. in linking users' activities to the featural and functional product characteristics to be designed. For such findings, the concept of reliability as a measure of dispersion (i.e. random variation in measurement results) is especially untenable, since this notion is based on repetition (Kanis, 1997): of being able to copy events, given a stable world (see above) rather than a reality emerging from interaction. Conceptually, the notion of validity, understood as the level of deviation from an adopted independent reference (Kanis, 2000), may to some extent be maintained, i.e. in an approach involving the absence of inconsistencies or negative instances as falsifying evidence, as well as the failure to present plausible alternative explanations of findings in a questioning approach. However, the terms validity/validation are best dispensed with, since the positive side of being unable to demonstrate falsification is at best corroboration: validation is simply impossible.

Other approaches to strengthening the 'credibility' or 'trustworthiness' of qualitative research findings include the application of various methods ('triangulation'), detailed description of the applied approaches (in field work, in the analysis) in order to promulgate the amenability for others to repeat a study, the use of 'outliers' as incentives to question current propositions rather than eliminating them, and presentation of 'low inference descriptors' in order to make the analysis tractable for readers (Murphy *et al*, p.197).

In spite of all these considerations, measures and precautions, qualitative research circles nevertheless tend to present themselves as subordinate to the quantitative reliability/ validity canon (cf. Silverman, p.25), making believe that this canon would be sustained by consensus (Seale, p.43). Some kind of consensus may hold, or at the least be suggested or propagated in particular fields; see, for instance, the discussion on usability methods in Gray and Salzman (1998). Yet to take this as a sign of the sound application of research criteria in quantitative areas of E/HF research would be a serious misunderstanding, as would the promotion of current practices in E/HF as a lead for other resarch areas.

The trouble with reliability and validity in E/HF
To begin with, design oriented study of user-product interaction offers various possibilities of measurement and observation involving humans (cf. Kanis, 1997):

a 'at' human beings, i.e. physical characteristics with the passive/involuntary role of participants, e.g. head circumference, hand breadth, leg length;

b 'through' human beings, i.e. people actively involved in carrying out standardised tasks, for instance to establish exertable forces, joint flexibility, reaction time, memory capacity, eyesight;

c 'by' human beings actually operating a product, i.e. 'doing' in terms of use actions;

d self-reports of user activities, i.e. perception, cognition, experience at issue, effort.

For these categories a negative trade-off exists between measurability/observability and design relevance (cf. Kanis, 1998). Generally, measuring human characteristics (*a*) and capacities (*b*) can be conceived as rather straightforward and controllable (see some later side-notes), while design relevance is absent or limited to setting boundary conditions for actual usage: how users perform activities within these boundaries usually has little to do with their characteristics, limitations and capacities. Conversely, for use actions (*c*) and self-reports on user activities (*d*) in context, design relevance is high, provided that such observations can be linked to designable featural and functional product characteristics. In contrast, observation of use actions turns out to be both cumbersome and delicate, involving such problems as reactivity-effects, which may also be prominent in self-reports on user activities.

By and large, smoothness or difficulties in measurements/observations involving *a*, *b*, *c*, and *d* run parallell with the (un)specifiability of dispersion ('reliability') and deviation ('validity') of these measurements/observations. For human characteristics in terms of *a*, as a rule both dispersion and deviation can be specified as is usual in the technical sciences, i.e. without additional difficulties.

For the other types of measurement/observation, the specification of *dispersion* tends to be problematic for *b* and *c*, where possible carry-over is often anticipated by a test-retest. In a study of E/HF literature (Kanis, 1997), several deficiences were encountered regarding this procedure: the correlation coefficient as a misleading, non-specifying criterion; a statistically non-significant t-test as a token of limited dispersion (it is rather just the opposite); the neglect of patterned dispersion, precluding its specification as a constant such as in standard deviations or as a variation coefficient (often human control works out proportionally); assessing intra-individual variation as 'unreliability' or even 'error', rather than conceiving this variability as a human condition in its own right. In a recent update of this study on the basis of electronically available research papers (1999- and 2000-issues of *Applied Ergonomics*, *Ergonomics*, and the *Proceedings of the Human Factors and*

Ergonomics Society), it was found that the same mistakes are still made. No example could be found of the application of proper procedures proposed by Kanis (1997). That paper, ironically, was never quoted in the data-bases searched.

As to self-reports ($\underline{d}$), specification of dispersion is simply non-applicable, since a reasonably argued repetition for this type of recording is illusory due to the irrevocability of perceptions, cognitions and effort in evolving, rather than stable, human conditions.

Specification of *deviation*, if an issue at all in measuring/observing $\underline{b}$ and $\underline{c}$, seems at least partly illusory due to interactivity. This not only involves the fact that, inter-individually, different phenomena may be observed (think of various interpretations of the same instruction, or diverse physiological reactions to the same task), but also that the sorting out of possible differences tends to be evasive. Finally, the specification of deviation for self-reports ($\underline{d}$) would have to rely on unquestioned references for comparison with internally referenced user activities. The only possibilities we have come across in field studies include conflicting evidence (e.g. participants contending that they have done something that recordings show they did not), and circumstantial evidence making sense as corroboration. Note that such considerations have nothing in common with mainstream, 'hard' testing of validity as has emerged from psychophysics traditions in psychology, and still dominates the E/HF research area. A recent study of E/HF literature (Kanis, 2000) shows that references to 'validity'/'validation' frequently tend to be only suggestive or even misleading, echoing a verificationist approach rather than a questioning one which looks for possible alternative interpretations and understanding of findings.

The conclusion seems obvious: application of research criteria in E/HF for specifying dispersion or deviation is deficient, and has nothing to offer to qualitative research. It is rather the reverse, particularly in studying human-product interaction in a design context.

References

Dourish, P. and Button, G. 1998, On "Technomethodology": Foundational Relationships between Ethnomethodology and System Design, *Human-Computer Interaction*, **13**, 395–432

Foddy, W.H. 1998, An Empirical Evaluation of In-Depth Probes Used to Pretest Survey Questions, *Sociological Methods & Research*, **27**, 103–133

Gray, W.D. and Salzman, M.C. 1998, Damaged merchandise? A review of experiments that compare usability evaluation methods, *Human-Computer Interaction*, **13**, 203–261

Green, W.S. and Jordan, P.W. 1999, The Future of Ergonomics, *Contemporary Ergonomics*, 110–114

Houtkoop-Steenstra, H. 2000, *Interaction and the Standardized Survey Interview, The living questionnaire*, Cambridge University Press

Kanis, H. 1997, Variation in measurement repetition of human characteristics and activities, *Applied Ergonomics*, **28**, 155–163

Kanis, H. 1998, Usage centred research for everyday product design, *Applied Ergonomics*, **29**, 75–82

Kanis, H. 2000, Questioning validity in the area of ergonomics/human factors, *Ergonomics*, **43**, 1947–1965

Kanis, H. and Green, W.S. 2000, Research for usage oriented design: Quantitative? Qualitative? In *IEA-HFES Proceedings*, 6-925–6-928

Moray, N. 1999, Culture, politics and ergonomics, *Ergonomics*, **43**, 858–568

Murphy, E., Dingwall, R., Greatbach, D., Parker, S. and Watson, P. 1998, *Qualitative research methods in health technology assessment: a review of the literature*, University of Nottingham

Seale, C.F. 1999, *The Quality of Qualitative Research*, Sage

Silverman, D. 1997, The Logics of Qualitative Research. In *Context and Method in Qualitative Research*, 12-25, Sage

Wilson, J.R. 2000, Fundamentals of ergonomics in theory and practice, *Applied Ergonomics*, **31**, 557–567

AUTOMOTIVE DESIGN

A COMPARATIVE ASSESSMENT OF THE USE OF CONVENTIONAL AND EXTENDED AUTOMOTIVE PEDALS

GE Torrens[1], G Williams[2], M Freer[3]

[1]Hand Performance Research Group, [2]Dexterity Research Limited, [3]Vehicle Ergonomics Group
Department of Design and Technology, Loughborough University of Technology,
Loughborough Leicestershire, LE11 3TU

There is current concern relating to the safety of drivers of smaller stature adopting a driving position that brings their upper body close to steering wheel mounted air bags. The aim of this pilot study was to identify differences in posture and movement that relate to reaction time in use between the two driving positions. This paper describes the comparative assessment of use of conventional and extended automotive foot controls by four subjects whilst undertaking a simulated driving activity. Four drivers, two female and two male, were selected who were all under the 25th percentile male stature and sitting height for the United Kingdom population. The results indicate that the pedal extension did not affect leg posture when added, but did increase extension at the elbow. Clearance between the sternum and steering wheel was increased by over 70mm when the pedal extensions were fitted.

Introduction

The use of air bags within a vehicle safety systems is widely used by all manufacturers. An air bag protects the driver or passenger through a rapid inflation of a lightweight sack. The sack has an opening within its casing to allow a controlled release of air. The inflated sack provides an air cushion to reduce the shock load to the body from high deceleration rates associated with a road traffic accident. However, there have been a number of cases reported in the press in the United Kingdom (U.K.) and United Sates of America (U.S.A.) of vehicle users being injured through close proximity to the air bag as it has inflated. Currently, there is no legislation that states a safe working distance air bag systems. Industry guidelines in the U.S.A. suggest a working distance of 10 inches (U.S.), approximately 250mm from the steering wheel hub to the driver forehead. The aim of this pilot study is to provide an initial assessment of the effect of using a foot pedal extension on driving posture and distance from the steering wheel.

Pedal extensions are used to provide access to foot pedal controls for those who may be pregnant, small in stature, or have a disability. There are a number of pedal extensions currently available. The authors know of only one product that is produced specifically for the European market. The foot pedal extension from Eze-Drive Limited, (Leicester, U.K.), caters for a range of European configured vehicles within an adjustable clamping system. The foot pedal extension effectively moves the pedal horizontally 120 mm forward of its initial position, towards the drivers foot.

Method

The assessments were carried out in a room of ambient temperature and limited direct sunlight. Four subjects, two female and two male, were chosen to provide a representation of people with a physical size that was the equivalent or under of a 25th percentile stature for males within the U.K. population. People of this stature perceived as being more vulnerable to injury, due to the proximity of air bag systems. The test procedure was discussed with each subject. A driving test rig was used that had adjustable dashboard and steering wheel, and foot pedal well. The adjustment of the two elements enabled ease of monitoring of the configuration favoured by each subject for driving. This test equipment had previously been used for perceptions of driver comfort (Porter and Gyi, 1998). The dimensions taken from the seating test rig were:

- Pedal seat (measured from the front edge of the pedal to the seat cushion front edge).
- Sternum to wheel distance (measured as a direct line between the centre of the steering wheel to the mid point along the sternum).

The distance between the hub of the steering wheel to the front of the foot pedals was kept constant. This was to simulate the seat moving forward or back. The configuration of steering wheel to foot pedal was based upon the dimensions of a Vauxhall Astra.

Subjects were asked to confirm that they had no history of neuromuscular or musculoskeletal injuries, or diseases. The assessment of each subject was divided into two phases; (i) physical characterisation, and (ii) performance characterisation.

(i) Physical characterisation

The physical characterisation of each subject was assessed using a previously documented series of methods (Torrens and Gyi, 2000). In addition, measurements were taken relating to seated posture that included:

- Stature
- Weight
- Sitting height (not slumped)
- Buttock popliteal
- Foot length
- Foot width
- Forward reach Sitting

- Fingertip arm length
- Fingertip elbow
- Shoulder breadth
- Shoulder fingertip
- Hand length
- Hand width (Metacarpophalangeal joints, digits 2-5)

The results from the measurement of each subject's anthropometrics were related to a United Kingdom population from the software package PEOPLESIZE (Open Ergonomics, 1999). The results are shown in Tables 1 and 2.

(ii) Task characterisation

The methods used in the recording of task performance were based on those documented in previous studies (Torrens and Gyi, 2000). The methods included: Motion capture using a CODA mpx30 system supplied by Charnwood Dynamics Limited, Leicester, UK. (http://www.charndyn.com). Photographs were taken of the postures taken up by each subject at every stage of the evaluation. Each subject was fitted with markers over anatomical reference points, laterally down the left side of the subject's body, prior to the start of the assessment. This process took approximately five minutes.

The motion-capture system employed used infrared emitting markers that were placed over anatomical reference points on the upper limbs and head, including:

- Supraorbital foramen (right and left)
- Mandible (at the midline)
- Acromium point (right and left)
- Sternum (suprasternal notch)
- Sternum (bottom of sternum body, above xiphoid process)
- Humerus (at the lateral epicondyle, left)
- Radius (at the styloid process, posterior, left)
- Digit 2 (at the Metacarpophalangeal joint, left)
- Digit 5 (at the Metacarpophalangeal joint, left)
- Hip (greater tochanter)
- Knee (lateral mid point patella)
- Distal lateral posterior femur
- Distal lateral anterior femur
- Proximal posterior lateral tibia
- Proximal anterior lateral tibia
- Ankle (lateral malleolus)
- Heel (lateral at the heel of the subject's footwear)
- Toe (lateral on the front of the subject's footwear, over toe 4)

Three markers were also placed on the edge of the foot pedal and steering wheel. The pedal markers provided an indication of the relative position of foot to pedal. The steering wheel markers provided a relative position of the trunk of the body from the wheel. Samples of each subject's posture were processed and an analysis of their performance involved the following elements:

- Position of body segments, specifically upper and lower limbs at the beginning and end of the task (with and without pedal extensions). The motion capture system was used primarily to record flexion/extension at the elbow and at the knee. References to ankle, hand/wrist, body trunk and head position were also measured.
- Reaction times to carry out an emergency stop (with and without pedal extensions)

Position of body segments

The subjects were asked to obtain a comfortable seating position. Each subject was asked to take up the driving posture he or she would normally use. The steering wheel and foot pedals were moved by an operator in to the positions that maintained the subject in their comfortable position. Each subject was asked to practice pushing the left-hand foot pedal control, (clutch in U.K.), as he or she would normally during driving over a period of five minutes. They were asked if their driving position was still comfortable and adjusted if necessary. Subjects comments and views were noted on their experience of driving postures with and without pedal extensions. Pedal and body segment positions were recorded using a CODA motion capture system. The capture time was four seconds at a sampling rate of 200Hz. This procedure was repeated following the addition of the foot pedal extension. This section of the assessment took approximately 20 minutes to complete. The results of this assessment are shown in Tables 3 and 4.

Reaction times

The subject was asked to repeat the process undertaken in setting up for the assessment of body segments. Only the lower limb markers were used in this assessment.

Following an initial pedal push task, each subject was asked to press the pedal as if part of an emergency stop. The subject was asked to start this task at the sound of a small single alarm bell. The tone from a digital stopwatch was used for this purpose. The alarm was used to avoid bias through verbal request by the operator and enhance repeatability. The actions of the operator were hidden from the view of the subject to avoid anticipation of the signal. The results are shown in Table 5.

Results and discussion

The results of each subject's physical characterisation are shown in Tables 1 and 2. The results of the task performance are shown in Tables 3 and 4. A summary description of the processed results from the motion capture recordings follows in Tables 4 and 5. The total time taken to process sections (i) and (ii) of the trials was calculated to be two hours per subject, involving three operators. The time taken to process the physical characterisation and task performance results from the four subjects was approximately 8 hours. The comparison of stature of the subjects and U.K. data through PEOPLESIZE (Open Ergonomics, 1999) showed that all four subjects were under a 25th percentile stature for U.K. males. However, subject 3 (male) was a 52nd Percentile in sitting height.

Table 1. Subject physical characteristics The peoplesize percentiles

Subject	Gender	Weight Kg	Stature mm	Peoplesize percentile equivalent for UK adults (F)	Sitting height mm	Peoplesize percentile equivalent for UK adults	Buttock popliteal Length mm	Foot length mm	Foot width mm
1	F	53.5	1664	(80)	877	86	484	223	73
2	M	79.9	1601	1	860	6	443	248	99
3	M	78.5	1699	23	917	52	461	264	88
4	F	58.0	1631	(62)	861	69	463	232	86

Table 2. Subject physical characteristics

Subject	Shoulder Breadth mm	Forward Reach Sitting mm	Finger Tip to Elbow mm	Hand Length mm	Hand Width MCP mm	Shoulder-Fingertip mm
1	253	693	415	161	70	680
2	300	740	428	172	85	670
3	307	723	467	189	90	745
4	285	647	424	175	77	685

In Table 3. it can be seen that the distance between the steering wheel and body is increased when using the pedal extension between 73mm and 78 mm, to between 426mm and 557mm, suggesting pedal extensions are useful in maintaining a perceived safe distance form an air bag system.

Table 3. A comparison of subject seating characteristics with and without using a pedal extension.

Subject	Seat Pedal 1	Seat-Wheel 1	Sternum -Wheel 1	Seat-Pedal 2 (with extn.)	Seat-Wheel 2 (with extn.)	Sternum-Wheel 2 (with extn.)	Distance between sternum -wheel (1-2)
1	404	-58	440	378	42	518	-78
2	464	2	483	456	121	557	-74
3	374	-88	365	330	-6	438	-73
4	375	-87	350	358	12	426	-76

Table 4. Flexion and extension angles at the elbow and knee during pedal pushing tasks with and without foot pedal extensions.

Subject	Conditions	Knee neutral angle°	Knee maximum extension angle°	Knee post extension angle°	Elbow neutral angle°	Elbow post knee extension angle°
1	Normal	107	126.8	107	111.2	108.1
1	Pedal extension	97.7	125.9	97.6	136.4	133.7
2	Normal	113.6	135.8	100.6	114.4	114.3
2	Pedal extension	103.4	132.7	99.9	110	112.8
3	Normal	93.6	112.3	98.5	111.3	106.4
3	Pedal extension	84.7	102.7	80.9	115.5	115.3
4	Normal	79.4	120.1	97	97.9	98.3
4	Pedal extension	101.3	125.2	98.9	117.7	117.4

Table 5. Performance characteristics of 3 of the 4 subjects using a foot pedal under normal use (gear changing) and when used during an emergency stop.

Subject	Conditions	Knee neutral angle°	Knee maximum extension angle°	Knee post extension angle°	Pedal movement time s	Foot movement time s	Reaction time s
2	Normal	86.6	122.4	91.8	0.389	0	0.389
2	Normal-extn.	94.7	130.1	93	1.438	0.693	0.745
2	Emergency stop	87.6	121.9	92.7	0.584	0.335	0.249
2	Emergency-extn.	96.1	126.6	91.1	0.943	0.741	0.202
3	Normal	85.8	132.2	84.2	1.076	0.54	0.536
3	Normal-extn.	97.4	133.4	96.2	1.178	0.837	0.341
3	Emergency stop	86.1	132.6	84	1.127	0.895	0.232
3	Emergency-extn.	97	133.5	96.4	0.711	0.526	0.185
4	Normal		120.6	81.5			0
4	Normal-extn.		139.8	83.7			0
4	Emergency stop	78.2	126.7	85.4	0.82	0.588	0.232
4	Emergency-extn.	90.7	147.7	77.2	0.663	0.338	0.325

The performance results indicate an increase in extension at the elbow when using the foot pedal extension. From the subjects comments and pictures taken during the task performances it was identified that the shoulders have displaced and a more upright sitting posture adopted to maintain as neutral an elbow posture as possible.

There was no conclusive indication from the reaction times collected to indicate if the pedal extension had made any difference to the use of the control pedal. One subject was missing and another's data was corrupted during recording. More investigation is required with a larger sample group and a review of the recording method (mainly marker positioning).

Through analysis of the limb movements, it was noted that each subject's leg did not move in a linear manner, with the knee acting as a hinge, but more a combination of rotations including the hip and ankle joints. It was also noted that the angle of the lower limb of the two female subjects, in the sagittal plane, was more acute in relation to the foot pedal than male subjects limb. The more acute angle can be attributed to a broader and shallower pelvis in females than males. These rotations and the difference in male and female driving task performances are likely to influence a drivers perception of seat discomfort and require further investigation.

Acknowledgements

The authors would like to thank Mr C McAllister, Defence Logistics Organisation, for his considerable support in the running and processing of results within this study. Also thanks to Eze-Drive Limited for the loan of their product. This study was, in part, funded by the Defence Logistics Organisation.

References

Open Ergonomics Limited, 1999, PEOPLESIZE software, Loughborough

Porter, J.M. and Gyi, D.E., 1998, Exploring the optimum posture for driver comfort, *International Journal of Vehicle Design* , 19(3), 255-266

Torrens G.E. and Gyi D., 1999, Towards the integrated measurement of hand and object interaction. *7th International Conference on Product Safety Research*, European Consumer Safety Association, U.S. (Consumer Product Safety Commission, Washington D.C.), 217-226

FERRARI AND ERGONOMICS IN DESIGN

Marjella Kamerbeek[1]
Heimrich Kanis[1]
Bill Green[1+2]

[1] *School of Industrial Design Engineering, Delft University of Technology, Jaffalaan 9, 2628 BX Delft, the Netherlands.*
[2] *School of Environmental Design, Division of Science and Design, University of Canberra, P.O.Box 1, Belconnen, A.C.T. 2616, Australia.*

In the process of adapting and re-designing the Ferrari Formula 1 gear changing system to their GT cars, the changes are tested by Ferrari test drivers. This user trial shows that the test drivers act in a very similar way. Given the flexibility of humans, their various preferences and their ability to adjust, it is not likely that the interior alone, as a working space, forces the drivers to behave in this specific way. Presumably, the test drivers have learned to drive the car in a particular way, which is, according to Ferrari, the only right way! It remains to be seen whether the ergonomic design policy adopted by Ferrari can be sustained in an expanding market for their sports cars.

Introduction

Ferrari S.p.A., manufacturer of racing and GT sports cars, is redesigning a gear changing system for their GT cars, which incorporates Formula 1 technology and experience. This system is meant to give the client a real racing car experience. In 1997, the system was first introduced in the F 355. Since then, all improvement has been in mechanical and software applications. The layout and design of the system is based on "rule of thumb" and "trial and error". Ferrari believes the only way to drive the car is the way their own test drivers do and therefore all changes are based on their assessments. In contrary, Delft University curriculum teaches its students to generate products that allow users as much freedom as possible in using the product without undermining functionalities, i.e. no particular way of usage is forced upon users (cf. Kanis, 1998).

After the introduction of the system, Ferrari has paid no real attention to the ergonomics of the system – placement, usage, etc. in spite of the express intention of the development. As more and more car manufacturers developed a way of changing gear at the wheel (buttons in several places on the wheel, paddles in different sizes behind the wheel, etc.), the question arose, which system is the best for Ferrari. After an evaluation process involving ergonomics, marketing and the design traditions of Ferrari, it was decided to concentrate on the ergonomic assessment and possible redesign of the existing paddles behind the wheel.

Objectives

The prime goal of this study was to find out what area behind the steering wheel should be covered by the levers, or 'paddles'. To get a clearer picture on how the wheel and the gear system is used, the following questions arose: what adjustments do drivers make to the cockpit after having entered the car and are these adjustments related to the anthropometrics of the drivers? In addition, what are the positions and actions of the hands on the wheel during driving?

Method

Subjects
The subjects for the user trial were four male test drivers of Ferrari. They were randomly chosen. Three of them are right handed, one is left-handed.

Tracks
For the tests, a racing track on the premises of Ferrari was used. The test track of Ferrari provides a balance between right and left-hand bends, bends with different radii and bends with different features e.g. having one or more centres. Each driver did two laps of this particular track.

A route through Maranello (including the hills of Maranello) was used to do the tests on normal roads. All four drivers took that same road but encountered different traffic situations.

Cars
A 360 Modena with F1 gearshift was used for the first eight tests (two tests for each driver: on the racing track and through Maranello). A camera was installed in the car to record all hand movements of the drivers during the tests. For the next eight tests, an F 360 Modena was used with a manual gear changing system.

Tests
The anthropometrical measurements of all four drivers were taken together with their seat and steering wheel adjustments in both cars. Hand movements and positions of the four test drivers from Ferrari were digitally recorded. The drivers did the two tests (track and route) in one car on the same day. After all four drivers had done the tests in one car, the same tests were repeated in the other car.

Analysis
The video recordings of the position of the driver's hands, relative to the steering wheel and relative to the dashboard in a Ferrari with Formula 1 gear change and in a Ferrari with a manual change, were analysed on a second by second basis. Every second the position of the hands, relative to the dashboard and relative to the steering wheel is given a numeric value in degrees between $-180°$ and $180°$ [$0°$ is top centre, $180°$ bottom centre, positive to the right, negative to the left]. All values were represented graphically, making it possible to compare different situations, different drivers, or, for instance, the left and the right hand. This comparison could be made for certain moments during the test or for the whole test (two laps together).

Results

Driving behaviour intra-individually

Figure 1a and figure 1b show that the differences between the two laps are minimal, exemplifying the professional behaviour of the Ferrari test drivers.

Figure 1a shows the position of the hands of a driver during the whole lap. For each lap he took about 94 seconds and he did the lap twice. The positions in the second lap practically overlap the positions in the first lap. Figure 1b shows the moments of changing gear of the same driver. Again, the moments that he chose to change gear in the second lap are almost the same as in the first lap.

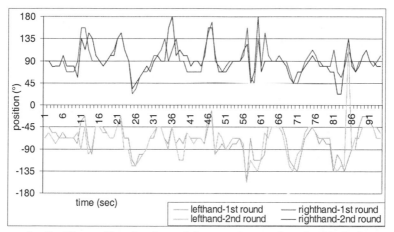

Figure 1a. Positions of hands of one driver on the steeringwheel, relative to the dashboard, during the test

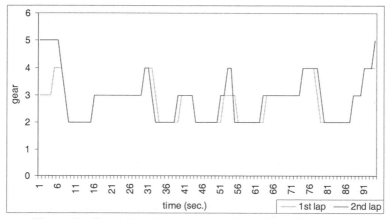

Figure 1b. Gear changing moments of one driver during tests

The peak for the position of both hands on the wheel occurs around 90°, see Figure 2. Apparently, the shape of the steering wheel (the drivers tend to 'hang' their thumbs in the wheel), together with the fact that the driver has to be able to reach the levers, causes the drivers to keep their hands in this particular area to the wheel.

From the average positions during gearing in the car with F1 gear and in the car with regular gear, see Figure 3, it is obvious that practically the same areas are used in both cars. The left hand is placed between –90° and –45°, the right hand is placed lower in the wheel compared to the left hand, between 67,5° and 112,5°.

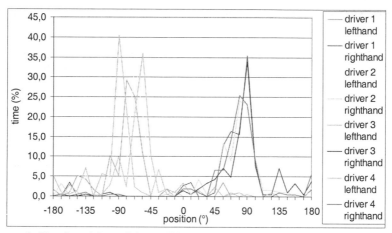

Figure 2. Hand positions drivers, F1 gear system, relative to the steering wheel

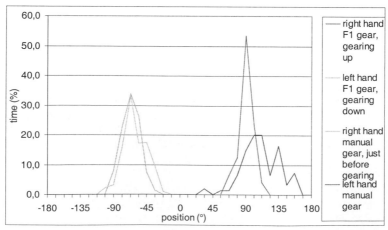

Figure 3. Average hand positions while changing gear or just before changing gear

The position of the right hand while driving the manual gear was actually the position of this hand immediately before the actual change. This might be a reason why the position of this hand is rather low, compared to the rest: the driver is getting ready to reach for the shift stick and therefore keeps his hand low on the wheel.

The fact that the areas used are practically the same could indicate that the drivers are either not really influenced by the position of the levers, or the drivers are so used to the F1 levers that they put their hands in that same position in the car with the manual gear. This in turn indicates that they are probably comfortable with this position.

There was a strong correlation between the left and the right hand positions.

Anthropometrical measures and the geometry of the working space
The anthropometrical characteristics and the geometry of the working space are given in Table 1.

Table 1. Anthropometrical characteristics and geometry of the working space

	Driver 1	Driver 2	Driver 3	Driver 4
Body length	1.78 m	1.81 m	1.78 m	1.61 m
Upper leg (right)	0.64 m	0.63 m	0.65 m	0.55 m
Arms (left/right)	0.78/0.78 m	0.78/0.78 m	0.75/0.76 m	0.73/0.73 m
Up. dist. wheel-seat (F1/manual gear)	0.70/0.69 m	0.76/0.76 m	0.76/0.75 m	0.68/0.67 m
Low. dist. wheel-seat (F1/manual gear)	0.54/0.53 m	0.59/0.59 m	0.56/0.56 m	0.48/0.48 m
Dist. wheel-floor (F1/manual gear)	0.43/0.43 m	0.43/0.43 m	0.44/0.43 m	0.40/0.41 m

As this table shows, the drivers neutralise the biggest differences in their anthropometrics by carefully adjusting the settings of the seat and the wheel. This adjusting is done very precisely and was almost identical for each driver every time he had to set his seat. Given these adjustments, there are still differences between posture and position of the different drivers (driver one is relatively close to the wheel in spite of his long arms, driver three is relatively far from the wheel in spite of the length of his arms).

Driving behaviour inter-individually
In spite of distinct body measurements and different settings for the workspace, the drivers' behaviour was very similar. This is particularly clear when one looks at their actions in the bends during the track test. To compare this second by second, only small sections of the laps can be taken at a time and the results need to be adjusted a little, since one driver might take e.g. 10 seconds to make a bend, another driver might take 11 seconds and so on. This would influence the shape of the graphs and make them not

comparable. To compare the position of the drivers, a particular section of track was selected (bends 11 and 13, together with the semi-straight piece 12). After stretching the total time of the fastest driver and compressing the total time of the slowest driver, it is significant that the drivers actually have their hands in the same position relative to the dashboard, at exactly the same points in the track, see Figure 5.

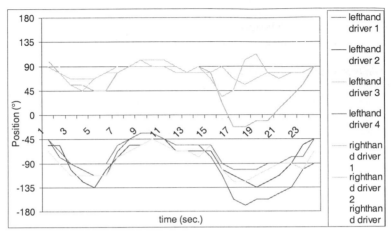

Figure 4. Hand positions of the four drivers, rescaled; 1 to 8 sec.: bend 11, 8 to 15 sec.: semi-straight 12, 15 to 24 sec.: bend 13

Discussion

On the basis of the observations, no major changes in the design of the gear system are necessary. In other words, the current design proved to be adequate and this is exactly what Ferrari wants. This result originates from the uniformity of the drivers, meaning that Ferrari has chosen a good solution for themselves. On the other side, this exclusive, rather then inclusive design method ignores the fact that less skilled drivers are the ones that actually buy and drive the cars. The question that remains is whether the current solution is good for them? On the basis of the tests that were done, this question cannot be answered. It seems unlikely that the lay-out of the cockpit makes drivers act uniformly, considering the flexibility and ability of humans to adjust to different circumstances. Presumably, learning and mimicking each other are important factors in test drivers ending up doing the same things. It remains to be seen whether the design policy adopted by Ferrari can be sustained in a world of sports cars which is gradually tending to become less exclusive, as a rapidly growing number of people can afford a Ferrari.

Reference
Kanis, H. 1998, Usage centred research for everyday product design, *Applied Ergonomics*, **29**, 75-82

DESIGNING FOR THE IN-CAR SAFETY AND SECURITY OF WOMEN

[1]Andree Woodcock, [2]James Lenard, [2]Ruth Welsh,
[3]Margaret Galer Flyte and [3]Sarah Garner

[1]*VIDe Research Centre, School of Art and Design, Coventry University,Coventry,*
A.Woodcock@coventry.ac.uk
[2]*ICE Ergonomics, Loughborough University, Loughborough,*
JLenard@ice.co.uk, RWelsh@ice.co.uk
[3]*Formerly Department of Human Sciences, Loughborough University, Loughborough.*

The need for automotive manufacturers to consider more seriously the requirements of female drivers and passengers is well established. The research summarised in this paper was commissioned by the Mobility Unit of the Department of the Environment, Transport and the Regions (UK) specifically to address the in-car safety and security needs of women drivers and their passengers. The research was multifacetted considering firstly, the injury patterns sustained by differently gendered vehicle occupants; secondly, the social, psychological and behavioural issues surrounding car usage; and thirdly the ways in which the results could be made more accessible (see also Woodcock and Galer Flyte, 2001). This paper presents an overview of the aims and methodologies adopted and the key findings of the research.

1 Introduction

Research into car safety, crash survival and accident analysis has been conducted both in the automotive industry and academic institutions. These studies have indicated that women are injured more in cars and sustain a different injury pattern to their male counterparts (Hill and Mackay, 1997). This might be due to the type of journeys undertaken, the type of car driven (e.g. women, on average, drive smaller cars than their male counterparts), or car design

Research on social and behavioural issues has tended to emphasize young, male drivers, the effectiveness of different forms of deterrents in reducing car related accidents, social, aging and cognitive effects on driving. In most of the studies the primary focus has been on the male driver, with women mentioned in comparison to men.

However, the AA (1993) conducted a survey of 600 women and 200 men which highlighted the attitudes of women towards driving, and established that different groups of female drivers had different concerns. A follow up report in 1997 included surveys with motorists on issues such as road tax, petrol duties and MoT tests and results from member surveys. This report showed, for example, that 19% of the women surveyed had never driven on a motorway alone, despite these being statistically the safest roads in the UK. The 1998, DETR report 'Focus on Personal Travel' suggested little difference in the number of car journeys made by male and female drivers. However, there was a significant difference in the length of these journeys, with men travelling approximately 45% further than women and access to cars (4 in 5 men have access to a car, but only 2 in 3 women have similar access).This research consolidates and extends these earlier findings by investigating how safe women are and feel themselves to be in their car and driving environment.

The actual safety and security of car occupants depends on the design of the car, its safety systems and the effective use of these by the occupants. Perceived safety and security may be affected by many factors such as media coverage, experiences, ability and the design of the car and the traffic environment.

Although specific issues in car design are routinely addressed (such as smart seating and occupant protection) the experiences of female motorists are rarely touched upon. This more user centred design approach to automotive design and usage is necessary if vehicles and environments are to be designed to suit the needs of all transport users.

Aims

The aims of this research were fourfold:

1. To consolidate data relating to women's safety and accidents;
2. To develop a series of 'best practice' guidelines relating to female occupant safety issues derived from national and international data, recommendations by learned bodies, and strategies which have been developed by women in relation to their own experiences;
3. To provide qualitative information regarding patterns of female car usage;
4. To disseminate this information in a manner which will heighten awareness of women's issues in relation to driving and the design of cars.

This led to a bipolarisation of the research into two areas; firstly an analysis of the causes and severity of injuries sustained by women drivers and passengers during crashes and secondly the social and behavioural issues effecting the in-car safety and personal security needs of women drivers and passengers. The following section considers the methods employed to capture this information.

Method

Analysis of Accident Statistics

The Birmingham report (Hill and Mackay, 1997) identified a number of safety issues relevant to female car occupants. For example, they are more likely to be killed in car accidents, they have more neck, kneecap, ankle and certain types of foot injury than men. Women tend to sit closer to the steering wheel which may expose them to potential injury from a deploying airbag. Osteoporosis in post-menopausal women increases their vulnerability to injury. Seat belts are beneficial to pregnant women but the advice to wear the belt low down across the abdomen is not necessarily widely given. However, no analysis has been reported regarding gender differences in car accidents. To this end accident injury data from two British databases were examined for differences between men and women in accident circumstances and injury outcomes.

Stats 19 Database

This is compiled from police reports for any accident on the highway resulting in human injury or death. The driver need not be a casualty in the accident, but the vehicle and driver record will appear in the data base provided someone involved in the accident, for example a passenger or a cyclist, was injured. The data used for the analysis covered three years (1996 -1998) totaling 944,638 drivers. Of these 324,215 were female and 620,423 male. The analysis considered firstly the measure of exposure (who is more likely to be involved in an accident); secondly, the circumstances surrounding the accident and injury outcome; and thirdly female vulnerability.

Cooperative Crash Injury Study (CCIS)

The CCIS data is gathered mainly from the examination of crashed vehicles (away from the crash scene) and hospital records or autopsy reports. It is based on retrospective, in-depth investigations. All vehicles from accidents that are classified as serious or fatal in the police notifications are eligible for inclusion. As it is not feasible to put this level of effort into every crash that occurs in the UK so CCIS takes a sample of accidents from certain regions around England. It therefore represents a stratified sample of accidents from the regions covered including virtually all fatal accidents, 80-90% of serious accidents, and around 20-30% of slight accidents. The sampling criteria is very well suited for identifying differences in the accident and injury circumstances between men and women.

It has been operating since 1983 and currently investigates around 1500 cases per year and now contains about 8000 drivers. The criteria for inclusion in the study is that (a) at least one occupant from a vehicle involved in the accident is injured and (b) at least one vehicle is less than seven years old and towed away from the accident scene.

The goal of the CCIS analysis was to identify differences in the accident circumstances and injury outcomes between men and women. An emphasis was placed on situations for which it is most likely to be possible to design countermeasures for the benefit of women: i.e. restrained occupants in frontal impacts, struck-side impacts, rear impacts and rollovers. In addition to this general survey, three other topics of special interest to women were also included: airbags, over-50 year olds, and pregnancy. Together the two databases provide a very strong basis for examining road safety issues.

Analysis of social and behavioural issues

The investigation of the social and behavioural issues complimented the accident data by considering the manner in which cars were used, the circumstances under which women were or felt themselves to be less safe in the car (e.g. incorrect positioning of the seat), and the road environment. Five methods were used to elicit information from different groups of women (such as the elderly and disabled, professional women, women with young and old children); focus groups, diary studies, stakeholder analysis, analysis of best practice and Internet analysis

Key findings

The main conclusion to the research was that women are significantly disadvantaged in the road and traffic environment. More specifically:

1.Women are more vulnerable to injury in crashes than men. This is a key finding which emerged from the accident data analysis.

2.Cars are not designed to suit women's anthropometric characteristics nor their journey requirements.

3.Many women lack confidence in driving and feel vulnerable in the road and traffic environment. This was evidenced in the focus groups and the Internet analysis.

4. Women feel that they would benefit from more information on choosing, using and maintaining their car.

These are examined in more detail in the following sections.

Ergonomics, social and behavioural issues

In terms of vehicle design the fit of the car to the female driver or passenger was not good. Typical problems related to difficulty positioning the seat so that all controls and displays could be accessed and the seatbelt remain comfortable. The elderly, pregnant women, babies and young children were especially not catered for in terms of the car design.

When driving the car women felt that modern cars had been designed to reach and 'sit' at high speeds, and lacked maneuverability at low speeds. Central locking systems did not cater for having different sets of access routes open. Emergency egress was worrying. Most women felt vulnerable when parking at night or when they had broken down and would have liked a panic or emergency system based around GPS technology.

Both female and male drivers were worried about aggressive driving and felt clearly threatened by the behaviour of other motorists. Women were more likely to avoid using the motorway, driving at night and on unfamiliar routes. Female drivers also said that they would not know what to do if they were broken down especially on motorways (evidence was found of contradictory advice being offered). Where there was more than one car in the household women used the smaller one as this was seen as being more easy to manoeuvre, more economical and easier to drive. Where the smaller car was an older one, this might not necessarily be the case. One female respondent to the Internet survey asserted that the partner who did the school run should place her safety and that of the children first, and always have the better car.

The literature review showed evidence of gender differences in driving skills and attitudes towards driving. For example, women on average take longer to pass their test and exhibit errors relating to lack of control in the performance of vehicle manouvres and low errors that do not endanger anyone (Reason *et al*, 1991). Having passed their test women were less likely to commit motoring offences (McKenna *et al*, 1998) and are more likely to adhere to driving regulations and traffic dominance hierarchies.

In-car and road safety issues

Most importantly, the analysis of the crash data showed that women were either equally or more vulnerable to injury than men especially in front, side and rear impacts. This difference cannot be necessarily attributed to women's use of smaller, lighter cars (which they drive more frequently than men). Soft tissue neck injury is more frequent among women across all impact types; in frontal impacts restrained female drivers have more skeletal chest and neck injuries than men; in side impacts women seated on the struck side of the vehicle have a higher incidence of skeletal pelvic injury than men. Female passengers are also more vulnerable to injury than male passengers, and form a higher percentage of this type of occupant.

Women constitute about 40% of car occupants involved in crashes and outnumber men in the passenger seats. Female drivers aged 17-29 are as likely to be involved in an accident as those aged 30-59, with those over 60 having a lower accident involvement rate. Women continue to be more vulnerable to injury than men in the older age group but the gap between the sexes in terms of injury does not appear to widen in the over 50s. Male drivers' accident involvement rate decreases with age through the three age bands. Female drivers are more likely to have a crash during the day (8am to 6pm), during weekends and the months of October, November and February. Women drivers tend to be involved in slightly more rear end collisions than men and less frontal impacts. Seat belt use is lower amongst pregnant women.

Recommendations and design implications

Only the major outcomes of our research have been outlined here, but the issues we have identified are of such significance that further investigation and action are warranted and. Clearly the data show that great benefits would be derived if the vehicle package was designed to fit the anthropometric characteristics of the female drivers and passengers, including those who are pregnant, elderly or who have reduced mobility. Likewise the development of smart occupant protection has the potential for significant benefits for women in crashes as does improvements in vehicle compatibility.

In terms of the road environment a combined engineering, education, design and enforcement strategy needs to address the increased levels of aggression on our roads. The Internet survey showed that this is an international issue. Although our work emphasized female drivers it is quite possible that men may have similar experiences, especially when driving smaller cars. That such aggression is tolerated and even accepted is forcing motorists off the road.

Lastly, and most importantly, we would strongly advocate a removal of the gender bias in automotive design. Exclusive reference to male anthropometric specifications in regulations and research programmes is not justified by real world data. for example, sex bias in crash test programmes and computer modeling is sub-optimal. Crash test criteria should be satisfied by both 50th percentile male and female dummies; research and experimentation to ensure that average females benefit equally from male oriented programmes and regulations should be encouraged.

Acknowledgements

The work reported herein was carried out under Contract PPAD9/72/36 funded by the Department of Environment, Transport and the Regions. Any views expressed are not necessarily those of the Secretary of State for the Environment, Transport and the Regions.

Project members: Department of Human Sciences: Sarah Garner and Margaret Galer Flyte: ICE - Ruth Welsh and Jim Lenard; Graphic Design Team, Coventry School of Art and Design - Paul Barrett, Lee Barrett, Richard Clark and Matthew Fish; Automotive Concept Design Team, Coventry School of Art and Design- Leia Anastasiou, Joonkian Chua and Anna Louise Clough, Advisor Sam Porter; Project Advisors: Anna Humpherson and Mike Bradley, Ford, UK

References

Automobile Association 1993, *Women and Cars: Emancipation, Enrichment and Efficiency.*

Automobile Association 1997, *Living with the Car*

Department of the Environment Transport and the Regions 1998, *Focus on Personal Travel*, The Stationary Office, London

Hill, J. and Mackay M. 1997, *In Car Safety of Women*. Birmingham Accident Research Centre, Report prepared for the Mobility Unit (DETR)

McKenna, R.P., Waylen, A. E. and Burkes, M.E. 1998, Male and female drivers; how different are they? AA Foundation for Road Safety Research

Reason, J.T., Manstead, A.S.R., Stradling, S.G., Parker, D. and Baxter, J.S. 1991, *The Social and Cognitive Determinants of Aberrant Driving Behaviour,* Laboratory Report LR253, TRRL, Crowthorne, Berks.

DRIVERS AND DRIVING

THINKING OF SOMETHING TO SAY: WORKLOAD, DRIVING BEHAVIOUR & SPEECH

Alex W Stedmon & Steven Bayer

HUSAT Research Institute
The Elms, Elms Grove,
Loughborough, Leics. LE11 1RG
tel: +44 (0)1509 611088
email: a.w.stedmon@Lboro.ac.uk

The success/application of in-car Automatic Speech Recognition (ASR) relies on designing interfaces to match the expectations, preferences and abilities of various user groups. Driver workload (underload or overload) is a primary factor affecting the integration of in-car systems. Using multiple measures of workload, it is possible to assess relationships between actual task difficulty (objective measures), perceptions of task difficulty (subjective measures) and how individuals react to their perception of task difficulty (psychophysiological measures). This experiment investigated a range of driver workload factors in a simulator, when using a speech interface, by manipulating traffic behaviour (density, flow, speed changes, etc) and road layout/conditions (geometry, speed restrictions, fog, etc). The findings illustrated that whilst increased workload affected driver performance using a speech interface did not appear have such an impact on the driving task.

Workload, Driving Behaviour & Speech

Automatic Speech Recognition (ASR) allows systems to be operated by speech input. Whilst little attention has been given to ASR in the driving domain (Graham, et al, 1998), it could improve the usability and safety of in-car systems, including voice-dialling of mobile phones, operating entertainment systems and Intelligent Transportation Systems (ITS) such as route guidance or travel/traffic information services.

Numerous studies (reviewed by Graham & Carter, 2000) have highlighted the dangers of using manual systems whilst driving and the potential benefits of speech recognition. Manual dialling of mobile phones has been shown to disrupt vehicle control activities such as lane-keeping and speed maintenance whilst voice-operated phoning resulted in less deviation and better control. Although real-road studies have generally shown smaller disruption effects compared to artificial environment studies the findings still indicate that ASR may offer some benefit to safer driving.

In-car use of ASR is fundamentally different from a number of other domain applications. In particular, the task of using ASR is secondary to the primary task of safe driving and, therefore, the driver's attention to the ASR process may be limited. The in-car environment is typical of a 'hostile environment' (Baber & Noyes, 1996), characterised by high workload and stress that can affect the speech produced and subsequent recognition process.

"Workload is more than merely doing a task, it encompasses an individual's perception of its complexity in relation to their ability to perform it" (Stedmon, et al, 2001) and this makes workload a highly problematic concept to define (Finch & Stedmon, 1998).

Ideally people operate at the peak of their arousal/performance (Weiner, et al, 1984). Underlying this hypothesis is a concept of arousal expressed as an inverted U-curve, that assumes there is an optimal level of arousal that yields an optimal level of performance. Driver workload (underload or overload) is a primary factor affecting the integration of in-car systems. Whilst driver underload will result in a deviation from the top of the inverted U-curve back towards the 'low arousal - low performance' end of the scale; driver overload will cause a deviation across the scale towards 'high arousal - low performance'.

As Graham, et al, (1998) suggest, sources of driver workload include:
- the driving task itself (e.g. lane-keeping, speed choice, keeping a safe headway and distance from other vehicles);
- the driving environment (e.g. traffic density, poor weather, road geometry, etc.);
- the use of in-vehicle systems (e.g. the presentation, amount and pacing of information to be assimilated and remembered).

In order to assess the effects of workload, it is useful to combine measures that take account of an individual's perception and response to their ability to perform a task. A number of studies have sought to correlate different measures of workload, suggesting that psychophysiological measures such as heart rate and heart rate variability serve as an index of mental workload and time-on-task (Aasman, et al, 1987). Wilson & Eggemeier (1991) observed that heart rate variability also shows a negative correlation with subjective ratings of workload. Objective workload may be measured as a function of the task difficulty/environment. In the driving domain workload can be manipulated through driving conditions (traffic, weather, road surface conditions, time of day, etc). Measures for assessing driving performance might include accidents, speed, lane deviation, speeding tickets, and traffic light violations (Stein, et al, 1987). Accidents are a clear measure of traffic safety and can be due to lapses in attention, excess speed, poor speed control or poor lane keeping. Excess speed may cause a driver to lose control of their vehicle (due to road geometry or hitting obstacles) whilst driving faster or slower than other road users (poor speed control) may increase the likelihood of an accident. Lane deviation is another indicator of driving performance. If a driver's ability to maintain lane position is impaired, then the likelihood of exceeding the lane boundaries and hitting another obstacle/vehicle also increases (Stein, et al, 1987). Speeding tickets and traffic light violations may be taken as indicators of driver Situational Awareness (SA). These parameters provide some indication of driver vigilance over the speedometer (inside the vehicle), and road signs, traffic lights and the behaviour of other traffic (outside the vehicle).

The success/application of ASR in the driving domain relies upon the careful design of the interface to match the expectations, preferences and abilities of various user groups. What might be a potential aid could just as easily prove to be hazardous by distracting drivers from the control of their vehicles (Stein, et al, 1987).

Driver Workload & Speech Trials

This experiment investigated a range of driver workload factors and the use of an in-car ASR interface, within a driving simulator. Within a series of scenarios traffic behaviour and road conditions were manipulated to assess the validity of different workload levels. The simulator allowed for strict experimental control of the workload variables between participants, whilst ensuring driver safety and an ease of data collection.

Method
  Participants 20 male participants were recruited via HUSAT's participant database (20-58 years, mean 33.3 years). All participants had normal, or corrected to normal, vision, did not wear pacemakers and were not taking any prescribed medication. All participants held full UK driving licences, drove at least 2-3 times per week and 6,000 miles a year.
  Apparatus Driving scenarios were generated and displayed using the full-size, interactive, HUSAT Driving Simulator running STI-Sim experimental software. Heart rate data were collected using ADI-Instruments MacLab/8 & Bio-Amp hardware and Chart v3.5 software. 2020Speech Mediator 6 software was used on an Aurix speech recogniser with a head mounted microphone and Press-To-Talk (PTT) switch on the dashboard. NASA-TLX and Bedford-Harper subjective workload questionnaires were administered.
  Design A repeated measures, within-subjects, design was used. The independent variable (workload) was manipulated across 5 experimental conditions as detailed in Table 1 below. The workload levels were evaluated in a baseline study (Stedmon, et al, 2001) and Condition 3 acted as a control so that comparisons could be made. Discrete differences between the conditions are highlighted in the table.

CONDITION	WORKLOAD VARIABLES			
	TIME	TRAFFIC	FOG	SPEECH
1	Low	Low	No	Yes
2	High	High	No	Yes
3	Low	High	No	Yes
4	Low	High	Yes	Yes
5	Low	High	No	No

Table 1: Workload Variables in the Driving Scenarios

  To minimise any order or carry-over effects, the conditions were counterbalanced across subjects using a Latin Square. The driving task was designed so that it represented, as far as possible, a real route. Changes were made to the layout so that aspects of road geometry (junctions, traffic lights, curves, hills, etc) were matched in all cases. Two variations of the road layout were used to minimise learning or fatigue effects. These were counterbalanced so that the driving task content remained homogenous, across subjects and conditions. This strengthened the analysis between scenarios and minimised any error variance from potential differences in the scenarios.
  Dependent variable measures were collected for psychophysiological workload (heart rate & heart rate variability), subjective workload (NASA-TLX & Bedford-Harper scores), and objective workload (accidents, lane deviation, speed, speeding & traffic light violations).

Procedure After signing a consent form, participants received written and verbal instruction on the experimental procedure. Electrodes were connected for the heart rate analysis and a baseline measurement was taken. Participants underwent familiarisation and training in the simulator and operation of the speech interface before running the experimental conditions. Prior to each condition participants received further instructions on time pressure. During the trials, at random intervals, participants were instructed to initiate inquiries using the ASR interface (user initiated interactions). They also received information from the interface which they were required to respond to (system initiated interactions). Immediately after each condition subjective workload scores were collected. At the end of the experiment participants were de-briefed and paid for their time.

Results

Mean data scores for psychophysiological, subjective, and objective workload, and speech task performance were obtained for each condition and analysed using a 1x5 (workload) within-subjects ANOVA. Post hoc analyses, where applicable, were carried out using a Bonferroni T-Test. Only simple main effects for single workload variables are discussed in this paper. Any interactions are stated but not described further. Due to the complexity of their nature they warrant further analysis and shall be covered in more depth in a future paper.

Heart Rate (HR) No significant main effect was observed for workload ($p>0.05$) illustrating that HR did not alter as a function of time pressure, traffic density, fog, or speech.

Heart Rate Variability (HR-V) Raw data were analysed using derivative signals between $0.02 - 0.13$ Hz. No significant main effect was observed for workload ($p>0.05$) illustrating that HR-V did not appear to alter as a function of time pressure, traffic density, fog, or speech.

NASA-RTLX Scores A significant main effect was observed for workload [$F (4,76) = 5.049$; $p<0.001$]. Post hoc (2-tailed) comparisons illustrated higher subjective ratings with increased time pressure [$t(19) = -3.163$, $p<0.01$]. Interactions were observed for time pressure and traffic density, traffic density and fog, time pressure and fog, and time pressure and speech.

Bedford-Harper Scores A significant main effect was observed for workload [$F (4,76) = 7.643$; $p<0.001$]. Post hoc (2-tailed) comparisons illustrated higher subjective ratings with increased time pressure [$t(19) = 4.465$, $p<0.001$]; and increased fog [$t(19) = -2.557$, $p<0.05$]. Interactions were observed for time pressure and traffic density, traffic density and fog, time pressure and speech, and fog and speech.

Road Traffic Accidents (RTAs) A significant main effect was observed for workload [$F (4,76) = 7.437$; $p<0.05$]. Post hoc (2-tailed) comparisons illustrated that more accidents occurred with increased time pressure [$t(19) = -4.156$, p <0.001]. Interactions were observed for time pressure and traffic density, traffic density and fog, and traffic density and speech.

Lane Deviation (LD) A significant main effect was observed for workload [$F (4,76) = 28.836$; $p<0.001$]. Post hoc (2-tailed) comparisons illustrated that lane deviation decreased with increased time pressure [$t(19) = -6.063$, p <0.001]; and increased with fog [$t(19) = -5.143$, $p<0.001$]. Interactions were observed for time pressure and traffic density, traffic density and fog, time pressure and fog, time pressure and speech, traffic density and speech, and fog and speech.

Vehicle Speed A significant main effect was observed for workload [F (4,76) = 24.173; p<0.001]. Post hoc (2-tailed) comparisons illustrated that speed increased under high time pressure [t(19) = 5.676, p <0.001]; and decreased under increased traffic density [t(19) = 3.796, p <0.001]; and fog [t(19) = 5.445, p <0.001]. Interactions were observed for time pressure and traffic density, traffic density and fog, time pressure and fog, time pressure and speech, traffic density and speech, and fog and speech.

Speeding Offences A significant main effect was observed for workload [F (4,76) = 3.493; p<0.05]. Post hoc (2-tailed) comparisons illustrated that the number of offences increased under high time pressure [t(19) = -3.493, p <0.01]; and speech [t(19) = 3.136, p <0.01]. Interactions were observed for time pressure and traffic density, and traffic density and fog.

Traffic Light Violations (TLVs) A significant main effect was observed for workload [F (4,76) = 4.442; p<0.05]. Post hoc (2-tailed) comparisons illustrated that the number of violations increased under high time pressure [t(19) = 2.517, p <0.05]; and decreased under increased traffic density [t(19) = 2.854, p <0.01]. No interactions were observed.

Discussion

Although the lack of significant data for HR and HR-V appear to undermine the subjective measures of workload that consistently rated the higher workload scenarios as being more demanding, the finding supports the earlier baseline study (Stedmon, et al, 2001). It would appear, therefore, that even though drivers perceived the driving task to be more demanding as workload increased, the psychophysiological basis of their behaviour did not alter.

This finding also supports a study by Wilson & Eggemeier (1991), that compared real and simulated flying exercises and found no reliable heart rate variations in the simulator even though they were apparent in real flight. They concluded that "a subject's physiological responses in a simulator task could be different from those during actual flight, due to differences in the responsibilities and ... mental workload" (Wilson & Eggemeier, 1991).

Although participants drove faster and had more accidents under increased time pressure and traffic density, the frequency of these events was still very low. In a similar study (Stein, et al, 1987) found that, "despite high workload, accidents and speeding offences were infrequent events".

As expected vehicle speed increased and more speeding offences were committed under increased time pressure. Whilst more speeding offences were committed with speech it may not be this factor that caused the offences. Speeding offences were only logged on the number of times a driver broke a particular speed limit, rather than the length of time spent speeding. As such, a driver who constantly decelerated and accelerated could invoke more penalties (whilst driving safely) than someone who broke a speed limit and remained at that speed (Whilst driving recklessly). As a speech effect was only observed for this measure it is difficult to ascertain the significance of this finding.

From the lane deviation data it appeared that participants deviated more in the fog (without all the visual cues of other conditions) and less with increased time pressure (even though they might have been tempted to overtake traffic).

The results for traffic light violations show that TLVs increased as a function of Time but decreased as a function of Traffic. This would seem to indicate that under increased time pressure drivers either could not slow down in time for the lights or consciously decided to drive through them. Under this condition of increased workload attention may have been diverted (for different reasons) from looking at the speedometer (inside the car) or looking at road signs (outside the car) with subsequent effects on driver SA. The significant finding for traffic density would appear to indicate that other road users provided cues/reference points for stopping at traffic lights.

The interactions that were observed indicate, in general, how workload factors may compound their affects on task performance over simple main effects. Traffic and fog would appear to have the most impact across a number of measures illustrating that when fog was present and traffic density increased, the driving task was perceived as being more difficult than when either of the workload variables were manipulated.

As with the earlier baseline study (Stedmon, et al, 2001), the findings support the hypothesis that workload affects driver performance in a simulator and provides a valuable basis for assessing the impact of in-car ASR on driving behaviour.

Acknowledgements

This research was carried under the UK Government LINK Inland Surface Transport (IST) programme, funded jointly by the Economic and Social Research Council (ESRC) and the Department of the Environment, Transport and the Regions (DETR).

References

Aasman, J, Mulder, G., & Mulder, L.J.M., 1987, Operator Effort and the Measurement of Heart Rate Variability. *Human Factors* 29, 161-170.

Baber, C., and Noyes, J., 1996, Automatic Speech Recognition in Adverse Environments. *Human Factors* 38(1), 142-155.

Finch, M.I., & Stedmon, A.W., 1998, The Complexities of Stress in the Operational Military Environment. In, M. Hanson (ed). *Contemporary Ergonomics 1998. Proceedings of The Ergonomics Society Annual Conference, Cirencester. 1998.* Taylor & Francis Ltd. London.

Graham, R., Aldridge, L., Carter, C., & Lansdown, T., 1998, The Design of In-Car Speech Recognition Interfaces for Usability and User Acceptance. In, D. Harris (ed). *Engineering Psychology & Cognitive Ergonomics - Vol.4.* Ashgate.

Graham, R., & Carter, C., 2000, Comparison of Speech Input and Manual Control of In-Car Devices whilst on the Move. *Personal Technologies* 4, 155-164.

Stedmon, A.W., Carter, C., & Bayer, S.H., 2001, Baselining Behaviour: Driving Towards More Realistic Simulations. In, D. Harris (ed). *Engineering Psychology & Cognitive Ergonomics.* Ashgate

Stein, A.C., Parseghian, Z., & Wade Allen, R., 1987, A Simulator Study of the Safety Implications of Cellular Mobile Phone Use. *Proceedings of 31st Annual Meeting of the American Association for Automotive Medicine.* New Orleans, USA, 28-30 September, 1987.

Weiner, E.L., Curry, R.E., & Faustina, M.L., 1984, Vigilance and Task Load: In Search of the Inverted U. *Human Factors* 26, 215-222.

Wilson, G.F., & Eggemeier, F.T., 1991, Psychophysiological Assessment of Workload in Multi-Task Environments. In, D.L.Damos (ed). *Multiple Task Performance.* London: Taylor & Francis.

ACTIVE: A FUTURE VEHICLE DRIVER CONTROL INTERFACE

Sally Lomas and Alastair Gale

Institute of Behavioural Sciences, University of Derby,
Kingsway House, Kingsway, Derby, DE22 5GX, UK
s.m.lomas@derby.ac.uk

A proportion of road traffic accidents are caused by driver inattention, linked to the use of secondary in-vehicle controls. Consequently the overall design of controls must strive to ensure that their operation and monitoring does not place excessive demands on the driver, whose primary task is the safe operation of the vehicle. In recent years the number and variety of non-safety critical controls in vehicles has increased with the implementation of new technologies and driver information systems. However, the amount of space to locate in-vehicle controls has remained fairly constant often resulting in small controls within cluttered consoles. An alternative approach to the use of conventional in-vehicle controls is described, together with the human factors issues involved.

Introduction

Driving imposes a high level of demand on the manual and visual resources of the driver who is required to obtain, process and monitor information from a wide range of sources (Rockwell, 1988; Zwahlen *et al*, 1988; Wierwille, 1993). To perform the task both safely and efficiently the driver must be able to divide attention successfully between the external road scene and the information provided by the vehicle's secondary controls and information systems.

Driver distraction or inattention away from the primary driving task can have serious consequences both to the driver and occupants of the vehicle, as well as to other road users. For instance, the late detection of traffic conflicts (Rumar, 1990) and driver attention problems (Shinar, 1978) have been cited as causal factors in a large proportion of road traffic accidents. Wang *et al,* (1996) and Wierwille and Tijerina (1994) have detailed that a number of road traffic accidents are caused by driver inattention linked to the operation of secondary in-vehicle controls such as the In Car Entertainment (ICE) and climate control (HVAC) systems. Wang *et al.* analysed the Crashworthiness Data System (CDS) and found that 2.5% of all accidents in 1995 involved driver inattention caused by distraction from secondary controls. It is highly likely that this figure is very conservative as some 46% of crash causes in the CDS were classified as 'unknown'.

The number and complexity of in-vehicle controls and displays has greatly proliferated in recent years due to the implementation of in-vehicle driver aids and communication devices. This is evidenced by the once simple car radio which rapidly evolved to incorporate; station presets, RDS, cassette, CD, and multiple CDs, MD, graphic equaliser and various esoteric displays - all still located within the same original small dashboard footprint. Devices such as mobile phones are now owned by a large proportion of the driving population and their use whilst driving has raised concern over the demands placed on the driver's attentional resources (Cain and Burris, 1999; Lamble et al, 1999a; Brookhuis et al, 1991). More recent communication systems under development potentially enable a driver to surf the web and access email using a single control.

There is large variability in the location and form of a particular control, both between manufacturers and within different car models and specifications produced by the same manufacturer. This can lead to delays in operating a control for drivers of unfamiliar vehicles, such as a hire car, and controls which are not generally used on a regular basis, such as fog lights. Laux and Mayer (1991) found that driver's experience an 'expectancy lag' when a control is not located where the driver expects it to be. To locate a control, however, the driver must firstly be able to recognise it. This may sometimes prove difficult since the driver can be unaware of the specific type of control being located. For instance; a sliding lever, rotating knob or push button, could all have the same function.

Complex controls affect driver behaviour affecting the amounts of time spent observing the road scene and on in-vehicle devices (Wierwille et al, 1991; Rockwell, 1988). If the driver has difficulty obtaining the necessary information required of a particular control in an average glance then they need to make more, or longer, glances towards that control. This difficulty is confounded by the fact that the stored amount of visual road traffic information decreases as the number of looks required to acquire the information from a control or display inside the vehicle increases (Senders et al., 1967). It is thus highly desirable for controls to be operated using the minimal number of glances towards them. Consequently the increased complexity of controls has given rise to new operational methods, such as driver voice activation.

The ACTIVE System

As part of the Foresight Vehicle LINK Initiative, we are developing a new control system based on novel technology. The system, ACTIVE (Advanced camera Technology In Visual Ergonomics), uses an advanced graphical user interface and driver gesture detection system. The latter uses miniature video cameras to sense driver hand positions and predictive technology to anticipate driver intentions. Further descriptions of the technical aspects of the system may be found in Cairnie et al. (in press).

ACTIVE replaces all non-safety critical controls by a Graphical User Interface (GUI), which is operated by a driver pointing. Figure 1 illustrates the basic principle, although the implementation will be very different from this. The system offers several advantages to the driver namely:

1. The ability to transfer the GUI layout between different vehicles. Thus, once a driver has aquainted themselves with the layout of the new system they will no longer have to adapt to a different layout of controls in an unfamiliar vehicle such as a hire car.
2. The display can be located closer to the driver's line of sight, thus reducing the reach distance and the time spent looking away from the road scene.
3. The location and pointing action required to operate the interface will, possibly, allow drivers to keep both hands on the steering wheel whilst activating a control.

Additional benefits are likely to include reduced wiring costs and increased flexibility in the styling of the dashboard.

Figure 1. The ACTIVE concept

Human Factors Issues to be addressed

There are many human factors which need to be addressed in the development of a novel system such as ACTIVE. The primary concerns are safety and ease of use. Three key issues are considered here.

Position of the ACTIVE system

ACTIVE controls have the potential to be located in a variety of places in the car. The position of the system is a very important factor which will affect the driver's ability to locate, operate and monitor individual controls accurately. To operate the controls successfully they will need to be located within convenient reach and be easily visible by the driver.

Several researchers have studied the effect of control location on their operation. Controls which are frequently used, or are important, should be placed higher on the dashboard, near the driver's line of sight. This reduces the need for the driver to remove their gaze and attention from the road (Zwahlen & Kellmeyer, 1991; Diffrient, Tilley & Harman , 1993). Popp and Farber (1991) studied where to place route guidance systems in cars and recommended a centrally mounted display because the driver is only required to make small eye movements away from the traffic scene.

Stalk mounted controls within the driver's fingertip reach (determined as within 17.5cm of the steering wheel) were researched by Moussa-Hamouda and Mourant (1981). They found that subjects' reactions to stalk mounted controls were faster compared to panel mounted controls. However, when two or more stalks were located on the same side of the steering wheel, a large number of inadvertent control operations happened.

Lamble et al (1999b) studied the effects of an attentional demanding in-vehicle task on the driver's ability to detect the approach of a decelerating car ahead. An LED display was placed in nine different locations within the car. They found that detection thresholds, of a decelerating car in the forward scene, were high when controls were located at the centre of the steering wheel and in the rear view mirror position and thus advised against placing in-car devices in these positions. They recommended that displays were located on top of the dashboard adjacent to the steering wheel, within 15-20° of eccentricity. In our approach we are testing a system mounted high on the dashboard.

Size of the controls

It is important that the GUI controls are an appropriate size so that the driver will be able to locate and operate them with ease and accuracy. The increasing number of in-vehicle devices in recent years has resulted in an increase in the number of associated controls and displays. This results in controls which are multifunctional or are smaller in size and more closely spaced, in order to fit all of the required controls in the available space. To operate the correct control the driver may have to devote a greater amount of time, more glances and/or longer duration of glances, and an increase in concentration (Zwahlen, 1991).

Recommendations regarding control size tend to focus on pushbuttons and are often specific to the particular application. Deininger (1960) studied different keyboard arrangements, varying pushbutton size in order to achieve faster keying in times. It was found that increasing the dimensions of a pushbutton from 9.5mm to 17.4mm the keying time decreased from 6.35 to 5.83 seconds and the errors hitting adjacent incorrect buttons) decreased from 7.1% to 1.3%.

Chambers and Stockbridge (1970) state that button size should be related to the size of the limb actuating it, fingertip width in the case of a pushbutton. Similarly Zwahlen and Kellmeyer (1991) believe that the dimensions of the fingers are the limiting factors regarding pushbutton size. They studied pushbuttons in vehicles and point out that a lot of the literature related to specific applications and did not take into account pushbutton location, or motion and vibration, which is important to in-vehicle applications. With our approach no physical contact is made with the 'button' itself and so functional control size has to be determined from empirical studies.

Control Feedback

Control feedback is important to allow the driver to monitor the control's 'state' (e.g. 'on' or 'off') and to provide information to the driver regarding the success of his/her action. Certain controls, such as the heating fan, windscreen wipers and headlights give immediate feedback to the driver. Popp and Faerber (1993) term these types of controls 'transparent' whereas controls which are not able to give instant feedback are regarded as being 'non-transparent'. An example of a non-transparent control would be the air conditioning temperature, the driver having to wait to see if their control action was successfully implemented.

Typically, to complete the task of, for instance, switching the fan to 'on' the driver must visually locate the control, reach towards and operate it (e.g. turning the dial or sliding a lever). This control gives proprioceptive and tactile feedback to the driver's muscles and limbs, and auditory feedback of the fan turning on. The ACTIVE GUI can not give tactile feedback to the driver, as it is not operated by touch. It is therefore important that the feedback provided by the controls (e.g. auditory) provides the driver with the relevant information without distracting from the driving task. Popp and Faerber (1993) investigated four different feedback modes: a tone, visual feedback (with and without auditory announcement) and voice feedback. They recommended visual feedback information for non-transparent driver actions and concluded that voice output is a good solution for messages which are of a high degree of interest. However, they noted that if voice feedback is used for simple feedback messages the result is an increase in driver irritation and subjective load as well as a decrease in driving performance. We are investigating a combination of auditory and visual feedback.

Discussion and Conclusion

There are numerous human factors considerations to be addressed in the design and development of this vehicle control interface and the main ones are outlined here. The development of ACTIVE aims to minimise the amount of time spent activating in-vehicle secondary controls, and thus the amount of time drivers spend looking away from the forward road scene. Empirical studies are currently being performed to address how drivers will interact with such a system.

Acknowledgements

This work was funded by the Foresight Vehicle LINK Programme in collaboration with the Applied Computing Department, University of Dundee, Daewoo Motor Company Ltd, and Vision Dynamics Ltd.

References

Brookhuis, K.A., De Vries, D. & De Waard, D. 1991, The effects of mobile telephoning on driving performance, *Accident Analysis and Prevention*, **23**, 4, 309-316

Cain, A. Burris, M. 1999. Investigation of the use of mobile phones while driving, Center for Urban Transportation Research, *University of South Florida*, Tampa, Florida.

Cairnie, N., Ricketts, I. W., McKenna, S. J. and McAllister, G. 2001, A prototype finger-pointing interface for safer operation of secondary motor vehicle controls, (Awaiting publication), *Personal Ubiquitous Computing.*

Deininger, R. L. 1960, Human factors engineering studies of the design and use of pushbutton telephone sets, *The Bell Systems Technical Journal*, **39**, 996-1012

Diffrient, N., Tilley, A. R. and Harman, D. 1993, *Humanscale 4/5/6*. (MIT Press. Massachusetts.)

Lamble, D., Kauranen, T., Laakso, M. and Summala, H. 1999a, Cognitive load and detection thresholds in car following situations: safety implications for using mobile (cellular) telephones while driving, *Accident Analysis and Prevention*, **31**, 6, 617-623

Lamble, D., Laakso, M. and Summala, H. 1999b, Detection thresholds in car following situation and peripheral vision: implications for positioning of visually demanding in-car displays, *Ergonomics*, **42**, 6, 807-815

Laux, L. F. and Mayer, D. 1991, Locating vehicle controls and displays: effects of expectancy and age, *AAA Foundation For traffic Safety Report*

Moussa-Hamouda, E. and Mourant. R. R. 1981, Vehicle fingertip reach controls-Human factors recommendations, *Applied Ergonomics,* **12**, 2, 66-70

Popp, M. M. and Farber, B. 1991, Advanced display technologies, route guidance systems, and the position of displays in cars, in A. G. Gale, (ed.) et al, *Vision in Vehicles III*, (North-Holland, Oxford), 219-225

Popp, M. M. and Faerber, B. 1993, Feedback modality for nontransparent driver control actions: why not visually? in A. G. Gale, (ed.) et al, *Vision in Vehicles IV*, (North-Holland, Oxford), 263-270

Rockwell, T. H. 1988. Spare visual capacity in driving-revisited, in A. G. Gale, (ed.) et al, *Vision in vehicles II*, (North-Holland, Oxford), 317-324

Rumar, K. 1990, The basic driver error: late detection, *Ergonomics*, **33**, 10/11, 1281-1290

Senders, J. W., Kristofferson, A. B., Levison, C. W. and Ward, J. L. 1967, The attentional demand of automobile driving. *Highway Research Record* 195

Shinar, D. 1978, *The Psychology of the Road*, (Wiley, New-York)

Wang, J. S., Kniping, R. R. and Goodman, M. J. 1996. The role of driver inattention in crashes: new statistics from the 1995 crashworthiness data system, *Annual conference of the Association of the Advancement of Automotive Medicine*, Des Plaines, IL, USA, 377-392

Wierwille, W. W., Hulse, M. C., Fischer, T. J. and Dingus, T. A., 1991, Visual adaptation of the driver to high-demand driving situations while navigating with an in-car navigation system, in A. G. Gale (ed.) et al, *Vision in Vehicles III*, (North-Holland, Oxford), 79-87

Wierwille, W. W. 1993. Visual and manual demands of in-car controls and displays. in B. Peacock and W. Karwowski (ed.), *Automotive Ergonomics*. (Taylor and Francis, London), 299-320

Wierwille, W. W. and Tjerina, L. 1994. An analysis of driving accident narratives as a means of determining problems caused by in-vehicle visual allocation and visual workload, in A. G. Gale et al (ed.), Vision in Vehicles V, (North-Holland, Oxford), 79-86

Zwahlen, H. T., Adams, C. C. and DeBald, D. P. 1988, Safety aspects of CRT touch panel controls on automobiles in A. G. Gale, (ed.) et al, *Vision in vehicles II*, (North-Holland, Oxford), 334-344

Zwahlen, H. T. and Kellmeyer, D. 1991, Position accuracy when pushing pushbuttons in a car as a function of car speed and location: Implications for design, *International Conference on Strategic Highway Research Program and Traffic Safety on Two Continents*, Gothenburg, Sweden, September 18-20. 212-248

AN ANALYSIS OF MOVEMENTS OF THE VISUAL LINE DURING DRIVING WITH NAVIGATION DEVICES

Hidehiro Yoritaka[1] and Kageyu Noro[2]

Media Network Center, Waseda University[1]
1-104 Totsuka-machi, Shinjuku-ku, Tokyo, 169-8050, JAPAN
yori@mn.waseda.ac.jp

School of Human Sciences, Waseda University[2]
2-579-15 Mikajima, Tokorozawa, Saitama, 359-1192, JAPAN

The present paper describes movements of the visual line in the act of driving with navigation devices. Subjects drove the car that was equipped with a navigation system. The navigation system was composed of Crystal display, Head Up Display and speaker. These three different devices were combined to present the information. The movements of visual line are measured in non-invasive method. They were calculated from the pictures from two CCD cameras set up on the different locations. Findings from this study are as follows: The location of information devices has strong influence on the movements of the visual line. The quantity of information is irrelevant to them when entering the crossing. It is suggested that the desirable location of information device is within the visual field.

Introduction

Recent traffic systems become more and more intelligent to provide certain traffic information, under the aim at presenting a route to a driver's destination, and sorting out a traffic jam. The more intelligent the traffic systems become, the more conformation need on displays both visually and auditory in the course of driving. Generally, visual information during driving occupy above 80% of all information collections. However, the research on human factors related the information receptivity is insufficient.

Practically, cars equipped a navigation system with Liquid Crystal Display (LCD) are on the increase. This means drivers can obtain useful information. On the other hand, they are also in the misgiving condition of excessive visual information in their driving by these instruments. To avoid this condition, a device that displays primary visual information in front of a driver has been developed. The device names Head Up Display (HUD), which was born in the field of aeronautics and applied it to cars. It is typical instruments for safety to decrease unnecessary eye movement to check information through devices.

The present study examined the location of devices and the quantity of information influence on eye movements in the act of driving.

Methods

Subjects
The subjects who participate in the study must fulfill the following criteria.   The driving experiences of subjects' were for more than three years, and usually drive a car, but do not usually use navigation systems.   Their sights were more than 0.7, and they do not wear glasses.   Four university students (3 male and 1 female) served as subjects.   Their mean age was 24.0 years old (*SD*=2.83, range 22 to 28).   They were not previously known the route, which they drove.

Experimental task
The distance of a driving route reached about 5 km.   The distance from a starting point to a particular crossing for measurement was about 3 km.   The sampling zone was from the 3 sec. before the first guidance where about 500 m before the near side of the particular crossing to the 3 sec. after the end of the guidance.   The guidance from LCD and HUD was always presented and guidance from the speaker was presented at 500m (36 sec. to crossings), 200m (14.4 sec.), and 100m (7.2 sec.) on near side of crossings.

Procedures
Subjects drove a standard-sized car that was equipped a navigation system (Sumitomo Electric Industries; SNV-AD10), LCD monitor (Sharp corporation; 6E-DK3), and HUD (Sumitomo Electric Industries; SHD-50-B).

The location of LCD was at 43° left and 33° down from the driver.   HUD was set up at 6° down in front of the driver.

A destination and route was set up and calculated by a navigation system beforehand. So a navigation system guided in accordance with this result.   LCD, HUD and speaker, these three types of devices presented the information provided by a navigation system (Figure 1).   Table 1 shows that navigation devices and the provided information by them. These devices are permuted including at least two kinds, the combinations of presenting information were 4 condition.

Figure 1. Examples of provided guidance; HUD (left) and LCD (Right)

Table 1. Navigation devices and the provided information

Navigation devices	Provided information
LCD	Map with route, the present location, enlarged view of the crossing and surrounding areas of the present location, compass direction, scale, distance of destination, etc.
HUD	Course, distance of destination, shape of the crossing and the traveling direction, etc.
Speaker	Beep and voice guidance

Measurement of eye movements
Stimulations may obstruct driving even if they are very slight, subjects should not equip with any instruments for measurement. Therefore non-invasive measurement of eye movements was used in this study.

Two CCD cameras were set up on the different locations and recorded the upper body of the driver. These two different pictures were synchronized on the time base, upper body movements were measured by the three dimensional measurement of two specific points (Figure 2, Yoritaka and Noro, 1997a). Eye movements were measured by the measurement method (Sotoyama *et al*, 1995) with which the distance of lower to upper eyelid (horizontal) and the ocular surfaces area (vertical). Subsequently, eye movements were corrected using upper body movements and converted to the visual line. Errors of this method are about 5 to 10 degrees (Yoritaka and Noro, 1997b).

The measurement frequency and the duration of fixation pauses of eye movements (Unit: 1/5 sec.) were also recorded.

Results and Discussion

Distribution of visual line
Figure 2 shows that the distribution of visual line at 3 sec. before and after 200m before the crossing under the condition of "HUD and speaker" and "LCD and speaker". Generally, under the condition using LCD, movements of the visual line tend to be scattered. On the other hand, under the condition using HUD, they tend to be gathered.

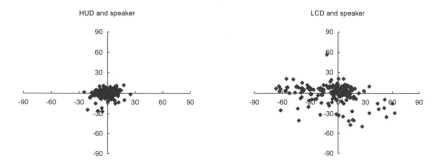

Figure 2. Distribution of the visual line at 3 sec. before and after 200m

Movements of visual line on devices

Frequency of the visual line movement on each device is judged as follows because of measurement errors: On LCD, the range of visual line is judged horizontal –43 ± 5°, vertical –33 ± 5°.   The range of visual line on HUD is judged horizontal 0 ± 5°, vertical –6 ± 5°.   The movement of visual line on LCD was not observed.   Accordingly, it is suggested that driver watched the LCD at more than 500m before the crossing, or after the end of guidance.   The frequency of visual line on HUD increased at 200m on this side of the crossing.   From the result of the frequency of visual line on devices, it is assumed that when approaching the crossing, drivers prior to the device placed narrower visual angle.   It is suggested that the desirable location of information device is within the visual field.

Figure 3 shows that the relation between movements of the visual line and time (3 sec before and after the guidance).   These results revealed that movements of the visual line under the condition using HUD reduce most at 10 sec. point (about 200m) on near side from the crossing, however, when using LCD, they were not changed.   The visual line, especially vertical direction, converged as approaching the crossing.   This tendency appears strongly under the condition using HUD that movements of the visual line are short.   The information acceptance depends considerably on drivers' size of chunk (competence of information acceptance).

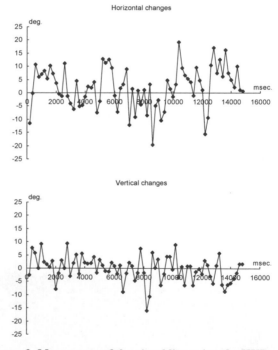

Figure 3. Movements of the visual line using the HUD

The average duration of fixation pauses of eye movements to HUD are shown in Table 2. From a result of the duration of fixation pauses of eye movements and the number of times on devices showed the distinct difference between them. The judgment of fixation pauses was defined as the same as the close observation previously mentioned. The distance from the crossing does not strong influence on fixation pauses.

Table 2. Average duration of fixation pauses of eye movements to HUD

	HUD and speaker (msec.)	LCD and speaker (msec.)	LCD, HUD and speaker (msec.)
Maximum duration	400.0	400.0	1200.0
500m-200m	254.5	200.0	400.0
200m-100m	266.7	200.0	375.0
100m-end	266.7	333.3	400.0

The average duration of fixation pauses of eye movements to HUD are shown in Table 2. Frequency of eye movements tends to reduce as approaching the crossing. Its tendency was remarkable under the condition of "HUD and speaker".

Table 3. Frequency of eye movements to HUD

	HUD and speaker	LCD and speaker	LCD, HUD and speaker
500m-200m	13	1	10
200m-100m	3	3	8
100m-end	3	3	5

One of the distinct differences between LCD and HUD was the quantity of the information. LCD provides much information such as map with route, the present location, enlarged view of the crossing and surrounding areas of the present location, compass direction, scale, and so on. HUD, however, simply provides the course, distance of destination, shape of the crossing and the traveling direction. It is exceedingly long duration of fixation pauses and many frequency of eye movements under the condition of "LCD, HUD and speaker" in spite of plenty of information. It is assumed that drivers must take more time to extract required information. For example, quantity of changes of information on HUD is exceedingly less than that of LCD. This suggests that drivers do not seek the quantity of information when entering the crossing.

The devices and feature of visual line

The feature of eye movements are compared "LCD and HUD", "LCD and Speaker", with "HUD and speaker". There is no special feature of utilizing auditory guidance. Visual line the use of LCD scattered vertically 30 ~ 50 deg., and horizontally 20 deg. Visual line by HUD converged about 30 deg. radiuses.

The most frequent visual line dispersion is observed on "LCD and speaker"; no glance at LCD was observed from the beginning to the end of the guidance. Inversely, the combination made the visual line most convergent was "Speaker and HUD" condition. From the result, the combination "LCD and speaker" that typical navigation systems employ made the visual line more scattered. It suggests that location of information display device is more predominant than the plenty of information displayed by LCD.

The result revealed that movements of visual line on devices began at about

100msec. after the auditory information under the condition of "Speaker and HUD". Moreover, under the condition of " LCD and speaker", movements of visual line on LCD were not observed. The role of LCD as information display device is suggested to confirm the present location in the map. The occasion for get information concerning the crossing depended on auditory guidance. The condition with no auditory guidance, duration of fixation pauses of eye movements and frequency of visual line movement were on the increase as approaching the crossing. Thus, auditory guidance is suggested to use as a cue in renewal of information.

Conclusions

The present paper examined the eye movements in the act of driving following guidance by a navigation system, through measuring movements of visual line. Findings from this study were as follows.
• Movements of the visual line tended to be gathered using HUD, on the other hand, LCD made it scattered when the information is presented.
• When the guidance is presented about 200m on near side of crossing, movements of the visual line reduced mostly. The HUD contributes decreasing unnecessary movements of the visual line for driving.
• As approaching a crossing, frequency of movements of the visual line on each device decreased.
• An auditory guidance is supposed to cue in renewal of information.
• The period of information acceptance delayed in the case of no auditory information.
It is considered from the above that the location of information devices has strong influence on the movements of the visual line. In addition, movements of visual line are irrelevant to quantity of information when entering the crossing. It is suggested that the desirable location of information devices is within the visual field, and the information provided by devices need to be extracted simply.

Acknowledgments

This study has been carried out in a joint study with Sumitomo Electric Industries.

References

Sotoyama, M., Villanueva, M.B.G., Jonai, H., and Saito, S. 1995, Ocular Surface Area as Informative Index of Visual Ergonomics, *Industrial Health*, **33**, 43-5
Yoritaka, H. and Noro, K. 1997a, A method of measurement without equipment for the visual angle 1 - Measurement of body motion -, *The Japanese Journal of Ergonomics*, **33** (Supplement), 328-329
Yoritaka, H. and Noro, K. 1997b, A method of measurement without equipment for the visual angle 2 - Measurement of visual angle -, *The Japanese Journal of Ergonomics*, **33** (Supplement), 330-331

Hierarchical Task Analysis of Driving: A New Research Tool

Guy H. Walker, Neville A. Stanton, & Mark S. Young

Brunel University, Department of Design,
Egham, Surrey, TW20 0JZ, UK
guy.walker@brunel.ac.uk   neville.stanton@brunel.ac.uk   mark.young@brunel.ac.uk

Much is known about automotive engineering, vehicle dynamics, cognitive psychology, and even what drivers are actually doing as they drive. Comparatively little is known about the specific nature and structure of the driving task itself. This paper presents an overview of a recently constructed Hierarchical Task Analysis of Driving (HTAoD), and discusses its utility in terms of inferring the cognitive information processing that underpins driving task enactment.

Introduction

Adaptive cruise control, drive by wire technology, collision avoidance systems, and sophisticated driver monitoring are indicative of the increasing power of proposed in-vehicle technology. These technologies, amongst others, not only increasingly automate many of the functions previously performed by the driver, but they are realistically expected to enter vehicles in the coming 15 years (Walker, Stanton & Young, 2001). Given the dramatic effect that this is to have on the specific nature of the driving task, it is interesting to note that the only attempt at a systematic and exhaustive task analysis of driving quoted in contemporary literature (for example, Michon, 1993) remains the work of McKnight and Adams (1970). McKnight and Adam's work was prepared for the U.S. Department of Transportation in order to "identify a set of driver performances that might be employed as terminal objectives in the development of driver education courses" (McKnight & Adams, 1970, p. vii). Whilst providing some extremely useful insights into the range and quantity of tasks enacted by drivers, it's stated purpose severely limits its research applicability. A sizeable corpus of knowledge exists about what drivers are actually doing whilst they drive (for example Tijerina, et al, 1998; Lechner & Perrin, 1993) but thus far very little is actually known about the specific nature and structure of the driving task itself. Therefore to date, driving research lacks an important and valuable research tool.

Hierarchical Task Analysis of Driving (HTAoD)

To address this shortfall, a systematic and exhaustive task analysis of driving has recently been designed and constructed in order to meet the needs of present and future driving research. The task analysis follows the rubric of Hierarchical Task Analysis (HTA) as pioneered by Annett and Duncan (1967). It begins by defining the driving activity (drive a car), setting the conditions in which this activity will take place (a modern, average sized, front wheel drive vehicle, equipped with a fuel injected engine, and airbag, on a British public road), and the performance criteria to be met (drive in compliance with the Highway Code, and the Police Driver's System of Car Control). This is the highest level task goal, and is completely specified by 7 first tier tasks. These are comprised of 5 definable behaviour levels, which altogether are completely specified by 1600 individual operations and tasks. They are bound together by 400 plans of a logical form. These plans define how task enactment should proceed, often contingent upon specific conditions being met or events being present. A summary of the HTAoD, and its defining features and structures is shown in figure 1 below.

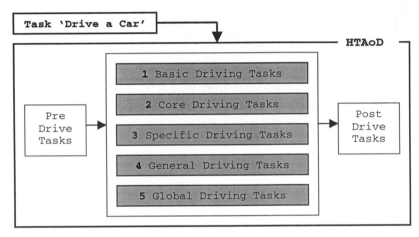

Figure 1. Overall structure of HTAoD

It is interesting to note the wide range of assumptions currently made about the driving task, especially given the fact that none of these assumptions appear to be based on any kind of systematic task analysis of driving. In particular, a recurring contention made in the literature is one originally made by McRuer, Allen, Weir, and Klein (1977), whereby driving is stated as being comprised of a 3 level hierarchy; navigation, manoeuvring, and control. On the basis of the current HTAoD it can be argued that driving is in fact made up of 5 behaviour levels. These being Basic Driving Tasks, Core Driving Tasks, Specific Driving Tasks, General Driving Tasks, and Global Driving Tasks.

The driving task falls readily into these 5 behaviour levels, each one representing a point along a continuum defined by the different levels of complexity embodied by the context, and the nature of the task. In general terms, the behaviour levels move from very basic low level physical driver tasks through to high level cognitive tasks; from tasks and operations, through rules and procedures, finishing up with overarching goal oriented driving strategies. Basic driving tasks are concerned with fundamental physical vehicle control tasks performed by the driver. Core driving tasks are still basic physical operations applied within a basic contextual setting of controlling vehicle speed and trajectory. Specific Driving Tasks cover strategies and operations for coping with specific road environment situations, such as junctions or crossings. General driving tasks represent increasingly goal orientated, overarching strategies for adapting to different road types, or general procedures for dealing with other traffic.

Global Driving Tasks are concerned with goal orientated operations related to high level driving strategy, such as rule compliance or navigation. The context, in a dynamic manner, demands the enactment of all 5 behaviour levels from the driver. Figure 2 summarises this classification.

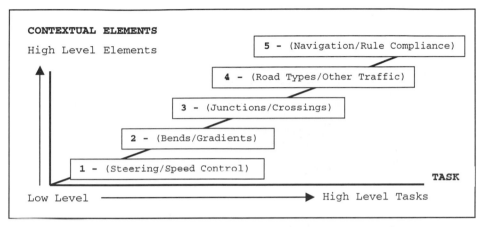

**Figure 2. HTAoD behaviour levels and their relation to task
and context**

These 5 behaviour levels hint at an inherent hierarchy with regard to the cognitive information processing employed by drivers. However, the reader should be warned that this Hierarchical Task Analysis of Driving (HTAoD) is neither a theory of driving, or necessarily a description of how the driving task is actually performed by all drivers in all situations. It represents a description of normative driving, albeit a relatively stylised 'ideal' case of normative driving, and is therefore a subtractive model. Arguably it is easier to subtract from an ideal model of driving as opposed to adding to an imperfect model. Furthermore, a descriptive ideal model of driving is likely to have increased practical applied value.

Fundamentally, the HTAoD is an event driven model. It is interesting to speculate that in theory, a computer or some form of expert system could indeed be programmed with this HTAoD, along with every possible combination of contingent driving event, such that it would then be able to drive as well as a human driver. Although potentially of some use in a practical domain, this brute force, rule based 'IF-THEN' approach is unlikely to represent the way that human drivers process information as they drive. The HTAoD is offered as a tool for research, and thus the intention is to employ it as a means of uncovering and understanding in more detail the nature and structure of cognitive processing enacted by drivers.

HTAoD: Modelling Cognitive Processes in Driving

A comprehensive HTA is a necessary prerequisite for a host of ergonomics methodologies that were previously not able to be applied to the driving task. Therefore at the outset, specifying the driving task within a structured HTA format makes it amenable to a wide variety of Cognitive Task Analysis (CTA) methods. This factor alone offers significant applied value. Of more interest within the present article is the notion that the HTAoD contributes greatly towards our understanding of driving and drivers, and offers a new and informative perspective on the modelling of underlying cognitive processes in driving.

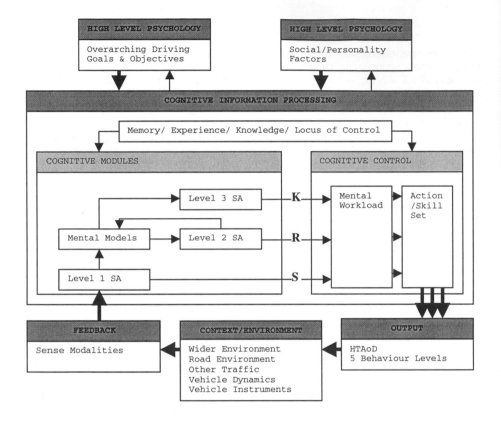

Figure 3. Feedback model of driving

The feedback model of driving presented in figure 3 above, is a closed loop circular arrangement, representing a schematic for the flow of information around the various functional elements of the driving system. The model captures the constant flow of information presented to the driver, and the constant adjustments and monitoring performed by the driver congruent with overall driving goals.

The enactment of the various tasks and operations specified in the HTAoD arise as a direct consequence of driver enacted low level cognitive processing, in conjunction with higher level psychological factors. Closed loop conceptions of driving are not new in themselves (Milliken & Dell'Amico, 1968; Verwey, 1993) however, this model is enhanced through its rigorous grounding in all the functional aspects of driving. These functional aspects range from psychological cognitive mechanisms and specific driving tasks, through to automotive engineering and vehicle dynamics. Furthermore, the model is contextually based, and offers a platform for welding the elements of cognitive control with specific cognitive mechanisms or modules.

It is possible to reconcile the HTAoD with existing knowledge about cars, drivers, and driving, whilst at the same time explaining in more detail the workings and theoretical grounding of the feedback model. Higher level psychological processes define the overall driving goals, such as why the journey is being undertaken, the underlying driving style of the driver, and the various other social and personality factors pertinent to the driving scenario. The lower level cognitive processing defines how the constant flow of information presented to the driver from the context is superimposed upon high level driving goals, leading to the selection of specific combinations of driving tasks. In turn, the outcomes of driving task enactment occur within a context and cause changes to external factors, such as the behaviour and status of the vehicle in its environment. The behaviour of the vehicle can be captured from an engineering psychology point of view in terms of system and control dynamics (Wickens, 1992), and within the domain of automotive engineering, vehicle dynamics is also well understood.

The consequences of these driving tasks, and the resulting behaviour of the environment and control/system dynamics of the vehicle are fedback, through various sensory channels, to the driver. This feedback represents context dependent 'knowledge of results' (Annette & Kay, 1957), and is used as input to the information processing system. To a great extent the nature of feedback made available to the driver (not just through the vehicle), but from the road and wider environment is also relatively well understood, or is at least measurable.

Significantly less well understood is what the driver does with this constant flow of information, and exactly how they put it to use within the driving task. Exploring the model in more detail draws upon Rasmussens taxonomy of Skill, Rule, and Knowledge (SRK) based processing (Rasmussen, Pejtersen, & Goodstein, 1994). These dimensions represent three levels of cognitive control. The SRK taxonomy of cognitive control is highly contingent upon expertise, and upon the nature of the context (Rasmussen et al, 1994). The feedback model represents an initial attempt, in light of the HTAoD, at bringing together functional cognitive mechanisms with the notion of cognitive control.

In terms of cognitive mechanisms, Stanton and Young (2000) relate the SRK taxonomy to the three levels of Situational Awareness (SA), level 1 (perception of elements in the environment), level 2 (comprehension of those elements), and level 3 (projecting the understanding of those elements onto future states). Level 1, 2, and 3 SA are linked to S, R, and K processing respectively. A combination of memory, experience, and locus of control, in conjunction with attention and mental workload, defines what mode of cognitive control is entered for the selection and enactment of driving tasks. The three levels of cognitive control interact with the 5 behaviour levels defined in the HTAoD. Further in-depth simulator based studies and research will confirm or refute this model more completely. However, at the current level of analysis it is not inconceivable that equipped with knowledge of the types of processing occurring in the drivers mind, coupled with an understanding of the behavioural groupings of driving tasks, will ultimately enable such processing to be fully supported via appropriate feedback to the driver. Powerful new in-vehicle technology offers the opportunity to deliver this appropriate feedback.

Conclusions

The domain of human factors is lagging behind the likely implementation of a wide range of powerful new vehicle technologies. Without proper human factors evaluation these new in-vehicle technologies represent a potential threat in terms of driver-vehicle interaction and a potential threat in terms of driver safety, efficiency, and enjoyment. But equipped with the appropriate knowledge, tools, and design solutions, human factors offers the potential to make this same technology an opportunity. An opportunity to design out problems, and moreover, to enhance and optimise the car-driver interface. The HTAoD presented in this article represents a new tool that speaks towards this goal. It achieves this through enabling a complete, systematic and highly structured description of the driving task. Representing the theme of this paper, the HTAoD greatly contributes to our knowledge of the driving system as encapsulated in the presented feedback model of driving. If the functional characteristics and interrelationships of these cognitive processes can be captured and understood, then new technology is eminently capable of fully supporting them through appropriate feedback. Although not a theory of driving in itself, the HTAoD is an invaluable tool for inferring the cognitive processes underpinning driving. A tool that the domain of driving research has until now been lacking.

References

Annett, J. and Kay, H. 1957, Knowledge of results and 'skilled performance', *Occupational Psychology*, **31**, (2), 69-79

Annett, J. and Duncan, K.D. 1967, Task analysis and training design, *Occupational Psychology*, **41**, 211-221

Lechner, D. and Perrin, C. 1993, The actual use of the dynamic performances of vehicles, *Journal of Automobile Engineering*, **207**, 249-256.

McKnight, J.A. and Adams, B.B. 1970, *Driver Education Task Analysis: Volume I - Task Descriptions*, (NHTSA, Virginia)

McRuer, D.T., Allen, R.W., Weir, D.H., and Klein, R.H. 1977, New results in driver steering control models, *Human Factors*, **19**, (4) 381-397

Michon, J.A. 1993, *Generic intelligent driver support: A comprehensive report on GID*, (Taylor and Francis, London)

Milliken, W.F.Jr., and Dell'Amico, F. 1968, Standards for safe handling characteristics of automobiles. In P. G. Ware (Chair), *Vehicle and road design for safety*, **183**, (3A). Symposium conducted by the Institution of Mechanical Engineers, London.

Rasmussen, J., Pejtersen, A.M., and Goodstein, L.P. 1994, *Cognitive systems engineering*. (Wiley, New York)

Stanton, N.A. and Young, M.S. 1999, Psychological factors in driving automation. Unpublished manuscript

Tijerina, L., Gleckler, M., Stoltzfus, D., Johnstone, S., Goodman, M.J., and Wierwille, W.W. 1998, *A Preliminary Assessment of Algorithms for Drowsy and Inattentive Driver Detection on the Road*, (NHTSA, Virginia)

Verwey, W. 1993, How can we prevent overload of the driver? In A.M. Parkes and S. Franzen (Eds.) *Driving future vehicles*, (Taylor and Francis, London), 235-244

Wickens, C.D. 1992, *Engineering psychology and human performance*, (2nd ed), (Harper Collins, New York)

Walker, G.H., Stanton, N.A., and Young, M.S. 2000, Where is computing driving cars? A technology trajectory of vehicle design, *International Journal of Human Computer Interaction*, In Press

SUPPORTING THE NAVIGATION TASK: CHARACTERISTICS OF 'GOOD' LANDMARKS

Gary Burnett[1], Darren Smith[2], Andrew May[3]

1. *School of Computer Science and Information Technology, University of Nottingham, Jubilee Campus, Wollaton Road, Nottingham, NG8 1BB*
Gary.Burnett@cs.nott.ac.uk
2. *Vosper Thornycroft (UK) Ltd, Victoria Road, Woolston, Southampton, SO19 9RR*
DarrenSmith@vtis.com
3. *HUSAT Research Institute, Loughborough University, The Elms, Elms Grove, Loughborough, LE11 1RG*
a.j.may@lboro.ac.uk

Landmarks (e.g. traffic lights, churches, monuments) have great potential to support travellers in navigation tasks. Concern has been expressed regarding the current generation of vehicle navigation systems and their reliance on distances, rather than landmark information, particularly within their voice instructions. This paper describes a direction-giving study which aimed to establish which landmarks are valued for navigation and their salient characteristics. Participants (n=32) provided written directions for three urban routes, based on either a single experience of the routes (via a video), or their long-term experience of the area. The study revealed the significance of 'road furniture' landmarks, such as traffic lights, pedestrian crossings and petrol stations. Prospective attributes of these landmarks include permanence, visibility, location in relation to a decision point, uniqueness and brevity.

Introduction

Finding one's way (whether in real or electronic worlds) is a complex human activity. Difficulties with this everyday task often cause stress and frustration for the individual and inefficiency within travelling networks. For road vehicles, the navigational uncertainty that arises from current methods (e.g. handwritten notes, paper maps) can affect drivers' behaviour in a number of undesirable ways. For instance, at the manoeuvering/tactical level of the driving task, there may be an inappropriate use of indicators, sudden lane changes or late/sharp braking. In contrast, at the strategic level of driving, drivers may make poor route choices or may avoid unfamiliar environments. Older drivers in particular have been shown to consciously adapt their driving habits as a result of navigation demands (Burns, 1997).

Advancements in computing, communications and map technologies have led to the development of GPS-based vehicle navigation systems (or 'SAT NAV'). This technology has been available to drivers in Europe since the mid 1990s, and the prediction is that it will be commonplace as the market matures and costs fall (Rowell, 1999). Many of the current systems provide instructions (using symbols and/or text and often voice messages) that indicate the location and direction of each turning. For systems that utilise a large LCD, map-based information can also be accessed.

The ergonomics/human factors issues for this technology have constituted an area of considerable activity for researchers in the last 15 years or so (see Srinivasan, 1999 for a recent review). This focus is not surprising given that a) this technology is arguably the most sophisticated with which drivers have had to interact within vehicles, and, b) much of the system functionality is of potential use when the vehicle is in motion. One clear finding of research has been the importance of voice messages for safe and effective interactions (Kishi and Sugiura, 1993; Burnett and Parkes, 1993). An outstanding issue of concern, however, is the *content* of voice messages. Current systems tend to emphasise distance-to-turn information, using either exact figures (e.g. "right turn in 200 metres") or informal, time-based terms (e.g. "right turn coming up"). Research has indicated that a much wider range of information types are used in navigation tasks, in particular landmarks, e.g. traffic lights, petrol stations, churches (Burnett, 1998). Furthermore, studies have revealed how the inclusion of landmarks within vehicle navigation systems can increase the effectiveness, efficiency and satisfaction of the system for users (see Burnett, 2000 for a review).

However, if such features of the environment are to be included in future systems, it is imperative that there is a detailed and practical understanding of what constitutes a 'good' landmark. The inappropriate use of landmarks (e.g. those which are poorly visible or cannot be readily identified) may lead to driver confusion and reduce, rather than increase, the usability of a vehicle navigation system. On this issue, there are recommendations in the human factors literature regarding good types or classes of landmark, and a consistent finding across several studies is the potential for traffic lights, petrol stations and bridges as landmarks for navigation (see Burnett, 1998 for a review). Unfortunately, such guidance does not account for the fact that landmark quality varies considerably from one situation to another (even for the best overall landmarks). For example, it can be difficult to identify a correct set of traffic lights associated with a turning in contexts when there are several sets close to one another. What is needed ideally is a generic understanding of the characteristics that *define* a good landmark, so that the best landmarks can be chosen on a case-by-case basis. It is believed that such knowledge will be beneficial for the analysis of landmarks in other navigational contexts (for pedestrians, within information space, virtual worlds, etc.).

Method

A study was devised in which 32 participants (16 male, 16 female; age range 22 to 60; all experienced drivers) were asked to write down the information they felt an unfamiliar traveller would need to drive and navigate three linked routes successfully. All routes were within an urban driving environment, covered approximately ten miles in total distance, involved 19 distinct decision points (i.e. junctions where navigational uncertainty would be expected), and took approximately 24 minutes to drive.

In a factorial design experiment, two conditions were adopted, whereby participants provided directions based on either A - a single experience of the routes (via a video) (n=16), or B - long-term experience of that driving environment (n=16). For A, participants had no prior experience of the routes, whereas for B, participants had lived and/or worked in the test area for at least five years.

Each of these conditions is associated with its own merits and problems (see Table 1). By examining which landmarks are commonly referred to by *both* groups (i.e. the degree of overlap), it is argued that one has the best set of potential navigational landmarks.

Table 1. Pros and cons of different information sources for direction giving studies

Condition	Advantages	Disadvantages
Single experience (via video)	Based on direct observation – the view of an unfamiliar traveller	A 'snap-shot' experience of routes – therefore, limited by specific views available
Long-term experience (via memory of locals)	Based on repeated exposure to landmarks – gives overall view of merits	Individual's memory for landmarks prone to subjective biases

Initially, participants were informed of the general nature of the study (concerning the directions people use for navigation). To avoid bias, no mention of landmarks was made. For participants in the single experience condition, the routes were filmed in an experimental car using a micro 'lipstick' camera mounted in front of the rear view mirror and forward facing. The focal length of the lens was 7.5mm representing a viewing angle of 45 degrees. These participants were shown this video and were permitted to rewind and review any part of the video until they felt happy with their route descriptions. Participants in the long-term experience condition were initially provided with a 'minimal' schematic map of the routes. These maps were carefully designed to provide sufficient information to enable participants to 'realise' the intended route, without affecting the directions they would give.

Once the participants had written their directions for all three routes, they completed questionnaires concerning their overall strategies and perceived abilities for navigation. The experimenter used this time to highlight all landmarks contained within their directions. Participants were then interviewed in a semi-structured style regarding their reasons for choosing landmarks. Specifically, for each landmark within their directions, participants were asked why they had referred to such an object/feature of the environment, what made it stand out from other potential information, etc.?

Analysis and results

In the first instance, the written directions provided by participants were examined and a frequency count was conducted for each specific landmark (e.g. the traffic lights at the junction of Lee Way and Park Drive). Figure 1 plots the number of participants who referred to a specific landmark in their directions for both conditions. The most valued landmarks across the two groups are those with data points located towards the top right of the graph. The figure demonstrates the relative importance/value attached by individuals to designed objects within the road infrastructure, or 'road furniture' landmarks (e.g. traffic lights and petrol stations). Indeed, over a third of all landmark references made were for traffic lights.

Wilson and Chisel (1996) assessed, for some 15000 vehicles and 800 pedestrian groups, driver assistance or obstruction rates at three pedestrian crossings of Laurentian University, Sudbury, Canada during autumn and winter sessions. At Laurentian, most of the 4000 students and 700 faculty and staff drive to university and use one or more pedestrian crossings on a 3 km private road network on campus. One third of drivers interacting with pedestrians were driving 50% or more above the speed limit and they obstructed 20 times as often as they assisted pedestrians. Those driving at speeds ranging from 5 km/h to 50% above the speed limit obstructed 7 times as often as they assisted. Finally those driving at or below the speed limit obstructed twice as often as they assisted. Surprisingly, public service vehicles (100%), university security/maintenance vehicles (80%), and general drivers (90%) obstructed at similar rates which, in turn, were much than those found in earlier studies in England and in Canada.

A locale based driving improvement intervention was subsequently introduced on campus. The nature of the intervention was the installation of five 25 km/h speed bumps, a three-way flashing red stop signal and the painting of a cross-hatched crossing at the education building crossing. One set of three speed bumps consisted of a central bump beside the main cross-hatched pedestrian crossing from the large on-campus arts building parking lot and two flanking bumps 60 and 80 metres away respectively beside two other pedestrian cross-hatched crossings. A second pair of speed bumps 70 metres apart defined either end of the day care facility on the road to the physical education centre. The three way flashing stop signal was installed a few hundred metres from the triple speed bumps at the major egress from the administration and cafeteria buildings to the student residences. Two questions were asked. Could any difference in obstruction rates or driving speeds be detected at crosswalks near and far from the speed bump-traffic light interventions? Was there any evidence of change in obstruction rates after the interventions?

Method

Three pedestrian crossing areas were chosen on the 3 kms of private roads of Laurentian university in Sudbury, Ontario. Two crossings were at the same locations as used in the Wilson and Chisel study: Willet, a two metre wide crosshatched crosswalk, located on a wide curve in the road, connecting the Willet-Green mining/research centre with the physical science buildings and Education, a cross-hatched crossing site (unmarked in the original Wilson and Chisel study) at the major intersection dividing physical education buildings from other academic faculties. The third site, the central marked pedestrian crossing at the three speed bumps, was chosen because it was the major intervention. The Wilson and McArdle method was used for recording, in detail, events surrounding interactions between pedestrians and vehicles at the crossings. In two separate 30-minute sessions in which a second observer was used to record events, observer reliability was found to be 100%.

Results

About 575 vehicles and 155 groups of pedestrians traversed all three crossing sites during the 8 hours of observations. Of these events, 111 (71%) of pedestrians did not interact with any vehicle. Of the 44 pedestrian groups interacting with a vehicle 6 (4%) were assisted by the vehicle while 38 (25%) were obstructed (binomial probability < .001). These interactions involved forty general vehicles of which the assistance/obstruction ration was 6/34 and four public service vehicles of which all obstructed ($\chi^2 = 0.13$, df = 1, n. s.). The relative frequency of assistance to obstruction in the earlier Wilson and Chisel study was 25/167. No significant differences were noted between the two studies ($\chi^2 =$ n. s.).

Consider location. Sixteen interaction crossings occurred at the Willet site, 24 at the central site, and four at the education site. Assistance was observed six times at the central site and nowhere else ($\chi^2 = 27.3$, df = 2, p < 0.001).

Consider vehicle speeds. The speed limit throughout the campus was 40 km/h or less. Of the 575 vehicles, 301 (52%) were measured as travelling between 60 and 89+ km/h, of which 51 vehicles were near or above the upper value of this range, 144 (25%) between 45- 59 km/h, and 130 (23%) less than 45 km/h. Of the latter group, only two travelled less than 30 km/h and three between 30 and 35 km/h in the immediate vicinity of the crossings. The corresponding frequencies for the three speed categories in the earlier Wilson and Chisel study were 63, 90, and 36 vehicles, respectively. A chi square test of the results of the two studies was significant ($\chi^2 = 22.5$, df = 2, p < 0.001). Examination of the data showed a relatively greater proportion of high speed drivers throughout the campus after introduction of safety interventions.

At the three-speed bump area, the frequencies of the three driving speeds 60+, 45-60, and less than 45 km/h were 61, 20, and 25, respectively, while at the two areas without speed bumps the speed frequencies were 238, 124, and 105, respectively. No significant difference was observed in the distribution of speed at the two types of sites ($\chi^2 = 2.8$, df = 2, p > 0.2).

Discussion

The overall rate of assistance relative to obstruction at pedestrian crossings was not seen to change after introduction interventions. In both cases, obstruction was 7 times as likely as assistance. However, now all of the assistance was concentrated at the area of three speed bumps. One possible inference is that drivers became less considerate at other locations when strongly encouraged by speed bumps at one location. Moreover, overall the fraction of drivers travelling at the highest speed category increased from one third to one half. No significant difference in proportions of drivers at different speeds at bump sites compared with other sites was noted. Were drivers going faster to make up for slowing down at speed bumps? Further driver improvement interventions to encourage assistance of pedestrians at crossings appear warranted. Meanwhile pedestrians should exert great caution when crossing roads at Laurentian University.

References

Wilson, Tay, 1991, Locale driving assessment – A neglected base of driver improvement interventions. In *Contemporary Ergonomics*, (Praeger, London), 388-393.

Wilson, Tay and McArdle, G., 1992, Driving style caused pedestrian incidents at corner and zebra crossings. In *Contemporary Ergonomics*, (Praeger, London),388-393.

Wilson, Tay and Godin, Marie, 1994, Pedestrian/vehicle crossing incidents near shopping centres in Sudbury, Canada. In *Contemporary Ergonomics*, (Praeger, London), 186-192.

Wilson, Tay and Chisel Christine, 1997, On-campus pedestrian crossings: Opportunity for locale based driver improvement. In *Contemporary Ergonomics*, (Praeger, London), 86-91.

AIR TRAFFIC CONTROL

INITIAL EVALUATION OF A RADICALLY REVISED EN-ROUTE AIR TRAFFIC CONTROL SYSTEM

H. David[1], J.M.C. Bastien[2]

[1]*EUROCONTROL Experimental Centre*
91222 Bretigny-sur-Orge, CEDEX, France

[2]*Université Paris V – René Descartes*
45 Rue des Saints-Peres
75270 Paris, France

25 postgraduate students (native French speakers) were given an overall briefing on Air traffic control, followed by individual familiarisation with the keyboard system and 20 training examples.

They were then presented with traffic corresponding to an entry rate of 250+ aircraft per hour in random direct flight for one-hour nominal duration. (When no action was required, the simulation rate was accelerated by a factor of six, so that the average length of one exercise was about 24 minutes.)

19 students controlled this traffic correctly, with no unresolved conflicts. The total of potential conflicts was 1322 of which 1304 (98.6%) were solved correctly.

Introduction

An examination of the 'En-Route' Air Traffic Control (ATC) system, summarised in Dee (1996) and in more depth in David (1997a), led to the publication of EEC Report 307, (David, 1997b). This report describes how En-Route ATC can be re-designed to provide a human-oriented task (primarily conflict resolution) supported by computer-based facilities, employing satellite communication and navigation systems. Briefly, after a careful task definition, a set of task analyses assuming different levels of human involvement were carried out, and a semi-automated system was adopted. Sub-tasks were allocated to define a satisfying, functional human controller task. Displays and control operations were then devised to provide simple (for the controller) displays and control options. The system was adapted, for demonstration purposes, to use a conventional low-precision colour display and a standard (QWERTY) keyboard. Although general design principles suggest that a three-dimensional image would be most appropriate, a co-planar display is in practice preferable (May et al 1996, Wickens 2000).

The RADICAL Interface

The 'co-planar' display provides a conventional 'map' display, with an additional height/time and distance profile for a selected aircraft. These two windows are the only ones provided in the demonstrator – additional information would be made available in any real system. In normal operation, only aircraft involved in potential future conflicts would be shown on the map display, linked to show which aircraft are in potential future conflicts.

Aircraft are shown by icons representing their relevant characteristics, Colour indicates flight level, size indicates size, position and angle represent position and direction of flight, and wing sweep represents speed. The system presents the profile for an aircraft in the most urgent conflict, although an experienced controller may choose to over-ride this choice. The profile shows all conflicting aircraft by blocks representing the duration of the conflict, in solid colour for conflicts with the present flight plan, or in outline for aircraft that would be in conflict if a level change were made. The full trajectory of the aircraft is shown on the map display, with the trajectories of conflicting aircraft up to the point of closest approach – which may be after the tracks have crossed. A white line, indicating the distance and direction of the problem links the positions of the target and problem aircraft.

All displays are entirely symbolic, using no letters or digits. The extensive use of colour to differentiate aircraft and other information makes it impossible to provide a useful monochrome illustration. Interested readers are referred to EEC Report 307 (David 1997b) or to the EEC website (www.eurocontrol.fr).

The controller constructs a potential solution to the problems of the target aircraft by specifying manoeuvres via the keyboard. As each keystroke is entered, the display shows its consequences. Additional conflicts may appear as a consequence of a proposed manoeuvre, or the aircraft may leave at a wrong position, height or time. Red lines joining the actual and required exit indicate these differences. Some manoeuvres may be unacceptable, for example, requiring excessive speeds to exit correctly, and are indicated by a red final track segment. The controller may cancel all or part of a set of manoeuvres, or may accept this aircraft and operate on the other aircraft involved.

(It is assumed that aircraft communication is by data-link. The manoeuvres decided by the controller are coded and sent to the relevant aircraft for action, and to adjacent aircraft for information. A speech message would be generated at the controller's position, and in the pilots' cockpits, in their language of choice. A 'message repeat' facility would be available at each position.)

Aims

The aim of this study was to demonstrate that the RADICAL interface could be learned and operated efficiently without the expert skills required for conventional ATC operations.

Method

Twenty-five postgraduate students (All French-speaking, none familiar with air traffic control) were given an initial general briefing on Air Traffic Control, followed by individual familiarisation with the QWERTY keyboard, and twenty short coached training examples. They were given a (one-page) guide to the preferable solution strategies.

They were then presented with traffic building up to 40 aircraft simultaneously present, in random direct flight in a space representing 125 x 90 NM corresponding to an entry rate of 250+ aircraft per hour for a nominal one hour. (The system replaced each aircraft as it left, providing an initial 2 minutes of conflict free flight. Since different solutions provided slightly different flight paths, samples diverged during successive exercises.)

Because the system was set to show only aircraft in conflict, it was completely blank when there were no conflicts. To save time, the simulation rate was increased to six times real time during these periods, and automatically returned to real time when a new conflict was detected.

Results

Table 1 presents the mean performances of the 25 participants. (These are unweighted means of individual performance means, not overall mean values.)

Table 1. Mean Performance

Parameter	Arithmetic Mean	% Nil Scores
Conflicts	52.88	
Resolved (correctly)	52.16 (=98.64%)	
Unresolved	0.72 (= 1.36%)	76%
Time to resolve Conflicts	24.85 seconds	
Time before Conflict start	20.71 Minutes	
Number of Control Orders	47.56	
Time to Construct Orders	13.97 seconds	
No. of Modified Orders	1.84	48 %
Time to construct	19.41 seconds	
No of Cancelled orders	21.52	
Time to construct	19.63 seconds	
Total Exercise Duration	23.68 Minutes	
Position Exit errors	0.08	92 %
Height Exit errors	0.62	68 %
Time Exit Errors	3.2	12 %
Total Exercise duration	23.68 minutes	

Nineteen out of 25 students correctly resolved all the conflicts presented to them. Two students failed to solve one conflict in time. One failed to solve two and one three conflicts. One failed to solve 5 out of 50 conflicts presented, probably because she hit the keys so hard that they bounced, and produced double entries. One student missed 6 out of 61 conflicts, apparently because she became 'locked' on one problem, and could not cope with the traffic that accumulated while she solved that problem. (For technical reasons, the model does not update the display during the solution of a potential conflict.)

The mean time taken to resolve conflicts was 25 seconds, measured from the first appearance of the conflict on the screen to the acceptance of an order resolving the conflict. The mean time to construct an order was 14 seconds, and a mean of 47.6 orders was required to solve a mean of 52.2 conflicts. (One order may solve several conflicts, but a less urgent conflict will not be considered until all more urgent ones have been solved.)

Cancelled and modified orders do not necessarily imply errors by the student. They usually represent the exploration of a potential solution that did not in fact work. (Solutions involving changes of flight level only can be checked visually on the height profile display before entry, but those involving height or speed changes can only be checked for conflicting traffic by entering the solution and observing any consequences.)

The total exercise duration of about 24 minutes corresponds to about 16.8 minutes running in real time and 7.2 minutes running at 6 times real time, during which no unsolved conflicts were present. The longest exercise took 35 minutes, corresponding to 30 minutes in real time and five minutes at 6 times real time.

The students were not specifically required to avoid 'exit errors' – defined as differences in exit position of 1 NM, in time of 1 minute and in height of 1 Flight Level (FL= 100 feet). Most errors can be attributed to 'finger trouble' since the 'Y' key, which accepted the order being developed, was next to the 'T' key which was used to instruct the aircraft to maintain the order for a fixed time before returning to track.   (A means of correcting this error is available in the model, but it is cumbersome and was not taught to the students. They did not use it.)

A few height errors were caused where an aircraft was unable to return to its prescribed height after a manoeuvre. The considerable number of time errors was generated by a difference in the time definitions used in the algorithm signalling errors to the controller and that scoring actual exit errors.

Discussion

En-route air traffic control is currently a two-stage process. A Planning controller, who compares the planned flights of aircraft along fixed routes, defined by ground 'beacons'. He may identify potential conflicts and solve these by deciding on a change of flight level. The Planning controller talks to adjacent controllers, but not with the aircraft. The Executive controller talks to aircraft, acknowledging aircraft entry to the sector, monitoring their flight through the sector, including any level changes decided by the Planning controller and any heading or speed changes, and returning the aircraft to track, and finally handing the aircraft to the next sector. This system constrains the aircraft to follow fixed routes, which not only increases the cost to the airlines, but also increases the number of potential conflicts. Although sector capacity varies according to the complexity of the sector, the capacity of a sector may be 50-60 aircraft per hour at most. A conventional Planning Controller relies on strips for his planning procedures and cannot cope with more than five or six 'direct route' aircraft at the best. Controllers take three years of general training, and an extra six months of 'learning the sector' to reach this level of performance.

The capacity increase provided by the RADICAL interface is such that no formal comparison can be made – the least efficient novice performs better using this interface than the most experienced team of professional controllers using conventional methods.

It should be born in mind that this demonstrator is not a complete system. In such a system, controllers would monitor the performance of the aircraft, being informed by suitably derived symbology of any deviations from plan and taking the necessary action to return the aircraft to their plan or adapt the plan to changing circumstances. System planning would include procedures and software tools for coping with 'foreseeable emergencies' such as loss of pressurisation, where the nature of a problem and the necessary response can be predicted, but its actual occurrence cannot. Procedures would also be needed to cope with temporary or permanent restrictions on airspace available (Military training areas, major weather problems) and to optimise traffic routes (by making use of jet streams, for example.)

The existing model is capable of coping with continuous 'slow climb' procedures, which may afford fuel economies for long-distance flights.

Air Traffic Control is, for the most part, a 'tidal' activity, experiencing peaks of workload and periods of low activity. Future systems must take account of the possibilities of overload (by providing safety-oriented fallback systems) and of underload (by increasing safety margins, or reviewing non-optimal flight paths).

The workload in this system is proportional to the incidence of potential conflicts. A complete solution to a potential conflict takes about 14 seconds, and may solve more than one potential conflict. Conflicts are detected and resolved within about a minute of the system becoming aware of the flight. In a 'real' system, conflicts would be resolved and solutions transmitted to the aircraft at entry, so that no accepted aircraft would have a potential conflict. In the event of system failure, the system is inherently safe.

It should also be noted that ease of learning an interface does not necessarily imply de-skilling or 'dumbing down'. As in the present system, controllers would rapidly learn the semi-repetitive traffic in their area, and develop methods of 'streaming' or otherwise optimising the traffic flow.

Finally, it should be recognised that the practice of real Air Traffic Control, as opposed to more-or-less elaborate simulation, requires mental and moral stamina, patience and professional conscience, which, in the long run, may be more important that the short-term cognitive and motor skills examined here.

Conclusions

1. The RADICAL interface for En-Route Air traffic control can be learned and efficiently operated with one morning's training by non-ATC personnel.

2. The interface is independent of the language spoken by controllers and/or pilots, and can be organised to enhance pilots' mutual awareness. (The students participating in this study were French-speaking.)

3. The capacity of the interface is at least 40 aircraft simultaneously present per hour, corresponding to about 250 aircraft per hour, for approximately 20 minutes workload per hour.

Acknowledgements

The authors gratefully acknowledge the co-operation of the students of the Laboratoire de l'Ergonomie de l'Informatique. They also acknowledge the encouragement of the director of the EUROCONTROL Experimental centre, M. J-C Garot. This study does not form part of the official EUROCONTROL work program.

References

(Paper copies of EEC Notes and Reports are available from the address above. Recent Notes and Reports are available at the EEC Web-site (www.eurocontrol.fr))

David, H., (1997a) Deep Design – Beyond the Interface, In K-P. Holtzhausen *Advances in Multimedia and simulation. Human-Machine-Interface Implications* (pp 65-77) Bochum, Germany: Fachhochscule Bochum
[1]David, H. (1997b) *Radical Revision of En-route Air Traffic Control*. EEC Report No. 307. EUROCONTROL Experimental Centre, Bretigny-sur-Orge, France
Dee, T.B. (1996), Ergonomic Re-design of Air Traffic Control for increased capacity and reduced stress, In Robertson, S.A. (Ed.), *Contemporary Ergonomics 1996* Taylor and Francis.
May, P.A., Campbell, M. & Wickens, C.D. (1996) Perspective displays for air traffic control: Display of Terrain and Weather, *Air Traffic Control Quarterly* 3(1) 1-17
Wickens, C.A. (2000) The When and How of using 2-D and 3-D Displays for Operational Tasks. In Proceedings *of the IEA2000/HFES 2000 Congress*, .San Diego California USA

---

[1] This report includes a 3.5"disc with compiled code, source code and documentation for demonstrations of traditional (DEMOLD), intermediate (DEMON) and fast interfaces (DEMFAST)

Compiled code, source code and documentation for the improved version (DEMFAST1) and for analysis programs are available from the author.

Finding ways to fit the automation to the air traffic controller

Barry Kirwan & Judith Rothaug

Human Factors and Manpower Unit, EUROCONTROL HQ
96 Rue de la Fusée, B-1130 Brussels, Belgium.
barry.kirwan@eurocontrol.be & *judith.rothaug@eurocontrol.be*

Abstract

The next fifteen years will probably see more than a doubling of current air traffic levels in European airspace. Since air traffic controllers are already very busy, some degree of automation will be needed to enable such desired safe increases in air traffic capacity. Given the safety critical nature of Air Traffic Management (ATM), there is significant effort being undertaken to harmonise the human and the automation. One project that is aimed at such harmonisation is the SHAPE project (Solutions for Human-Automation Partnerships in European ATM) being carried out by EUROCONTROL. This project is outlined, and some preliminary findings from the areas of human error, ageing, and the future role of the controller, are discussed.

Background

This paper outlines a project in the domain of Air Traffic Management (ATM), specifically concerned with the introduction of computerised assistance (i.e. automation) for certain air traffic controller functions (e.g. detection of aircraft 'conflicts'; the conveyance of certain routine messages to and from aircraft crew; etc.). Experience in other related domains such as cockpit automation has shown that failure to take account of Human Factors issues early on in the development of such automation endeavours, can lead at the least to 'start-up' problems, and at the most 'automation-assisted accidents' (Billings, 1997). A new project has therefore been launched at EUROCONTROL headquarters in Brussels, which aims to provide assessment techniques or tools, and guidance to ensure that the implementation of future automation will not negatively impact on human (and therefore system) performance and safety.

The project is called Solutions for Human-Automation Partnerships in European ATM (SHAPE) and runs from 2000 – 2002. It does not assume that the ATM system and automation must be designed completely around the human (Human Centred Automation), but instead recognises the critical role of both human and automation, and the need for a joint or co-operative system. Such a system will be one which is reasonably intuitive and not placing excessive (nor insufficient) demands on the controller, whilst keeping the controller 'in the picture'. It will also be a system where the current error detection and recovery capabilities of the controller are retained, and where the human has a good enough knowledge of the system and skill set to be able to optimise ATM system performance and recover from system faults and failures. The controller will still be 'in control', albeit supplemented and assisted to handle increasing levels of traffic predicted in the next fifteen years. This vision of the controller's role continues up until the year 2015, after which new ATM strategies may emerge.

There have been a number of recent reviews of the potential impacts of automation on ATM system performance (eg Wickens et al, 1998; Parasuraman and Mouloua, 1996), and these have outlined the major areas and Human Factors/Automation issues that need to be resolved. The SHAPE programme of work has therefore synthesised a number of requirements for tools or assessment techniques that need to be developed, some of which are described below:

1. **New error forms** – although automation will reduce some error types, other new forms are likely to arise (eg errors associated with different modes of equipment configuration). Additionally, currently the very 'hands-on' nature of ATM facilitates error detection and recovery – many errors are actually made in today's ATM, but these are almost all recovered, the net result being a 'high reliability system'. The solution is to be able to evaluate system vulnerability to error and to ensure the system design still maintains error detectability and recoverability characteristics.

2. **Skill-set change** – Automation can lead to skill degradation, and to the need for new skills. Accurate skill/knowledge requirements are key solutions in automation, as well as 'human-sensitive design', i.e. design which is intuitive and highly usable, and which matches system and user models. A means of identifying what new skills need to be trained (and how), and what skills can be relinquished, is needed.

3. **Trust** – trust (confidence in the automation) is a key trait to try to control when implementing automation – it is possible to gain and maintain trust early on in the system design, but trust is easily lost, and once lost, becomes very hard to regain. Additionally, over-trust in automation can be as damaging as under-trust. A tool is therefore needed to evaluate trust in a tool or automated function, as well as guidance principles for developing and maintaining a correctly calibrated level of trust.

4. **Team Impacts** – Automation inevitably results in changes of task allocation and will therefore have impacts on the current controller team-working roles and responsibilities. There may be loss of some interactions, some new interactions, and it will be necessary to evaluate the impacts of such changes on system performance. This will require a new form of team-based task analysis. There may also be Team Resource Management (TRM) aspects relevant to automation, e.g. in terms of reluctance to rely on automation (or over-reliance). A key solution will be a better understanding of the impacts of automation on team performance and effectiveness, and guidance principles on how to integrate automation into team functions.

5. **Recovery from System Failure** – inevitably automated support will suffer from failures. Once controllers have become used to the automation support, and traffic capacity has increased, former skills will inevitably degrade or at the least become less fluent. When system faults or failures occur, therefore, recovery will be critical but difficult. Solutions for supporting controllers in such events rely on achieving a

better understanding of the impacts of sudden loss of automation on human performance, with methods for investigation of this phenomenon. Such studies will result in guidance on how to support the controller should such failures occur.

Other aspects of human system performance that SHAPE will consider are aspects of Mental Workload, Situation Awareness measurement, and the general role of the controller for future. The aim with all of these areas is to gain a better understanding of how these issues are affected by certain planned automation tools, and how the issues themselves interact with other issues, and then to develop assessment approaches that can be applied to new systems tools development programmes.

At this early stage in SHAPE, the work is focusing on the development of a better understanding of the issues themselves, with some preliminary work on development of tools. Some early results are presented below.

Human error and automation

For current systems, one can learn about human errors from incident analyses, such as the formal systems most ATM centres use nationally. However, for new systems, this is not the best way to learn – it is better to learn about errors before automation tools are installed into new systems, so that they can be eradicated or else contingencies can be made for their occurrence. Ideally therefore, what is needed is a system for identifying errors at the design stage, and also a system for recording errors that occur during the various training and test simulations that will occur with real controllers and simulated traffic. Whilst various techniques exist for identifying errors with new system designs, few assessment approaches exist for use during real-time simulations such as system validations (proof of concept) or training work with simulated traffic. Therefore, a proformer was developed that can be filled out rapidly by a controller after a simulation session, documenting any errors that occurred, and classifying their significance and causes. The proformer is shown in Figure 1, and the lower half is based on the developing EUROCONTROL HERA (Human Error in ATM) incident analysis system (Isaac et al, 2000).

Such a system has been applied with some success in a real-time simulation at the EUROCONTROL Bretigny, involving controllers using a medium term conflict prediction tool. The approach recorded problems with the tool as well as errors, e.g. the prototype tool failing to detect a conflict it should have detected. Errors included losing the picture due to false traffic alerts, input errors with the mouse leading to selection of wrong information, problems in coordination between adjacent controllers relating to different 'pictures' of the tool, and a read-back error from the pilot. The controllers found it useful because they could encode what had happened to them, and also could make comment on aspects of the prototype system that should be improved. This can of course happen currently, but the sheet is more formal and by stating the potential consequences, the comments can be prioritised and analysed accordingly. The benefits are then a more efficient and effective learning mechanism during such simulations, maximising the knowledge gained from them, and potentially allowing all controllers to learn from all other controller's errors. Such information can be fed back into the design of the system, and also into ongoing training media development, warning controllers of certain errors, and showing how to recover should they occur.

ERROR RECORDING SHEET

Date   Time   Shift   Sector
                                      Approach ☐   Area ☐

Description of what happened:

How detected & recovered?

Self   Prompted by system   Other controller   Pilot   Not recovered

☐   ☐   ☐   ☐   ☐

Actual consequences: None   Efficiency   Safety
                          ☐   ☐   ☐   *Loss of*
Potential consequences: ☐   ☐   ☐   *separation*? ☐

Brief description of actual or potential consequences:

Fail to detect   ☐   Mis-hear   ☐   Receive info late ☐   Mis-read   ☐

Forget info   ☐   Forget to act ☐   Mis-remember   ☐

Mis-judgement   ☐   No decision   ☐   Wrong decision   ☐   Late decision ☐

No planning   ☐   Late plan   ☐

Keying error   ☐   Selection error ☐   Positioning error ☐   Timing error ☐

Omit information ☐   Send wrong info ☐   Other ☐........................

Interface....................legibility   input devices   label overlap   layout   *other:*

Traffic & Airspace.........high traffic load   low traffic   unusual traffic   *other:*

Pilot-ATCO comms.......confusable callsigns   accents   non-RT phrases   *other:*

Procedures..................ambiguities   complex   recent changes   *other:*

Training & Experience....inadequate training   experience   on-job-training *other:*

Social & Team Factors...hand-over   co-ordination   team pressure   *other:*

Environment................noise   distraction   lighting   temperature   *other:*

Personal factors............tired   domestic issues   anxiety   health   *other:*

Organisational factors... .staffing   sector split   morale   pressure   *other:*

Figure 1: **Error recording sheet** (actual sheet is in colour)

Automation and Controller Ageing

Ageing has a significant impact on the cognitive functioning of human beings. Examples of cognitive functions impeded by ageing that are highly relevant for job performance of Air Traffic Controllers (ATCOs) are, amongst others, working memory (especially speed of processing), attention, and spatial visualization.

The few studies addressing ageing in Air Traffic Management (ATM) found a negative relationship between age and job performance. A recent study by Heil (1999) concluded a U-shaped relationship between age and job performance: performance increases until the age of 35, levels off until early to mid 40s and starts declining after the mid 40s.

However, there is a relative lack of data for European ATM as well as for the possibly compensating effect of experience. The question of how older controllers cope with new automated systems and how they could be supported in this process has not been addressed by research so far. The results presented here are the initial part of a longer study. This first phase consists of interviews with 37 ATCOs in three air traffic centres. The interviews were conducted on an anonymous basis.

The average age of the interviewed controllers was 50,32 years, ranging from 33 to 64 years. Only two subjects were under the age of 40, three subjects were female.

Three main impacts of ageing on controllers job performance could be identified: Firstly, the subjects recognize a general slowing down in information processing. This effect appears to be stronger in situations with high demand (i.e. high traffic load) and when working only part-time on operational tasks. Secondly, the ability to do multi-tasking is decreasing, dealing with complexity gets more difficult, the personal capacity for number of aircraft decreases; and the job in general gets more demanding. Thirdly, the learning of new tasks, concepts, or procedures becomes more demanding with age.

The impact of age appears to differ depending on the kind of task to be executed. They clearly depend on the type of position one is working. For one position, which is more associated with overall sector management and planning, ageing related declines can be negligible. However, on a radar position declines due to age are clearly recognized.

However, those statements must not be seen in isolation. Older controllers' rich experience to some extent counteracts the above mentioned impacts of age. Their experience supplies them with a broad memory bank of situations and appropriate solutions, helps to anticipate traffic constellations, to plan ahead, and to prioritize more efficiently. They have a "bag of tricks" they can fall back on. Furthermore, they can rely to a great extent on well-practised problem solving skills, and they are better in handling emergencies. However, there appears to be a limit of the beneficial effects of experience. The subjective perception of the performance curve of controllers confirmed the findings by Heil (1999): U-shaped with a definite decline after 45-50 years.

The forthcoming transition to a new system with an increased level of automation raised some concerns amongst the older controllers. A major concern is the move to electronic flight progress strips, and the related loss of paper strips as a backup system. The increased reliance on computers is as concerning as the enormous amount of technical background knowledge that is required to operate the new system.

The next steps in this project will be a questionnaire survey, addressing a broader sample of European air traffic controllers, followed by an experimental study, which will explore the relationship between age, experience and job performance in automated ATM systems, and will determine how significant this perceived decline is for system performance. This work will then develop guidance for future automation tool design to help compensate for ageing effects.

The Future Role of the Controller

With so much automation predicted for future ATM, concern has been expressed about the role of the controller in the future. Currently the controller is very active and has little automation. The job is very skilled and controllers take pride (rightly so) in their trade. The SHAPE project wants therefore also to look at the long term implications for controllers. As an initial exercise, two groups of Human Factors personnel from a range of European countries discussed the future potential roles of the controller. Part of the purpose was to set up an alternative vision to that of some designers who see the far future as being highly automated. It was considered that Human Factors should have its own vision about the future role, but that such a vision must be based on system goals of operability and safety, and not merely a vested interest in the subject matter. The tentative conclusion was that the controller would possibly have more of a role of being an optimiser (the term system manager was used, though it is recognised that this term has other and broader implications), managing and optimising the smooth flow of traffic, via the use and supervision of the automation tools, and acting as an optimising interface with aircraft crew.

In summary SHAPE has started and has begun to address core issues related to the introduction of significant amounts of automation over the next decade. The intention is to develop ways of harmonising automation and controller roles and abilities, via a better understanding of how automation issues affect the controller, and how to optimise future controller-automation 'partnerships'.

References

Billings, C.E. 1997, *Aviation Automation: The Search for a Human-Centred Approach,* (Lawrence Erlbaum, Mahwah)

Heil, M. C. 1999, *An investigation of the relationship between chronological age and job performance for incumbent Air Traffic Control Specialists,* Washington, DC: Federal Aviation Administration Report DOT/FAA/AM-99/18.

Isaac, A., Shorrock, S., Kirwan, B., Kennedy, R., Andersen, H., and Bove, T., 2000, Learning from the past to protect the future – the HERA Approach, European Aviation Applied Psychology Conference, Crieff, Scotland, September.

Parasuraman, R. and Mouloua, M. (Eds), 1996, *Automation and Human Performance: Theory and Applications* (Lawrence Elbaum Associates, Mahwah)

Wickens, C.D., Mavor, A.S., Parasuraman, R., and McGee, J.P., (Eds), 1997, The future of air traffic control: human operators and automation, (National Academy Press, Washington DC)

HUMAN FACTORS IN THE OFFSHORE INDUSTRY

FATIGUE OFFSHORE: A COMPARISON OF SHORT SEA SHIPPING AND THE OFFSHORE OIL INDUSTRY

Andy Smith[1] and Tony Lane[2]

[1]Director, Centre for Occupational and Health Psychology
Cardiff University
63 Park Place
Cardiff, CF10 3AS
Smithap@cardiff.ac.uk

[2]Director, Seafarers International Research Centre
Cardiff University
63 Park Place
Cardiff, CF10 3AS
Lanead@cardiff.ac.uk

Mounting concern with seafarer fatigue is widely evident among maritime regulators, insurers, ship owners, trade unions and welfare agencies. We are carrying out a research programme to investigate this topic the first phase of the research is concerned with specific comparisons between short sea shipping and the offshore oil industry. The overall objectives of the research are: to predict worst case scenarios for fatigue, health and injury; develop best practice recommendations appropriate to ship type and trade; and produce advice packages for seafarers, regulators and policy makers

Background to the programme

Global concern with the extent of seafarer fatigue and the potential environmental costs is widely evident everywhere in the shipping industry. Maritime regulators, ship owners, trade unions and P&I clubs are all alert to the fact that in some ship types, a combination of minimal manning, sequences of rapid turnarounds and short sea passages, adverse weather and traffic conditions, may find seafarers working long hours and with insufficient recuperative rest. In these circumstances fatigue and reduced performance may lead to environmental damage, ill-health and reduced life-span among highly skilled seafarers who are in short supply. A long history of research into working hours and conditions and their performance effects in manufacturing and process industries as well as in road transport and civil aviation has no parallel in commercial shipping. With a few exceptions, maritime research on work patterns and conditions has been conducted aboard, or in simulations of, warships. The best known and most recent merchant ship field studies (Colquhoun et al., 1988) were conducted aboard warship auxilliaries.

Given the absence of research on offshore fatigue we are carrying out a research programme which generally aims to:

- Predict worst case scenarios for fatigue, health and injury
- Develop best practice recommendations appropriate to shiptype and trade
- Produce advice packages for seafarers, regulators and policy makers

Specifically, the programme aims to provide advice on:

- incidence and effect of fatigue in terms of specific ship types and voyage cycles
- optimal shift patterns and duty tours to minimise fatigue
- identification of at risk individuals and of factors which affect fatigue/quality of rest
- significance of patterns of work and rest, and patterns of health and injury, in terms of seeking to improve health and safety of seafarers on board ship
- suggested ameliorative/preventative procedures for minimising the effects of fatigue
- appropriate guidance for seafarers on fatigue avoidance

These aims will be achieved by field studies using a battery of techniques to explore variations in fatigue and health as a function of the voyage cycle, crew composition, watchkeeping patterns and the working environment. The methods involve:

- A review of the literature
- A questionnaire survey of working and rest hours, physical and mental health
- Physiological assays assessing fatigue, rhythm adjustment and cardiovascular risk
- Instrument recordings of sleep quality, ship motion, and noise
- Self-report diaries recording sleep quality and work patterns
- Objective assessments and subjective ratings of mental functioning
- Pre- and post-tour assessments
- Analysis of accident and injury data

The present project

The present project has operationalised the central themes of the programme so that they can be applied to specific issues of current concern. This involves comparison of short sea shipping (shuttle tankers, offshore supply vessels, anchor handlers, daughter craft and diving support vessels) with the offshore oil industry. Our interest in short sea shipping comes from recent research by SIRC which has analysed mortality data and found supply vessels to have major problems. Research on fatigue on oil platforms is currently supported by HSE and the background and aims of that project are given below.

The health and safety problems of shift workers on offshore oil rigs are different to the onshore population, since complete biological adaptation to night shift is seen with some schedules offshore (Barnes et al., 1998). The advice of the European Work Directive is not appropriate in this case as it is based on the premise that night shift workers do not adapt. A parallel project (Arendt et al., 2000) aims to specify the psychological and physiological response to different work schedules on North sea oil rigs. Biological adaptation, risk factors for heart disease and performance efficiency and mood are being assessed. This multi-disciplinary and integrated approach will permit for the first time the simultaneous evaluation of psychological and physiological variables in the same individual while offshore. It would appear most appropriate to apply this approach to allow comparison with short sea shipping. This will allow us to achieve the objectives of advising the Industry and the Regulators as to a) work schedules which maximise performance and b) strategies to minimise risk to health and safety. This shows that it is most appropriate to conduct a first phase which compares the workers on the offshore installations with diving support vessels, supply vessels, anchor handlers, daughter craft and shuttle tankers. This range of comparisons allows assessment of similar shift patterns in different offshore industries (e.g. the installations with the supply vessels) and also the impact of different shifts (e.g. short sea tankers versus supply ships and installations). It also enables us to use the methodology being applied to the offshore installations on board ship and to also apply the techniques to be used in our overall fatigue programme to address this specific issue. Furthermore, access to these type of vessels has already been achieved through discussion with the owners.

Aims and objectives of present project

This project aims to collect the range of data and provide the level of information necessary to optimise working schedules for performance, safety and long term health. The objectives of the project are to:

- make recommendations for work schedules and duty tours
- identify characteristics of at risk individuals
- propose ameliorative/preventative procedures
- propose appropriate medical record keeping
- prepare a Seafarers Guide to Fatigue Avoidance
- prepare briefings for UK delegations to IMO, EC/EU, ILO

Methods

1. A review of the literature on offshore fatigue

This essential initial part of the research has critically evaluated the last review produced by Ivan Brown, considered studies conducted since then, and integrated the research on seafarers fatigue with the findings obtained on the oil rigs (Collins, Matthews & McNamara, 2000).

This is another extremely important part of the project. A current re-analysis of some 4000 incident reports from the UK offshore oil industry (Parkes & Swash, 2000) shows that shift and season explain the greatest proportion of variance in the occurrence and severity of reported injury. A similar analysis of seafarers injuries has now been conducted, and is briefly described in a following paper.

3. Questionnaire survey of the extent of fatigue and its consequences
We have been able to obtain the collaboration of NUMAST and MSF to conduct this survey. Three questionnaires have been administered simultaneously for completion by whole crews in the types of vessel and offshore installations being considered in this project. The first focuses on voyage cycle, duty tour and work and rest patterns. The second assesses sleep and fatigue using the Bristol Sleep Questionnaire and the profile of fatigue related states. The third questionnaire is the standard health screen SF-36 instrument designed to measure self-reported physical and mental health. Ample comparative data exist to allow comparison between offshore workers and other populations.

4. Physiological indicators of fatigue and rhythm adjustment.
Salivary cortisol has been measured as an indicator of stress/fatigue.

5. Sleep, alertness and cognitive performance
These have been assessed by sleep logs, actigraphy (wrist mounted accelerometers), alertness ratings and measures of cognitive performance (simple reaction time; focused attention and categoric search tasks).

6. Daily logbooks
Logbooks have been completed each day recording sleep, symptoms of fatigue and perception of human error.

7. Instrument recording of motion and noise
It is clearly very important to have precise information on the environmental factors. Motion and noise levels can now be recorded continuously over long periods. Information on the weather, sea state and other factors which may influence fatigue has also been recorded.

Procedures

Essentially the project consists of four main methods. The first involves analysis of the survey data and is designed to give an indication of the extent of chronic problems. The second involves the analysis of existing accident databases. The third assesses a range of functions before and after each shift at different time points in the tour. This technique has been widely used in fatigue research (Broadbent, 1979), the rationale being that pre-post shift differences reflect the magnitude of the fatigue during the day. All of the approaches will take into account the effect of season which is likely to be a major influence in the geographical location where the study is taking place.

reflect the magnitude of the fatigue during the day. All of the approaches will take into account the effect of season which is likely to be a major influence in the geographical location where the study is taking place.

Deliverables

The present project aims to provide the following deliverables:

- information that will further our understanding of fatigue in short sea shipping and the offshore oil industry
- advice on shift schedules and hours of work, with an aim to maximise performance and minimise health and safety risks both in the short and long term
- evaluation of ameliorative/preventative procedures

Conclusions

We believe the present project is justified for the following reasons. Regulation of working and rest hours beyond the measures already in place in EU Directives and ILO and IMO conventions is extremely likely. Without substantial research, however, debate and application of appropriate policy will not be informed by extensive and high quality scientific evidence. There is a risk that blanket rather than targeted measures will then be adopted. The present project will aim to deliver the following outputs:
- recommendations for work schedules and duty tours
- identification of characteristics of at risk individuals
- proposals for ameliorative/preventative procedures
- proposals for appropriate medical record keeping
- preparation of a Seafarers Guide to Fatigue Avoidance
- preparation of briefings for UK delegations to IMO, EC/EU, ILO

The objectives of the project will be achieved using multi-methodologies. Specifically, these are:
- A review of the literature on offshore fatigue
- A questionnaire survey of the extent of the problem
- Analysis of accidents and injuries
- Physiological assays indicative of fatigue and rhythm adjustment
- Recordings of sleep quality and performance efficiency
- Self-report diaries of subjective health and well-being

The preliminary findings from the project are reported in the subsequent papers in these proceedings.

Acknowledgement

The research described in this article is supported by the Marine and Coastguard Agency, the Health and Safety Executive, NUMAST and MSF. We would also like to acknowledge the contribution made by the ship and installation owners who have participated in the research.

References

Arendt, J., Hampton, S., Morgan, L., and Smith, A.P. 2000, Physiological and psychological markers for shiftwork offshore, Project funded by the Health & Safety Executive

Barnes, R.G., Deacon, S.J., Forbes, M.J., Arendt, J. 1998, Adaptation of the 6-sulphatoxymelatonin rhythm in shiftworkers on offshore oil installations during a 2 week 12-h night shift, *Neuroscience Letters*, **241**, 9-12

Broadbent, D.E. 1979, Is a fatigue test possible ? *Ergonomics, ***22**, 1277-1290

Collins, A., Matthews, V., and McNamara, R. 2000, Fatigue, health and injury among seafarers on offshore installations: A review, *Seafarers International Research Centre. Technical Report Series.* **No. 1**

Colquhoun, W.P., Rutenfranz, J., Goethe, H., Neidhart, B., Condon, R., Plett, R., and Knauth, P. 1988, Work at sea: A study of sleep, and of circadian rhythms in physiological and psychological functions in watchkeepers on merchant vessels. 1. Watchkeeping on board ships: A methodological approach, *International Archives of Occupational and Environmental Health,* **60**, 321-329

ACCIDENTS, INJURY AND FATIGUE IN THE OFFSHORE OIL INDUSTRY: A REVIEW

Alison Collins, Victoria Cole-Davies and Rachel McNamara

Centre for Occupational and Health Psychology
Cardiff University
63 Park Place
Cardiff, CF10 3AS

The contribution of fatigue to accidents has been extensively researched across a number of industries. However, findings cannot automatically be applied to the offshore oil industry because of the unique combination of working conditions including demanding work/rest patterns and extreme weather. The industry has also gone through major changes in recent years, leading to reduced manning of rigs and ships, increased automation and workload and decreased job security (e.g. NUMAST, 1992). A review by Brown (1989) into the work hours, fatigue and safety at sea highlighted the lack of objective evidence of the detrimental effect of fatigue at sea. Eleven years on from Brown's (1989) paper, the purpose of this literature review, was to identify the current research into fatigue, health and accident rates among seafarers and employees working in the offshore oil industry.

Introduction

This review is an extract from the original report (Collins, et al. 2000) published by the Seafarers international Research Centre (SIRC), as part of their technical report series.

In a review of work hours, fatigue and safety at sea, Brown (1989) found little objective evidence of the effects of fatigue, although he did find anecdotal evidence regarding personal fatigue experiences. Seafarers reported that they were often expected to work continuously, under conditions of task-induced or environmental stress for excessive (in relation to other industries) periods of time. Respondents' attributed a number of fatigue symptoms to their working arrangements that were in general agreement with research into fatigue effects (e.g. Bartlett, 1948 cited in Brown 1989). Thus the main objective in compiling this review follows from Brown's (1989) assertion that long hours are a major contributor to fatigue and accidents at sea. Although there is a great deal of research into the effects of fatigue in industry generally, these results cannot automatically be applied to the offshore oil industry because of the unique combination of working conditions. Thus, it is clear that in order to obtain an accurate picture of the extent of fatigue and its effects on rigs and support vessels, more specific research needs to be conducted in the field.

The Offshore Industry

The offshore oil industry can be generally divided into two occupational groups: personnel working aboard ships, and those working on installations. Although both groups are similar in terms of working away from home, and demanding work and rest patterns, many differences do exist between them, e.g. differing tour lengths, shift and rest patterns, vessel structure and physical conditions. Furthermore, the research areas covered within the two industries vary enormously. For the purpose of this review the sectors will be considered individually.

Accidents and Days into Tour

Offshore Installations

In general, offshore installation personnel work tours of two weeks on/off, though work duties can vary from one week to more than 2 weeks offshore. Such duty rosta's can affect accident rates. Forbes' (1997, cited by Parkes and Swash, 1999) found that drill floor accidents on an offshore installation decreased in the second week. There was also a trend for those working fixed shifts to show an increase in accidents on days 5 and 6. A report by Parkes and Swash (1999) for the HSE based on 3 large industry injury databases, showed an increased risk of serious injury relative to less serious injuries lasting 3 or more days, with increasing days into tour and the same was true for fatalities and 3 days plus injuries. Furthermore, where days into a tour exceeded 2 weeks in duration, the ratio of fatalities and serious injuries to injuries lasting more than 3 days, increased significantly. The study indicated the significance of working beyond the 2-week marker. However, the absence of exposure data means that absolute injury rates could not be calculated, thus, the results must be viewed with caution.

Ships

Similar systematic analyses of vessel accidents are hard to come by as the majority of sources detailing accidents aboard offshore support, or indeed any vessel types, are one-off, anecdotal case studies. Even where more thorough investigations have been carried out (e.g. Nielsen, 1999) information relating incident occurrence to days into tour, shift and injury type is noticeably absent. An exception is Raby and McCallum's (1997) study into working conditions that contribute to fatigue related incidents. They found that hours on duty prior to the casualty and hours worked in the 24, 48 and 72 hours preceding the casualty contributed to such incidents. In fatigue related personal injury cases mariners had worked an average of 7.7 hours prior to the incident in comparison to 3.2 hours in non-fatigue related incidents. In the 24 hours preceding the fatigue related incident seafarers reported working an average of 14.3 hours, compared to 8.4 hour.

Accidents and Time of Day

Installations

Most data from offshore installations relate to the 12-hour shift pattern. Laundry and Lees (1991) compared accident and employment records 10 years prior to and post a change from 8-hour to 12-hour shifts. They found that on-the-job injuries decreased with the 12-hour system. However, in both systems, accidents occurred more frequently during the day even though production and work rates remained at the same level. In contrast, Parkes and Swash (1999) found that twice as many fatalities and serious injuries occurred on the night shift, independent of days into tour. Miles (2000) cites an HSE research project investigating drill floor accidents with regard to shift patterns, time of day and time of year for the following shift patterns: 0000-1200/1200-0000 and 0600-1800/1800-0600. No difference in incident numbers was found between the two shift patterns, though season did significantly affect the number of incidents for both. Also incidents occurred around midday for both shift patterns and around crew changeover days (50% more incidents occurred around this time though the number of incidents was only 16), with an increase around midnight for the 12-12 shift in winter

Ships
Within the maritime industry Folkard (1997) found that collisions between ships at sea were more likely to occur during early morning hours with a peak between 0600 and 0700. These data were derived from a sample of 123 collision claims made between 1987 and 1991 (UK P & I Club, 1992, cited by Folkard, 1997). Marine pilotage accidents have also been found to show circadian variation, with two peaks occurring between 0400 and 1000, and 1600 and 2400 (Smith & Owen, 1989). Thus, it appears that high performance demands during the night may pose safety and occupational health hazards within the maritime industry. Although, reported accidents may be just the observable portion of a much greater number of unsafe behaviours and mishaps.

Survey report of fatigue, stress, health and illness

Offshore Installations
Working in the offshore oil industry has been associated with higher levels of self-reported stress due to factors such as social isolation (see the review by Parkes, 1998). However, recent changes within the industry have demanded more of the remaining workers in terms of efficiency and flexibility. In general, installation workers show 'stable extravert' personality characteristics (Parkes, 1993), a supposedly adaptable personality trait and have been seen as forming a 'healthy worker' group reflecting the high medical standards required of them, in comparison to onshore occupational groups. However, they reportedly have significantly higher levels of free-floating anxiety (Cooper and Sutherland, 1987). Parkes (1993) in a questionnaire survey of 172 control room operators found workers showed greater anxiety, sleep problems, dissatisfaction with shift schedules and higher perceived workload. This suggests "that psychosocial stressors in the offshore environment may play a significant role in mental health" (Parkes, 1998).

Ships
There is little evidence of detrimental effects of long and unsociable work hours in the short-sea and offshore shipping industries. Wigmore (1989) surveyed masters of offshore supply vessels and found they tended to work longer hours than other crewmembers, sometimes in excess of 19 hours per day. In a survey of over 1,000 officers across all sectors NUMAST (1995) concluded that reduced crew size (and therefore increased workload) was the main cause of fatigue in seafarers: shifts of between 12-20 hours (upwards of 85 hours per week) were commonly reported. NUMAST (1995) suggested that better work organization and closer monitoring of small crews might prevent

unnecessary stress and fatigue 'on the job'. However, it is not possible to ascertain from this data which ships and work systems would be optimum for achieving this aim.

Physiological Data

Offshore Installations

Physiological data from oilrig personnel working which reflect fatigue, stress and health also has been inadequately covered, although this is beginning to change. Circadian adaptation is one area that has been investigated to some extent and it has been shown that contrary to onshore workers, offshore platform personnel can adjust to night shift work (e.g. Arendt & Deacon, 1997) indicating that once offshore, circadian rhythms may adapt to the environment. Interestingly, although rig workers show best adaptation to 2-week shift schedules, this does not appear to have any performance benefits. Such adaptation does not appear to hold true for seafarers and reveals an interesting difference between the two groups of offshore workers.

Ships

Amongst seafarers several studies have examined the physiological status of ships' pilots in terms of stress and fatigue. Shipley (1978) examined heart rate as a stress indicator and found, broadly, that as job complexity increased, so did heart rate and therefore stress levels. Cook and Cashman (1982) studied ECG recordings of ships' pilots and the incidence of ectopic beats, thought to be activated by stress. They found the occurrence of ectopic beats was more common under demanding or hazardous pilotage conditions, although the magnitude of the effect is difficult to determine. Furthermore, whether pilots have a higher incidence of these irregular beats than the general population is difficult to ascertain.

Performance

Offshore Installations

Studies that have investigated the performance effects of working on offshore installations are limited. To an extent parallels as to the detrimental effects of nightwork and longer shifts on performance can be drawn from onshore studies. One of the few to specifically examine performance is McPherson (1999) who studied 20 personnel onboard an offshore platform to investigate the impact of fatigue as a result of tour length on cognitive processes. He found that 'fatigued' personnel were significantly slower at responding to both control and stroop conditions and made more errors than the 'alert' personnel. Although results were not reported in statistical form and details such as time of testing and number of tasks were omitted it does suggest that tour length does have an impact on attentional processes and warrants further research.

Ships

Amongst seafarers the relationship between fatigue and performance has also been neglected. Again, parallels can be drawn from onshore studies and it is highly likely that the same relationships would hold true for seafarers. Condon et al. (1986) in a study of watchkeepers, on a "4on/8off" routine and day-workers, found that the speed of a complex visual performance task, and subjective alertness ratings decreased slightly during the early hours and peaked during the day. Condon et al. (1988) also found that task speed, in relation to its peak level, is slowest at the beginning of watches starting at 0400 or after recent awakening. Thus they suggest that there should be a provision for an adequate "waking up" period before the start of the duty They also concluded that operational effectiveness variations could be reduced by watchkeeping systems, which allow a single long sleep per day.

A more substantial body of evidence details the effects of vessel motion, which may in turn induce fatigue, on performance, although, results differ depending ship type and experimental tasks employed. For example, Wilson et al. (1988, cited in Powell & Crossland, 1998) using a simulator found that cognitive processing was significantly slower as a result of motion, although no information regarding total motion exposure time was available. Furthermore, it is not possible to ascertain from these data whether the accuracy, as well as the speed of cognitive processing was affected. While, Pingree et al. (1987, cited in Powell & Crossland, 1998) found evidence to suggest that motion degrades performance on a psychomotor tapping task, although not on computer-based cognitive tasks. It would therefore appear that certain types of cognitive task are more sensitive to the effects of vessel motion than others.

Summary

It would appear that the majority of research into health and stress in the offshore oil industry has concentrated on oil-rig workers, although there is scarce mention of the role of fatigue. Similar studies in the maritime industry have tended to overlook offshore sectors altogether, in favour of more intuitively demanding occupations, such as fishing (e.g. Johnson, 1994). It is certainly evident that little has been done to determine the nature and causes of fatigue at sea since Brown (1989) concluded that excessive work hours were a major contributor to accidents. Furthermore, having identified the available literature relating to fatigue, health and injury in the offshore oil industry, it is apparent that large gaps currently exist within our knowledge of a number of areas, including accident causation, physiological state and cognitive performance.

Acknowledgement: Preparation of this review has been supported by the MCA, HSE and SIRC

References

Arendt J. and Deacon S. 1997, Treatment of circadian rhythm disorders – melatonin. *Chronobiology International*: **14** (2), 185-204

Brown, I.D. 1989, *Study into Hours of Work, Fatigue and Safety at Sea*. Cambridge: Medical Research Council, 119p

Collins, A., Matthews, V., and McNamara, R. 2000, Fatigue, health and injury among seafarers and workers on offshore installations: a review. SIRC Technical Report

Condon, R., Colquhoun, P., Plett, R., Knauth, P., Fletcher, N. and Eickhoff, S. 1986, Circadian variation in performance and alertness under different work routines on ships. In M. Haider, M. Koller and R. Cervinka (Eds). *Night and Shiftwork: Long-term Effects and their Prevention.* Frankfurt: Verlag Peter Lang. pp277-284

Condon, R., Colquhoun, P., Plett, R., De Vol, D. and Fletcher, N. 1988, Work at sea: a study of sleep, and of circadian rhythms in physiological and psychological functions, in watchkeepers on merchant vessels. IV. Rhythms in performance and alertness. *International Archives of Occupational and Environmental Health,* **60**, 405-411

Cook, T.C. and Cashman, P.M.M. 1982, Stress and ectopic beats in ships' pilots. *Journal of Psychosomatic Research,* **26** (6), 559-569

Cooper, C.L and Sutherland, V.J 1987. Job stress, mental health, and accidents among offshore workers in the oil and gas extraction industries. *Journal of Occupational Medicine.* **29** (2), 119-125

Folkard, S. 1997, Black times: temporal determinants of transport safety. *Accident Analysis and Prevention,* **29** (4), 417-430

Johnson, D.G. 1994, Job stress, social support and health amongst shrimp fishermen. *Work and Stress,* **8** (4), 343-354

Laundry, B.R and Lees, R.E.M 1991, Industrial accident experience of one company of eight and 12 hour shift systems. *Journal of Occupational Medicine.* **33**, 903-906

McPherson, G. 1999, Too tired to stay alert. *Safety and Health Practitioner,* **17**(5), 16-18

NUMAST 1992. Conditions for Change. London, NUMAST

NUMAST 1995 All in Good Time. London, NUMAST

Miles, R 2000, Developments in the understanding of working on extended nights offshore. Paper presented at the SPE International Conference on Health, Safety, and the Environment in Oil and Gas Exploration and Production, Stavanger, Norway, 26-28 June 2000

Nielsen, D. 1999, Deaths at sea-a study of fatalities on board Hong Kong-registered merchant ships (1986). *Safety Science*, 32, 121-141

Parkes, K.R. 1993, Human factors, shiftwork and alertness in the offshore oil industry. Health and Safety Executive, HMSO Books, London, Offshore Technology Report OTH 92 389, 105pp

Parkes, K.R. 1998, Psychosocial aspects of stress, health and safety on North Sea installations. *Scandinavian Journal of Work, Environment and Health*, 24 (5), 321-333

Parkes, K.R. and Swash, S 1999 Injuries on offshore oil and gas installations: An analysis of temporal and occupational factors. Report prepared by University of Oxford under HSE-OSD Contract MaTSU/8843/3516

Powell W.R. and Crossland P. 1998, A literature review of the effects of vessel motion on human performance. INM Technical Report No 98027

Raby, M. and McCallum, M.C. 1997, Procedures for investigating and reporting fatigue contributions to marine casualties. *Proceedings of the Human Factors and Ergonomics Society 41ˢᵗ Annual Meeting*

Shipley, P. 1978, A human factors study of marine pilotage. *Ship and Marine Technology Requirements Board, DTI*

Smith, A.P. and Owen, S. 1989, Time of day and accidents in marine pilotage. In: G. Costa, G.C Lesona,, K. Kogi and A Wedderburn (eds.) Shiftwork: Health, Sleep and Performance, Studies in Industrial and Organizational Psychology, Vol. **10**. (Frankfurt: Peter Lang), 617-622

Wigmore, L.M. 1989, Are our working hours safe? An independent look at working practices aboard British oil-rig supply vessels. *Seaways*, November, 7-8

AN ANALYSIS OF ACCIDENTS IN THE OFFSHORE OIL INDUSTRY

Rachel McNamara, Alison Collins and Victoria Cole-Davies

Centre for Occupational and Health Psychology
Cardiff University
63 Park Place
Cardiff, CF10 3AS

Two datasets containing accident-related information from short-sea vessels were analysed in terms of temporal and environmental factors, such as time of day, hours into shift, days into tour and motion. Accident frequency was found to demonstrate a time of day effect (incidence is higher between 09:00-16:00 hours), although there is no real evidence to suggest that fatigue was a causal factor. With regards days into tour, injury occurrence is greatest during the first week, however analyses of temporal factors in terms of injury severity and accident and injury type failed to yield additional information to that provided by overall incidence rates. In conclusion, the present analyses provide little evidence for a major role of fatigue in offshore accidents. This may be due to the problems inherent within the datasets e.g. lack of exposure information and large amounts of missing data within significant temporal and environmental variables.

Introduction

Little is currently known about the underlying causes of accidents at sea. Although accident-reporting systems do exist, systematic analysis of incidents has been somewhat lacking. The purpose of this report therefore, was to collect accident data from several existing sources and to determine the role of fatigue (as indicated by temporal, environmental and occupational factors) in causation. The report focuses specifically on short-sea vessels working in the offshore oil industry and examines similar variables and issues described in a recent analysis of injuries occurring on offshore oil and gas installations (Parkes & Swash: 1999). The main objectives can therefore be summarised as follows:

1.   To identify the extent to which accidents in short-sea shipping are attributable to temporal, occupational and environmental factors.

2.   To identify similarities and differences that exist in injury and accident causation between such vessels and offshore oil and gas installations.

There exists a significant body of self-reported data relating to working hours, shift work, the environment and occupational injury (e.g. Parkes; 1999, Sutherland & Cooper; 1991). However, studies that have been conducted tend to concentrate on the onshore population. It becomes apparent when reading the literature that the level of detail included in occupational accident reports varies widely. Williamson and Feyer (1995) examined Australian work-related fatalities within the context of time of day, specific injury details and contextual factors. Most accident databases allow for comparisons of this type, what is often missing however, are details relating to shift schedules – e.g. time into shift, and exposure rate information. Having said that, Williamson and Feyer (1995) were able to estimate exposure rates from a government survey. The results of the study indicated that fatalities were significantly more likely to occur at night than they were during the day.

Haneke et al. (1998) analysed a large sample of German work-related accidents in relation to time of day and hour into shift, using two separate exposure rate models in order to increase reliability. The results demonstrate a steady increase in incident frequency between 1-5 hours into shift, dropping off between 6-7 hours and increasing again between 8-9 hours. However, in the absence of appropriate exposure data, some studies have examined incidents in terms of comparative severity. For example, Jeong (1999) examined fatal accidents in relation to non-fatal ones, and found injury incidence to be highest during the first two hours of the day shift (N.B. this finding held true for both fatal and non-fatal accidents).Although useful in different ways, data in the above studies were collated from onshore workers. However, working in the offshore oil industry introduces very specific issues, both occupational and environmental, which will impact upon offshore personnel in addition to factors known to affect their onshore counterparts. There have been comparatively few studies examining such issues on oil and gas installations, although Forbes (1998) studied accident rates amongst drillers using exposure data supplied by the company whose records were examined, in terms of both hours into shift, and days into tour. Forbes' (1998) study demonstrates that accident rates are unusually high during the first hour of a shift, although total incident rates decline steadily over the course of a 12-hour shift. With regards days into tour, it would appear that a greater incidence of injuries occur in the first seven days of a tour than during the second week.

Parkes & Swash (1999) examined offshore accident and injury data from three sources: an industry wide database supplied the Health and Safety Executive, and records supplied by two large multinational oil companies. The main findings of the Parkes and Swash (1999) study can be summarised as follows:

1.   Days into tour: the ratio of fatalities and serious injuries to those lasting three or more days increases after the second tour week.
2.   Day versus night shifts: the distribution of injury severity is significantly different across both shift types. There is a significantly higher incidence of fatalities and serious injuries on the night shift (an effect found to be independent of days into tour).
2.   Hours into shift: injury severity was generally found to be independent of hours into shift, except when considered as a dichotomy: more serious/fatal injuries occurred after 12+ hours into shift than during the first 12 hours (this effect was found to be particularly evident amongst drillers).
3.   Clock hours: more 3+ day injuries than any other type occurred during all two-hour time periods (with the exception of 23:00-01:00). Due to a high rate of

serious/fatal injuries occurring between 23:00 and 01:00, the pattern of injury frequency in relation to severity during this period was reversed.

4.   Injured body part: injuries to the hand, arm or shoulder were found to be the most frequent, and accounted for the majority of crushes. Legs were the next most likely body part to be injured and accounted for the majority of sprains and strains. Injury type was found to be independent of both occupation and shift type.

It is important to note however, that exposure rates were not available for the data used by Parkes and Swash (1999) but rather that serious and fatal injuries were examined in relation to more minor ones. The current study therefore seeks to examine all of the variables described above specifically in relation to accidents and injuries occurring aboard vessels and to compare and contrast these findings to the pattern described for offshore platform workers. Specifically, all incidents were examined in terms of time of day, hours into shift, days into tour, motion (inferred from sea state and wind force), injury severity (i.e. fatalities, serious injuries) and injury type (i.e. burns/scalds, strains/sprains).

Data Used In The Study

This study analysed data from two sources. Dataset 1 refers to records obtained from a multinational oil company, and dataset 2 was provided by the MCA and refers to incidents reported to the MAIB. Both datasets include incidents occurring between 1989-1999. Both datasets include details of injuries incurred by personnel working on offshore oil support vessels. Unfortunately however, the datasets did not include identical information; dataset 1 is primarily concerned with relatively minor injuries, whereas dataset 2 deals in the main with major injuries. Nonetheless, comparisons between the two can be made in terms of time of day and shift effects.Some modifications were made to each dataset in order that only personnel working on merchant vessels were included in the analyses. Dataset 1 was modified to exclude all personnel working on installations, and in dataset 2, incidents occurring aboard fishing and passenger vessels were excluded. There are however, a number of problems inherent within both datasets. Largely as a result of inadequacies in original incident reporting systems, it was not possible to gain estimates of exposure rates, and large amounts of potentially useful data within temporal and injury severity variables was defined as missing.

Demographics

The following section of the report details information relating to injury type and ship area in addition to age and occupational status.

Dataset 1

Age of crewmembers in dataset 1 ranged from 18 to 67 years, with the highest incident rate occurring in those aged 34-44 years (50%). Recorded injuries extend over 108 different occupations. Contractors account for the greatest number of injured workers (27.7%) followed by marine personnel (23%) and divers (2.4%). The most common incident area was the open deck (accounting for 53.8% of total accidents). With regards to body part, arms were the most frequently injured (34.6% of incidents) followed by legs (20% of total incidents).

Dataset 2
The distribution of injuries across age categories in dataset 2 is shown in Table 2.

Table 2: Injury Distribution & Age (Dataset 2)

Age Category	% of dataset 2 (n=4145)
< 19	3.1
20-29	18.5
30-39	28.7
40-49	28.1
50-59	16.7
60 & over	4.9

56.6% of the sample consisted of ratings, 35.3% of masters/cadets/officers, and just 8.2% of catering staff/stewards.One-third (33.4%) of the accidents reported in dataset 2 were classified as collisions/contacts, 25.5% as strandings/groundings and 22.9% as machinery faults. 29.3% of incidents were classified as slips/trips or falls.

Summary of Main Findings

Accident Distribution as a Function of Time of Day and Hours into Shift
Chi-square tests were applied to all data in order to examine potential relationships between accident and injury rate, temporal, occupational and environmental variables. It is important to note however, that each dataset was considered separately, due to the differences between them highlighted earlier in the report. The following section is a summary of the general trends indicated by the both datasets, and not a detailed analysis of similarities and differences between them. The distribution of incidents across time shows a very similar pattern for both datasets. General distribution of accidents across time of day is shown in Figure 1. The majority of incidents were found to occur between the hours of 09:00-16:00, an effect found to be independent of whether personnel were on or off duty. Incident frequency was significantly greater during the first four hours of a shift. Although time of day effects were evident from these analyses, they do not correspond to natural troughs in circadian rhythms: accidents do not increase after lunch, or between 02:00-06:00. However, sleep inertia may explain in part why accidents tend to be higher at the beginning of a shift, although this is purely speculative as information regarding sleep patterns was unfortunately not available in this instance.

Accident Distribution as a Function of Days into Tour
Accident frequency was found to be greatest at the beginning of a tour, specifically during the first tour week, and then declined steadily over the course of a tour. Again, when examined in terms of on and off-duty incidents, the same pattern was evident for both groups.

Accident Distribution as a Function of Sea State
Accident distribution was found to differ significantly as a function of sea state. More specifically, a greater proportion of incidents occur in calm conditions (i.e. low − moderate wind force, and calm seas). However, this somewhat peculiar finding may simply be a reflection of work patterns: in other words, it is more than likely that a greater proportion of personnel are exposed to potential incidents in calm conditions, as they are more likely to be working.

Figure 1: Accident Distribution and Time of Day

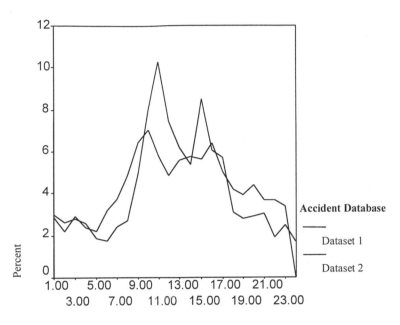

Time (Hourly intervals)

Injury Severity and Accident and Injury Type
Re-analysis of temporal and environmental variables in terms of injury severity and incident type did not shed any further light on the possible role of fatigue in accident causation. It might therefore be concluded that although injuries generally show time of day effects, subdividing these into severity categories does not yield any additional information.

Conclusion

Although the two datasets deal in the main with different injury types (the majority of incidents detailed in dataset 1 are relatively minor, whereas in dataset 2 they are, for the most part, major) they generate a very similar pattern of results. In both cases, accident frequency demonstrates a time of day effect (incidence is higher between 09:00-16:00 hours) although there is no evidence that fatigue is a causal factor. Accidents do not peak significantly during any time periods associated with natural circadian troughs (i.e. post lunch or in the early hours of the morning). However, some evidence is presented to suggest that sleep inertia impacts upon accident frequency, as accident rates are highest in the first few hours of a shift. With regards days into tour, (it was only possible to analyse information from dataset 1 in this instance) it seems that the occurrence of injury is greatest in the first week. However, it should be noted that database 1 contains a limited amount of data regarding fatalities and major injuries.

These results have however replicated some previous findings: for example, Jeong (1999) found the greatest number of accidents to occur in the first 2 hours of the day shift and Forbes (1998) demonstrated that there are significantly more injuries during the first seven days of a tour. It is not however possible to ascertain from the data analysed in the current report, whether serious injuries increase after the second tour week in relation to more minor ones (Parkes & Swash; 1999). Furthermore, the finding that accidents are generally more likely to occur at night (Williamson & Feyer: 1995) was not replicated. Unfortunately, due largely to inadequacies inherent within reporting systems, it was not possible to examine injury severity in terms of time into shift (not enough information relating to injury severity was available).

It is important to note that the absence of a link between accidents and fatigue indicated by these results does not mean that fatigue has no impact: rather that it is not possible to determine from data of this type. Specifically, information regarding days into tour and hour into shift was somewhat patchy, and no exposure rate information was available against which to qualify the findings. Neither was it possible to accurately match serious injures against more minor ones to provide a proxy measure of exposure. Therefore, if the role of fatigue in accident causation is to be accurately estimated, it is vital that reporting systems are standardised across the industry and information relating to days into tour, hours into shift, injury severity and exposure levels is recorded.

Acknowledgement: The research reported here was supported by MCA, HSE and the Seafarers International Research Centre.

References

Forbes, M.J 1997 *A study of accident patterns in offshore drillers in the North Sea.* Dissertation prepared for the Diploma of Membership of the Faculty of Occupational Medicine of the Royal College of Physicians

Hanecke, K, Tiedemann, S, Nachreiner, F, and Grzech-Sukalo, H 1998 Accident risk as a function of hours at work and time of day as determined from accident data and exposure models for the German working population. *Scandinavian Journal of work, Environment and health*, **24** (3), 43-48

Jeong, B.Y 1999 Comparison of variables between fatal and nonfatal accidents in manufacturing industry. *International Journal of Industrial Ergonomics*. **23**, 565-572

Parkes, K.R. & Swash, S. 1999 *Injuries occurring on offshore oil and gas installations: An analysis of temporal and occupational factors.* HSE OSD contract Ma TSU/8843/3516

Williamson, A.M. And Feyer, A.M. 1995 Causes of accidents and the time of day. *Work and Stress*, **9**, 158-164

FATIGUE, HEALTH AND INJURY OFFSHORE: A SURVEY

Victoria Cole-Davies

*Centre for Occupational and Health Psychology,
Cardiff University, 63 Park Place, Cardiff, CF10 3AS*

Surveys of the health and well-being of seafarers are few and far between. Previous studies relating to offshore installations have indicated that through changing circumstances, offshore workers often have to work longer hours, shifts and tours of duty, which may impact upon health and well-being. The present survey was designed to look for evidence of fatigue in both seafarers and offshore installation workers, and examined all possible contributory factors during both time spent offshore and time spent on leave. The first phase of the study was reported here and refers to crewmembers working in the short sea shipping industry. The results showed that there is very little evidence of fatigue or any detrimental effects caused either directly, or indeed indirectly by the offshore working environment.

Introduction

Research findings from other transport industries with regard to the role fatigue plays in the occurrence of accidents, cannot automatically be applied to the offshore industry. This is due to the unique combination of factors with which workers have to cope. Extreme weather conditions, noisy working environments, demanding work and rest patterns and motion, all contribute to the impact of fatigue in the individual (Parkes, 1997; 1998). Furthermore, major economic, structural and technological changes have taken place within the offshore industry in recent times, which have led to reduced installation and vessel manning, increased workload and job insecurity (e.g. Collinson, 1998; NUMAST, 1992). All the afore mentioned factors either alone, or in various combinations, are proposed to have an impact on the health and well-being of offshore workers.

Surveys of the psychological effects of working in the offshore industry have been in the main inconclusive. Although a number of studies have been conducted into stress and health offshore, they rarely refer to fatigue explicitly, or do not identify sufficient factors necessary to determine whether fatigue can have an impact in workers. Furthermore, the majority of surveys carried out in this area have applied to offshore installations only. There is some evidence that psychosocial stressors unique to the offshore environment do impact upon mental health (Parkes, 1997; 1998), yet other studies have either failed to examine the psychological state of participants, or concluded that they were no worse off than their onshore counterparts (Gann et al., 1990; Parker et al., 1998). However, it is apparent from the self-reported data that increased workload, excessive work hours, poor quality or lack of sleep and feelings of boredom and isolation all contribute to stress, poor mental health and fatigue offshore (Hellesoy, 1985; Parkes, 1997; Parker et al, 1997; 1998a 1998b).

More specific information relating to levels of anxiety and depression, the relationship between work hours, sleep and fatigue, and information regarding the numerous other contributory factors to fatigue, during both work and leave time is notably limited in the maritime sector. Therefore, in order to bring maritime research more in line with existing information on offshore installations, and to enable researchers to gain an accurate representation of the impact of working conditions in both sectors, the present survey was carried out. The survey was designed to identify all possible aspects of the offshore working environment, which may affect fatigue levels, and to determine the general health and well-being of offshore workers during both work and leave time. As a result of the survey, it is hoped that a comprehensive report of the incidence of fatigue, health and injury offshore environment is obtained, which pinpoints specific risk factors in offshore workers.

Aim of the Survey:
- To assess the work and rest patterns of Seafarers and Offshore installation workers
- To assess the extent to which working hours, shift patterns and time spent offshore are associated with fatigue, accidents and injuries, and poor physical and mental health in crewmembers
- To assess the impact of time spent offshore on leave time

The questionnaire (33 pages) was designed to encompass all aspects of an offshore worker's life, and assessed voyage cycles, work and rest patterns, fatigue and sleep, health-related behaviours and general health and well-being. It was divided into the following three sections:

1. *Offshore*: Questions refer to time spent offshore, encompassing measures of work and rest patterns and practices, and feelings about work.
2. *On Leave*: Questions relate specifically to time spent on leave/at home and include scales of health and well-being, fatigue, sleeping habits, and health-related behaviours such as exercise, smoking, eating and drinking habits.
3. *Life in General*: Refers to life in general, assessing accidents and injuries, general fatigue and general health and well-being using a number of standardised health scales such as the General Health Questionnaire (GHQ), Profile of Fatigue Related Stress (PFRS), the Cognitive Failures Questionnaire (CFQ) and the Moss Short Form Health questionnaire (SF-36).

The questionnaire was distributed to three target samples. Firstly, questionnaires (423) were distributed to crewmembers onboard offshore support vessels by visiting researchers. This reflected the initial phase of the project, which targeted short sea shipping crewmembers. The questionnaire will shortly be distributed in the same fashion to workers onboard other short-sea shipping vessels such as standby vessels and shuttle tankers. Several hundred questionnaires were also sent out to members of the officers union NUMAST (1600) and also to members of the offshore installation workers union MSF (1800), accompanied by a cover letter from the union leaders underlining the importance of the survey.

The initial phase of the project, which targeted short sea shipping crewmembers, yielded a response of 20% (85 out of a possible 423 respondents). The following section summarises the information gathered from a sample of crewmembers onboard offshore support vessels (a pipe-layer, and two dive support vessels) operating in the North Sea.

Demographics, work and rest patterns offshore

The average age of the short sea shipping respondents was 42.85 years, ranging from 22 to 63 years of age. The sample comprised mainly of men (96.5%) who were mostly either married (68.2%) or living with a partner (14.1%), only 13% were single or divorced. Nearly half of all respondents were officers (42.4%), with stewards and catering workers comprising 11.8% of the sample. 21.2% of the sample were project workers onboard the offshore support vessels, and 17.6% and 5.9% were managers and divers respectively. Crewmembers had spent an average of 18.59 (11.33) years at sea, though 52.9% had also worked for periods offshore, with 16.2 % having returned to sea for financial reasons. 61.2% of crewmembers worked 4 weeks on 4 weeks off tour duty, and whilst onboard 84.7% of the sample worked the 12 hours on and 12 hours off shift pattern, Crewmembers mostly worked on a fixed shift basis (71.8% of the sample), and 51.8% often worked nights. Reinforcing the claim that offshore workers work excessive hours, a large number of respondents (50.6%) claimed to work over 12 hours per day, the average longest period spent on continuous duty being 14.46 hours (5.64). 35.3% claimed to work more than 85 hours per week, with 50.6% working between 60 and 80 weekly hours.

Crewmembers' ideal sleep length was as an average of 7 ½ hours, whereas the average time spent sleeping was just over 7 hours, proving that there was not much discrepancy between the two. Furthermore, 80% claimed that they have the opportunity to gain 10 hours rest in every 24-hour period, and 90.6% of the sample also stated that they are able to obtain at least 6 hours of uninterrupted sleep (table 1). The general consensus appeared to be that sleep was not disrupted by environmental conditions such as motion and bad weather, although 29.4% of the sample claimed that noise disrupted their sleep quite a lot.

Table 1. Opportunities for sufficient sleep and rest periods onboard

	Opportunity for 10 hrs rest %	Opportunity for 6 hours sleep %
Yes	80	90.6
No	20	9.4

68.8% of short-sea shipping crewmembers believed that the effects of fatigue increases the longer they are at sea, and that the most difficult period with regard to fatigue was during the night shift, with 22.4% of the sample naming the period between midnight and 4am and 34.1% claiming the period between 4am and 8 am as the most difficult time. 24.7% still feel 'below par' on return to sea, even after a period of leave. Just over 50% of crewmembers said that it takes them 2-3 days to adjust to life onboard once more, after a period of leave, and 10.6% stated that it could take them up to a week to readjust. During the period of adjustment, 37.6% believe that their performance is adversely affected. However, only 8.2% of crewmembers admitted to having been involved in a fatigue-related incident or accident offshore.

According to their ratings of fatigue symptoms experienced whilst offshore (table 2), there did not appear to be a great deal of evidence that suggested fatigue was a wide spread problem for the crewmembers. Most respondents (around 70% in each case) rated the various symptoms such as confusion, poor sleep, depression, tension, concentration problems as ranging from 'hardly experienced' to 'not experienced at all'. Furthermore, the same was true for ratings of fatigue during leave time (table 2).

Table 2. Percentage of crewmembers who rated on a scale of 1 – 5 (1 = very: 5 = not at all), the extent to which they experienced fatigue symptoms offshore and on leave

Rating	Confusion		Lethargy		Poor quality sleep		Depression		Tension		Loss of concentration		Increased use of caffeine	
	Sea	Leave	Sea	Leave	Sea	Leave	Sea	Leave	Sea	Leave	Sea	Leave	Sea	Leave
1	4.7	2.4	8.2	5.9	11.8	2.4	2.4	0	8.2	1.2	7.1	2.4	11.8	1.2
2	2.4	2.4	7.1	4.7	9.4	4.7	8.2	8.2	14.1	7.1	8.2	8.2	9.4	2.4
3	4.7	4.7	25.9	14.1	29.4	10.6	9.4	8.2	25.9	15.3	24.7	9.4	18.8	7.1
4	28.2	7.1	24.7	23.5	21.2	24.7	29.4	10.6	27.1	20.0	27.1	14.1	20.0	10.6
5	50.6	74.1	24.7	44.7	21.2	48.2	40.0	62.4	20.0	48.2	25.9	57.6	31.8	67.1

In addition, in general, although some respondents claimed to suffer from a little tension (51.8%), mental and physical tiredness offshore (44.7% and 38.5%, respectively), and experienced sleepiness and tiredness 2-3 times per week (23.5% and 35.3%, respectively), 80% still claimed their typical state to be either alert (63.5%) or very alert (16.5%) during working hours.

General health and well-being

It appeared that crewmembers are very healthy, with 89.4% of respondents claiming not to have taken any day's sick leave whilst offshore in the previous 12 months. Whilst on leave, 78.8% stated that they hardly ever visit their GP, and 80% have not experienced sick-days during their leave period. When asked about how stressful they thought their job was, at worst, only 28.2% found it to be moderately stressful. The majority of the sample (63.5%) found work mildly or not at all stressful. In fact, only 6% of the seafaring sample fell into the stressed category (table 3). However, 62.4% of crewmembers stated that they suffer from the after-effects of fatigue from working at sea, in their leave time (table 4.) and 40% claimed that the effect lasts for up to 3 days, or for up to a week for 15.3% of the sample. Furthermore, 43.5% of respondents believed that their performance was affected during this period.

Table 3. Work stress

Work stress	%
Not at all stressful	23.5
Mildly stressful	40
Moderately stressful	28.2
Very stressful	4.7
Extremely stressful	1.2

Table 4. After-effects of fatigue from working at sea

After-effects	%
No	35.3
Yes	62.4

It was evident from the responses to the standardised questionnaire scales (table 5) such as the General Health Questionnaire (GHQ), Cognitive Failures Questionnaires (CFQ), Profile of Fatigue Related Stress (PFRS) and the Moss Short Form Health Questionnaire (SF-35), that short sea shipping crewmembers did not appear to demonstrate any detrimental of the effects of fatigue, or suffer from stress and ill health to any real extent. In fact, the short sea shipping sample compared favourably to the onshore normative group on all standardised measures of fatigue, health and well-being, suggesting that they are probably quite a resilient population, able to cope with the working and psychosocial demands placed on them by the offshore environment.

Table 5. Standardised scales for mood, health and well-being

Scale	Short sea shipping mean	Onshore group mean
SF-36 Physical functioning	93.05 (8.67)	77.6 (31.5)
SF-36 Role Functioning (physical)	94.94 (15.70)	75.0 (38)
SF-36 Role Functioning (emotional)	92.09 (19.34)	75.0 (38)
SF-36 Social functioning	90.01 (18.24)	82.1 (27.5)
SF-36 General health	75.04 (19.41)	65.6 (22.7)
SF-36 Mental health	77.98 (16.10)	74.2 (13.3)
SF-36 Vitality	69.82 (17.97)	61.0 (22.1)
SF-36 Bodily pain	84.33 (17.29)	73.9 (29.4)
PFRS Emotional distress	30.40 (17.17)	36.9 (16.1)
PFRS Fatigue	22.91 (11.55)	28.6 (14)
PFRS Cognitive difficulty	20.99 (10.72)	26.2 (11.5)
PFRS Somatic symptoms	22.72 (9.08)	26.9 (13.3)
GHQ	21.94 (4.13)	28.2
CFQ	35.38 (13.38)	43.5 (17)

SF-36: High score = better health/functioning. Standard deviations in parenthesis

Discussion

It was fairly apparent from the results of the offshore survey, which specifically targeted short sea shipping crewmembers, that there was very little evidence of any fatigue effects, or indeed any accidents or incidents which may have occurred either directly or indirectly as a consequence of fatigue. It was clear that working hours are long, shifts demanding, and there was evidence that the majority of crewmembers once on leave, claim to suffer from fatigue as an after-effect from working offshore, and believe that their performance is detrimentally affected during this period. However, it was also apparent from measures of self-reported alertness and attention, stress, health and well-being, that these factors and perceptions were not reflected. Offshore workers appear unaffected both whilst offshore and on leave. Then again, whilst offshore, reportedly there are opportunities for adequate sleep and rest periods. Sleep quality and quantity appear to be sufficient, with little disruptions reported due to the environmental conditions. It seems likely that a buffering process in place, whereby any possible detrimental effects of the working hours and environmental conditions are lost to the restorative benefits of sleep. For this particular sample, it is clear that fatigue was not cumulative and certainly not a chronic problem. The fact that only a small percentage of short sea shipping workers suffered from stress, and that overall, they compared favourably to an onshore comparison group on measures of health, depression, anxiety and cognitive dysfunction, also suggested that they were a perfectly healthy and undeniably 'normal' population. This also supports the concept that seafarers are more psychologically robust than other populations (Parker, 1997; 1998) in that they have to cope with the demanding conditions offshore, and do so quite effectively. Although the results must be viewed with caution and cannot be automatically applied to other seafaring populations due to the wide variation in working practises and conditions

Acknowledgement: The research described in this article was supported by MCA, HSE, NUMAST, MSF and SIRC.

References

Collinson, D.L. 1998, "Shift-ing lives": Work-home pressures in the North Sea oil industry. *Canadian Review of Sociology and Anthropology.* 35 (3), 301-324

Gann, M., Corpe, U., and Wilson, I., 1990, The application of a short anxiety and depression questionnaire to oil industry staff. *Journal of the Society of Occupational Medicine*, 40, 138-142

Hellesoy, O.H (1985). *Work Environment: Stratjord Field* (Universitetsforlaget, Oslo).

NUMAST, 1992, *Conditions for Change*, (NUMAST, London)

Parker, A.W. and Hubinger, L., 1998, On tour analyses of the work and rest patterns of Great Barrier Reef pilots: Implications for fatigue management, *Australian Maritime Safety Authority (AMSA)@ www.amsa.gov.au/sp/fatigue.90p*

Parker, A.W., Balanda, K., Briggs, L. and Hubinger, L.M., 1998, A survey of the work and sleep patterns of Great barrier Reef pilots, *Australian Maritime Safety Authority (AMSA)@ www.amsa.gov.au/sp/sleep.60p*

Parker, A.W., Hubinger, L.M., Green, S., Sargent, L., and Boyd, R, 1997, A survey of the health, stress and fatigue of Australian seafarers, *Australian Maritime Safety Authority (AMSA)@ www.amsa.gov.au/sp/fastoh.120p*

Parkes, K.R., 1997, *Psychosocial aspects of work and health in the North Sea oil and gas industry, part 1: review of the literature*, Health and Safety Executive, HSE Books, Sudbury, Suffolk, Publication No. OTH 96 523

Parkes, K.R., 1998, Psychosocial aspects of stress, health and safety on North Sea installations, *Scandinavian Journal of Work, Environment and Health*, 24(5) 321-333

GENERAL ERGONOMICS

ERGONOMIC EVALUATION OF A WEIGHTED VEST FOR POWER TRAINING

Philip Graham-Smith, Neil Fell, Gareth Gilbert, Joanne Burke and Thomas Reilly

Research Institute for Sport and Exercise Sciences,
Liverpool John Moores University,
Henry Cotton Campus, Webster St.,
Liverpool, L3 2E

Weighted vests are becomingly increasingly popular as resistance devices for power training. They provide additional resistance to the lower limb musculature for vertical jumping and sprinting drills, but this additional load may cause increased compressive forces on the spine. The aim of this study was therefore to quantify the mechanical loading of the spine by measuring changes in stature following unloaded and loaded jump training using a commercially available weighted vest. The PowerVest© has been uniquely designed to distribute the load in the lumbar region and over the shoulders. Spinal shrinkage was found to increase throughout the 5 sets of 10 vertical jumps, in both counter-movement and drop jump conditions. The addition of a PowerVest© with 10% body mass did not significantly increase spinal shrinkage over the period of testing compared to unloaded counter-movement jumps and drop jumps from a height of 20 cm. It was concluded that the PowerVest© provided a safe means of adding resistance to the body in power training activities under the conditions studied. Subjective responses also indicated that the PowerVest© was comfortable to wear, rarely impaired lower and upper limb movement and that performance enhancing benefits could be felt from wearing the PowerVest©.

Introduction

The ability of the muscles to generate power for explosive actions is critical for success in many sports. Explosive muscle actions are required in sports that involve both jumping and throwing movements. Equally, sudden bursts of power are essential for rapid changes in direction, or accelerating during various sports events, such as football, rugby, basketball, tennis and hockey (Newton and Kraemer, 1994).

The optimal training method to develop explosive speed and power is not always clearly established. For several decades sprint athletes have trained with various apparatus, such as pulling a tyre or sled, wearing a parachute or simply being held by a partner, to increase the external resistance and therefore overload the neuromuscular system to perform at a higher level (Dintiman *et al.*, 1998; Seagrove, 1996).

In addition to the implements described above for enhancing speed capabilities, the systematic loading of weight to the body in different forms (vest, pants, or suit) has perhaps received greater attention (Bosco et al., 1984; Bosco, 1985; Bosco et al., 1986; Sands *et al.*, 1996). This work has focused specifically on the enhancement of explosive power production in the vertical jump rather than horizontal speed development. Research by Bosco (1985) showed an average increase in vertical jump performance of 10 cm in male international athletes after wearing a weighted vest (at 11% of body mass) for only three weeks. With the additional benefits of being able to use a weighted vest for multi-directional activities, the use of weighted vests for sport specific training is likely to further increase their popularity as a training device.

Care needs to be taken in the implementation of adding weights to the body as it has been found that this increases the risk of injury. Research has demonstrated that wearing a weighted vest in jump training leads to higher than normal impact forces and greater stresses imposed on the spine (Fowler *et al.*, 1994). Compressive loading of the intervertebral discs increases their susceptibility to injury as they become narrow and stiff (Perey, 1957). The measurement of changes in stature due to compression of the intervertebral discs has been used as a reliable technique to study spinal loading in sports activities (Boocock, 1990; Leatt *et al.*, 1986). The technique has been used for evaluating spinal loading in a range of occupational and exercise training contexts.

The aim of this study was therefore to quantify the mechanical loading of the spine by measuring changes in stature following unloaded and loaded jump training using a commercially available weighted vest. The 'PowerVest©' has a unique design compared to other vests on the market with the load being distributed in pouches around the shoulders, lumbar spine and waist. An additional aim was to subjectively assess the design criteria, (comfort and fit) and performance enhancing effects of the PowerVest© with an international rugby union team.

Methodology

Six male subjects with previous experience of jump training volunteered for the study. The mean age, height, and body mass were 24 ± 3 years, 1.79 ± 0.07 m, and 88.6 ± 20.6 kg respectively. Each subject was required to attend the Human Performance Laboratory, Research Institute for Sport and Exercise Sciences, Liverpool John Moores University, on four separate occasions. A precision stadiometer was used to measure changes in stature (spinal shrinkage) (Althoff *et al.*, 1992), and on the first occasion subjects underwent training to familiarise themselves with this equipment.

On each of the three subsequent trials, subjects were required to attend the laboratory having participated in no physical activity 24 hours prior to testing. Subjects were instructed to rest with trunk supine and legs raised with knees flexed and ankles supported (Fowler's position) for 20 minutes prior to each test session to allow for a controlled period of spinal unloading. Subjects were required to perform counter-movement jumps, with and without the addition of a PowerVest© (10% of body mass), and drop jumps with no additional load, from a height of 20 cm. Counter-movement and drop jumps are widely used by athletic trainers to develop explosive speed and power.

For the counter-movement jumps, subjects were instructed to execute a downward movement at a self-selected speed and depth and immediately jump upward for maximum height. These instructions encouraged subjects to find their own optimum jumping conditions (Young *et al.*, 1998). Subjects were also encouraged to bend the knees upon landing to absorb the impact, then pause briefly to prepare mentally for the next repetition. This resulted in a pause between jump repetitions of approximately 2-4 seconds. For the drop jumps, the athletes were required to drop from the raised platform, and immediately on landing perform a maximal vertical jump (Bobbert, 1990). For both types of jump, all subjects were instructed to keep their hands on their hips to negate the effects of the arms, Figure 1.

Figure 1. Subject performing a counter-movement jump with the PowerVest.

Each session of testing consisted of five sets of 10 jumps, with 3 minutes recovery between each set. A 20 minutes standing recovery (with PowerVest© removed) followed all test sessions. The order of testing was performed in a counter-balanced design with at least 72 hours recovery between testing sessions. Subjects were also tested at the same time of day to control for circadian variation (Atkinson and Reilly, 1996).

Measurements of stature were taken pre-exercise, 2 minutes into recovery at the end of each exercise set, and after the 20 minutes standing recovery. The 2 minutes recovery enabled breathing to return to normal. The weighted vest was still worn during this period, being removed immediately prior to the measurement of stature. The pre-exercise data were obtained to elicit the individual's natural shrinkage (i.e. loss of height). These data were extrapolated to determine the predicted shrinkage over the session of testing. The final shrinkage value was the difference between this expected value and the observed value. At the time of each stature measurement, subjects were asked to rate their degree of back pain on a scale from 0 (no pain) to 5 (unbearable pain).

Questionnaires to examine athletes' reactions to the (i) design criteria (comfort and fit), and the performance enhancing effects of the PowerVest© were sent with an international Rugby Union Team on their summer tour of Australia in 1999. Each player who used the PowerVest© in training was asked to complete a questionnaire.

Results and Discussion

Following the final set of vertical jumps, spinal shrinkage increased by 1.84 mm ± 1.59 in the unresisted counter-movement jumps, 0.30 ± 1.60 in the drop jumps and 2.58 mm ± 1.39 in the counter-movement jumps with PowerVest©. The addition of the PowerVest© therefore does appear to elicit greater spinal shrinkage, but this was not statistically significant (F=3.46, P=0.058). Surprisingly there was a greater magnitude of spinal shrinkage for the unloaded and loaded counter-movement jumps, compared to the drop jumps from a height of 20 cm. This is an unexpected finding considering that drop jumps are associated with high impact forces upon landing (Fowler *et* al., 1994) and that subjects experience two landings in drop jumping compared to one in counter-movement jumps. Following qualitative analysis of video footage it was apparent that subjects had a tendency to flex the trunk further forwards in the counter-movement jumps. In the drop jump condition subjects adopted a more upright posture. This result may therefore be explained by the increased muscular activity of the erector spinae muscles required to counterbalance the pronounced forward flexion of the trunk in the counter-movement jumps. In all test sessions a small amount of the exercise-induced spinal shrinkage was reversed during the 20-minute standing recovery, although this was not statistically significant. Equally, none of the subjects tested complained of having any back pain, all giving a rating of zero for the three jumps conditions.

In relation to the design and performance aspects of the PowerVest©, the international rugby union squad used the PowerVest© in a range of training exercises, including speed, agility and plyometric drills in addition to sport specific drills like rucking and sheild contact drills. The responses from the questionnaires provided very positive comments to the design and performance enhancing benefits of the PowerVest©. The majority of the players (83%) found the PowerVest© 'very comfortable' to wear. Over half of the players (58%) reported that the PowerVest© 'rarely' restricted leg movement when sprinting or jumping, with 33% stating that 'no restriction' was incurred, i.e. to hip flexion. Two-thirds of the players (67%) reported that the PowerVest© 'rarely' restricted arm movement when sprinting or jumping, with 33% stating that 'no restriction' was incurred. Over half of the players (58%) agreed that the PowerVest© remained fixed in position during the sprint and jump exercises, although a minority (17%) did report movement of the vest when performing the different activities. Nevertheless, every player did report that they either felt faster, more agile or more powerful after removing the vest. The results lend support to the PowerVest© as an effective means of providing additional resistance to the athlete in speed, agility, plyometric and sports-specific training activities.

Conclusion

The additional load of 10% of the subject's body mass in the form of the PowerVest© did not significantly increase the magnitude of spinal shrinkage when performing counter-movement vertical jumps with a PowerVest© compared to jumps without. It can therefore be concluded that within a typical plyometric training session, i.e. 5 sets of 10 with 3 minutes recovery, the PowerVest© provides athletes and coaches with a safe device for resistance training. The results of this study lend support to the unique load distribution of the PowerVest© around the shoulders, lumbar region and waist. The same results may not necessarily be found for other commercially available weighted vests. As with all weighted vests caution should also be expressed for their over usage as the long term effects of wearing such devices has not been investigated.

Acknowledgement

The authors would like to thank PowerVest© for funding this project.

References

Althoff, I., Brinckmann, P., Forbin, W., Sandover, J. and Burton, K. (1992). An improved method of stature measurement for the quantitative determination of spinal loading: application of sitting postures and whole body vibration. *Spine*, 17, 682-693.

Atkinson, G. and Reilly, T. (1996). Circadian variation in sports performance. *Sports Medicine*, 21, 292-212.

Bobbert, M. F. (1990). Drop jumping as a training method for jumping ability. *Sports Medicine*, 9, 7-22.

Boocock, M.G., Garbutt, G., Linge, K., Reilly, T. and Troup, J.D. (1990). Changes in stature following drop jumping and post exercise gravity inversion. *Medicine and Science in Sports and Exercise*, 22, 385-390.

Bosco, C. (1985). Adaptive response of human skeletal muscle to simulated hypergravity condition. *Acta Physiologica Scandinavica*, 124, 507-513.

Bosco, C., Rusko, H. and Hirvonen, J. (1986). The effect of extra-load conditioning on muscle performance in athletes. *Medicine and Science in Sports and Exercise*, 18, 415-419.

Bosco, C., Zanon, S., Rusko, H., Dal Monte, A., Belotti, P., Latteri, F., Candelero, N., Locatelli, E., Azarro, E., Pozzo, R. and Bonomi, S. (1984). The influence of extra load on the mechanical behaviour of skeletal muscle. *European Journal of Applied Physiology*, 53, 149-154.

Dintiman, G., Ward, B. and Tellez, T. (1998). *Sports Speed*. Champaign, Illinois: Human Kinetics.

Fowler, N. E., Lees, A. and Reilly, T. (1994). Spinal shrinkage in loaded and unloaded drop jumping. *Ergonomics*, 37, 133-139.

Leatt, P., Reilly, T. and Troup, J. D. G. (1986). Spinal loading during circuit weight-training and running. *British Journal of Sports Medicine*, 20, 119-124.

Newton, R. U. and Kraemer, W. J. (1994). Developing explosive muscular power: Implications for a mixed methods training strategy. *Strength and Conditioning*, 16, 20-31.

Perey, O. (1957). Fracture of the vertebral end plate in the lumbar spine: an experimental biomechanic investigation. *Acta Orthapaedica Surgical Supplement*, 25, 1-100.

Sands, W. A., Poole, R. C., Ford, H. R., Cervantez, R. D., Irvin, R. C. and Major, J. A. (1996). Hypergravity training: Women's track and field. *Journal of Strength and Conditioning Research*, 10, 30-34.

Seagrove, L. (1996). Introduction to sprinting. *New Studies in Athletics*, 11, 93-113.

Young, W. B. (1998). Acute enhancement of power performance from heavy load squats. *Journal of Strength and Conditioning Research*, 12, 82-84.

THE DESIGN OF METHODS FOR CARRYING POLICE EQUIPMENT

Janette Edmonds and Glyn Lawson

InterAction of Bath Limited, 5/6 Wood Street, Bath, BA1 2JJ

This paper presents an ergonomic study to develop different methods for carrying police equipment. The Avon and Somerset Constabulary commissioned the study following complaints by female officers regarding the poor usability and discomfort caused by the use of the existing utility belt. The study was comprised of three stages, namely: user requirements analysis, concept development and user verification. It was found that the current utility belt did not meet the user requirements and that a single solution was unlikely to suffice. Three concepts were developed in collaboration with police officers and manufacturers. Prototypes were provided to support user trials for each of the three concepts. It is discussed how a seemingly straightforward design problem revealed several complexities due to the number and variation of user requirements

Introduction

Two significant legislative changes related to police equipment have occurred in recent years. These are the Police (Health and Safety) Act (1997), which stipulates the requirement to secure the health and safety of officers whilst on duty, and the Personal Protective Equipment at Work Regulations (1992), which requires that clothing and equipment are appropriate for the job to be performed. In particular, the equipment must suit the wearer, at least in terms of size, fit and weight. So, from the health and safety perspective, there is a duty of care to be afforded to police officers with regard to the provision of equipment, which naturally includes the method of carrying it.

The health and safety team from the Avon and Somerset Constabulary received several complaints from female officers regarding the utility belt. The problems included body aches and pains and the difficulty of use whilst performing their duties. In addition, it was recognised that there had never been a coherent approach to providing a method of carriage, and that its development, and indeed the issuing of police equipment, had been rather ad hoc. In the early years, police officers carried little, if any, equipment. Gradually, equipment such as personal issue radios, batons and handcuffs became standard issue. Latterly, more equipment is carried on the belt, some of which is police issue, such as CS incapacitant spray and first aid kits, and the remainder is personal supply, such as penknives, and personal torches. As could be expected, a number of problems have arisen.

The Avon and Somerset Constabulary commissioned this study to identify a solution to these problems. It was recognised that an ergonomic approach would allow the problem to be systematically solved from the perspective of the end user.

The aim of the paper is to discuss how ergonomic techniques were applied to the development of carriage methods, a seemingly straightforward design problem, which revealed several complexities due to the number and variation of user requirements.

User Requirements Analysis

The steps included in the user requirements analysis are described in the following sub-sections.

Literature Survey

A literature survey was conducted to review relevant research related to the design and evaluation of equipment carriage methods. Several issues were identified from the literature survey, including:

- The optimal positioning of weights (from studies of load carriage in the military)
- Physiological responses to equipment carriage and body armour
- The effect on the respiratory system of wearing a belt (from studies on back belts)
- Wearing utility belts in cars, particularly ingress and egress, and compatibility with the seatbelt

Questionnaire

A questionnaire was distributed to 440 female officers within the Force to elicit their views, experiences, requirements and design ideas. The questionnaire was developed with the support of a purposely-formed focus group, which included representation from the health and safety team, and the Police Federation and Welfare group. A total of 210 questionnaire responses were received and analysed (a response rate of 48%). In summary, 82% reported at least one problem. The problems included:

- Equipment commonly stuck into the body and became tangled, for example in the seat belt
- Discomfort was experienced whilst driving and sitting (including discomfort from wearing the seat belt)
- Back pain, hip pain, and general discomfort were experienced
- The utility belt was felt to be obstructive, bulky, poorly sized, and looked untidy
- The belt was reported to slip around and ride up the body.
- It was unstable whilst running and easily came undone during difficult arrests
- Equipment accessibility was poor
- Discomfort was experienced in different weather conditions
- The weight distribution was uneven and the belt was too heavy (1.3 – 3.8kg)

The other key findings were:

- There was large variance in the positioning of equipment items
- Many respondents were found to carry several ancillary items
- It was felt that the new police uniform (due to be issued at the end of 2000) would help with equipment distribution due to the provision of additional pockets and the shape of the jacket.

Interviews

Interviews were conducted with eight officers representing different job roles (for example, foot patrol, vehicle patrol, mounted section and plain clothes) to explore specific issues in more detail. They were asked to describe the tasks and duties they undertake, the equipment they carry for different circumstances, how the equipment is used, difficulties relating to different conditions and tasks, and their ideas and requirements for a future design. The information from the interviews provided a greater appreciation of the problems previously described in the questionnaire responses.

Task analysis

A task analysis was developed and recorded using InterWORK, the InterAction of Bath in-house task analysis tool. The task analysis was based on the information from the interviews and questionnaires, and included information relating to equipment used during different activities, and the frequency and importance of use. The most common activities associated with police duties were reported to be driving, walking or running and deskwork.

A list of user requirements was complied using the data gathered. This formed the basis of a "design requirements" checklist for usc during the design.

Concept Development

It became apparent that a single solution was unlikely to satisfy all the user requirements on the checklist. Therefore, three concepts were developed:

- A modified belt design
- A harness
- A vest

Different options were defined for placing equipment on the body, with the respective concept providing the solution for the primary equipment. For instance, the new police uniform includes pockets on the thighs and pockets in different positions on the 'bomber-style' jacket which allowed ancillary items to be placed away from the primary carriage concept. The options were developed, paying particular attention to weight distribution, frequency and importance of use of each equipment item, and user preferences.

The proposed equipment positions were then sketched on scale diagrams of the 1st percentile female adult to identify constraints posed by the smallest possible user.

The concepts were assessed against the "design requirements checklist" and presented to the focus group for comment and evaluation prior to further development.

The concepts were developed in collaboration with manufacturers of equipment holders and carriage devices, and commented on throughout by the user focus group. The key reasons for involving the manufacturers at this stage were:

- To ensure that the designs were practical and cost effective to manufacture
- To maximise the use of "off the shelf" items to reduce the cost to the Avon and Somerset Constabulary
- To gain support for the development of the prototypes
- To gain the benefit of the manufacturers' experience and design ideas

The following topics were discussed with the manufacturers:

- Fastenings and attachment of the equipment to the carriage device
- Sizing and shaping
- Orientation and access to equipment
- Space requirements between items
- Security and safety issues

As discussions proceeded, it became apparent that the primary focus of the design had to be on the critical items of equipment, i.e. ASP (baton), CS incapacitant spray, personal radio and 'quick' cuffs. This was because the number and variety of additional equipment were too large for any single solution. It was felt that these four items were critical to all officers, and carrying them in the trouser or jacket pocket was not feasible as this would present a health and safety hazard to the user, or would significantly reduce their usability. Moreover, these items were ranked the four most important items in the questionnaire results.

The aim was to design a device to satisfactorily accommodate these items, but with provision for other items to be carried as well. The concept design sketches are presented in Figures 1 to 3 below.

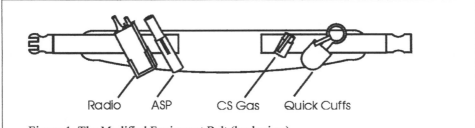

Figure 1. The Modified Equipment Belt (back view)

The modified belt concept incorporated a large padded section round the back and hips, designed to prevent the equipment from digging into the body and to distribute the weight more evenly. This would also provide improved lumbar support and shaping to the female form.

A shaped section was provided at the front to allow the officers to move their legs, to sit down easily and to allow the seat belt to be worn across the hips.

The harness and vest concepts were similar to current designs, although the positioning of the equipment was modified in response to the user requirements. It was agreed that the shaped section and the final positioning of equipment would be decided at the pre-user trials.

Within each concept, male and female docking ports were used to attach the equipment holder to the carriage method. The reasons for this were to allow flexibility in positioning of the equipment, easy removal, and the ability to angle the equipment to suit the posture, i.e. standing, sitting, crouching, and so on. The holder for the CS incapacitant spray, which was an 'off-the-shelf' design incorporated the following features:

- A slide mechanism to support rapid retrieval
- Single direction positioning (to prevent spraying gas into one's own face)
- A lanyard to prevent loss if it were dropped
- A hardened case to prevent damage to the canister

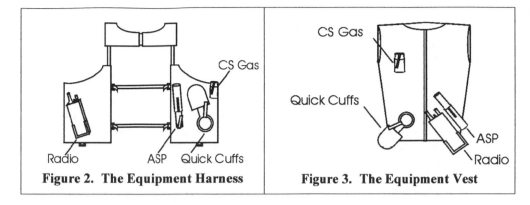

Figure 2. The Equipment Harness | Figure 3. The Equipment Vest

User Verification

Initial prototypes were developed for use during the pre-user trials. Officers tested each concept in three different conditions:

- Driving a police car (including, use with a seat belt, ingress and rapid egress)
- Simulating office work
- Dealing with a series of obstacles on an obstacle course, including climbing, running, high leg manoeuvres, and moving within confined spaces

Officers were asked to comment on the designs and a group discussion was held at the end of the trial.

Modifications were made to the designs based on the comments prior to the presentation of the main prototypes for use during the main trials. Some reservations were raised by the authors regarding the use of the equipment vest due to the distribution of the weight being primarily toward the front, which could potentially cause discomfort, particularly in the lumbar region. It was decided that the concept would still be presented for use during the trial, as several officers particularly favoured the design.

The modified prototypes were distributed to officers across the constabulary for trial. Officers were asked to wear the concept for a minimum of three shifts and to test it under "real" conditions. They were asked to complete an evaluation sheet for each concept. The final results were not available when this paper was in preparation.

Discussion

It is often the case that when applying an ergonomic approach to practical problems, the solution or solutions become relatively obvious due to the systematic methods used to understand the problems and user requirements.

It became apparent early on during this study that the detailed exploration of the user requirements revealed an enormous variety, not just in body size and shape, but also in the way the equipment is worn, and the way it is used for different tasks, job roles, circumstances, and conditions, even before personal preferences are considered. For example, covert operations require equipment to be accessible, but discretely positioned. Driving requires the equipment be positioned to avoid interference with the seat belt and vehicle controls. Public disorder situations may mean that drinks bottles need to be carried. Torches may be required during both day and night time duties (for example, if derelict buildings are to be entered). Rural policing may have different requirements to urban policing. The vehicle used (foot, car, horse, motorcycle, support van) and the clothing worn (e.g. body armour, public order personal protective equipment, plain clothes, uniform) also dictate what equipment needs to be carried on the person and the positions where equipment can be placed.

Compromises were inevitable for a design problem of this nature, and although solutions were identified, none of them was able to solve all the issues. For example, the harness is the best option for avoiding interference with the seat belt, but reduces the accessibility of the equipment when a jacket is worn over the top. The vest improves the visual accessibility of equipment, but places a greater weight toward the front of the body and is not favoured by officers with large breasts. The belt provides good weight distribution on the body, but is unlikely to be used as effectively by officers travelling by motorcycle or horse, or for covert operations.

It was, therefore, found that a seemingly straightforward design problem became difficult to solve due to the variation in the user requirements.

The benefits of the approach, however, enabled the user requirements to be understood in considerable detail so that the design difficulties and the required compromises could be properly understood. There were significant benefits from involving officers at all stages to support the development of solutions to meet their needs.

Acknowledgements

The authors wish to acknowledge the support of the officers involved in the study and the help given by Dennis Bray, Sarah Neate and Gemma Dye from the Health and Safety section of the Avon and Somerset Constabulary.

The authors also wish to acknowledge the support of the manufacturers during the study, particularly for the development of the prototypes.

References

Home Office. 1997, *Police (Health and Safety) Act.* (HMSO)

Health and Safety Executive. 1992, *Personal Protective Equipment at Work Regulations.* (HMSO)

InterAction of Bath Limited. *InterWORK: Software Work Analysis Tool.*

A PREREQUISITE FOR PROTECTION - WHAT PUNCH FORCES MAY POLICE OFFICERS FACE ON PATROL?

Robin Hooper and Dalien Cable[1]

Department of Human Sciences, Loughborough University,
Loughborough, LE11 3TU, UK.
r.h.hooper@lboro.ac.uk
[1] *(now at TNO Technische Menskunde, Kampweg 5, Postbus 23*
3769 ZG Soesterberg, Netherlands)

Police officers suffer personal attack and injury during routine patrol work, often by punches to the head causing linear and rotational accelerations, as well as bruising, laceration and fracture. Headwear subserves many functions, but in this context can be viewed as PPE. Few data exist about accelerations caused by punching, none in the context of police work. The types of punch used against officers 'on the street' were observed and reproduced in the laboratory. Volunteers struck an instrumented, model head. Combined (x, y & z) linear forces, depending on punch type, lead to accelerations of 181 g, giving a head injury criterion of 371, with the potential to cause rotation up to 4,500 rad.s^{-1}. The latter has a high potential for damage. The possible protection factor for PPE in the patrol setting is discussed. Allowing a 10% margin, absorption to reduce linear accelerations by 75g (frontal) and 90g (lateral) is one possible design/test criterion.

Introduction

More than 18,000 assaults on police officers were recorded during 1993 leading to injury (Brown, 1994). Concern about the risk of assault on officers whilst performing their duties has increased since then. The application of Health and Safety legislation to the Police Service in 1998 highlighted the issue to a greater extent and enhanced attention paid to the protection afforded officers on routine patrol by their uniform and equipment (see Hooper, 1999 and 2000), in contrast to the special riot provision.

Most assaults occur when dealing with public disorder and disputes. Other situations include carrying out traffic and foot stops as well as during arrest. 44% of the injuries inflicted were caused either by a punch, kick or headbutt, 38% of injuries resulting from general scuffles (e.g. being elbowed, kneed, pushed, or bitten) with 31% of all injuries being to the head (Brown, 1994). Clearly assailants get close to officers.

The most common attack is by punching. Injuries sustained include: cuts and bruises 65%, fractures, serious cuts and bruises, concussion and trauma 17%, pain, discomfort and sprains 7%, with no injury reported 11%. For serious injuries, the face, head and neck areas suffer most frequently at 42%, 33% of victims taking more than 8 days sick leave. The next most frequent 'site' of serious injury was a resultant of multiple areas being targeted.

Although there are a variety of approaches that may reduce attacks and/or injury, one facet is to provide effective PPE (personal protective equipment). Forces currently provide strengthened bowlers for female officers and either helmets or caps for male officers. Each of these has pros and cons, outside the scope of this paper. None can be judged (or reengineered/redesigned) from a standpoint of protection from assault, as there is a lack of data available detailing the forces exerted by members of the public during assaults on officers. This study set out to address this lack of relevant data and had the following aims:

Aims:
- to characterize those involved in assaults and the situations in which assaults occur
- to identify the main types of strikes and punches used against officers
- to simulate the various punches and measure the resulting accelerations
- to consider the likelihood for injury by comparison with published criteria
- to suggest tolerance levels protective headwear should withstand.

Approach and methods

Assaults and assailants

The first two aims were addressed (i) through discussion with 5 police officers, military police, club security staff and members of the public who have been involved in/observed fighting and (ii) direct observation of assault situations around night clubs / public houses.

The majority of assaults is by unarmed assailants, two thirds occurring in public places e.g. around pubs and clubs between 10pm and 3am at weekends often induced by alcohol. Two thirds of assailants were also in groups, so an officer may have to deal with multiple assailants with the possibility of attacks from different directions (15% of incidents end with multiple assaults). Assailants are generally 17 – 25 yr, male, unemployed or in low status jobs with a prior conviction. The 30 participants in this study were male in this age range.

The most usual assault involves punching, characterized by 5 styles: 'step-and-punch', 'straight-punch', 'swing-punch', 'jabbing-punch' – note these are not intended to convey precise, boxing definitions– and 'elbow-strikes'. The head can suffer trauma from a fall.

Punching forces and resulting accelerations

Several studies have measured such forces, using an instrumented, mechanical neck and head with an internal accelerometer. For example, in Kozey and Stanish (1985), the head was struck various times by trained boxers. Linear acceleration in one axis was measured giving peak linear headform accelerations around 90g (g = acceleration due to gravity). This is well within human tolerance limits reported by Goldsmith and Ommaya (1984) and also well below the levels at which sports' helmets are rejected. Even in unprotected heads these levels would not be expected to cause skull fracture or intra-cerebral lesions.

Impact forces on heads tend to have tangential as well as normal, compressive components that can produce rotation of the head. These can give rise to angular accelerations of the intra-cranial contents and result in diffuse brain injury. Smith, Bishop and Wells (1998) measured linear accelerations of 20 – 58 g with resultant angular accelerations from 275 – 675 $rad.s^{-2}$. In boxing matches, Pincemelle *et al* (1989) found higher values of 30 – 120 g and >3,500 $rad.s^{-2}$ respectively.

For head protection, angular accelerations must be assessed since it is the rotational accelerations as well as the accumulation of traumas from multiple punches that seem most likely to underlie injury.

Biomechanical simulation of the human head was achieved by the use of a Hybrid III head with an Eurosid neck unit mounted on a metal base (similar to that used by Smith et al, ibid). Within the headform, tri-axial accelerometers were mounted on an alloy sub-frame. A simple system was developed using Window's Workbench that allowed individual accelerations along the x, y, z axes each to be measured at 435 Hz. The three highest values were averaged in each axis and the calibration factor applied to convert the voltage output to corresponding multiples of g.

The main head unit was attached to a bracket on the laboratory wall, providing adequate clearance to prevent accidental injury. The head was padded and boxing gloves used to prevent injury to fists. The absorption of the padding layers was tested as follows. A weighted pendulum was swung ten times into the headform without padding. The swings were repeated with the padding in place. The average absorption of the padding was calculated. Finally the swings were repeated with a glove in place to measure its average absorption. The pendulum provided consistent impacts, so testing the reliability of the system as well. Accelerations of 5g (padding) and 4g (gloves) were absorbed and are added to the measured values as appropriate.

A single tangential acceleration

As stated earlier, previous experiments only noted the peak linear accelerations found in one axis for each punch. True accelerations may actually be higher as the other two axes were not considered. To combine the separate x, y, z, components into one single tangential acceleration (F) required application of Pythagoras' Theorem:

$$F = \sqrt{x^2 + y^2 + z^2}$$

The risk of resultant injury can be considered by establishing the head injury criterion for the maximal linear accelerations and making comparison with the Wayne state curve (see below). However, these risks are still concerned with maximal linear accelerations. Consideration also needs to be given to the possibility for rotational accelerations, since accelerations may not merely displace the head linearly.

Format of testing

Each of the 30 male participants gave informed consent that excluded anyone with previous injury or current condition likely to increase risk of injury during the experiment. The equipment and 5 'punch styles' noted above were demonstrated, any questions or doubts being clarified. The five attacks were then performed. Initial practice was not permitted as fights occur without a preparation period.

On completion, participants rated their perception of punching effort, as boxers have been found to hold back due to fear of injury. In a similar way, the participants believed they punched at 84% of maximum effort on average in these trials. Although open to debate (see discussion), the accelerations reported here have been increased by x1.19 (= 1/0.84) to reflect this under-performance.

A separate study was conducted to simulate the possibility of an officer being knocked over during a confrontation and hitting his/her head on the ground. The headform was removed from its base and dropped five times from each of two heights (5 cm and 20 cm). The two impact velocities tested were one metre per second and two metres per second. It is unlikely that a head would be going any faster from this type of fall, the deceleration being very brief and intense.

Results

Table 1   Single tangential accelerations for each type of punch
(g = acceleration due to gravity)

Step-and-punch	straight-punch	swing-punch
Lowest punch = 46.1g	Lowest punch = 50.4g	Lowest punch = 41.5g
Average punch = 83.7g	Average punch = 84.1g	Average punch = 90.4g
Highest punch = 121.0g	Highest punch = 125.4g	Highest punch = 181.4g
Standard Deviation = 15.8	Standard Deviation = 20.8	Standard Deviation = 27.2
Jab	Elbow strike	Falling impact
Lowest punch = 30.4g	Lowest elbow = 27.0g	highest acceleration:
Average punch = 60.9g	Average elbow = 78.4g	154.5g
Highest punch = 91.6g	Highest elbow = 122.4g	
Standard Deviation = 15.0	Standard Deviation = 31.5	

The data were transformed by finding the head injury criterion (HIC) (Goldsmith and Ommaya, 1984), using the highest accelerations measured in each axis for each of the punches:

$$\mathbf{HIC} = \left[\int \mathbf{a.dt}/_{(T2-T1)} \right]^{2.5} \times (T2-T1)$$

Where   a = the 25 highest tangential accelerations found for each punch
dt = the time over which these accelerations act (the sampling rate of 435 s^{-1} gives one measure every 0.0023 s).
and   (T2 – T1) = the pulse duration.

This calculation gives Head Injury Criterion values, based on the highest accelerations measured, of 301 for the step-and-punch, 300 for the straight-punch, 371 for the swing-punch and 19 for the falling impact.

The first two values can produce rotational accelerations up to about 3,500 rad.s^{-1}, while the swing punch could exceed 4,500 rad.s^{-1}.

Discussion

The Wayne State curve (Fig. 1) concerns the likelihood of injury by linear acceleration. The curve is the tolerance level that denotes the onset of concussion. It originated from experiments performed on embalmed cadavers. Rigid structures were struck into the foreheads with the duration range of one to six milliseconds. These results were correlated with concussive effects generated in animals. The curve suggests concussion occurs with values above/to the right of the curve.

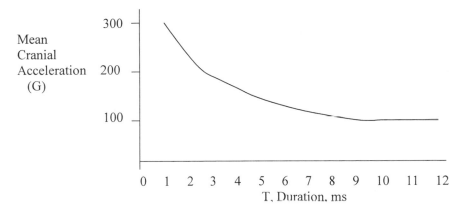

Figure 1: The Wayne State Curve

In the work reported here, the accelerations were increased by a factor of x1.19 to allow for the subjective reports from participants that they did not punch with full force. On average each participant perceived their effort to have been 84% of maximum. Such a rating is open to question but it is clear the measured forces did not reflect maximum effort despite the padding provided. It is only natural to be concerned about safety. Whether a true maximum effort is likely in a scuffle is also open to doubt, especially as in many cases those involved are influenced by considerable quantities of alcohol. It is a matter for speculation whether this would cause reduced punching power (interference with coordination) or increased power (inhibition release). It seems prudent to make this degree of increase.

Even with these allowances, the values found would have to act over a period of time approaching 1 ms to have a concussive effect – i.e. brief acting linear accelerations of 180 g and HIC of 400 found here are not expected to lead to concussion. There still will be risk arising from multiple blows. And these linear blows have the potential to cause bruising, laceration and fractures as accelerations of >90g can cause these outcomes.

In addition, the rotational accelerations need to be considered, since >1,000 rad.s^{-1} can lead to concussion. The blows reported here have the potential to create rotational accelerations well above this borderline and so pose a threat of injury to the brain.

It is feasible to provide protection from blows and strikes of many types – there is a fully-encasing riot helmet for example. The data collected here are intended to indicate the accelerations that might be encountered in the course of incidents during routine patrol work.

As the rotational accelerations are potentially damaging, PPE should provide a degree of protection by being able to prevent such accelerations. The approach would of necessity require reduction of the linear accelerations in such a way as to lower the risk of subsequent rotational acceleration and of fracture.

If the PPE provided were to reduce accelerations to 40g, most risk of injury would be prevented. Maximal frontal blows (see table 1) would require absorption of about 85g (frontal) and 140g (lateral) to reduce accelerations to this level. To provide this level of protection against the average punch +1 S.D. would require accelerations to be reduced by 75g and 90g respectively, allowing a 10% margin.

What this work did not cover was the head area struck. It is impossible to be fully prescriptive but a considerable proportion will occur to the face as well as other aspects of the head. Further work will be required to define the areas of protection needed. And of course, there are many other factors that need to be considered as the headwear serves other functions also. It provides protection from the elements, provides recognition, has to serve the patrolling officer well in other, non-violent aspects of the work and will, by its style, influence both the response of the public to the officer and vice versa. Protection afforded through risk assessment on a patrol, working arrangements and incident management are inevitably part of the decision making. The accelerations to which officers may be subject are just part of any such considerations.

Acknowledgement

The authors thank the Police Scientific Development Branch for the loan of the model head unit including the tri-axial accelerometer.

References

Brown, B. (1994) Assaults on Police Officers: An examination of the circumstances in which such incidents occur. *In Police Research Series paper no. 10 (1994)* Home *Office Police Research Group*

Goldsmith, W. & Ommaya, A.K (1984). Head and neck injury criteria and tolerance levels. (pp.149-187) *in B.Aldman & A.Chapon (Eds.). The biomechanics of impact trauma.* Elsevier Science Press, Amsterdam.

Hooper, R.H. (1999) Why do police officers leave their body armour in the cupboard? *Contemporary Ergonomics 1999,* pp358-362.

Hooper, R.H. (2000) Is British police uniform adequate for the task? *Proceedings of the IEA 2000,* pp6.125-6.128.

Kozey, J.W & Stanish, W.D (1985). The measurement of headform accelerations from boxing punches. *Medicine and Science in Sports and Exercise,* 17, 222.

Smith, T.A, Bishop, P.J, & Wells, R.P (1998). Three dimensional analysis of linear and angular accelerations of the head experienced in boxing. *Proceedings of the IRCOBI Conference on the Biomechanics of Impacts* (pp.271-285) Bron, France: IRCOBI

Pincemelle, Y., Trosseille, X., Moch, P., Tarrier, L., Breton, F., & Renault, B. (1989). Some new data related to human tolerance obtained from volunteer boxers. *Proceedings of the 33rd Stapp Car Crash Conference Philadelphia*: Society of Automotive Engineers.

SEX DIFFERENCES IN COGNITIVE FAILURES AND VIVIDNESS OF VISUAL IMAGERY

Neil Morris

University of Wolverhampton
Department of Psychology
Bankfield House
Wolverhampton WV1 4QL
UK

It has been argued that the Vividness of Visual Imagery Questionnaire (VVIQ) measures some cognitive skills associated with particular professions, for example, Isaac and Marks (1994) found that pilots, air traffic controllers and elite athletes had higher VVIQ scores than matched controls. The VVIQ may therefore be a useful addition to selection procedures in some cases. However a high VVIQ score may be a necessary but not sufficient requirement for some skills. In this study 40 male and 40 female students completed the VVIQ and Broadbent *et als.* (1982) Cognitive Failures Questionnaire (CFQ) to examine the proposition that vivid imagery may be associated with high levels of reported absentmindedness. The results showed that VVIQ and CFQ scores were significantly correlated in females (r=+0.40) but not in males (r=+0.11) although males and females did not differ in the magnitude of either their VVIQ or CFQ scores. It is concluded that women, but not men, show an association between level of cognitive failure and reported imagery vividness. This *qualitative* difference between the sexes suggests that the VVIQ should only be used for selection purposes with great caution.

Introduction

Galton (1880) arguably produced the first psychometric instrument when he developed his procedure for measuring the vividness of mental imagery. He constructed a questionnaire that he had completed by both students and by professional men. The questionnaire contained specifications for various situations for which they were to try to elicit images. One request was to call up from memory the scene of their breakfast table that morning. His subjects were to say whether the image they had was dim or clear, the objects ill or well defined, in colour or monochrome, and to rate the extent of the contents of the field of view etc. (Watson, 1963). Galton found that some individuals reported that they could evoke no visual images while others reported near veridical imagery. In particular, he found that many eminent scientists had poor vividness of imagery that he explained as perhaps the price of profound abstract reasoning. This procedure has long been associated with discriminating between different modes of thinking associated with exceptional levels of performance. Galton's procedure was formalised as the Vividness of Visual Imagery Questionnaire

(VVIQ) by Marks (1973). He used five items from the Betts (1909) formulation and 11 additional items. The VVIQ has good psychometric properties in terms of reliability and validity (Gur and Hilgard, 1975;Isaac and Marks, 1994). It is completed twice, once with eyes open and then with closed eyes. The two scores can be summed.

Subsequent studies have examined the properties of the VVIQ and other vividness questionnaires with similar content. Carnoldi *et al.* (1991), for example, distinguished at least two properties of vividness – the extent to which the 'image' was reported to phenomenologically approach visual experience and its luminosity, that is, the clarity/sharpness of the image.

Sex differences in visual imagery vividness have been found by Sheehan (1967) with females reporting more vivid imagery. This finding was replicated in Isaac and Marks (1994) using the VVIQ but many studies fail to find sex differences (White, Ashton and Brown, 1977) and it is not clear why this is except that sex differences tend to be confined to children and adults under 50 years (Isaac and Marks, 1994). Another source of sex differences ambiguities on this dimension may lie in the failure of factor analysis to provide a solution with a dominant vividness factor. Both Gur and Hilgard (1975) and Richardson (1977) suggest independent factors of vividness and imagery control.

That reflecting on the vividness of ones imagery is a multi-component process suggests that there may be more than one 'type' of vivid imager. Isaac and Marks (1994) examined a range of groups with the VVIQ and found that elite athletes, pilots and air traffic controllers, in particular, had very high VVIQ scores relative to matched controls. It is not surprising that good 'visualisation' skills are associated with these groups – they all require considerable visuo-spatial competency and very high levels of performance. The above groups may well demonstrate high vividness and good control of imagery. However, it is possible that some individuals might combine high vividness with somewhat lower levels of control and thus score relatively highly on the VVIQ but have a different set of skills to a pilot or air traffic controller. Thus it is likely that good visualisation is a necessary but not sufficient characteristic and that high VVIQ scores might also be associated with other features that are contra-indicated within the aviation industry. For example, a poet might well benefit from good visualisation but lack the vigilance of an air traffic controller. Indeed he/she might be sufficiently introspective to display considerable absentmindedness, reflected as a range of cognitive failures.

Reverie or 'daydreaming' "is a human function that chiefly involves resort to visual imagery" (Singer, 1975, p.55). The link between reverie and vivid visual imagery seems to be strong. Individuals scoring high on the Positive Vivid Daydream scale also report more vivid visual imagery than low scorers (Fusella, 1972). Perhaps people differ in their ability to control their experience of visual imagery such that some individuals fall 'victim' to reverie (perhaps using this creatively) while others actively use their imagery to perform visuo-spatially demanding tasks. This would be reflected in a relatively high correlation between reverie and vividness of imagery in groups with high imagery control (this would be positive or negative depending on the valency of the tests used – for example, a vivid imager has a *low* score on the VVIQ and an absentminded person has a *high* score on the Cognitive Failures Questionnaire so the relationship would be positive for a vivid imager who was not prone to absentminded reverie) and one would expect these dimensions to be *decoupled* in individuals with poor imagery control because whereas good imagery control would facilitate tasks requiring vigilance, poor imagery control would not necessarily produce levels of absentmindedness that would be problematic in situations with modest vigilance requirements. Extreme absentmindedness, or cognitive failure, is generally considered to be pathological and is more likely to correlate with cognitive decline associated with stroke or Alzheimer's disease. Thus in a 'normal' population sample there will be some who score very low on measures of absentmindedness but most other individuals will have modest scores, reflecting everyday adequate functioning, rather than high scores.

Reverie is one aspect of cognitive failures, a dimension measured by the Cognitive Failures Questionnaire (CFQ – Broadbent, Cooper, Fitzgerald, and Parkes, 1982). A large scale study of 2949 American navy recruits showed this instrument to measure, reliably, a general cognitive failures factor. No second order factors of any significance were produced (Larsen, Alderton, Niedeffer, and Underhill, 1997). It provides a good prediction of mood state in pregnant women, but not in non-pregnant women (Morris *et al.*, 1998). The CFQ correlates with vulnerability to stress (Reason, 1988) and the probability of causing a traffic accident (Larsen and Merritt, 1991) during the six months covered by the questionnaire (respondents are requested to report the frequency of cognitive failures in the last six months).

CFQ scores outside of the pathological range but nevertheless high would be contra-indicated in professions such as pilot or air traffic controller. It is distinctly plausible that this combination could arise and that an alternative career path would be advisable for such individuals. This study examines a group of undergraduates, the 'pool' from which future aviation personnel would be recruited, to examine this proposition. In particular, it compares males and females. As noted above, there is considerable ambiguity about between sex imagery vividness differences. Additionally, the Isaac and Marks (1994) studies of pilots and air traffic controllers only examined males. A very different pattern might emerge in females.

Method

Participants

40 male and 40 female participants, aged between 18 and 36 (mean age 21.7), took part in this study as part of an instructional session in psychometrics. All participants were first year undergraduates on a BSc Psychology degree at the University of Wolverhampton.

Materials

Marks (1973) Vividness of Visual Imagery Questionnaire and Broadbent *et als.* Cognitive Failures Questionnaire were combined with a cover sheet to create the booklet of psychometric tests. A cardboard storage box was used as a 'post box'.

Procedure

Participants were seated in a large teaching room in the University of Wolverhampton and were required to maintain sufficient distance from each other to ensure that no one else could observe their questionnaire booklet. They were told to fill in the front sheet that simply asked for their age and sex. It was stressed that they should not write their name on the booklet. They were instructed in filling out the VVIQ and the CFQ and it was stressed that this was to be carried out 'under examination conditions' with no communication allowed during the data collection phase of the study. It was also stressed that if they did not wish to participate then they should feel no obligation to do so and should read quietly during the study. When all participants had completed the questionnaires they were required to drop them into a sealed cardboard box with a 'letterbox' to ensure anonymity was maintained. The questionnaires were subsequently scored using the standard scoring conventions for these questionnaires and the participants were debriefed.

The VVIQ was filled out first and the 'eyes open' and 'eyes closed' scores were combined. The CFQ requires the participant to reflect on their cognitive failures over the last six months. It was pointed out to the participants that in this instance this meant that they should consider the beginning of this period to coincide with the first week of teaching.

Results

The means and standard deviations for VVIQ and CFQ scores for both males and females are shown in Table 1. Also in Table 1, for comparison, are the CFQ scores for a non-student female sample from Wolverhampton who participated in a different study. Independent samples t-tests showed that males and females did not differ on VVIQ scores $(t(38) = 1.76, p>0.05$, two tail) or CFQ $(t = 1.54, p>0.05)$. For males a Pearson Product Moment Correlation Coefficient revealed no significant correlation between VVIQ and CFQ scores $(r(39) = +0.11, p>0.05)$ but for females there was a significant correlation $(r = +0.40, p<0.01)$. Thus the two groups did not differ in the magnitude of their scores on either questionnaire but there was a relationship between VVIQ and CFQ in females but this correlation was lacking in males. The CFQ scores for these samples are very similar to those found in a more general sample of the Wolverhampton population.

Table 1: Means and standard deviation (in brackets) for VVIQ (max = 160) and CFQ scores (max = 96) for male and female students at the University of Wolverhampton. Also shown are means and standard deviations for a sample of female workers in Wolverhampton (CFQ scores only).

Males		Females		Female workers
VVIQ	CFQ	VVIQ	CFQ	CFQ
80.1 (22.9)	46.8 (16.0)	76.0 (15.5)	43.7 (12.6)	43.9 (11.9)

Discussion

These results show that although males and females do not differ in reported vividness of imagery or cognitive failures they differ radically in the association between these. For males vividness and cognitive failure seem to be unrelated whereas for females one could argue that vividness is a modestly good predictor of cognitive failure (and *vice versa*). This is consistent with the view, but perhaps not persuasive, that for females vividness and cognitive failure share a component of cognitive control. If this proves to be the case then imagery vividness is qualitatively quite different in women and men – vivid imagery has some association with low incidence of cognitive failure in women but there is no such relationship in men. In this case this is true for a sample that very closely match, in terms of VVIQ scores, the samples of air traffic controllers and pilots used in the Isaac and Marks (1994) study.

In Isaac and Marks (1994) table 4 (p.494) mean ratings for the two groups are reported. When the VVIQ scores in table 1 above are converted to this scale air traffic controllers have slightly more vivid imagery (mean = 2.0) than the pilots and Wolverhampton male and female students (after rounding all three groups fall in the range 2.4-2.5) suggesting that there is a *prima facie* case for considering these four groups to be comparable in terms of reported vividness of imagery. One might also reconsider the idea that the pilots and air traffic controllers have exceptionally vivid imagery – the Wolverhampton sample were not chosen from a group requiring particularly well developed visuo-spatial skills and there is no reason to suppose that their scores are remarkable.

The Wolverhampton student samples are also comparable with Broadbent *et als.* (1982) samples with respect to CFQ scores although they score, on average, ten points higher than the navy recruits used in the Larsen *et al.* (1997) study. This, taken together with the data from the Morris *et al.* (1998) study suggests that the student sample is typical of an adult UK population. The Larsen *et al.* (1997) sample may be atypical in as much as even mildly absentminded individuals may not be selected for service in the US navy.

Whether the distinction between the two modes of using vividness of imagery outlined in the introduction is valid remains to be seen. These data are compatible with this idea but do not provide a strong test of it. Further studies homing in on imagery control should address this problem. What is clear is that a female sample, with a similar distribution of VVIQ scores to a sample of male students and samples of professional pilots and air traffic controllers, is *cognitively quite distinct* from at least one of these groups (but in no way superior or inferior to this group) suggesting that the Isaac and Marks study has limited validity for making statements about the skills underpinning elite professional performance within this domain until the qualitative differences have been addressed. Great caution is needed if the VVIQ is to be used, even in combination with other instruments, at any stage in the selection process.

Acknowledgment

Many thanks to Claire Wilkinson for bibliographic help.

References

Betts, G.H. (1909) *The distribution and function of mental imagery.* New York: Teachers College, Columbia University.

Broadbent, D.E., Cooper, P.F., Fitzgerald, P. and Parkes, K.R. (1982) The Cognitive Failures Questionnaire (CFQ) and its correlates. *British Journal of ClinicalPpsychology,* **21,** 1-16.

Cornoldi, C., De Beni, R., Giusberti, F., Marucci, F., Massironi, M., and Mazzoni, G. (1991) The study of vividness of images. In R.H. Logie and M. Denis (eds.) *Mental images in human cognition.* New York: Elsevier. 305-312.

Fusella, V. (1972) *Blocking of externally generated signal through self-projected imagery: the role of inner acceptance personality style and categories of imagery.* PhD thesis, City University of New York.

Galton, F. (1880) Statistics of mental imagery. *Mind,* **19,** 303-318.

Gur, R.C. and Hilgard, E.R. (1975) Visual imagery and the discrimination of differences between altered pictures simultaneously and successively presented. *British Journal of Psychology,* **66,** 341-346.

Isaac, A.R. and Marks, D. (1994) Individual differences in mental imagery experience: Developmental changes and specialization. *British Journal of Psychology,* **85,** 479-500.

Larsen, G.E., Alderton, D.L., Neideffer, M., and Underhill, E. (1997) Further evidence on dimensionality and correlates of the Cognitive Failures Questionnaire. *British Journal of Psychology,* **88,** 29-38.

Larsen, G.E. and Merritt, C.R. (1991) Can accidents be predicted? An empirical test of the Cognitive Failures Questionnaire. *Applied Psychology: An International Review,* **40,** 37-45.

Marks, D. (1973) Visual imagery differences in the recall of pictures. *British Journal of Psychology,* **64,** 17-24.

Morris, N., Toms, M., Easthope, Y. and Biddulph, J. (1998) Mood and cognition in pregnant workers. *Applied Ergonomics,* **29,** 377-381.

Reason, J. (1988) Stress and cognitive failure. In S. Fisher and J. Reason (eds.) *Handbook of Life Stress, Cognition, and Health.* New York: John Wiley.

Richardson, A. (1977) The meaning and measurement of mental imagery. *British Journal of Psychology,* **68,** 29-43.

Sheehan, P.W. (1967) A shortened form of Betts' Questionnaire Upon Mental Imagery. *Journal of Clinical Psychology,* **23,** 386-389.

Singer, J. (1975) *Daydreaming and fantasy.* Oxford: Oxford University Press.

Watson, R.I. (1983) *The great psychologists: From Aristotle to Freud.* Philadelphia: Lippincott.

White, K.D., Ashton, R., and Brown, R.D.M. (1977) The measurement of imagery vividness: Normative data and their relationship to sex, age, and modality differences. *British Journal of Psychology,* **68,** 203-211.

ENHANCED MOTION PARALLAX AND TELEPRESENCE

Paul B. Hibbard[1], Mark F. Bradshaw[1], Simon J. Watt[1], Ian R.L. Davies[1], Neil S. Stringer[1] and Andrew R. Willis[2].

[1] *Department of Psychology,*
University of Surrey
Guildford
Surrey GU2 7XH

[2] *Centre for Human Scienes,*
Defence Evaluation & Research Agency,
Farnborough
Hampshire, GU14 0LX, UK

Abstract

Three geometrical tasks were used to assess whether operators in telepresence and virtual reality environments can benefit from enhanced motion parallax (i.e. increasing the effect of head movements to enhance depth information) and the effects of this enhancement on the interpretation of visual information. Sensitivity to depth improved with increasing parallax gain, especially if the line of sight was rotated to keep the object of interest centred. Depth and width settings were affected by motion gain, consistent with the predicted change in perceived distance. Performance in a shape matching task, which does not require information about distance, was considerably better than performance for the setting task. These results demonstrate that enhancing motion parallax increases sensitivity to depth, but that the choice of parallax gain should be considered carefully if absolute size and shape estimates are important.

Introduction

Telepresence and virtual reality systems aim to provide operators with sufficient information to perform specific tasks, and to give the sense of presence in the remote or virtual environment. Typically, this requires the provision of information about the three-dimensional structure of the environment, which is often achieved using binocular cues. This introduces the need for specialist hardware for stereoscopic viewing, and often leads to discomfort arising from the conflict between the operator's states of vergence and accommodation. An alternative cue, providing information that is geometrically equivalent to that available from binocular disparity and an equally compelling perception of 3D structure (Rogers & Graham, 1979), is motion parallax. This is a potentially useful cue since in a visually coupled system no additional hardware is required, and the discomfort arising from accommodation/vergence conflict suffered in binocular systems is avoided.

An important issue in the design of a system that relies on motion parallax to provide depth information is how to link the motion of the operator to that of the vantage point of the system. It would appear most sensible to maintain a one-to-one link between the motion of the operator and the vantage point, so as to preserve the geometrical relationship between the operator and the viewing system. However, it has been suggested that a gain of less than one (such that the vantage point moves through a smaller distance than the observer) is felt to be more "natural" by most observers (Runde, 2000). Alternatively, in some applications it might be desirable to use a gain of greater than one, so as to enhance the relative motion of different parts of the scene which should, in principle, enhance the detection of depth. This may be seen from the geometry in figure 1a. If the operator's head moves through a distance x, the change in location of the image of a point at a distance D on the retina, θ, is given by:

$$\tan\theta = \frac{x}{D} \qquad (1)$$

and the difference between the change in position of two points, ω, is given by:

$$\tan\omega = \frac{x\delta}{D^2} \qquad (2)$$

where δ is the depth separation to the points and D is the distance from of the points from the observer.

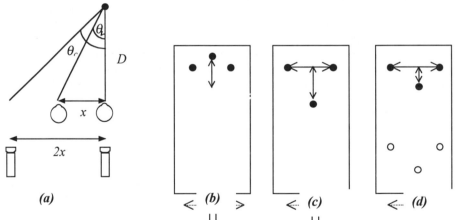

Figure 1. (a) increasing the gain of motion parallax increases the displacement of a point on the retina for a given head movement (in this case, from θ_h to θ_c). (b) the depth nulling task (c) the shape setting task (d) the shape matching task.

Increasing the gain will therefore increase the difference between the movement of two points that are separated in depth. For a gain g, and a head movement x_h, the vantage point will move through a distance $x_c=gx_h$, giving a difference in the movement of the points:

$$\tan\omega = \frac{gx_h\delta}{D^2} \qquad (3)$$

This increase in image motion should increase sensitivity to differences in depth. One potential disadvantage of changing motion gain is that the geometrical relationship between the operator and the scene is altered, leading to predictable distortions of visually perceived space. To see this, consider how a depth interval might appear to an operator using a system with a movement gain g. From (2), assuming that the operator interprets image motion using proprioceptive information about the change in their head position, the perceived distance D_p to an image point and the perceived depth interval δ_p between two points are given by:

$$D_p = \frac{D}{g} \quad (4) \qquad\qquad \delta_p = \frac{\delta}{g} \quad (5)$$

respectively. Thus, while using a motion gain of greater than one may increase the operator's sensitivity to a depth difference between two points, those points will appear closer to the operator, and the depth difference will appear smaller. Additionally, if distance information provided by motion parallax is used in estimating an object's width w, the perceived width w_p will also decrease with increasing motion gain:

$$w_p = \frac{w}{g} \qquad\qquad (6)$$

While equations (4) to (6) predict systematic distortions of perceived space, many of the perceptual judgements that we make do not require full, metric three-dimensional information (Glennerster *et al*, 1996). Rather, it may be possible to perform tasks on the basis of relative information, thus avoiding the problem of distortion described here. In the current study, a series of tasks were used to investigate the effect of altering motion parallax gain on perceptual judgements, to determine (i) whether changing the motion gain changes sensitivity to depth differences (ii) whether the perception of object shape, size and distance is affected by changes in gain and (iii) whether tasks requiring only relative information can be completed successfully regardless of the gain of motion parallax.

Methods

Observers (n=5) viewed the stimuli monocularly on a computer screen, while moving their heads back and forth through a distance of 11cm at a rate of 1Hz to generate motion parallax. Each participant performed 3 tasks (figure 1b-1d):

1. **Depth nulling** Observers viewed two small light sources, at a distance of 2m, presented at eye height 7.5cm on either side of the centre of translation. A third light was positioned midway between these lights, and observers moved this light back and forth until it appeared conlinear with the other two lights. Ten settings were made at each of 4 motion gains (0.5, 1, 2 and 4).
2. **Shape setting** Observers viewed two lights, at a distance of 2m, and adjusted their separation until it appeared to be 15cm (in comparison with a hand-held T shape that served as a reference). A third light appeared midway between these points, and it was moved back and forth until it appeared to be 15cm in front of the other two lights

(again in comparison with the handheld standard). The task was therefore to position the lights in a triangle with a base and height of 15cm. Five settings were made at each of the 4 motion gains used in the nulling task.

3. **Shape matching** Observers viewed two sets of three lights, one of which formed a triangle with a base and height of 15cm, with the base at a distance of 1.41m. The other set consisted of two lights at a distance of 2m with a lateral separation of 15cm, and a third light that appeared midway between these lights, whose position was adjusted to form a triangle with the same shape as that of the closer set of lights. Again, 5 settings were made at each of the above motion gains.

Results

Standard deviations of the depth nulling settings decrease with increasing motion gain (figure 2a), indicating increased sensitivity to depth differences. However, sensitivity decreases for the highest gain of 4. This is most obvious when the results are plotted as the relative displacement of the three points on the retina (figure 2b). Clearly, sensitivity falls below that predicted for the highest gain. This decrease in performance is most likely attributable to the increased motion of all image points at higher gains decreasing operators' sensitivity to the relative motion between points. This may be avoided if the line of sight is rotated as the head moves so as to keep the object of interest centred on the screen, thus avoiding the need to make eye movements to keep it fixated. The experiment was repeated with this manipulation. In this case, sensitivity continued to increase for all motion gains (figure 2c), and was constant when expressed in terms of retinal displacement (figure 2d).

Results for the shape setting task are shown in figure 3a. Width settings were relatively constant across changes in motion gain, while depth settings decreased with increasing gain. This is not consistent with the geometrical prediction that both settings should increase with increasing gain, to compensate for the expected decrease in perceived size. Depth settings are expressed in terms of relative retinal displacements in figure 3b. Settings increase with increase motion gain, consistent with the prediction that objects will appear closer with increasing gain. The results are expressed in terms of scaling distances (i.e. the distance at which the setting made by the observer is consistent with a width or depth of 15cm), in figure 3c. While scaling distances decrease with increasing motion gain, this increase is less than predicted (as shown by the dotted line). This is consistent with results for similar tasks in which shape is defined by binocular disparity, where it is found that compensation for changes in viewing distance is considerably less than perfect (Glennerster *et al*, 1996). The effect of this incomplete scaling for changes in apparent viewing distance is to alter the perceived shape of objects. This can be seen by plotting the depth:height ratio of the set triangles (figure 3d). This ratio decreases with increasing gain, indicating that objects appear progressively more stretched in depth as the gain increases (thus requiring less depth to maintain a constant perceived shape). Results for the matching condition are also shown in figure 3d. Here, the change in perceived shape with changing gain is less pronounced. This is as expected, as the task does not require knowledge of object distance, but requires one shape to be matched to another, both of which will be equally affected by changes in gain. Perceived shape also appeared to vary less with changes in gain when the line of sight was

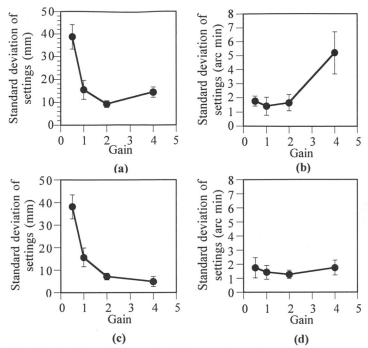

Figure 2: Standard deviations of nulling settings plotted (a) in mms and (b) in arc minutes of visual angle. (c) and (d) show settings for the case in which the line of sight was rotated to keep the object centred.

rotated, such that the objects would always appear to be at the same distance.

Conclusions

Enhancing motion parallax information increases sensitivity to depth differences, as predicted. This improvement is particularly pronounced when the line of sight is rotated so as to avoid the observer having to make large eye movements to keep the object of interest fixated. However, the perceived shape of objects is affected by changes in motion gain, consistent with geometrical predictions if incomplete scaling for changes in apparent object distance is taken into account. These results have clear implications for the choice of motion gain for tasks in which absolute shape, size and distance judgements are important.

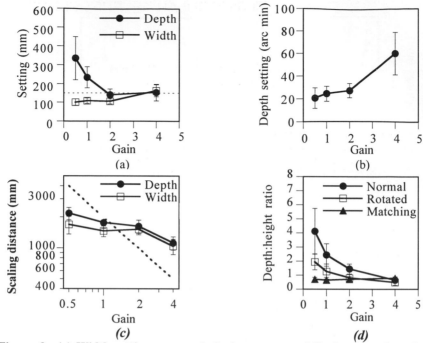

Figure 3. (a) Width settings were relatively constant, while depth settings decreased with increasing gain. (b) Depth settings expressed in terms of retinal displacement, demonstrating some compensation for changes in gain. (c) Scaling distances show that scaling for perceived distance is less than predicted. (d) The apparent shape of the object is affected by this incomplete scaling. Objects appear stretched in depth as gain is increased. Shape settings are less affected by motion gain for the matching task and also when the cameras are rotated to maintain the apparent distance to the object.

This work was carried out as part of Technology Group 5 (Human Sciences and Synthetic Environments) of the MoD Corporate Research Programme.

References

Glennerster, A., Rogers, B.J. & Bradshaw, M.F (1996) Stereoscopic depth constancy depends on the subject's task, *Vision Research*, **36**, 3441-3456.

Rogers, B.J. & Graham, M.E. (1979) Motion parallax as an independent cue for depth perception, *Perception*, **8**, 125-134.

Runde, D. (2000) How to realize natural image reproduction using stereoscopic displays with motion parallax, *IEEE transactions on circuits and systems for video technology*, **10**, 376-386.

THE STATION POINT AND BINOCULAR TELEPRESENCE.

Ian R.L. Davies[1], Mark F. Bradshaw[1], Paul B. Hibbard[1],
Simon J. Watt [1] and Andrew R. Willis [2].

[1] Department of Psychology,
University of Surrey
Guildford
Surrey GU2 7XH

[2] Centre for Human Sciences,
Defence Evaluation Research Agency,
Farnborough
Hampshire, GU14 0LX, UK

Abstract

Telepresence systems have the potential to allow remote control of manual operations such as vehicle guidance, or fine mechanical movements. To support such tasks effectively, the operator must perceive the remote scene indirectly, in the same way they would perceive it directly. Television images can provide equivalent percepts to direct viewing, provided the observer is at the station point. Here we investigate the effects of viewing the monitor displaced from the station point when the stimuli are defined stereoscopically or by motion or by both. For stereoscopic stimuli, displacement causes perceptual distortions consistent with the optical geometry. However, motion defined stimuli are less affected by displacement, and when the stimuli include both cues, judgements are veridical. Thus, it appears, that provided there is motion in the scene, displacement from the station point may not be important.

Introduction

Telepresence and virtual reality systems can both provide high fidelity 'indirect' images of a scene. The core principal of these technologies is the same as for photography: projective isomorphism (Gibson, 1971). The indirect image viewed from the station point (SP — the centre of projection) produces a retinal image that is geometrically equivalent to the image produced by direct viewing of the real scene. The principal of retinal equivalence can be extended to moving scenes and to the binocular case. In the former, relative motion within the scene produces equivalent retinal changes to direct viewing, and this additional information can enhance perceptual judgements (Rogers and Graham, 1979) In the latter case, if two cameras view the scene separated by the average interocular distance and the images are sent separately to the eyes, then viewed from the SP the scene disparities will be equivalent to those generated in direct viewing. Binocular information adds to the sense of immersion in the scene, and enhances accuracy of spatial judgements (Asbery and Pretlove, 1995; Lippert et al, 1982).

Binocular and motion information are not absent from conventional photographs. Rather, they specify a flat surface at a particular viewing distance — the picture surface — while the perspective or pictorial cues specify the spatial layout of the depicted scene. Perceived layout is then a compromise between the depicted scene and the flat surface due to some kind of weighted sum of the various cues to depth (e.g.. Landy et al., 1991). Thus, part of the benefit of including binocular and motion information is due to reduction in 'cue-conflict'.

The projective correspondence account of picture projection, although appealingly simple, cannot be the whole story. If pictures are not viewed from the SP, then the indirect image falls out of correspondence with the direct image and so the depicted spatial relations will no longer be veridical. The extent of this departure will vary as a function of the degree of displacement from the SP. Optically, visual angles representing distances along the observer's line of sight are compressed if the observer is closer than the SP, and expanded if the observer is further away than the SP. Thus, if perceived virtual space was solely determined by the optic pattern, then the former leads to magnification: objects will appear closer, but compressed along the line of sight. Equivalently, viewing from further from the SP should produce minification: objects should appear further away and expanded (Lumsden, 1980).

However, perspective is robust (Kubovy, 1986). We are not aware of perceived space changing as we move relative to a picture. Furthermore, experimental studies often find little effect of deviation from the SP (Rossinski and Farber, 1980; Yang and Kubovy, 1999). This stability of picture perception implies to some (e.g. Kubovy, 1986; Pirenne, 1970) that some kind of compensation mechanism is at work yielding this invariance. Somehow we know how far we are from the SP and we compute what the image properties would be when viewed from the SP.

The compensation account is perhaps plausible for oblique displacements from the SP, but less so for displacements along the normal to the surface. In the former case, the assumption that the SP is on the normal to the surface is usually valid, and can be used to drive compensation. But in the latter case there is no obvious information to specify where along the normal the SP lies. Yang and Kubovy (1999) argue that evidence for compensation is strongest for oblique displacements, and evidence for perceptual distortions strongest for displacements along the normal.

Our concern here is that in many telepresence systems little control is exercised over the observation point. For many purposes or tasks this may not matter, but if the visual information is being used to plan fine motor movements, in say surgery, or in guiding a remote vehicle, then even small perceptual distortions may have great significance.

Most studies of the effects of non-station point viewing have used static mages without valid binocular information. The geometry of displacement from the SP is essentially equivalent for the conventional and binocular or motion cases. If perception is driven entirely by geometrical optics, then, as shown in Figure 1 (dashed line), perceptual distortions in perceived depth should occur as a function of the extent of the displacement. The magnitude of the distortion is different for the binocular and motion cases as each cues scales with distance in a different way: binocular disparity scales inversely with the square of the viewing distance whereas relative motion scales inversely with distance.

To determine the effect of non-station point viewing observers performed a variant of the ACC task devised by Johnston (1991) where they had to adjust the amount of depth in a stimulus until it matched the 0.5 width (i.e. to create a *circular* cylinder). The stimuli were defined by (*i*) binocular disparity, (*ii*) relative motion, or (*iii*) disparity and motion and were simulated to appear at 50, 100 or 200 cm. Three distances were used to determine how each cue scales with changes in simulated distance. The ACC task was completed in three viewing conditions: when the observer was positioned at 0.5SP, SP and 2SP.

Methods

Subjects
Six undergraduates took part in the study, and they were paid for their time. All had normal or corrected to normal vision and stereoacuity better than 40 arc sec.

Stimuli
Stimuli were computer generated with a SP of 100 cm from the monitor. They were elliptical half cylinders, of which only the surface texture was visible. The texture on the surface of each cylinder consisted of 400 Gaussian blobs, each with a spatial standard deviation of 0.7mm, which were distributed randomly over the surface. Cylinders were presented at 3 simulated distances 50, 100 and 200 cm. The cylinders were defined by (i) binocular disparity (ii) multiframe motion or (iii) binocular disparity and multiframe motion. In the motion condition, the cylinders rotated through an angle of $\pm 8.4°$, at a rate of $2.1°s{-1}$. Stimuli were oriented horizontally and rotated around a vertical axis. Stimuli were presented on a 19" computer monitor, with a spatial resolution of 800x600 pixels and a refresh rate of 100Hz. Binocular disparities were generated by viewing the stimuli through CrystalEyes liquid crystal shutter glasses that were synchronised to the display rate of the monitor. Alternate images were presented to the left and right eyes so that each eye received a new image every 20ms (50Hz). Only the red gun of the monitor was used.

Procedure
The task was to adjust the depth of a cylinder until its profile appeared circular (the half-height was fixed at 3cm). Each trial began with the presentation of a small fixation cross, presented at the distance (defined by vergence) of the back of the cylinder to be presented on that trial. When this fixation cross was fused, observers pressed a button on a keypad to start the trial, and the cross was replaced with the experimental stimulus. The depth of the cylinder was initially set to a random value (between 0 and 12 cm), and could be increased or decreased by pressing either of two buttons. When the observer was satisfied that the cylinder appeared circular, they pressed a button to terminate the trial, and the fixation cross for the next trial appeared. Observers viewed the stimuli from either 100 cm (the station point), 50cm or 200 cm.

Results

Figure 1 shows the depth settings for the 3 conditions. In each figure the dotted line is the prediction for perfect scaling — that is when the changes in station point is taken into account completely. The dashed line indicates predicted settings if the change SP is ignored. Note how the dashed line differs in Figure 1a,c and 1b,d. This is because disparity and motion scales differently with changes in viewing distance (or station point). Clearly, under binocular conditions (1a), large distortions in perceived depth are evident whereas under monocular motion (1b) performance remained relatively veridical. In the latter case SP changes may not be significant because the information to judge shape is available solely from the motion information. That is, information about viewing distance is not necessary. In the combined cue conditions shown in 1c and 1d the observer in the non-station point conditions are faced with a cue-conflict stimuli so we plot the depth relative to the disparity in the stimuli (1c) and the depth relative to the motion in the stimulus (1d). Clearly when both cues are available observers can compensate for SP violations completely when the magnitude of the motion in the stimulus is considered (1d).

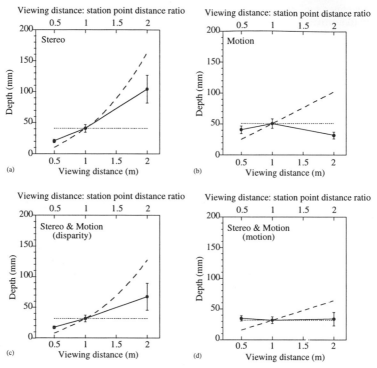

Figure 1 The amount of depth set (a) disparity alone (b) motion alone and (c and d) disparity and motion defined stimuli in the 0.5SP, 1SP and 2SP viewing conditions.

The addition of disparity in this condition clearly contributes to this which is most probably due to the fact that an estimate of size is possible (which is unknown on the basis of motion alone).

Figure 2 shows the depth settings made from 0.5SP, SP and 2SP for three simulated distances in the disparity only condition. It is known that changes in viewing distance are not taken into account completely when disparity information is scaled for depth. This graph shows that the degree of scaling that takes place is approximately constant for station point violations although there is some evidence that scaling is greater for 0.5SP.

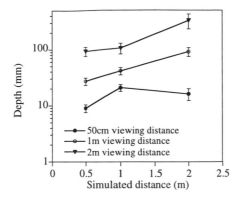

Figure 2: Depth set for stimuli defined by disparity and displayed at three simulated viewing distances (50, 100 and 200 cm). Stimuli were viewed from 0.5SP, SP and 2SP.

In conclusion, we have shown that displacement from the station point produces perceptual distortions in line with the geometrical optics, if the stimuli are defined stereoscopically. However, if there is motion in the scene this may be sufficient to correct the stereoscopic error. It remains to be tested whether observer motion rather than object motion has the same effect. If it does, then the cost of providing binocular and head linked motion information would be justified for certain tasks.

Acknowledgement

This work was carried out as part of Technology Group 5 (Human Sciences and Synthetic Environments) of the MoD Corporate Research Programme.

References

Asberry, R. and Pretlove, J., 1995, Telepresence for remote viewing, *Service Robot: An International Journal*, **1**, 20-23

Lippert, TM, Post DL and Beaton RJ A study of direct distance estimates to familiar objects in real-space, two-dimensional and stereographics displays. In Proceedings of the human Factors Society 26th annual meeting, 324-328, 1982.

Gibson, J. J. 1971, The information available in pictures. *Leonardo*, **4**, 27-35

Huber, J. and Davies, I. 1997, Pictures as surrogates: the perception of slope in the real word and photographs. *Contemporary Ergonomics*

Kubovy, M. 1986, *The psychology of perspective and Renaissance art*. (Cambridge University Press, New York)

Johnson, E. B., 1991, Systematic distortions of shape from stereopsis, *Vision Research*, **31**, 1351-1360.

Landy, M. S., Maloney, L. T,, Johnson, E. B. and Young M. J., 1991, In defense of weak fusion: Measurement and modelling of depth cue combination, *Mathematical Studies in Perception and Cognition* (New York University Press, New York).

Pirenne, M. 1970, *Optics, painting and photography*, (Cambridge University Press, London).

Lumsden, E. A. 1980, Problems of minification and magnification: an explanation of the distortion of distance, slant, shape and velocity. In M. Hagen (Ed.) *The Perception of pictures* Vol. I, (Academic, New York).

Rossinski, R. R and Farber, J. 1980, Compensation for viewing point in the perception of pictured space. In M. Hagen (Ed.) *The perception of pictures* Vol. I, (Academic, New York).

Rogers, B.J & Graham, M.E. (1979). Motion Parallax as an Independent Cue For Depth Perception. Perception 8 pp 125-134.

Yang, T, and Kubovy, M., 1999, Weakening the robustness of perspective: evidence for a modified theory of compensation in picture perception, *Perception and Psychophysics*, **61**, 456-467.

ASSESSMENT OF THREE GLOVES WHEN USED BY A WHEELCHAIR-RACING ATHLETE IN THE ISMWSF T1 CLASS.

GE Torrens[1], S Briance[1] and W Butler[2]

[1] *Hand Performance Research Group, Department of Design and Technology,*
[2] *Department of Physical Education, Sports Science and Recreational Management,*
Loughborough University, Loughborough Leicestershire, LE11 3TU
g.e.torrens@lboro.ac.uk

This paper describes the design specification and assessment of a new wheelchair-racing glove for athletes with C6 lesion of the spine. There are limited choices of commercially available glove designs due to the small number of racing athletes. The aim of this paper is to provide recommendations for the design specification and assessment of gloves that will enable those involved in wheelchair racing to produce and test their own customised glove design. Three glove designs were assessed in a laboratory and on a racing track: (i) a leather British military CS95 combat glove used in conjunction with a high-tack resin material; (ii) a modified chemical workers glove, and (iii) a prototype glove incorporating a high friction pad, and wrist support. The prototype glove was found to produce less slippage per revolution in use and stayed on the athlete's hands more effectively than the other gloves evaluated.

Introduction

There are approximately 100 wheelchair athletes registered with the British Disabled Athletics Association, of those only around fifty are actively involved in racing. The wheelchair athlete group can be further divided into 45 male paraplegic athletes, 5 female paraplegic athletes, and 5 quadriplegic athletes. Since the market worldwide is so small development of the athlete's racing equipment has been limited. The types of glove currently used by British athletes include open digit gloves, based on mountain bike gloves; full digit gloves, often made of leather and from military origins and purpose; and, special purpose gloves, often in the form composite material mitts that enclose digits two to five, with digit one (thumb) separately sheathed. Because of the small market size existing gloves are used or modified by athletes. Some one-off or small batch produced designs are available, but are expensive. An experienced wheelchair athlete approached the Department of Design and Technology, Loughborough University, to help improve the design of his racing gloves. Athletes of all abilities have a number of attributes that make up their physiological performance: acceleration, peak power generation, endurance and recovery. The contact points between the human body and the sports equipment being used can be an area where the athlete's performance may be inhibited. During wheelchair racing the main area for force transferral is through the hand. To achieve an optimum transferral of forces through the hand to the sports equipment requires grip. Reducing slip during a performance will optimise control and power output through to the sports equipment, improving grip.

The aim of the paper is to provide recommendations for the design specification and assessment of gloves that will enable those involved in wheelchair racing to produce and test their own customised glove design. The objectives of this communication are:

- To provide a specification for the glove material and the configuration of glove holding straps based on a wheelchair racing athlete's requirements;
- To assess the ease of use of three designs of racing glove; and,
- To test the performance of three designs of racing glove.

Three glove designs were assessed in a laboratory and on a racing track: (i) a military combat glove; used in conjunction with a high-tack resin material; (ii) a modified chemical workers glove, and (iii) a prototype glove incorporating a high friction pad, and wrist support.

Method

The study described here involved an initial evaluation of current racing glove designs through interviews with the subject and assessment of the subject's racing technique. Subsequently two designs of glove were compared with the current racing glove through rolling road trials and track performances. These assessments provided more detailed information on which to base the recommendations for a racing glove design stated in the conclusion of this study. The rolling road and track performance measurements taken in this study have been slippage at the glove-pushrim interface and the overall performance of the athlete over a set time duration or distance.

A three wheeled track and road chair used in this trial, designed by Bromakins, Loughborough, U.K., was the subject's own racing wheelchair. A VHS video camera , Panasonic CCD video camera, model number F 15, and a Panasonic, AG-7350 video recorder, was used to record the rolling road trials. Each frame recorded 1/250th of a second of event time. The high frame speed was used to capture a sharp image suitable for analysis of the racing technique of the subject. It was also used to analyse the different glove performances when used with the wheelchair on the rolling road.

The three gloves were tested: a leather glove, at that time used by the subject in conjunction with a tacky resin, (Klister); a new prototype design; and a modified nitrile rubber -dipped cloth work glove. The leather glove was a United Kingdom military issue combat glove, CS95, Size 9. Each glove was held in place on the hand using a 15mm wide Velcro strip. The Velcro was looped around the gloved hand at the wrist. Once in place over the wrist crease, the loop was tightened by the athlete. The low-cost chemical workers glove had been modified by using a synoacrylate adhesive, (trade name 'Superglue'), to stick the first and second digits of the glove together along their adjacent sides. This modification kept the first digit in contact with the main part of the hand and avoided the first digit being caught in the wheel rim during the athlete's racing action.

The prototype glove was based on the recommendations of the wheelchair athlete. The following features were incorporated into the new glove design, based upon the subject's comments:

- The main body of the glove was made of thin leather, with a cotton/nylon mix liner to enable the subject to get the glove on and off more easily.
- The wrist area of the glove had an extended wrist cover, with an outer cover patch, made of Ghana rubber, that ran from the tip of the glove thumb around to and over the glove knuckle, covering a width that was level with the fifth digit of the hand.
- To help support the wrist two plastic splint supports were inserted into the glove between the liner and the leather sections of the glove. The size and shape of the plastic mouldings were based on casts taken from the subject's hands some weeks earlier.

To assist in keeping the hand in a fist shape the glove tread was extended to the same length again as the glove length on the hand. At the end of the glove the tread had two Velcro strips attached, to enable the subject to over-fasten the Velcro strips.

Laboratory assessment of athlete performance

The laboratory-based trial was to investigate the preparation for racing of the athlete and their racing technique. The biomechanics laboratory of the Department of Physical Education, Sports Science and Recreational Management (PESS&RM), Loughborough University was used to undertake the rolling road trials in this study. The subject was asked to comment freely during the trial on any matters relating to his performance and that of the equipment he was using. The trial was over a period of 90 minutes to reduce the influence of fatigue on the results. The overall test duration was base on recommendations from other researchers (Macdougal, Wenger and Green, 1991, Dallmeijer, Kappe, Dirkjan, Veeger, Janssen, and van der Woude, 1994, Wang, Beale, Moeizadeh, 1996). The subject was asked to go through their normal preparation for racing e.g. settling into their wheelchair, muscle warm up and stretching exercises. The recordings were also used to define the path of contact between the pushrim and glove. Measurements of slippage at the glove and pushrim interface and a time performance was also measured, some weeks later. The test set-up was identical to the previous assessment except that the video camera was set at right angles to the front-to-rear axis of the wheelchair and rolling road. It was noted by Dallmeijer *et-al* (1994) that power output and therefore speeds varied between athletes with Cervical injuries and those with Thoracic. The subject was consulted to confirm the racing speeds chosen were appropriate for his level of fitness. The trial involved two phases: assessment of slippage at two racing/training speeds and acceleration to three racing speeds from a standing start. The speeds were based upon the subject's racing experience, reflecting sprinting, middle distance and long distance speeds and accelerations.

Measurement of slip

Slip occurs when the glove slides across the surface of the rim but appears to stay in firm contact with it. In this session slippage was also seen as comparative movement between the markers placed on the glove and those on the rim. Slippage between the hand and the glove was reported through subject comments. The subject, who had a consistent pull stroke, was fitted with reflective markers, for contrast during filming, in the form of white adhesive circles. These were attached to the glove over the wrist. There were also 6 reflective markers fitted at regular intervals to the spokes of the wheel within the pushrim. The glove was marked around the wrist with white tape, and a wheel spoke marker was coloured to act as a reference spot. The smaller the slippage the higher the number of turns needed for the hand and wheel reference markers to coincide again. The schedule was split three five-minute periods: one warm-up and two slippage test periods. Before setting the speed targets it was checked that the subject could achieve them over the given time periods.

To warm up the subject settled into a preferred rhythm. Following the warm-up period of five minutes two slippage tests were carried out.

- Slippage test one: the subject worked for 5 minutes at a constant speed of 2.68 m/s, and represents hill climbing. The speed chosen was around of 40 - 50 % of maximum racing speed achievable by the subject.
- Slippage test two: The final 5 minutes of warm up was conducted at 4 m/s, 70% of maximum race speed and reflects a cruising speed. This period was filmed also with the intention of monitoring levels of slip. The athlete was then allowed to rest for five minutes before the next assessments.

The acceleration performance of the athlete was recorded as the time taken to reach a given speed. The subject was asked to accelerate as fast as possible to reach a target speed, as shown on a cycle speed indicator. Each test was started from stationary position to eliminate inaccuracies due to inconsistent rolling starts. Overcoming inertia from a stationary start was considered to be enough to simulate the kind of forces encountered on the track.   The operator monitored the speed from a cycle computer. The cycle computer was mounted on the wheelchair, which was also visible to the subject.

The acceleration test was repeated three times, for each target speed, for each of glove design, giving nine results per glove. Wang *et al* (1994) suggested a three-minute rest between repetitions, but following discussions with the subject a shorter rest time of one minute was taken. The longer rest time was considered to add to the overall duration of the test increasing the likelihood of results being affected by fatigue. The table of results can be seen in the results section of this paper.

Athlete racing track performance assessment

Racing track assessments were undertaken to confirm the results found in laboratory testing. A digital stopwatch was used to monitor the subject as he sprinted from a standing start over a straight 100metres. A hand held video camera was used to record the event. The facility used was a 400m synthetic rubber covered track used for international grade athletics events. Once the subject had finished his warm up routine he was asked to take a position at the start and accelerate along the 100metres-sprint section of the track. The subject continued around the track to recover and cool down. The procedure was to be repeated three times with each glove. The results of this test are shown in Tables 1 and 2.

Results

The subject involved in this assessment was a 30 year old male. He had sustained a lesion in the C6, thoracic region of his spine, producing functional characteristics classified as tetraplegia.

Laboratory assessment of glove requirements

During the trials of the three types of gloves in the laboratory and on the track it quickly became clear that the modified chemical workers glove was not suitable for this purpose. Therefore, the results refer only to the military glove, used with Klister resin, and the prototype racing glove. A comparison of percentage slippage per pushrim revolution produced by a wheelchair athlete when using a military glove with Klister resin and a prototype racing glove.

Table 1. Percentage slippage per push stroke cycle of a wheelchair athlete at a range of speeds (based upon the number of revolutions of a reference marker to return to zero)

Speed and acceleration values m/s	% slippage per revolution of reference marker to military glove marker	% slippage per revolution of reference marker to prototype glove marker
4 m/s	17	20
0-3.57m/s	14	14
0-3.57m/s	14	11
0-3.57m/s	14	13
0-4.47m/s	8	10
0-4.47m/s	10	13
0-4.47m/s	7	13
0-5.36m/s	7	11
0-5.36m/s	8	11
0-5.36m/s	8	

The result for the 0-5.36m/s acceleration repetitions using the prototype glove was void due to an incomplete video recording. The military glove currently used provided consistent results over the period of the tests. It can be seen that slippage increased in the results taken later in the series of tests. Fatigue would seem to be the likely cause of the increase in slippage. The increase in slippage in the result 0-4.47m/s b may be due to an increased effort put into the sprint by the athlete or operator error in the processing of the recording. The difference between a constant sub-maximal speed and a sprint acceleration speed may be explained by the effort required being greater for the athlete to attain the sprint speed against that required maintaining a constant speed. It can be seen that less consistent results were obtained using the prototype racing glove. The inconsistency may be due to the difference in size and configuration of the glove requiring the athlete to adjust their racing technique. Table 2. shows a comparison of time taken to reach a specified speed by a wheelchair athlete on a rolling road when using a military glove and a prototype racing glove.

The track sprint results show that the prototype glove consistently out-performed the currently used military glove during acceleration sprint testing. The athlete was between 0.5s and 5.2s faster using the prototype glove than the military glove during testing. The results show that in the earlier sprint tests the athlete was 1.83s and 0.98s faster over 100m when using the prototype glove. However, in the later repetitions the military glove was seen be faster by 0.24s and 0.81s in the last test of each glove.

Table 2. Acceleration of wheelchair athlete to terminal velocity (in Seconds) using two different types of glove

Glove type	Rep.1	Rep2.	Rep.3	Average
Military glove 0-8mph	6.2	4.8	4.5	5.17
Military glove 0-9mph	7.7	6.5	8.8	7.67
Military glove 0-10mph	7.4	7.1	7.3	7.27
Prototype glove 0-8mph	4.3	3.8	4	4.03
Prototype glove 0-9mph	4.9	3.7	3.6	4.07
Prototype glove 0-10mph	4.9	4.8	5.1	4.93

Table3. 100m Sprint times by a wheelchair athlete when using two different types of racing glove (tail wind 10-15 mph)

Track test 100m repetitions	1	2	3	4	Average
Military glove and Klister resin	33.1	30.58	30.06	30.9	31.160
Prototype glove	31.27	29.6	30.3	30.9	30.518

The subject's comments indicated that aspects of prototype glove worked well, but towards the end of the testing flaws in the glove design were highlighted resulting in an increase in sprint times. The subject indicated they perceived the friction material Ghana had worked well on the prototype gloves. However, other parts of the glove specification required revision and further development. Under commercial constraints this product development would not have taken place. The authors would welcome further discussion about the needs of wheelchair athletes or the needs of the wider community.

Acknowledgements

This study was sponsored by the Nuffield Foundation, undergraduate research bursary and supported by Mr G Dalzell, Defence Logistics Organisation and Dr B Bennett, Bennetts Safetywear Limited. The authors would like to thank Mr C McAllister for his help in data processing.

References

Macdougal D.J., Wenger H.A., Green H.J., *1991, Physiological testing of the high performance athlete.* (Human Kinetics, Champaign,Ill).

Dallmeijer A.J., Kappe, Dirkjan, Veeger, Janssen, van der Woude, 1994, Anaerobic power output and propulsion technique in spinal cord injured subjects during wheel chair ergometry. *Journal of Rehabilitation Research and Development*, **31**, 120 - 128.

Wang Y.T., Beale D., Moeizadeh M.N.A., 1996, An electronic device to measure drive and recovery phases during wheelchair propulsion-a technical note., *Journal of Rehabilitation Research and Development*, **33**, 305 - 310.

EXAMINATION GLOVES FOR NURSES
A PROTOCOL FOR USABILITY TESTING

Laura Norton[1], Sue Hignett[2],

[1]*Institute for Occupational Ergonomics, University of Nottingham,
University Park, Nottingham, NG7 2RD, UK.
Laura.Norton@nottingham.ac.uk*
[2] *Nottingham City Hospital NHS Trust, Nottingham, NG5 1PB, UK*

Powdered latex gloves are recognised to contribute to the development of latex hypersensitivity. Nottingham City Hospital responded to official guidance, encouraging the removal of these gloves, by reviewing the provision of gloves throughout the Hospital. A study was therefore undertaken to develop a protocol for comparative testing of sterile and non-sterile gloves used by nurses. The design features of gloves that may impact upon their usability were identified. Tests representing typical nursing tasks and capable of reflecting changes in usability were developed and the performance of six subjects with seven types of gloves analysed. Subjective comments about each glove were also received via a questionnaire. The study provided an ergonomic approach to ensuring the provision of gloves that will not hinder the practice of nurses.

Introduction

The guidance on latex allergy recommends that powdered gloves be removed from hospitals as recognised to be a major contributor to latex hypersensitivity (Medical Devices Agency, 1996; 1998). Nottingham City Hospital responded to these recommendations by drafting guidelines on the usage of gloves (Nottingham City Hospital, 1998) and information on alternative products. The provision of gloves throughout the hospital was also reviewed with the intention of standardising the purchasing across specialties. A review of relevant literature suggested very little research had considered the ergonomic factors related to glove design with consideration to nurses, who represent a large population of users. To ensure that the design of the sterile or non-sterile gloves used by nurses, at Nottingham City Hospital, facilitate rather than hinder their tasks a comparison of various glove types was considered appropriate. Therefore a pilot study was undertaken to develop a protocol for the comparative testing of both sterile and non-sterile gloves and assist in developing a method capable of providing an evidence-based decision for the recommendation of a suitable supplier. Therefore only a small sample of experienced nurses was used during this trial.

An ergonomic perspective was adopted to ensure that all important design features were considered when comparing gloves. The ergonomic considerations to the design features that may affect usability were identified and are illustrated in figure 1.

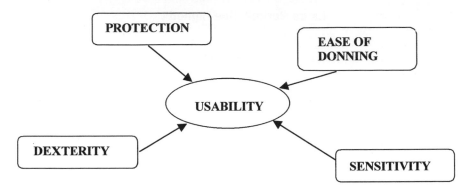

Figure 1. Design features that may affect usability

The most common materials used in glove construction are latex, vinyl and synthetic materials, including nitrile, dupraprene and triflex (O'Gilivie, 1998). Nurses commonly use vinyl gloves; even though they have had poor results for puncture resistance and permeability (Johnson, 1997). Nurses sometimes use double gloving (two gloves on one hand) to increase the level of protection (Mahony, 1998). However, wearing one glove of a thickness at least 0.83mm provides greater protection (Nelson and Mital, 1995). The factors influencing puncture resistance also include the amount of time the glove is worn exposure to high levels of stress and hydration (Kornewicz and Kelly, 1995).

The dexterity of latex gloves with a thickness of 0.83mm is similar to that of the bare hand (Nelson and Mital, 1995). However comparative studies of latex and nitrile gloves suggest that nitrile gloves are less elastic with surgical tasks (Chua et al, 1996, Newsom et al, 1998). The impact on nursing tasks is not documented.

The ease of donning a glove is also important to any user (Fisher et al, 1996). Research into a replacement for powder, previously used to aid donning, has suggested hydrogel-coated gloves to be a suitable alternative, when donning a glove with either dry or damp hands (Robert and Brackley 1996, Fisher et al, 1996).

The Pilot Study

The study incorporated tests that were sensitive to any changes in tactility and reflect the gross and fine motor activities associated with nursing. The aim of developing a protocol for comparative testing of sterile and non-sterile gloves.

The gloves were provided from four companies asked to provide samples of sterile and non-sterile gloves in latex and synthetic materials. Gloves not fulfilling the criteria of being powder free or with a high protein level, greater than 50µg/g/glove (Medical

Devices Agency, 1996), were omitted as both are recognised as contributing to latex hypersensitivity. There were seven gloves included three sterile and two non-sterile latex gloves, one non-sterile nitrile and one non-sterile vinyl.

Six subjects were recruited using convenience sampling. Anthropometric dimensions were measured including: hand length, breadth including thumb, hand spread and maximum hand depth. All were within a normal distribution range of the 5th- 95th percentile, except for maximum spread and hand depth, which were under the 5th percentile dimension in some subjects. All subjects were right handed and did not declare latex hypersensitivities or upper limb disorders. The sample included nurses from medical wards, intensive care and theatres with a mean number of 14.2 years experience.

The tests devised for the pilot were developed from the literature (Chua et al, 1996, Fisher et al, 1996, Nelson and Mital, 1995, Roberts and Brackley, 1996, Thompson and Lambert, 1995, Woods et al, 1996) and in consultation with local clinical nurse specialists. There were five tests involving tasks commonly associated with nursing activities, figure 2.

Objective Tests	Description of Test	Measurement
Tactility	Two pieces of fine, medium and coarse-grained sandpaper in a box. The subject was blindfolded and asked to identify and extract one of each type.	Pass/fail
Gross Motor	Open a sealed dressing pack, remove three pieces of gauze using tweezers and place them in separate marked squares. Then to cut two pieces of tape with scissors and stick them on two marked rectangles.	Time taken
Fine Motor	Five three way taps secured to a cork board and placed on a table. Two black marks of the same width and length were made on the base and spoke of each tap. Subjects lined up the black marks on each tap.	Time taken and a cumulative five second penalty imposed for inaccuracy
Fine Motor	Move as many metal drawing pins as possible, point first, from a pot with tweezers in thirty seconds	Number of pins moved
Gross Motor	A specimen bottle filled with water and sealed. Open the bottle, draw 5ml in a 10ml syringe and re-secure the lid.	Time taken and a five second penalty imposed for spillage

Figure 2. Objective tests associated with nursing tasks

Subjective testing was also completed with a questionnaire that used five point likert scales followed by an open question to gain subjects views on the following issues:

- Ease of donning
- Ease of handling small objects
- Glove fit
- Hindering effect on ability to carry out task
- Inhibition of sensation

The gloves were awarded points according to how well they performed in each test; the higher the ranking, the higher the score. The results were presented by grouping the objective test data, subjective questionnaire data and the comparative ranking for the sterile and non-sterile groups. One of the fine motor tests, in which tweezers were used to pick up metal drawing pins, was found to be lacking in sensitivity, with minimal differentiation between gloves. The results were therefore omitted from the results. A new fine motor task to test a specific nursing skill, such as removing stitches, would be necessary in any future study. The sample was not large enough to draw robust conclusions about the comparative benefits of the gloves tested but the following observations were made.

Sterile Gloves
One latex glove performed better than others in the tactility test and subjectively with regard to fit and sensation. The glove that did not achieve as good a result was found to be difficult to put on.

Non-sterile
A latex glove was found to perform very well in one of the gross motor tests and provided the greatest tactility. A positive result regarding this glove was confirmed by the questionnaire that indicated staff found it to fit well and be easy to put on.

The Nitrile glove also performed well with the fine motor test but some difficulties were reported with handling smaller objects and putting the glove on. The vinyl glove was ranked last in this group and had low scores on both the tactility and gross motor testing. Staff referred to this glove as 'restricting' and 'inhibiting sensation'.

Discussion

The use of ergonomic principles allowed sterile and non-sterile gloves, a critical piece of equipment for nurses, to be tested by methods that reflect their intended use. The study aimed to develop a protocol for comparative testing of the usability of this product and it is inappropriate to draw firm conclusions from such a small sample size. A larger sample involving a representative population considering all variables such as anthropometry, nursing specialities, grade, sex and age would be needed. However the study did suggest the protocol to be sensitive enough to distinguish between gloves that provided greater sensitivity, dexterity and were easier to don.

Conclusion

This pilot study trial led to the development of a protocol, which can be used for the comparative testing of sterile and non-sterile gloves used for nursing tasks. The use of ergonomic principles would be of great benefit in ensuring that large-scale purchasing provides the nurse with a product that assists their practice.

References

Chua, K.L., Taylor, G.S., Bagg,J. 1996, A Clinical and Laboratory Evaluation of Three Types of Operating Gloves for Use in Orthopaedic Practice. *British Journal of Orthodontics. ,* **123**,: 15-20

Fisher, M.D., Neal, J.G., Kheir, J.N., Woods, J.A., Thacker, J.G., Edlich, R.F. 1996, Ease of Donning Commercially Available Powder-Free Surgical Gloves. *Journal of Biomedical Materials Research (Applied Biomaterials).* **33**,:291-295

Johnson, F. 1997, Disposable Gloves: Research Findings on Use in Practice. *Nursing Standard. ,* **11**,:16, 39-40

Kornewicz, D.M. and Kelly, K.J. 1995, Barrier Protection and Latex Allergy Associated with Surgical Gloves. *AORN Journal.* **61**,:6, 1037-1040

Mahony, C. 1998, The Need for a Clear Policy on Glove Use. *Nursing Times.* April 29, 94, **17**, 52-54

Medical Devices Agency. 1996, Latex Sensitisation in the Health Care Setting (Use of Latex Gloves). *Device Bulletin MDA DB 9601.* London: MDA

Medical Devices Agency. 1998, Latex Medical Gloves (Surgeons and Examination), Powdered Medical Gloves (Surgeons and Examination), *Safety Notice SN 9825.* London: MDA

Nelson, J.B. and Mital,A. 1995, An Ergonomic Evaluation of Dexterity and Tactility with Increase in Examination/Surgical Glove Thickness. *Ergonomics. ,* **38**,: 4, 723-733

Newsom, S.W.B., Shaw, P., Smith, M.O. 1998, A Randomised Trial of the Durability of Non-Allergenic latex-free surgical gloves versus latex gloves. *Annals of The Royal College of Surgeons of England. ,* **80**,:288-292

Nottingham City Hospital. 1998, Latex Sensitivity: Policy and Procedures (Draft). *Internal Document.* Nottingham City Hospital NHS Trust.

O'Gilivie, W. 1998, Medical Gloves and Glove Materials. *Professional Nurse.* **14**,:3, 205-211

Roberts, A.D., and Brackley, C.A. 1996, Comfort and Frictional Properties of Dental Gloves. *Journal of Dentistry. ,* **24**,: 5, 339-343

Thompson, P.B. and Lambert, J.V. 1995, Touch Sensitivity Through Latex Examination Gloves. *Journal of General Psychology. ,* **122**,:1, 47-58

Woods, J.A., Leslie, L.F., Drake, D.B., Edlich, R.F. 1996, Effect of Puncture Resistant Surgical Gloves, Finger Guards and Glove Liners on Cutaneous Sensibility and Surgical Psychomotor Skills. *Journal of Biomedical Materials Research (Applied Biomaterials). ,* **33**,: 47-51

An evaluation of ergonomic effects during repetitive farm working in vineyards

Takeshi SATO[1], Takeshi KINOSHITA[2], Masami MIYAZAKI[2], Kazuyoshi SEKI[2], Shoji IGAWA[1]

1) Lab. Information & ergonomics, Nippon sport Science University, 7-1-1 Fukazawa, Setagaya, Tokyo 158-8508 Japan

2)Waseda University 2-579-15 Mikajima Tokorozawa 359-1164 Japan

Vineyard farmer's work is characterized by repetitive raising and lowering of the neck/shoulder/arm and many possibly localized muscular fatigue. The purpose of this study was to investigate working tasks during real farm working by making a comparison between on the standard horizontal shelf and on the modified Y-typed shelf of Japanese grape vine. This study analyzed the work load-reduction effect of using cultivate installation that were improved in comparison to another conventional shelf of grape vine employed by expert farmers in Japan. A method for recording and analyzing repetive up-down picking movement is also presented. Eight experienced volunteers from our laboratory performed a standardized repetitive real farm working tasks, on the standard horizontal and the modified Y-typed shelf of grape vine. Electromyograhic activities(EMG) were recorded from sis part of neck/shoulder/arm muscles, trapezius, deltoid, biceps and triceps brachii, brachioradialis, sternocleidomastoid, and form working performances were Video-filmed as posture was checked for standardization. For our choice of three muscles, we expected that these muscles were functioning and fatiguing during this real field farm working, were fitted to study this investigation. Objective parameters for localized muscular fatigue or activity were derived from the time course of the rectified EMG and it's mean power frequency(MPF) of the EMG recording. Reduction of MPF shows about 27% in brachioradialis muscle on each typed shelf, and trapezius muscle about 33% on standard shelf and 22% on Y-typed shelf. Experimental finding showed that posturing hands above shoulder level significantly increased the risk of localized muscular fatigue($p<.05$). The use of modified Y-typed shelf of vine resulted in 8% reduction in trapezius muscular activity.

Introduction

The mechanism of occupational fatigue and disorders, especially in the neck/shoulder region, is a chief matter of discussion. Studies of muscle fatigue represent one of the oldest research fields in Ergonomics and work physiology. Disorder and complaints in the neck/shoulder region have been described as a common problem in general populations and many occupations(Makela et al. 1991), for example carpenters(Hammarskjoid et al. 1992) and sitting work(Hagberg and Sundelin 1986). These articles indicated especially the case during the work task demanded the arms to be lifted. Many occupational situations involve highly repetitive dynamic contractions, which have been identified as a risk factor for musculoskeletal complaints. However, reports in the agricultural workers are not as many as those in the industrial workers. In the farming work load, mechanical overstress is thought to cause localized muscular fatigue. The cultivation of grapes and keeping high quality have to relay on the manual work by well trained workers. Farming works often do heavy physical movements, and are endurable to muscle fatiguing work. Since 1975 the shape of the vineyard shelf was improved to Y-type for experimental farming. The purpose of this research is to evaluate the grape shelf from the human skeletal muscle load in the origin for the data of the labor for farming in actual fields.

Methods

Eight healthy right-handed males, mean age was 36.2 years(ranged 19 to 56 years) and mean height was 169.4cm (ranged 155.7 to 177.7cm), participated in the experiment. Two of them were well trained farmers whom the technical assistants belongs to Department of Agriculture, The University of Tokyo, the rest were untrained volunteers from Waseda University. None of the subjects had specific diseases and symptoms in the neck/shoulder/arm region. All the subjects were fully informed of the nature and possible risk of the various procedure, all of which had been approved and followed the procedures of the local Ethics Committee.

It is well known that Electromyography(EMG) is one of reliable methods measuring task performance in human. The EMG data are most often used as the indicator of evaluation for the level of muscle activity or the level of muscular fatigue during or after task performance. Muscular fatigue is the result of sustained exertions of a specific muscle or muscle groups. And the term of localized muscular fatigue(LMF) was used to characterize fatigue experienced in regional muscles in response to postural or focused exertion stress.objective parameters for localized muscular fatigue or activity were derived from the time course of rectified EMG and its mean power frequency of the EMG recording. The reduction of mean power frequency was evaluated as indicator of LMF. The myoelectric signals were recorded by surface EMG electrodes on the six part of neck/shoulder/arm muscles, trapezius, deltoid, biceps and triceps brachii, Brachioradialis and sternocleidomastoid at a sample frequency of 1.5KHz. The surface electrodes were used(NT-615U Nihon Kohden, Japan), in a bipolar configration and were placed along the music fibres with 25mm distance on the each tested muscled. The repetitive farming work was performed at about 30 minutes both on the standard horizontal shelf and on the modeified Y-typed shelf of grape vine. The EMG signals were analyzed by computer of a 1024-point fast fourier transformation and mean power frequency and other objective parameters were computed. Results are presented as mean values with standard deviation and standard error. Comparisons were tested using by Student's t-test for paired observation with a significance level of p<0.05.

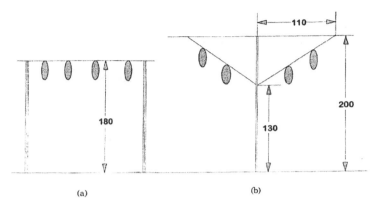

(a)
(b)

Figure 1. The shape and measure of two-typed grape vine shelf.(cm), (a) the standard horizontal shelf. (b) the modeified Y-typed shelf in Japan.(a) is spread widely and introduced many farmers cultivated grapes. However, farmers used type (b) shelf is still a minority.

Results

An increase of EMG amplitude showed in a scene of each farm working motion, e.g. arm elevation or lift at an overhead level, up-ward posturing, gathering grapes and it's transport etc. The motion of picking grapes has increased EMG amplitude in Brachioradiallis muscle. The posturing hands at or above shoulder level has increased in trapezius and deltoid muscular activity. The EMG amplitude increase of biceps and triceps brachii muscles has shown throughout motion of real farm working. These amplitude increases has indicated that recorded specific muscles were actually functioning and fatiguing. In all muscles, the frequencies in EMG power spectrum showed significant decrease on the two-typed shelf of vine. Compared with the statistically difference, however, the difference on the standard shelf was greater than that on the Y-typed shelf. And we obtained same tendency among each subject. Experimental findings also shows that the MPF values bradually shifted lower in all muscles on the two typed shelf of vine. In comparison with Y-typed shelf, the MPF curve in horizontal shelf showed a obvious downturn. The reduction of MPF values in the standard shelf was significantly higher than that on the Y-type shelf in all muscles. It was showed MPF value of brachioradialis muscle reduced finally about 28% on each shelf. As a result of above-mentioned two objective parameters, repetitive farm working in two typed shelf has regarded specific muscles or muscle groups as having been fatigued. Above all, in trapezius muscle it's parametric decrease was indicated statistical difference both on each shelf and among each subject. The degree of fatiguing had differences between on the standard shelf and Y-typed shelf. The ratio of decrease on the standard shelf was higher than that on Y-typed shelf. The tendency has also been shown among each subject.

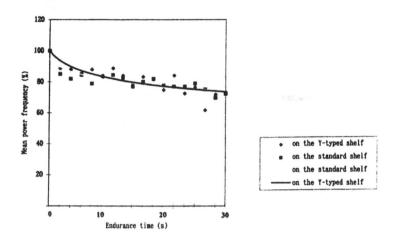

Fugure 6. The MPF values vursus endurance time in brachioradialis muscle.

Discussion

In this present study, the purpose was to investigate working tasks during real farm working by making a comparison between standard horizontal shelf and modified Y-typed shelf of vine. The EMG signals were usually one of several methods available for evaluation of muscle fatigue. Upper trapezius, deltoid, biceps and triceps brachii muscles are often adopted for evaluation of disorders and complains in the neck/shoulder region in many previous articles. It is well known that mechanical loads imposed on the neck-shoulder structure due to repeated or sustained elevations of the arm are a major risk factors(Nieminen et al. 1993). It was expected that farmers are able to grow grapes in lower working tasks. This result indicate that the cultivation on the modified Y-typed shelf of vine is suitable than that on the standard shelf. However the use of standard shelf is popular and the modified Y-typed shelf has not spread yet. Farm working, especially keeping high quality of pomiculture, is still needed to reley on the manual work by well trained farmers. Since this article deals with working tasks in the just harvest time, futuremore, we will expect the study throughout a year. And we will need to study not only about the working posture but also the working environment and the tools used by farmers.

References

Nieminen H, Takala EP and Viikari-Juntura E., 1993, Nomalized of electromyogram in the neck-shoulder region. European Jounal of Applied Physiology, Vol67, pp.199-207

ADDITIONAL PAPERS

A PROFILE OF PROFESSIONAL ERGONOMISTS

RB Stammers & EJ Tomkinson

Centre for Applied Psychology, University of Leicester
Leicester, LE1 7RH, UK

The results of a survey on UK ergonomists is reported. The sample was derived from the 1997/8 Ergonomics Society Membership Directory and a return rate of 56% from a postal questionnaire was obtained. There is a wide variety of professional specialisms within the respondent sample, with the practitioner appearing as the most numerous sub-group. Researchers make up the next largest group. A variety of qualification routes to membership are indicated, with the MSc being the most common one. Information on years of experience and on other topics is presented.

Introduction

There is a need to know more about the practice of ergonomics. This paper will focus on the UK, but the issue is an international one. Many individuals and groups in the UK, from time to time, make statements on the practice of the subject, but all too often the information on which this is based is sketchy. As part of a European project, a study has been carried out to assess the current situation with regard to university courses in ergonomics (Stammers & Tomkinson, 2001).

A postal survey was used to gather information from qualified ergonomists working in the UK. Whilst the main role of the questionnaire was to collect information on courses, the opportunity was also taken to collect information to help in building up a picture of contemporary professional ergonomists.

The Ergonomics Society (ES), as a collaborator in the research, could have given access to up-to-date names and addresses of members. It was thought important, however, to contact, (a) people some time after graduation, and, (b) people who might have left the ES and/or the profession. The 1997/1998 ES Membership Directory was therefore chosen as the source of potential participants. This was published in October

1997 and at the time of the survey was over 2 years old. Since its publication, a number of people would have left the Society and a number would have joined (these numbers would be roughly equal). Of the latter group, most would have recently finished their degrees and were of less interest for the survey. Of those who had left the Society, it was possible that they were still practising as ergonomists and their responses were therefore of interest for the survey.

The final version of the questionnaires was sent, via the post, to people who had been randomly chosen from the 1997/8 Ergonomics Society Membership Directory. Members were rejected if they lived outside the UK or were not graduate, registered or fellow members. Eighty-two questionnaires were returned out of the 156 sent out. However, 9 were also returned due to wrong addresses. The response rate (82/147) was therefore calculated as 56%.

The results were categorised using the respondent's professional area. In the first section of the questionnaire, respondents were asked what their major area of activity was. They could choose from, Management, Research, Consulting/Advising or Teaching. There was also the option of specifying their major area of activity if the other options were not suitable.

Forty-one respondents chose Consulting/Advising, 19 chose Research. Only 7 chose Management and only 3 chose Teaching. Also, 12 people fell into a General group, these were those that had chosen more than one category and those that did not fit into the other categories e.g. unemployed. Due to the small numbers for Management, Teaching and the General group, these were all placed into one group labelled 'General'. Three groups of respondents were thereby created, Consulting/Advising with 41 people, Research with 19 and General with 22.

These groups were of most use in relation to the course survey, this is described in an accompanying paper and in the full report. The focus in this paper will be on information that was collected that will help build up a picture of ergonomists at work in the UK in the year 2000.

Findings

Before discussing the findings, some comments on the sample and on possible bias is necessary. The numbers chosen to send the questionnaire to were approximately a 25% sample of the graduate, registered and fellow members. However, members not living in the UK were rejected from the random sample and replaced. A higher percentage of the UK-resident members was therefore sampled, estimated at over 30%.

It is possible to speculate on possible sources of bias. Ex-members are more unlikely to respond, especially if they are no longer working in ergonomics. Against this background a return rate of 56% was felt to be a healthy rate, even if it was possibly biased in its makeup towards working ergonomists.

The first finding has already been covered above. This was the division of the sample depending on professional specialism. Fifty percent of the sample saw themselves as consultants/advisors, 23% described themselves as researchers and 27% of the respondents could be described as generalists.

Thus practitioners, as opposed to those based in ergonomics in other ways, make up a sizeable majority of professionals in the field.

One of the questions asked was whether the respondent's work directly involved the use of Ergonomics knowledge. Thirty-six out of 41 Consultants (87.80%) indicated that it did. Eighteen out of the 19 Researchers (94.74%) and 19 out of 22 Generalists (86.36%) also indicated that it did. Thus 11% of the overall sample were not directly using ergonomics knowledge in their work.

The respondents were also asked if they are currently members of the Ergonomics Society. The following breakdown was obtained.

Consultants	*Researchers*	*Generalists*	*Overall*
90%	100%	86%	91%

The obvious question is what is the overlap between ex-members and those no longer using ergonomics in their work? Of the 9 respondents who indicated non-use, 4 were no longer in the Society. This leaves 5 who have kept their membership despite no longer using ergonomics in their work.

Although a concern to the ES, it is hard to determine how widespread non-membership of the ES by practising ergonomists is. The small number of 3 professionals who have left the Society, but still use ergonomics knowledge is too small to draw any conclusions.

There was also a question on membership grade. The numbers did not differ greatly across group categories and are given below as overall percentages, but are contrasted with the latest membership figures taken from the ES Annual Report 2000. These latter percentages refer to these membership categories only, as these were the groups sampled from. The odd 3% of 'others' in the sample row refers to two people who had changed their membership grade.

Given that the 2000 Report figures will include overseas members, then the sample might be considered to be fairly representative, although graduate member numbers appear somewhat low.

	Graduate members	*Registered members*	*Fellow*	*Others*
Sample	15%	59%	23%	3%
2000 Report	37%	48%	15%	-

Another question asked fellows and registered members how many years it was since they qualified for a 'professional' grade of membership. This did not apply to graduate members, who would not normally have reached this stage of experience yet. Again, there were no substantial differences between the groups, so the overall findings are presented below.

Years	0-5	5-10	10-15	15-20	20+
Percentage	33%	26%	13%	13%	15%

Another question asked whether respondents were qualified as "European Ergonomists". The overall figure of 21% indicates a sizeable uptake of this qualification by UK ergonomists. An notable point here was there was a higher percentage of the Researchers (26%) than the Consultants/Advisors (17%) holding this qualification. The Generalists come between the other two samples (23%).

Data on membership of other professional societies was also collected. The following percentages of respondents were members of the societies indicated. Some of the respondents were not members of the ES but they did not indicate membership of any other society. These figures do not seem very high and indicate a fair degree of single society membership status for the ES members.

Society	BPS	HFES	IOSH	BOHS
Percentage	20%	10%	10%	4%

Another area of questioning concerned the route to qualification. Over the years a number of routes to membership have been in place. The majority of the sample qualified with an MSc in the subject (54%), those qualifying with a BSc numbered 18% and those coming though the experience and/or research degree route totalled 20%. Of those qualifying with an MSc, about half had some work experience before taking the degree.

A question that focused just on those that had completed a taught degree course in ergonomics, enabled an overview on number of years since completing the course. This reinforced the findings given above on years since qualification (by any route) and showed a wide spread of experience in both the sample, and presumably in the overall membership.

Years since grad.	4-5	6-10	11-15	16-20	21-25	26-30	N/A
Percentage	8	18	16	16	21	15	6

The remainder of the questionnaire focused on course related issues which are being reported on separately.

Discussion

Given that there is no existing clear profile of the ergonomics-qualified people in the UK, then the results presented have no real points of contrast. Some usable suggestions emerge, however. These give reasons for both satisfaction and concern for the ES.

Practitioners are largest sub-group within the Society, with researchers being the only other sub-group of substantial size. Whilst it must be a concern that there are some ex-members who are still involved in ergonomics, there are also existing members who are not now using their ergonomics professionally. The Society remains an "interest" and a professional society, but not one that can claim exclusivity for its activities.

The sample exhibits what could be called a healthy spread across age and experience categories. This a broad-based membership in terms of its experience bodes well for the Society in the future.

Most members do not have "divided loyalties" with other societies. The MSc is the most common route to qualification, although it is notable that there have been a variety of ways of achieving membership.

Reference

Stammers, R.B. & Tomkinson, E.J. 2001, **A study of ergonomics training in the UK.** (Leicester: Centre for Applied Psychology, University of Leicester).

RESULTS OF A SURVEY OF ERGONOMICS TRAINING

RB Stammers & EJ Tomkinson

*Centre for Applied Psychology, University of Leicester
Leicester, LE1 7RH, UK*

The results of a study on UK ergonomics courses is reported. Course directors were interviewed and a postal questionnaire was used to survey qualified ergonomists. Course directors report on pressure from an number of sources and courses have changed in a variety of ways. The survey results indicate a general level of satisfaction with course content and teaching methods. One area of concern is the poor rating that skills development received. Whilst, with the passage of time, respondents indicate that knowledge gained from their courses is of less use to them, there appears to be a relatively low level of demand for continuing professional development activity.

Introduction

As part of a European project, a study has been carried out to assess current university courses in ergonomics in the UK. Advantages and disadvantages of current practice should be of value not only in the UK should help in course development internationally.

The problem with any study of this kind is the collection of sufficient valid information. Previous experience in the ES had shown a low response rate to questionnaires distributed with Society newsletters. In the light of this it was decided to draw on two very different sources of expertise. One was the directors of current MSc courses. There was also a postal survey of qualified ergonomists. With the assistance of the ES, it was possible to get direct access to these two groups.

There are both BSc and MSc courses in ergonomics in the UK. There is only one BSc course that is fully in ergonomics, a second, combined, degree, is no longer taking any students. It was decided not to explore the BSc degrees in detail, but to restrict the study mostly to the MSc degrees. Whilst this excluded one unique area of provision, it did allow for a focus on similar degree programmes from a number of universities.

The postal survey was used to gather information from a second source of informants. The details of the sampling procedures and sample are included in an accompanying paper and in the main report (Stammers & Tomkinson, 2001). In distributing the questionnaire, efforts were made to get a good response rate. This involved sending reminder letters to non-respondents and by providing return envelopes with postage stamps on them.

The first section of the questionnaire asked respondents about their current work and their ES membership status. Another section asked questions about the respondents' training in Ergonomics. The final section asked about training that respondents had subsequently undertaken and about their opinions on different forms of continuing professional development.

The findings of the survey were detailed and only the main points are summarised here. Eighty-two questionnaires were returned out of the 156 sent out. However, 9 were also returned due to wrong addresses. The response rate (82/147) was therefore calculated as 56%. The results were categorised using the respondent's professional area. Three groups of respondents were thereby created, Consulting/Advising with 41 people, Research with 19 people and General with 22 people.

Findings

Interviews
The course directors of five Ergonomics courses in the UK were interviewed. The interviewees were from the following universities: Birmingham, Loughborough, Nottingham, Surrey, University College London. It is clear that courses have undergone changes over the years in response to a variety of circumstances.

A number of sources of pressure have led to changes in programmes. Some come from within the discipline itself, leading to some topics having less coverage in the past and some having more. Another shift has been towards more interactive teaching and less formal lectures. This shift has had three consequences. One consequence is that students are expected to do more self-directed learning. Another is that there is probably less breadth in coverage. Thirdly there is a move towards a problem-solving and information retrieval model of learning rather than a recall of specific facts one.

Some changes have come about because of pressure on staff time and changes in the management of university activities. The suggestion has been that the courses should be more "cost-efficient". It was felt that there were a number of negative consequences of these developments. One example would be that modularisation has tended to put artificial fences around subject areas, making it harder to examine the interactions between topics.

Another impetus for change arises from difficulties that students have had in obtaining funding for attendance. The criteria used to award grants to course is the quality of the course content and the kind of jobs graduates obtain afterwards. This, together with uncertainty over whether the grant-awarding bodies will continue their support, have been additional sources of pressure. Related to this is the fact that most students nowadays have to be self-supporting, therefore there is a demand for part-time courses and other ways of making courses accessible.

There have also been demands from students and potential employers for courses to become more specialised, it was felt that there was now less of a place for general ergonomics programmes. There are contrasts between different courses in the number and variety of modules that can be studied, but the scope for specialisation is rather limited, with courses offering from 1 to 4 options. Practical work is an important part of all courses.

Students are recruited from a large range of disciplines and backgrounds. Courses mostly recruit graduates, but there are also people without first degrees, but with relevant experience. There is a high proportion of mature students. There appears to be differences between people from a technology background and those from a human sciences background. Those from a technology background tend to be less competent at discursive, thoughtful writing. It was also suggested that this group are less accepting of a strictly scientific approach, resisting the notion of theory-based approaches. Human scientists were thought to be less competent at technical topics and it was commented that these two, typified, groups tend to not work well together at first.

However, it would be wrong to over-generalise and one interviewee suggested there can be differences within groups as well as between them. Mature students can relate what they learn to the experience they have had, new graduates, on the whole, cannot do this. It was also thought that some students, particularly mature ones, who have been out of education for a while, find the transition back into education difficult.

Survey
Turning now to the membership survey, respondents were asked rate the importance and frequency of use of the following areas of ergonomics knowledge. These had been determined to be the key topics in all courses:

Anthropometry	Methods of analysis and evaluation
Anatomy	Statistics/data handling
Biomechanics	Workspace design
Work physiology	Systems ergonomics
Cognitive psychology	Human reliability/error
Social/organisational psychology	Health and safety
Physical environment	Human-computer interaction

For "importance", a five point scale ranging from "critical to current work", through "very important, "of some importance", "of marginal importance" to "not important at all" was used. A five point scale was also used for "frequency of use", this had the range, "daily", "weekly", "monthly", "rarely" to "never". There was evidence of use of the full scale of responses for each of the knowledge categories. Thus certain topics were seen as critical for some participants, others were not rated as important at all. There was no strong consensus on the critical importance or irrelevance of any area. Similar patterns of responding emerged for ratings of frequency of use.

More interesting were those topics where there was a strong bias towards one area of the scale and where patterns of difference emerged for different participant groups. A summary of responses for each area follows:

Anthropometry: For this topic, both the consultant group (C) and the generalists (G) gave a wide spread of importance ratings. For the researchers there was a central clustering. On frequency, a similar pattern emerges.

Anatomy: Alone of all the topics, this one gets a clustering to the low end of both importance and frequency.

Biomechanics: For all groups a fairly flat profile of responses indicates a variety of importance ratings for this topic. The frequency ratings cluster at the low end.

Work physiology: Within each group there is a fairly even range of ratings of importance for this topic, but the C group tend to rate it at the low end. There as a similar pattern for frequency.

Cognitive psychology: There is a clustering towards to high end of ratings of importance for this topic, with the G and R groups giving generally high ratings. For frequency, there is a flat profile for all groups.

Social/organisational psychology: There was a fairly even spread of scores across the ratings for all groups for both importance and frequency.

Physical environment: This topic gets quite a high rating for all groups, although it would appear that it is not very frequently used, the modal response being "monthly".

Methods of analysis and evaluation: There is a strong clustering towards the high end of the importance scale, the frequency ratings were also at the high end.

Statistics/data handling: The R group's ratings were exclusively at the middle to high end of the scale for importance, the G group were similar. This topic was seen as less important for the C group. On frequency, there was a similar pattern.

Workspace design: The C and G groups' responses cluster at the high end, whilst the R group's rating are evenly spread. There is a clustering around the central rating for all groups on frequency.

Systems ergonomics: This topic was of "critical" importance for many of the C group, but the other groups' ratings were evenly spread. The frequency ratings were generally spread out, although the R group were rare users of this topic.

Human reliability/error: There was a fairly even spread of responses for both scales for all groups for this topic.

Health and safety: A clustering towards the high end for both importance and frequency was found for all groups here.

Human-computer interaction: For this topic, for both scales, and for all groups, there was a pattern of high end responses, but not a strong one.

Respondents were asked to add other subject areas that they had studied and to rate them. Only a few extra topics were added by some participants. Another question asked about any topics that respondents thought should have been studied in more depth and any that could have been omitted. No strong indications emerged here. HCI was mentioned as a topic that needed more coverage by around 12% of respondents.

Another question referred to topics that were not available that respondents would have liked to have studied. Again HCI is the only topic that gets any indication of interest here.

Another area of questioning focused on the respondent's impressions of the usefulness of their course of study immediately on completion, in subsequent years and in the future. A pattern emerges, with courses being seen as immediately useful and then becoming less useful with the passage of time.

There were also questions on teaching methods used. For most methods (lectures, lab classes etc.), ratings on a 5-point usefulness scale indicated a positive, but not strongly positive response.

Another question explored the extent to which "skills" were gained from the course studied. Participants were asked to give a yes/no response to a list of practical topics. The findings here were quite striking in that they indicated some quite low levels of positive response (a mean of 49%). At the high end, 74% felt that they had gained skills in anthropometry, but at the low end, only 15% felt they gained "research/consultancy proposal writing" skills. The other results are listed below:

Laboratory experimentation	59%	Interviewing	41%
Retrieving information	51%	Questionnaire design	59%
Analysing data	67%	Spoken presentation	48%
Field experiments	46%	Research report writing	64%
Physiological measurement	59%	Consultancy report writing	33%
Anatomical measurement	41%	Writing	28%

One question explored the extent to which respondents felt that their knowledge about ergonomics had became more integrated over time. They were asked to reflect on the extent to which such knowledge was integrated before the course and then at various stages after the course. It could be argued that the question, as posed, suggested that knowledge "should" become more integrated over time. It is interesting, however, that there was a strong positive indication that this was so.

The final section of the questionnaire explored views on continuing professional development (CPD). The first question required a yes/no answer on whether respondents have ever undertaken any further study (formal or informal) of ergonomics. Some interesting differences emerged here, with 68% of the R group positively responding and 54% of the C group and 55% of the G group similarly answering "yes". Asked to indicate reasons for further study, "relevance to current/future project" and "self-initiated to update knowledge" emerged as the most common reasons. When those who had undertaken some CPD were asked about the type of activity that was used, independent reading was most commonly used (83%) followed by conferences (66%) and then short courses of 1-3 days (43%).

In terms of media used for CPD, familiar media such as reading and face-to-face teaching were positively rated. There were less positive ratings for less familiar media such as computer-based learning. A final question asked whether the respondents thought that they would be undertaking CPD in the future. A small majority thought that they would (59%), the C group indicating a less positive response (49%) than the R group (74%) and the G group (64%).

Discussion

It is hoped that the findings of this study will be of use to both the ES and those responsible for courses. The ES has had an interest in the national training situation for some time. It has also had discussions on the concerns of the course directors and has tried to help. The ES has some influence on course content through its accreditation activities, but little pressure is needed as there appears to be a broad consensus on course content, with some specialisms reflected in the whole or part of individual courses. The interviews reveal concerns, but also demonstrate that different courses have responded to pressures in different ways.

The survey results could be taken to vindicate much of what has gone on in courses over the years. The knowledge needs of ergonomists appear to be covered. Some specialisms would seem to need more or less of certain topics. However, as career routes are not always predictable in advance, there are limited possibilities for this.

However, there should not be complacency. The greatest area of concern is with skills. The relatively poor showing of this area reveals a dilemma for courses. The pressure is to be cost-efficient, but skill learning involves intensive teaching. Also, in relation to the demand for more flexible modes of teaching, there is the issue that whilst theoretical topics can be studied in, eg, distance learning modes, practical work largely requires face to face teaching.

The findings in relation to CPD also present interesting contrasts. Whilst the respondents indicate that their course-acquired knowledge becomes less useful with the passage of time, there is also a low level of participation in CPD. This is at a time when professional bodies are increasingly requiring evidence of knowledge updating. Changes in attitude towards, and also in resourcing of, such activities appears needed.

Reference

Stammers, R.B. & Tomkinson, E.J. 2001, **A study of ergonomics training in the UK.** (Leicester: Centre for Applied Psychology, University of Leicester).

HUMAN FACTORS INTEGRATION IN THE PETRO-CHEMICAL INDUSTRY

Jayne Elder[1], W Ian Hamilton[1] and Peter Martin[2]

[1]Human Engineering Limited
Westbury On Trym
Bristol, BS9 3AA

[2]M W Kellogg Limited
Greenford
London

Human factors integration (HFI) is a systematic method of incorporating all aspects of human factors into the appropriate stages of an engineering or system design process. The concept was first applied in the defence industry but its success has meant that it is now being applied in other sectors. This is particularly the case in industries with a major hazard potential.

M W Kellogg is a prime contractor to the oil and gas industry. They have taken a proactive approach to ensure that human factors are considered from the very earliest stages of design. Along with Human Engineering, the Company has developed a human factors integration process specifically tailored to the support requirements of their engineering design process.

The HFI process comprises a three tier hierarchy of documents that provide increasingly detailed levels of guidance on the activities and methodologies to be applied.

This paper reports on the development and delivery of these guidance materials and their integration with the established engineering design procedures operated by M W Kellogg.

Human Factors Integration

Human Factors Integration is a systematic method of incorporating all aspects of human factors and ergonomics into the appropriate stages of the engineering process.

The integration of human factors into engineering design was originally made systematic by defence procurement, first in the US and shortly afterwards in the UK. Contractors are required to consider human factors in the design and modification of defence systems. This is to ensure total system effectiveness by recognising the capabilities and limitations of the system operators, maintainers, and repairers.

The success of these programmes, together with generally increased awareness of human factors issues and their impact on the potential safety and efficiency of a system, has given rise to a recognition within non-defence sectors of the need for human factors integration. This is particularly the case in safety critical industries such as petro-chemical and rail.

Benefits

The objective of human factors integration (HFI) is to ensure that human factors methods and principles are applied appropriately and consistently during systems development in order to achieve a safe and effective design for the users.

Given the increasing emphasis on safety, particularly from the UK HSE, it is ever more important that human factors are taken into account in the design process. Adherence to good human factors principles in design not only improves the comfort and productivity of users, but also reduces the risk of injury to users and damage to equipment. The inclusion of HFI in a project may also provide a commercial/competitive advantage.

To cover the full range of human factors issues, HFI encompasses the following six elements, known as domains. They are:

- Manpower
- Personnel
- Training
- Human Factors Engineering
- System Safety
- Health Hazard Assessment

Early human factors integration means that human reliability can be designed into a system while under development. This is much more effective than addressing human reliability issues in a completed system. Ultimately, the benefits of human factors integration can be realised in terms of improved safety and reduced life cycle costs.

Failure to apply human factors and ergonomics may result in an increase in the likelihood of accidents occurring. A recent example of ergonomics deficiencies contributing to an accident is the explosion and fire at the Milford Haven Texaco refinery in 1994 (HSE, 1997).

Texaco Refinery

In July 1994 an explosion and a number of fires occurred at the Texaco Refinery in Pembroke, where hydrocarbon fuels are produced from crude oil. A severe electrical storm caused plant disturbances and resulted in all but one of the cracking units being shut down. Approximately 20 tonnes of flammable hydrocarbons were released forming a drifting cloud of vapour. The vapour was ignited and the resulting force of the explosion was equivalent to four tonnes of high explosive.

However, it was a combination of failures in safe working procedures, equipment and control systems during the plant upset that directly caused the explosion some five hours after the initial storm:

- a control valve was in fact shut when the system indicated that it was open,
- a modification had been carried out on the flare drum pump-out system without assessment of all its consequences,
- control panel graphics did not provide necessary process overviews,
- attempts were made to keep the unit operating when it should have been shut down.
- inefficient classification of alarms overwhelmed the operators as the incident developed.

The occurrence of the incident figured prominently in the media and was notified to the European Union. The Health and Safety Executive brought legal proceedings against the Partners involved, which resulted in a fine of £200,000 plus payment of £144,000 costs. The consequent business interruptions seriously affected the UK refining capacity and the damaged refinery cost approximately £48 million to rebuild.

The use of HFI before the incident could have identified most of the contributory factors, whereupon the necessary ergonomics intervention could have been applied. For instance, when considering the HFI domain of human engineering, the requirement for the process overview screens, mass and volumetric balance summaries and efficient alarm design would have been identified. The training requirements for the operators during unplanned events would also have been addressed when considering the HFI domain of training.

Development Of Guidance Material

Review

M W Kellogg had already recognised the importance of human factors and had established some basic procedures as part of the quality assurance system for engineering work. It was recognised that these were incomplete and a review was commissioned to determine how they might be improved.

The main finding of this review was that the existing human factors procedure did not tie-in effectively with the Company's engineering process. What was needed was a set of procedures that explicitly linked HFI activities and techniques to the goals of the various stages of the engineering design process.

Engineering Process

The established engineering design process comprises a number of phases. Typically a project lifecycle will be:

- Response to enquiry
- Front-end engineering design (FEED)
- Detailed design
- Procurement
- Construction & pre-commissioning
- Commissioning & beneficial life

The main impact of human factors on design occurs early in the engineering design process (FEED and early Detailed Design). At later stages in the process, designs become fixed and cannot be so easily influenced. Consequently, at later stages the focus of HFI shifts from analysis and modelling techniques to assessment/audit and testing, with particular emphasis on safety management.

Hierarchy of Guidance Documents

This work will ultimately deliver a set of guidance documents to M W Kellogg that are based on five steps to successful HFI. These are:

1. Understanding HFI,
2. Identifying the requirements for a project (supported by human factors checklists),
3. Identifying what human factors studies need to be carried out,

4. Carrying out the studies as appropriate,
5. Acting on the findings.

Three levels of documentation are being created. The levels comprise:

Level 1 - An overview document providing background information on human factors, HFI, HFI in the M W Kellogg engineering process, and describing the structure of the entire document set.

Level 2 – A set of guidance documents to support the implementation of HFI at each stage of the engineering design process.

Level 3 – A set of desk manuals providing information and instructions on human factors techniques for all relevant activities such as: task analysis, human reliability analysis, workspace modelling, etc.

All of these documents will sit within the existing quality system documentation and will be accessible to all project personnel.

HFI In The Engineering Process

Human Factors Integration Plan

The guidance material is premised on the assertion that effective human factors integration relies on three key elements. The first of these is the Human Factors Integration Plan (HFIP). This is a crucial management plan that is developed early in the engineering process and serves to guide the allocation of resources and the timing of human factors studies throughout each stage of the project's engineering programme. The HFIP is created by applying a human factors checklist that has been developed to tease out the HFI elements required for the project.

Management Organisation

The second key element is management organisation. In order for ergonomics to be systematically included in the project design process, 'Ergonomics Champions' should be assigned to each project. The objective for the Ergonomics Champions is to promote the cause of HFI within the project team and ensure it is adequately addressed and implemented. This function splits into two roles:

- HFI Co-ordinator
- HFI Champion

The HFI Co-ordinator may be a full or part time role depending on the scope of the ergonomics requirement. This person should usually be drawn from the Health, Safety and Environmental (HSE) Engineering Group and should report to the Project HSE Manager.

An HFI Champion should be assigned for each of the engineering disciplines involved in the project (e.g. Piping, Instrumentation and Control, etc).

Depending on the size and requirements of the project, there may also be a requirement for an administration assistant.

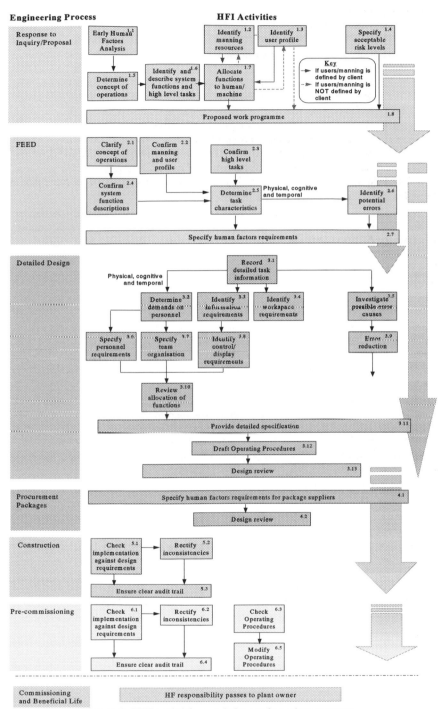

Figure 1. HFI activities and the engineering process

HFI Issues Register

The recommendations / findings from the HF studies should be stored in an Issues Register. This will provide a method for managing the results in order to ensure that none are overlooked and all are acted upon satisfactorily. It will also help provide a project audit trail.

The Issues Register would typically be developed in the form of a database. The HFI Co-ordinator would manage the Issues Register and the HFI champions would identify the issues to be entered into it.

Activities During the Engineering Process

Figure 1 illustrates the activities required to integrate human factors with the engineering design process. It shows where in the engineering process HFI activities should occur and the numbers on the diagram provide cross-references to the Level 3 documents that describe the techniques that should be applied.

For example, in order to Identify Workspace Requirements (Diagram Reference 3.4) the relevant techniques that are cross referenced are: HTA, workspace modelling, anthropometry, and link analysis.

Conclusion

The work to deliver the full set of HFI guidance documentation is still in progress. Nevertheless the principles have already been applied and proven within a wide range of safety critical engineering projects in the petro-chemical and other industries.

References

Health & Safety Executive 1997, *The explosion and fires at the Texaco Refinery, Milford Haven, 24th July 1994*, Her Majesty's Stationery Office, Norwich.

AUTHOR INDEX

SUBJECT INDEX